Advances in

BOTANICAL RESEARCH

incorporating *Advances in Plant Pathology*

VOLUME 32

Advances in

BOTANICAL RESEARCH

incorporating *Advances in Plant Pathology*

Advances in

BOTANICAL RESEARCH

incorporating *Advances in Plant Pathology*

Plant Protein Kinases

Volume editors

M. KREIS

Institut de Biotechnologie des Plantes,
UMR/CNRS 8618, Bâtiment 630,
Université de Paris-Sud,
F-91405 Orsay Cedex, France

J. C. WALKER

Division of Biological Sciences,
University of Missouri,
308 Tucker Hall,
Columbia, MO 65211, USA

Series Editor

J. A. CALLOW

School of Biological Sciences,
University of Birmingham,
Birmingham, UK

VOLUME 32

2000

ACADEMIC PRESS

A Harcourt Science and Technology Company

San Diego San Francisco New York Boston
London Sydney Tokyo

This book is printed on acid-free paper.

Academic Press
A Harcourt Science and Technology Company
Harcourt Place, 32 Jamestown Road, London NW1 7BY, UK
http://www.academicpress.com

Academic Press
A Harcourt Science and Technology Company
525 B Street, Suite 1900, San Diego, California 92101-4495, USA
http://www.academicpress.com

ISBN 0-12-005932-0

A catalogue record for this book is available from the British Library

Typeset by Gray Publishing, Tunbridge Wells, Kent
Printed in Great Britain by MPG Books Limited, Bodmin, Cornwall

00 01 02 03 04 05 MP 9 8 7 6 5 4 3 2 1

CONTENTS

Light and Protein Kinases

JOHN C. WATSON

Calcium-Dependent Protein Kinases and their Relatives

ESTELLE M. HRABAK

Receptor-Like Kinases in Plant Development

KEIKO U. TORII and STEVEN E. CLARK

A Receptor Kinase and the Self-Incompatibility Response in *Brassica*

J. M. COCK

Plant Mitogen-Activated Protein Kinase Signalling Pathways in the Limelight

S. JOUANNIC, A.-S. LEPRINCE, A. HAMAL, A. PICAUD, M. KREIS and Y. HENRY

Protein Phosphorylation and Dephosphorylation in Environmental Stress Responses in Plants

K. ICHIMURA, T. MIZOGUCHI, R. YOSHIDA, T. YUASA and K. SHINOZAKI

Protein Kinases in the Plant Defense Response

GUIDO SESSA and GREGORY B. MARTIN

SNF1-Related Protein Kinases (SnRKs) – Regulators at the Heart of the Control of Carbon Metabolism and Partitioning

N. G. HALFORD, J.-P. BOULY and M. THOMAS

Carbon and Nitrogen Metabolism and Reversible Protein Phosphorylation

D. TOROSER and S. C. HUBER

Protein Phosphorylation and Ion Transport: A Case Study in Guard Cells

JIAXU LI and SARAH M. ASSMANN

CONTRIBUTORS TO VOLUME 32

S. M. ASSMANN *Department of Biology, 208 Mueller Laboratory, The Pennsylvania State University, University Park, PA 16802, USA*

J.-P. BOULY *Laboratoire de Biologie du Développement des Plantes, Institut de Biotechnologie des Plantes, UMR CNRS 8618, Bâtiment 630 Université de Paris-Sud, 91405 Orsay Cedex, France*

S. E. CLARK *Department of Biology, University of Michigan, 830 N. University, Ann Arbor, MI 48109-1048, USA*

J. M. COCK *Reproduction et Développement des Plantes, UMR 9938 CNRS-INRA-ENSL, Ecole Normale Supérieure de Lyon, 46 allée d'Italie, 69364 Lyon Cedex 07, France*

N. G. HALFORD *IACR-Long Ashton Research Station, Department of Agricultural Sciences, University of Bristol, Long Ashton, Bristol BS41 9AF, UK*

A. HAMAL *Institut de Biotechnologie des Plantes, Laboratoire de Biologie du Développement des Plantes, Bâtiment 630, UMR/CNRS 8618, Université de Paris-Sud, 91405 Orsay, France*

D. G. HARDIE *Biochemistry Department, Dundee University, MSI/WTB Complex, Dow Street, Dundee DD1 5EH, UK*

Y. HENRY *Institut de Biotechnologie des Plantes, Laboratoire de Biologie du Développement des Plantes, Bâtiment 630, UMR/CNRS 8618, Université de Paris-Sud, 91405 Orsay, France*

T. P. HOLTSFORD *Division of Biological Sciences, University of Missouri–Columbia, MO 65211, USA*

E. M. HRABAK *Department of Plant Biology, University of New Hampshire, Durham, NH 03824, USA*

S. C. HUBER *United States Department of Agriculture, Agricultural Research Service and Departments of Crop Science and Botany, North Carolina State University, Raleigh, NC 26795, USA*

K. ICHIMURA *Laboratory of Plant Molecular Biology, Tsukuba Life Science Center, The Institute of Physical and Chemical Research (RIKEN), 3-1-1, Koyadai, Tsukuba, Ibaraki 305–0074, Japan*

E. R. INGHAM *Division of Biological Sciences, University of Missouri–Columbia, MO 65211, USA*

JIAXU LI *Department of Biology, 208 Mueller Laboratory, The Pennsylvania State University, University Park, PA 16802, USA*

S. JOUANNIC *Laboratoire de Biologie du Développement des Plantes Institut de Biotechnologie des Plantes UMR/CNRS 8618, Bâtiment 630, Université de Paris-Sud, 91405 Orsay, France*

M. KREIS *Institut de Biotechnologie des Plantes, UMR/CNRS 8618, Bâtiment 630, Université de Paris-Sud, F-91405 Orsay Cedex, France*

A.-S. LEPRINCE *Institut de Biotechnologie des Plantes, Laboratoire de Biologie du Développement des Plantes, Bâtiment 630, UMR/CNRS 8618, Université de Paris-Sud, 91405 Orsay, France*

G. B. MARTIN *Boyce Thompson Institute for Plant Research and Department of Plant Pathology, Cornell University, Ithaca, NY 14853, USA*

T. MIZOGUCHI *Laboratory of Plant Molecular Biology, Tsukuba Life Science Center, The Institute of Physical and Chemical Research (RIKEN), 3-1-1, Koyadai, Tsukuba, Ibaraki 305–0074, Japan*

A. PICAUD *Institut de Biotechnologie des Plantes, Laboratoire de Biologie du Développement des Plantes, Bâtiment 630, UMR/CNRS 8618, Université de Paris-Sud, 91405 Orsay, France*

G. E. SCHALLER *Department of Biochemistry and Molecular Biology, University of New Hampshire, Durham, NH 03824, USA*

G. SESSA *Boyce Thompson Institute for Plant Sciences, Tower Road, Ithaca, NY 14853, USA*

SHENG LUAN *Department of Plant and Microbial Biology, University of California at Berkeley, Berkeley, CA 94720, USA*

K. SHINOZAKI *Laboratory of Plant Molecular Biology, The Institute of Physical and Chemical Research (RIKEN), Tsukuba Life Science Center, 3-1-1, Koyadai, Tsukuba, Ibaraki 305-0074, Japan*

M. THOMAS *Laboratoire de Biologie du Développement des Plantes, Institut de Biotechnologie des Plantes, UMR CNRS 8618, Bâtiment 630 Université de Paris-Sud, 91405 Orsay Cedex, France*

K. U. TORII *Department of Botany, University of Washington, Seattle, WA 98195, USA*

D. TOROSER *United States Department of Agriculture, Agricultural Research Service, and Departments of Crop Science and Botany, North Carolina State University, Raleigh, NC 26795-7631, USA*

J. C. WALKER *Division of Biological Sciences, University of Missouri, 308 Tucker Hall, Columbia, MO 65211, USA*

J. C. WATSON *Department of Biology, Indiana University – Purdue University Indianapolis, 723 West Michigan Street, Indianapolis, IN 46202-5132, USA*

R. YOSHIDA *Laboratory of Plant Molecular Biology, Tsukuba Life Science Center, The Institute of Physical and Chemical Research (RIKEN), 3-1-1, Koyadai, Tsukuba, Ibaraki 305–0074, Japan*

T. YUASA *Laboratory of Plant Molecular Biology, Tsukuba Life Science Center, The Institute of Physical and Chemical Research (RIKEN), 3-1-1, Koyadai, Tsukuba, Ibaraki 305–0074, Japan*

CONTENTS OF VOLUMES 20–31

Contents of Volume 20

Contents of Volume 21

Contents of Volume 22

Contents of Volume 23

PATHOGEN INDEXING TECHNOLOGIES

Contents of Volume 24

Contents of Volume 25

Contents of Volume 26

Contents of Volume 27

Contents of Volume 28

Contents of Volume 29

Contents of Volume 30

Contents of Volume 31

Plant Protein–Serine/Threonine Kinases: Classification into Subfamilies and Overview of Function

D. G. HARDIE

Biochemistry Department, Dundee University, MSI/WTB Complex, Dow Street, Dundee DD1 5EH, UK

I. INTRODUCTION

Although there were reports of protein kinases in higher plants based on biochemical assays prior to 1989 (e.g. Harmon *et al.*, 1987), in that year the first two plant protein kinase *sequences* were reported by Lawton *et al.* (1989). From this modest and recent beginning, the number of higher plant protein kinase sequences in the European Molecular Biology Laboratory (EMBL) database (excluding expressed sequence tags, ESTs) has mushroomed, there currently being about 1000 in all (Fig. 1), of which about 200 are from

Advances in Botanical Research Vol. 32
incorporating Advances in Plant Pathology
ISBN 0-12-005932-0

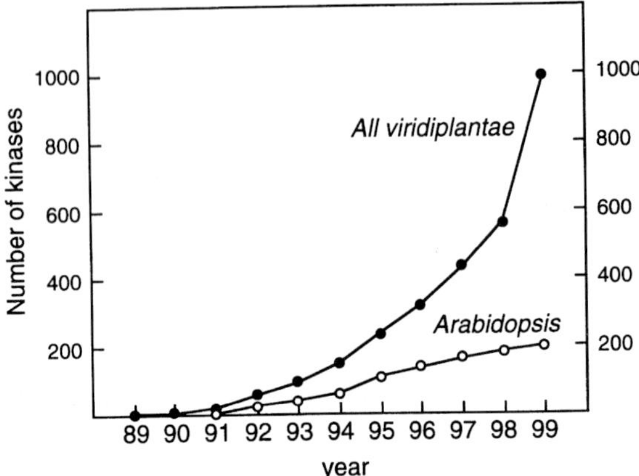

Fig. 1. Cumulative total of plant protein kinases in the EMBL database plotted against year of deposition. In July 1999 the EMBL/EMBLNEW databases were searched with the term 'kinase' in the 'description' field, *viridiplantae* or *Arabidopsis thaliana* in the 'organism' field, and the year of first entry in the 'date' field. Proteins other than protein kinases (usually around 10–20% of the total hit by the 'kinase' search term) were removed by manual inspection. This search will miss protein kinase sequences if the word 'kinase' does not appear in the 'description' field [e.g. the two kinase sequences deposited by Lawton *et al.* (1989), which have been added, and many ESTs]. No attempt was made to delete multiple entries of the same sequence. The large increase in non-*Arabidopsis* kinase sequences in 1999 appears to be due to the deposition of large numbers of sequences from cotton (*Gossypium hirsutum*).

Arabidopsis thaliana. We will know the exact number of protein kinases in the *Arabidopsis* genome within the next year or two, but reasonable estimates can already be made. In a 1.9 Mb sequence on chromosome 4 (Bevan *et al.*, 1998), there were 11 open reading frames (ORFs) encoding protein kinases out of a total of 389, representing 3% of all ORFs in this segment of chromosome. Of 3360 *Arabidopsis* ORFs in the PEDANT database (http://pedant.mips.biochem.mpg.de/index.html) in July 1999, 44 (1%) are reported to have homology to protein kinases. For comparison, the frequency of protein kinase genes in the genome of the nematode worm *Caenorhabditis elegans* is about 2% (400/19 000) (*C. elegans* sequencing consortium, 1998). Taking 1% and 3% as the lower and upper estimates of the frequency of protein kinase genes, and assuming a total complement of 19 000 genes (Bevan *et al.*, 1998), one would expect there to be between 250 and 500 protein kinase catalytic subunits in *Arabidopsis*.

We may therefore already have at least partial sequences of more than half of all *Arabidopsis* protein kinases. While such predictions are always risky, it may also be that representatives of every major protein kinase subfamily have already been sequenced. The corollary of this is that no new subfamily of protein kinase may remain to be discovered, and that the remaining gaps will merely provide additional members of known subfamilies, particularly those for which either the spatial or temporal expression is restricted. It is therefore not unreasonable to survey the plant protein kinases at this time, even though no higher plant genome sequence is yet complete.

An important caveat is that this review will only cover the members of the classical 'eukaryotic' protein kinase family, which have the 11 conserved subdomain motifs defined by Hanks and Hunter (1995a, b). Higher plants also contain intriguing members of the 'prokaryotic' two-component histidine kinase family (see the chapter by Schaller). It remains possible that other protein kinases unrelated to the classical 'eukaryotic' protein kinase family remain to be discovered.

DNAs encoding higher plant protein kinases have generally been cloned by one of four routes:

1. Cloning by sequence homology, which can be termed 'fishing' for kinases. This can either be performed utilizing polymerase chain reaction (PCR) with degenerate primers, based on the sequence motifs known to be conserved in mammalian and yeast protein kinases, or by screening complementary DNA (cDNA) libraries at low stringency with protein kinase DNAs derived from non-plant species. The first approach, which was pioneered by Hanks (1987), was used in the cloning of the first four plant kinase DNAs to be reported (Lawton *et al.*, 1989; Feiler and Jacobs, 1990; Walker and Zhang, 1990), and is largely responsible for the explosion of sequences evident in Fig. 1. However, as the various genome projects near completion, genome sequencing will tend to supersede the 'fishing' method, as the latter will be less and less likely to find novel sequences. Both of these approaches provide the predicted amino acid sequences of kinases, but give no information about their function other than by virtue of sequence similarity with kinases of known function, usually of animal or fungal origin. The 'functional' approaches [(2) through to (4) below] will therefore become increasingly important.
2. Cloning by functional homology, e.g. selection of cDNAs by their ability to complement mutations in other species, especially in the yeast *Saccharomyces cerevisiae*. Although many plant DNAs have been subsequently shown to complement yeast mutations, surprisingly this approach does not appear to have been widely used for the initial cloning. However, it was used to clone DNA encoding the LAMMER kinase AFC1 (Bender and Fink, 1994) and a cyclin-dependent protein kinase (CDK)-activating kinase (Umeda *et al.*, 1998).
3. Cloning from a partial amino acid sequence of a purified protein. Although this can be regarded as the 'classical' approach for cloning of a known protein, in practice it has been rarely used in the case of plant protein kinases, probably because very few have been purified to homogeneity from plant sources in the first place. A rare example was the original cloning of a calmodulin-like domain protein kinase (CDPK)

by Harper *et al.* (1991). The availability of complete or nearly complete genome sequences allows a new variant of this method, which lies within the realm of 'proteomics'. In this approach a polypeptide (e.g. a spot on a 2D gel, corresponding to a kinase purified by functional assays) is cleaved into fragments (e.g. using trypsin or cyanogen bromide) and the masses of the resultant peptides are determined by mass spectrometry. If a complete or extensive genome database is available, the polypeptide can usually be assigned unequivocally as a specific gene product by matching the spectrum to the masses of the peptides predicted from the DNA sequence. This method can be performed with very much smaller amounts of protein than are required to obtain amino acid sequences by conventional chemical sequencing.

4. Cloning of mutants. This approach involves selection of mutants defective in some physiological process, and then cloning of the mutant gene, e.g. by genetic mapping and chromosome walking. The latter approach is laborious, but if successful the rewards can be great, because the mutant phenotype immediately gives direct clues as to the function of the gene. A significant short cut is possible if the mutants are generated by insertional mutagenesis using *Agrobacterium tumifaciens* transfer DNA (T-DNA), since it is then much easier to locate the mutated gene. Although this approach is not specifically aimed at cloning of protein kinases, given their important role in cellular regulation, it is not surprising that many protein kinases have emerged from these genetic screens. Interesting examples include the disease-resistance genes *Xa21* from rice (Song *et al.*, 1995) and *Pto* from tomato (Martin *et al.*, 1993), and the genes giving rise to the developmental mutants *CLAVATA1* (Clark *et al.*, 1997) and *TOUSLED* (Roe *et al.*, 1993), and the brassinosteroid-resistant mutant *BRI1* (Li and Chory, 1997), all from *Arabidopsis*. This method also allows the function of specific genes identified by genome sequencing to be addressed: libraries of insertional mutants can be screened by PCR to find lines where the mutagen has inserted into the gene of interest.

II. CLASSIFICATION OF PROTEIN KINASES

Protein kinases can be classified by various biochemical criteria, such as by their substrate specificity or their regulatory properties. However, an inherent problem with a purely biochemical approach is the difficulty in comparing results obtained by different laboratories, especially when different species or tissues are used: it is often difficult to know whether the protein kinases under study correspond to the same molecular entities. A protein kinase is only rigorously defined when its DNA and/or amino acid sequence is established. Similarities in amino acid sequence would therefore seem the most logical method to classify them.

Figure 2 shows a phylogenetic tree derived from sequences of *Arabidopsis* protein kinases that were in the EMBL database in July 1999. Although the analysis was based on 111 kinase sequences, only 48 are shown in Fig. 2 to make the diagram more legible. Certain of the major protein kinase subfamilies (indicated by bold type at extreme right) are represented by two members only. More complete phylogenetic trees for these subfamilies are presented in later figures. This analysis is restricted to *Arabidopsis* partly to reduce the scale of the task. However, this also obviated an inherent difficulty in comparing protein kinases from different species, i.e. deciding whether two closely related sequences represent true interspecific homologues (i.e. orthologues) or just close relatives (i.e. paralogues). Answers to this question may become possible when the order of genes along chromosomes in different species can be compared, but at present it remains a difficulty. A second restriction was to only consider kinases where the available sequence was complete or nearly complete, since it is sometimes difficult to be certain about classification when this information is incomplete. Therefore EST sequences have not been considered. A final restriction was that, with one exception discussed later, the phenograms were produced using only the sequences of the core kinase domains [as defined by Hanks *et al.* (1988)]. The reason for this was that, when analysing large numbers of sequences, the alignment programs perform better with shorter sequences of approximately the same length. However, kinases that cluster together according to their kinase domain sequences do tend to be related in other regions as well.

Figure 2 shows that the initially confusing array of 111 kinases can be grouped into about 14 subfamilies, i.e. the seven subfamilies represented in bold type, plus the CK2, CDC2, LAMMER, NPH, PVPK1, S6 kinase, CK1, NAK and CTR1/Raf-like subfamilies. In the remainder of this review each of these subfamilies will be consided and our current knowledge of their structure and function will be briefly discussed. There are also a few interesting outliers on the tree, such as the TOUSLED kinase, that will be considered under 'Miscellaneous kinases'. Hopefully this treatment will provide a useful introduction to the more detailed discussions in later chapters. The next section briefly discusses the general functions of protein kinases in animals and fungi.

III. FUNCTIONS OF PROTEIN KINASES IN ANIMALS AND FUNGI

The study of protein kinases by biochemical approaches has a much longer history in animal cells (dating back to the 1950s), and there are of course much larger numbers of researchers working in that field. We also have a rather detailed knowledge of protein kinase function in the yeast *Saccharomyces cerevisiae* (and to a lesser extent in *Schizosaccharomyces pombe*) owing to the amenability of these organisms to genetic manipulation, and the fact that the genome sequence is already complete in the former case. Not surprisingly, therefore, protein kinase function is better understood in animal cells, and in yeast, than in plants. Although one needs to maintain an open mind and should not stick to this approach too slavishly, animal and yeast protein

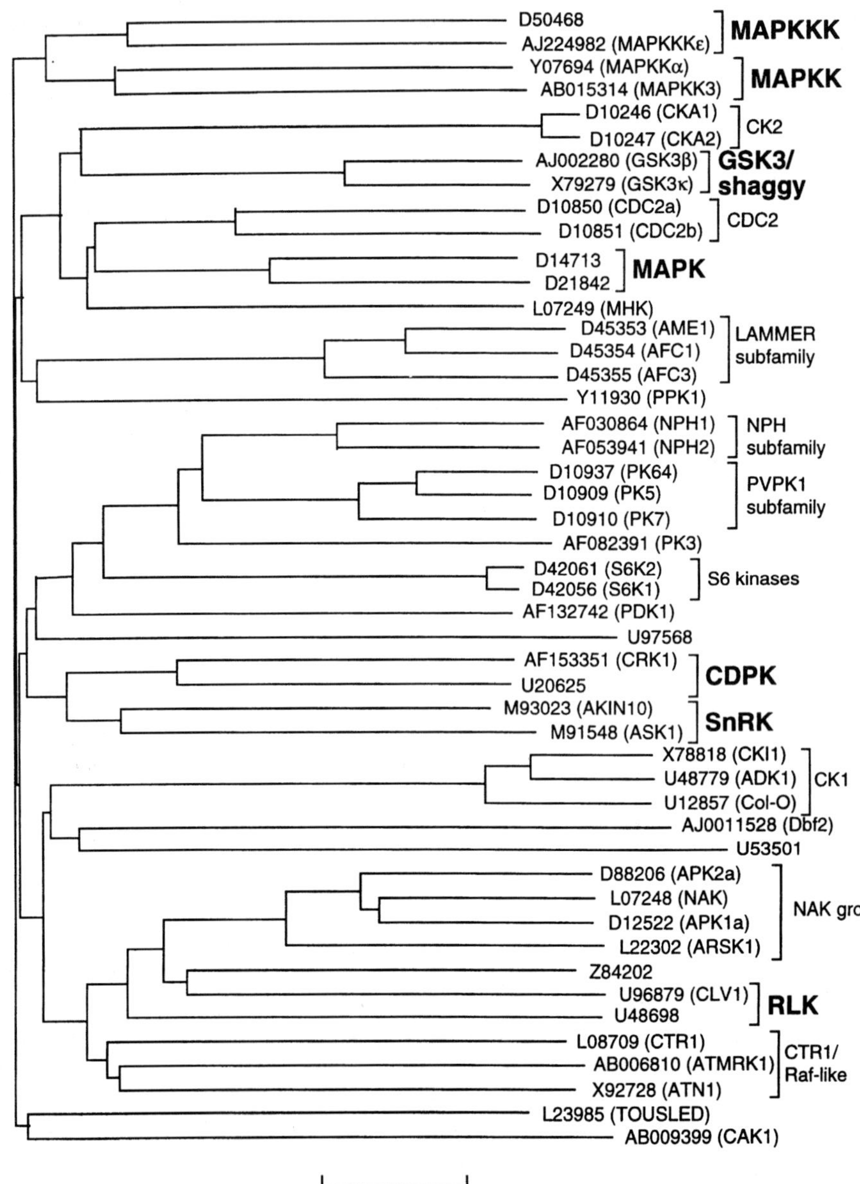

kinases provide good models for the functions of plant systems, particularly where they appear to be homologues. This approach has validity because the more insight we gain into signal transduction pathways, the earlier they appear to have evolved and the more conserved they appear to be. In this section a brief overview will be provided of the functions of animal and yeast kinases. In later sections, when discussing the different subfamilies of plant kinases, the section will often begin by comparing them with their animal or fungal homologues.

As pointed out previously (Hardie, 1999), an analogy can be drawn between the complex network of protein kinases and phosphatases in a eukaryotic cell and the central processor unit (CPU) of a computer. Like a CPU, the kinase/phosphatase network receives *input* information from receptors and sensor proteins, processes and integrates it, and then converts it into appropriate *output*, i.e. changes in metabolism, ion fluxes, gene expression, or cell growth and division. The analogy is imperfect in that most kinase–phosphatase switches work in analogue fashion (i.e. being capable of intermediate states between 'off' and 'on') rather than being purely digital. Nevertheless this remains a useful general way of thinking about the functions of kinases and phosphatases, i.e. that they are involved in processing of *information*. To make this concept more concrete, some of the key inputs and outputs that are interconnected by the protein kinase/phosphatase network in animal and fungal cells are considered below.

A. INPUTS

1. *Soluble extracellular messenger molecules*, i.e. hormones, growth factors and neurotransmitters (animals), or mating factors (yeast). The receptors for these messengers may themselves be protein kinases, particularly transmembrane proteins with protein kinase domains on the intracellular face, or they may interact with cascades of protein kinases via protein–

Fig. 2. Phylogenetic tree (phenogram) of kinase domains derived from full-length kinase sequences from *Arabidopsis thaliana*. The tree shows a selection of 48 sequences from the total set of 111 analysed for this chapter. Seven major protein kinase subfamilies shown in bold, i.e. receptor-like kinases (**RLK**s), the **GSK3/shaggy** subfamily, the MAP kinases (**MAPK**s), MAP kinase kinases (**MAPKK**s), MAP kinase kinase kinases (**MAPKKK**s), the calcium-dependent protein kinases (**CDPK**s) and Snf1-related kinases (**SnRK**s), are represented in this parent tree by one or two examples only: subtrees showing the whole subfamily are shown in later figures. Individual kinases are labelled with their EMBL/GENBANK accession number, followed by the commonly used acronym, where available. The tree was created using the ClustalX (Thompson *et al.*, 1997) graphical multiple alignment package using a neighbour-joining method. The branch lengths are proportional to the distance between the sequences. The scale bar at the bottom corresponds to 0.1 evolutionary distance units.

protein interactions. Classic examples include growth factor receptors of the protein–tyrosine kinase class, the mitogen-activated protein (MAP) kinase cascades that lie downstream of many receptors, and protein kinases that respond to second messengers such as cyclic adenosine monophosphate (cAMP; protein kinase A), phosphatidylinositol-3,4,5-triphosphate (protein kinase B), diacylglycerol (protein kinase C) and calcium (calmodulin-dependent protein kinases).

2. *Immobilized messenger molecules*, i.e. messengers anchored in the surface of neighbouring cells, or in the extracellular matrix. In these cases protein kinases play important roles in development and differentiation by conveying information about the identity of neighbouring cells. A good example is the Eph family of protein–tyrosine kinase receptors, and their protein ligands the ephrins (Zhou, 1998). These are involved, for example, in guiding axons of nerve cells to grow towards appropriate targets by inhibiting inappropriate cell–cell interactions. It could be argued that responses to soluble and immobilized messengers [categories (1) and (2)] are not intrinsically different. However, they are listed separately because perception of positional information would appear to be particularly crucial in plant development, and receptor-linked kinases may be playing important roles in this process.
3. *'Foreign' molecules*. Although the antibodies on B cells and the receptors on T cells that recognize 'foreign' molecules are not themselves protein kinases, they do activate cascades of downstream kinases to initiate immune responses. These are analogous, although apparently not homologous, to the intriguing systems by which plants respond to pathogen invasion.
4. *Stressful environmental conditions, including starvation for key nutrients.* Examples of protein kinases activated by environmental stress or starvation for specific nutrients include the Hog1 kinase, involved in the response to osmotic stress (Maeda *et al.*, 1994), and the Snf1 kinase (Hardie *et al.*, 1998), involved in the response to glucose deprivation, both in *S. cerevisiae*.

B. OUTPUTS

1. Changes in Metabolism

Protein kinases in animal cells were discovered in the 1950s and 1960s through their ability to phosphorylate and regulate metabolic enzymes such as phosphorylase and glycogen synthase. Although the kinases that regulate them are only just being defined in molecular terms, regulation of plant metabolic enzymes such as pyruvate dehydrogenase (Thelen *et al.*, 1998), phosphoenolpyruvate carboxylase (Hartwell *et al.*, 1996; Vidal and Chollet, 1997) and sucrose phosphate synthase (Huber and Huber, 1996) has been studied for several years (see also the chapter by Toroser and Huber in this volume).

2. Changes in Membrane Transport

Ion channels and pumps are important classes of targets for mammalian protein kinases, and this is also the case in plants (see the chapter by Li and Assmann).

3. Changes in Gene Expression

Transcription factors are another very important class of targets for mammalian and yeast protein kinases. Once again, this is likely to be true for plants, and Ca^{2+}-stimulated protein kinases phosphorylate the bZIP transcription factor HBP-1a(17) (Meshi *et al.*, 1998), while CK2 phosphorylates CCA1, a Myb-related transcription factor (Sugano *et al.*, 1998). Changes in gene expression can of course profoundly influence any of the other outputs listed here, such as metabolism and membrane transport (see above), cell division and development/differentiation (see below).

4. Changes in Cell Division (Cell Cycle)

This is the particular province of the cyclin-dependent protein kinases (CDKs) that control entry into S phase and mitosis, being regulated through their association with cyclins and inhibitor proteins, and via phosphorylation via other kinases and phosphatases.

5. Changes in Development and Differentiation

Although there are few, if any, cases where developmental pathways have been worked out in complete molecular detail, many mutations that lead to abnormal development occur in genes encoding protein kinases. Examples in animals include the *sevenless* and *shaggy/zeste-white3* genes in *Drosophila melanogaster*, which encode members of the receptor protein–tyrosine kinase and GSK3 subfamilies, respectively. Examples in plants include *ERECTA* and *CLAVATA1* from *Arabidopsis*, both encoding members of the receptor-like kinase subfamily (see the chapter by Torii and Clark elsewhere in this volume).

The remainder of this review will provide an overview of the structure and function of plant protein kinases, utilizing the subfamilies identified in the phylogenetic tree of *Arabidopsis* protein kinases as the basis. They are considered in a somewhat arbitrary order, commencing with those for which the most information about biochemical function is available from plant systems. The phylogenetic trees were constructed using *Arabidopsis* sequences only, which inevitably introduces a bias in coverage towards that species, but where appropriate examples from other species will also be covered. Coverage of each subfamily is intended to be introductory rather than comprehensive in nature, especially because many of them will be considered in more detail in later chapters.

IV. THE CALCIUM-DEPENDENT PROTEIN KINASE (CDPK) SUBFAMILY

Figure 3 shows a phylogenetic tree for the calcium-dependent protein kinases (CDPKs) and the SNF1-related kinases (SnRKs). These two subfamilies cluster next to each other in the parent phenogram shown in Fig. 2 and will be dealt with in the next two sections of the chapter. In all living cells, free Ca^{2+} must be maintained at levels well below the multimolar concentrations that occur in sea water or the extracellular medium, due to the relatively high concentrations of cellular phosphate and the low solubility of calcium phosphate. This requires very active mechanisms for calcium homeostasis, and it has been pointed out (Sanders *et al.*, 1999) that this would have made Ca^{2+} a good candidate during evolution to become a signalling molecule. Many different stimuli acting on animal cells increase cytosolic calcium from its normal resting level of around 10^{-7} mol l^{-1} to the 'excited' level of $\geq 10^{-6}$ mol l^{-1}. Most, although not all, effects of increasing calcium in this concentration range in animal cells appear to be mediated by its binding to the small, ubiquitous calcium-binding protein, calmodulin. The Ca^{2+}–calmodulin complex in turn binds to and activates a large number of different proteins, with an important subset of these being the calmodulin-dependent protein kinases (CaMKs). The CaMKs have a variety of different functions, phosphorylating targets that range from metabolic enzymes to transcription factors. Some CaMKs are specific for individual targets (e.g. myosin light chain kinases, phosphorylase

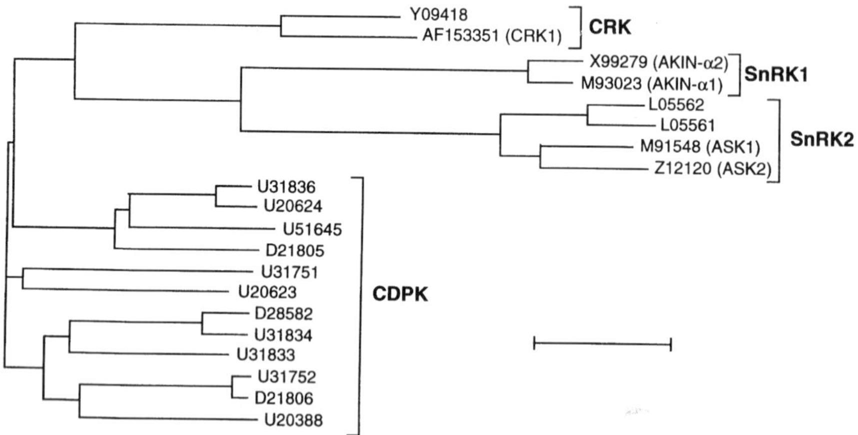

Fig. 3. Phenogram derived from sequences of the kinase domains of the Ca^{2+}-dependent protein kinase (CDPK), CDPK-related kinase (CRK), and SNF1-related kinase (SnRK1 and SnRK2) subfamilies. The tree was made using the methods described for Fig. 2. The scale bar at the bottom right corresponds to 0.1 evolutionary distance units.

kinase), whereas others have multiple targets (e.g. CaMKI, CaMKII, CaMKIV). In general, they are maintained in an inactive state in the absence of calmodulin via autoinhibitory regions, located just C-terminal to the kinase domain, that associate with the active site cleft and thus block access to exogenous substrates (Hu *et al.*, 1994; Goldberg *et al.*, 1996). In some CaMKs (e.g. CaMKII), the autoinhibitory region contains an autophosphorylation site that is phosphorylated upon binding of Ca^{2+}–calmodulin. In most other cases, the autoinhibitory region resembles a phosphorylation site but does not have a phosphorylatable amino acid, when it is referred to as a *pseudosubstrate* (Hardie, 1988).

Calcium is also a ubiquitous intracellular messenger in plants, increasing in response to numerous stimuli such as abscisic acid, gibberellin, red light, salinity/drought, osmotic and oxidative stress, touch, cold or heat shock, fungal elicitors and nodulation factors (Sanders *et al.*, 1999). Calmodulin, and calmodulin-activated enzymes, are also present. However, when a calcium-dependent protein kinase was extensively purified from soy bean it was found not to require the addition of calmodulin (Harmon *et al.*, 1987). This unexpected behaviour was elegantly explained when the DNA was cloned and was found to encode a protein kinase catalytic domain followed by a C-terminal domain with 39% identity to calmodulin, with the four 'EF-hand' Ca^{2+}-binding motifs of calmodulin being conserved (Fig. 4; Harper *et al.*, 1991). The plant Ca^{2+}-dependent protein kinases can also therefore be termed 'calmodulin-like domain protein kinases', the acronym remaining the same in either case. Although mammalian phosphorylase kinase contains calmodulin as a tightly bound regulatory subunit, the situation where a calmodulin-like domain is fused on the same polypeptide as the kinase domain has not been

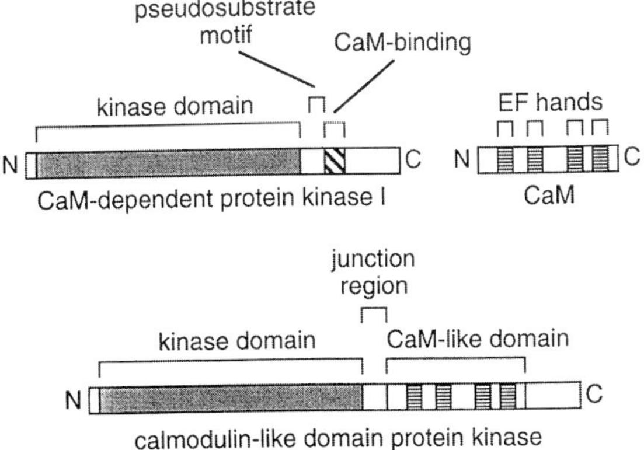

Fig. 4. Domain structure of a typical mammalian calmodulin-dependent protein kinase (CaMK1), a calmodulin, and a typical plant CDPK, showing how the latter might have arisen by gene fusion events. Redrawn from Hardie (1999).

found in animals and fungi. However, examples similar to plant CDPKs have been found in protists such as *Paramecium tetraurelia* and the malarial parasite *Plasmodium falciparum* (Son *et al.*, 1993; Zhao *et al.*, 1993).

The CDPKs will be considered in more detail in the chapter by Hrabak, but an overview is provided here. At least 12 complete CDPK sequences have been obtained from *Arabidopsis*, plus two CDPK-related kinases (CRKs; Fig. 3). DNAs encoding multiple forms have also been cloned from various other plant species. The CDPKs appear to be the largest subfamily of plant kinases (although there is a larger total number of receptor-like kinases, the latter are more divergent and fall into several distinct subfamilies). Why such a large family of CDPKs is required is not clear, although it seems likely that they are expressed in different cellular and subcellular locations, and perhaps also at different times. In agreement with this, antibodies against *Arabidopsis* CDPK6 and CDPK9 showed that their temporal and developmental expression was different (Hong *et al.*, 1996). With respect to subcellular localization, CDPK isoforms have been found to be associated with the plasma membrane in tobacco (Iwata *et al.*, 1998; Yoon *et al.*, 1999) and plasmodesmata in maize (Yahalom *et al.*, 1998). One potential cause of membrane localization is N-terminal myristoylation, which appears to occur in at least some CDPKs (Ellard-Ivey *et al.*, 1999). There are also indications that the substrate specificity of CDPK isoforms may differ. Different isoforms from soy bean have been reported to differ in their Ca^{2+} sensitivity and in their specificity for peptide substrates (Lee *et al.*, 1998). Also, activation of a stress-inducible promoter in maize leaf protoplasts was only obtained on expression of two out of four CDPKs tested (Sheen, 1996).

Although in most cases it has not been proven that they are physiological substrates *in vivo*, a large variety of proteins is phosphorylated and regulated by CDPKs in cell-free assays. These include nitrate reductase (Douglas *et al.*, 1998), sucrose synthase (Huber *et al.*, 1996), a bZIP transcription factor (Meshi *et al.*, 1998), the maize actin-depolymerizing factor ZmADF3 (Smertenko *et al.*, 1998), nodulin-26 (Lee *et al.*, 1995) and tonoplast intrinsic protein (Johnson and Chrispeels, 1992) (aquaporins in the membranes of root nodule symbiosomes and seed storage vacuoles, respectively), K^+ (Li *et al.*, 1998) and Cl^- (Pei *et al.*, 1996) channels from guard cell vacuole membranes, and plasma membrane H^+-adenosine triphosphate (ATP)ases (Camoni *et al.*, 1998; Lino *et al.*, 1998). It has also been found that activation of a stress-inducible promoter in maize protoplasts was obtained by expression of constitutively active CDPKs (Sheen, 1996). This indicates that these kinases can regulate gene expression, although in this case the identity of the actual target for CDPK was not clear.

The sequence similarities between CDPKs and mammalian CaMKs (Fig. 4) suggest that the former might have arisen by fusion of genes encoding an ancestral CaMK and a calmodulin. CDPKs contain a 'junction region' between the kinase domain and the calmodulin-like domain (CLD). There is now extensive evidence that this represents a pseudosubstrate sequence analogous to that of the CaMKs, whose inhibitory effect is relieved by binding of Ca^{2+} to the CLD (Harmon *et al.*, 1994; Harper *et al.*, 1994; Huang *et al.*, 1996; Yoo and Harmon, 1996).

Some plant Ca^{2+}-dependent protein kinases that are not members of the immediate CDPK family have been reported. Perhaps the best characterized of these is the 'chimeric Ca^{2+}/calmodulin-dependent protein kinase' (CCaMK). Poovaiah and co-workers have cloned cDNAs from *Lilium longiflorum* (Patil *et al.*, 1995) and tobacco (Liu *et al.*, 1998) encoding representatives of this subclass. CCaMKs have a kinase domain followed by a calmodulin-binding region related to that of mammalian calmodulin-dependent protein kinase II (CaMKII), then a domain with three EF-hand motifs related to the neural Ca^{2+}-binding protein visinin. When expressed in bacteria, the *Lilium* kinase was almost completely calmodulin dependent for phosphorylation of an exogenous substrate, although the endogenous visinin-like domain did bind Ca^{2+}, and appeared to be responsible for a Ca^{2+}-activated autophosphorylation (Takezawa *et al.*, 1996). A curious feature of these kinases is that they only appear to be expressed in developing anthers, although biochemical analyses have identified what appear to be similar kinases that are expressed more widely (Pandey and Sopory, 1998). Watillon *et al.* (1993) have cloned a somewhat different cDNA from apple using screening of an expression library with radiolabelled calmodulin. The full-length clone encoded a protein with a kinase domain followed by a calmodulin-binding region related to that of CaMKII, but no calmodulin-like or visinin-like domain. Another class (CDPK-related kinases) has a sequence equivalent to a CLD but which is rather divergent from calmodulin, including non-conservative substitutions in the Ca^{2+}-binding motifs (Lindzen and Choi, 1995). There are two representatives of this class shown in Fig. 3. Furomoto *et al.* (1996) expressed an example of this class (from maize) in bacteria and found that it was active in the absence of calcium.

V. THE SNF1-RELATED PROTEIN KINASE (SnRK) SUBFAMILY

Members of this subfamily, which are considered in more detail in the chapter by Halford *et al.*, are characterized by their sequence similarity with the catalytic (α) subunits of the mammalian AMP-activated protein kinase (AMPK) and yeast (*Saccharomyces cerevisiae*) SNF1 protein kinase complexes. They are also quite similar to the CaMKs/CDPKs within their kinase domains (Fig. 3), but are unrelated in their regulatory domains and are not sensitive to Ca^{2+}. Halford and Hardie (1998) have divided the SNF1-related kinases (SnRKs) into two groups:

1. The SnRK1 subfamily, which is related to AMPK-α and Snf1p in both the N-terminal kinase domain and the C-terminal regulatory domains.
2. The SnRK2 subfamily, which is related only in the N-terminal kinase domains and have smaller, unrelated C-terminal domains.

Mammalian AMPK was discovered via biochemical assays owing to its ability to phosphorylate and inactivate biosynthetic enzymes such as 3-hydroxy-3-methylglutaryl-coenzyme A (HMG-CoA) reductase and acetyl-CoA carboxylase (Hardie *et al.*, 1998). The yeast *SNF1* gene (encoding Snf1p) was discovered genetically via mutants that would not grow on sucrose or non-fermentable carbon sources (SNF = sucrose non-fermenting; Hardie *et al.*, 1998). The SNF1 kinase complex is required for derepression of genes that are repressed by glucose, e.g. those required for growth on sucrose, or growth on non-fermentable carbon sources. Although the functions of AMPK and SNF1 might at first sight appear to be different, it is now clear that AMPK also regulates gene expression in mammals (Foretz *et al.*, 1998; Leclerc *et al.*, 1998), while SNF1 also phosphorylates acetyl-CoA carboxylase in yeast (Woods *et al.*, 1994). Both can be regarded as 'metabolic master switches' (Winder and Hardie, 1999) that effect major alterations in metabolism in response to nutritional and/or environmental stress. In the case of mammalian AMPK, the signal that switches it on is a rise in 5′-AMP coupled with a fall in ATP (Hardie *et al.*, 1999), and/or a fall in phosphocreatine (Ponticos *et al.*, 1998). The kinase therefore responds to a fall in cellular energy charge, acting like a 'fuel gauge' (Hardie and Carling, 1997). In the case of the yeast SNF1 complex, starvation for glucose is the initial stimulus that switches the system on. This correlates with an increase in the cellular AMP:ATP ratio (Wilson *et al.*, 1996), but the intracellular signals remain unknown. Both AMPK and SNF1 appear to exist *in vivo* as heterotrimeric complexes comprising a catalytic α subunit (AMPK-α1, -α2 in mammals, and yeast Snf1p) and non-catalytic β subunits (AMPK-β1, β2, and yeast Sip1p, Sip2p and Gal83p) and γ subunits (AMPK-γ1, -γ2, -γ3, and yeast Snf4p). The β subunits appear to be the scaffold on which α and γ subunits assemble, and have conserved KIS and ASC domains that are the sites of interaction with the α and γ subunits, respectively (Fig. 5). The N-terminal regions of the β subunits are very divergent, and it has been suggested that these might have a targeting role (Hardie *et al.*, 1998). The γ subunits are composed of four tandem CBS domains (Bateman, 1997) and appear to be required, in the presence of the activating signal, to relieve autoinhibition by a region of the α subunit (Hardie *et al.*, 1998).

A plant gene related to *SNF1* was first cloned by Halford's group in 1991, and was shown to rescue a *snf1* mutation in yeast (Alderson *et al.*, 1991). Members of this SnRK1 family have now been cloned from a large variety of higher plant species (see the chapter by Halford *et al.*). Halford and Hardie subdivided the SnRK1 subfamily into two subclasses, SnRK1a and SnRK1b; the latter appear to be expressed exclusively in developing seeds (Halford and

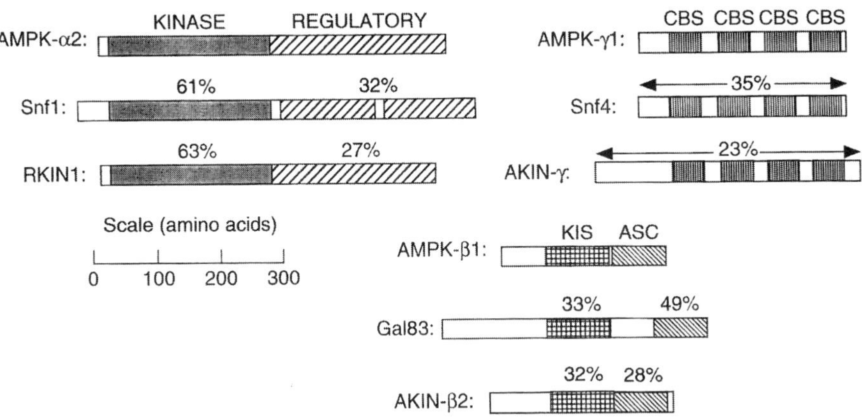

Fig. 5. Domain structures of the catalytic α (top left) and regulatory β and γ subunits (top right, bottom) of mammalian AMP-activated protein kinase (AMPK), *S. cerevisiae* SNF1 and higher plant SnRK1 protein kinases. A single isoform is shown as an example for each of the animal, fungal and plant kingdoms. The nomenclature of the domains is explained in the text. Figures above each domain refer to the percentage identity to the homologous mammalian domain.

Hardie, 1998; Takano *et al.*, 1998). In 1992 Hardie's group had independently identified by biochemical approaches an activity with properties very similar to those of mammalian AMPK (MacKintosh *et al.*, 1992), and tentatively named it HMG-CoA reductase kinase (HRK) after one of its potential substrates. This kinase had a catalytic subunit of only 58 kDa (Ball *et al.*, 1994) which was identified using antibodies as a product of a *SnRK1* gene (Ball *et al.*, 1995).

Two genes encoding *Arabidopsis* members of the SnRK1 subfamily are currently in the EMBL database (Fig. 3). Originally called AKIN10 and AKIN11, these are now renamed AKIN-α1 and -α2 to establish consistency with the nomenclature of catalytic subunit isoforms of the mammalian kinase (AMPK-α1 and -α2). At the messenger RNA (mRNA) level, SnRK1-α1 and -α2 appear to be ubiquitously expressed in *Arabidopsis*.

Although it has a catalytic subunit of 58 kDa that appears to be encoded by a SnRK1 gene, the AMPK-related kinase identified by Hardie's group in cauliflower (Ball *et al.*, 1994) had a native molecular mass of 150–200 kDa, suggesting that it was an oligomer like its mammalian and yeast counterparts. Very recently, homologues of the mammalian and yeast β and γ subunits have been cloned from *Arabidopsis* (Bouly *et al.*, 1999) and a γ homologue from potato (Lakatos *et al.*, 1999). These appear to interact with the SnRK1 catalytic subunits, at least in the yeast two-hybrid system. As expected, the β subunits have conserved KIS and ASC domains, while the γ subunits have four CBS domains (Fig. 5). These findings indicate that the plant SnRK1 kinases may exist as heterotrimeric $\alpha\beta\gamma$ complexes, like their mammalian and yeast counterparts.

The specificity of the cauliflower and spinach SnRK1 kinases for peptide substrates is very similar to those of their mammalian (AMPK) and yeast (SNF1) homologues, with the motif $\Phi(X,\beta)XXSXXX\Phi$ [Φ = hydrophobic (M, L, I, F, V); β = basic (R, K, H)] being recognized (Dale *et al.*, 1995b; Sugden and Hardie, 1998). At least three important metabolic enzymes are phosphorylated and inactivated by the SnRK1 kinases *in vitro*, i.e. HMG-CoA reductase (Dale *et al.*, 1995a), and sucrose phosphate synthase and nitrate reductase (Sugden and Hardie, 1998). In each case the site phosphorylated conformed to the recognition motif described above. Although it has not yet been proved conclusively that these enzymes are substrates *in vivo*, they are the key regulatory enzymes of isoprenoid synthesis, sucrose synthesis and nitrogen assimilation, respectively, implying that activation of the SnRK1 kinase would switch off these important biosynthetic pathways. It is notable that one function of mammalian AMPK is to switch off biosynthesis, and HMG-CoA reductase is regulated by AMPK in mammals by phosphorylation at a serine exactly equivalent to that phosphorylated by SnRK1 on *Arabidopsis* HMG-CoA reductase (Dale *et al.*, 1995a). It may also be noted that sucrose phosphate synthase and nitrate reductase are phosphorylated at the same sites by CDPKs (see above), introducing the possibility that other CDPK substrates might also be targets for SnRK1 kinases. The similarity in specificity between CDPKs and SnRK1s is not surprising given that the sequences of their kinase domains cluster together in a phylogenetic analysis (Figs 2 and 3).

The SnRK1 kinases also appear to be involved in regulating gene expression, because transgenic potato plants in which SnRK1 expression was reduced in tubers or leaves by an antisense approach exhibited abnormal expression of sucrose synthase (SuSy). In tubers, SuSy mRNA expression was abolished, whereas in leaves the induction of SuSy mRNA by high sucrose was prevented (Purcell *et al.*, 1998). This is intriguing, since SuSy (despite its name) is thought to be involved in sucrose catabolism, and fulfils a metabolic role analogous to that in yeast of invertase, the latter encoded by *SUC2*, which is one of the classic genes regulated by *SNF1* in that organism. Although the results of Purcell *et al.* (1998) strongly suggest that the SnRK1 kinases regulate gene expression, the targets for phosphorylation by the kinases responsible for this remain unknown.

The signals that regulate the SnRK1 kinases, and the physiological circumstances under which they are switched on, also remain unclear at present. The mammalian AMPK cascade is switched on by an increase in 5′-AMP and a fall in ATP, which acts by no fewer than four mechanisms (Hardie *et al.*, 1999): (i) allosteric activation of AMPK; (ii) binding of AMP to AMPK, making it a better substrate for an upstream kinase; (iii) binding of AMP to AMPK, making it a worse substrate for protein phosphatases; and (iv) allosteric activation of the upstream kinase. The upstream kinase (AMP-activated protein kinase, AMPKK) phosphorylates AMPK at threonine-172 (Thr172) within the 'activation loop' (A loop) between the conserved DFG and APE motifs. Like AMPK, the SnRK1 kinases are also activated by phosphorylation (an effect mimicked by mammalian AMPKK *in vitro*) and inactivated by dephosphorylation (MacKintosh *et al.*, 1992; Ball *et al.*, 1994; Sugden and Hardie, 1998). It has recently been demonstrated using a phosphopeptide-specific antibody that the phosphorylation of the SnRK1 kinases occurs at a site equivalent to Thr172, corresponding to Thr175 on *Arabidopsis* AKIN-α1 (Sugden *et al.*, 1999). Although the regulation by phosphorylation is therefore highly conserved between the mammalian AMPKs and the plant SnRK1s, the regulation by 5′-AMP is not completely conserved. In particular, AMP does not appear to activate allosterically the SnRK1 kinases [mechanism (i) above]. Because the upstream kinase has not been characterized from plants, it is not yet feasible to test mechanisms (ii) and (iv). However, AMP does inhibit dephosphorylation of spinach SnRK1 by protein phosphatase-2C [mechanism (iii)], an effect that appears to be specific since GMP has no effect at the same concentrations (Sugden *et al.*, 1999). This raises the possibility that AMP does activate the SnRK1 system *in vivo* by at least some of the same mechanisms that activate the AMPK cascade, and that the former might be a sensor of cellular energy charge, like its mammalian counterpart.

Although the physiological circumstances that cause activation of the SnRK1 cascade *in vivo* remain unclear there are indications that, as in yeast, the availability of carbohydrate might be a key factor. This proposal comes from observations that antisense expression of SnRK1 expression in potato leaves prevents sucrose- or glucose-induced induction of sucrose synthase (Purcell *et al.*, 1998), and from a recent observation that incubation of *Arabidopsis* plants with high sucrose causes two- to three-fold activation of a SnRK1 kinase (Bhalerao *et al.*, 1999). This would be analogous to the role of the SNF1 kinase in yeast, except that the yeast kinase appears to be required for the response to *low* rather than high carbohydrate.

There appear to be four members of the SnRK2 subfamily in the *Arabidopsis* sequences currently in the EMBL database (Fig. 3). Members of this subfamily have kinase domains related to those of the SnRK1 family, but their C-terminal regions are shorter (giving an overall molecular mass of ≈45 kDa) and are not related to those of the SnRK1 subfamily. A characteristic feature of these C-terminal domains is highly acidic patches, consisting of runs of either aspartic or glutamic acid residues. The SnRK2 kinases have not yet been identified biochemically: Ball *et al.* (1994) resolved a cauliflower kinase (termed HRK-B) of native molecular mass 45 kDa, which had a specificity for peptide substrates very similar to that of the SnRK1 kinase (HRK-A) from the same species. In view of the similarity in the kinase domains between the SnRK1 and SnRK2 subfamilies, HRK-B is a good candidate to be a product of a *SnRK2* gene, although this remains unproven.

The functions of the SnRK2 subfamily also remain unclear. The 'founder' member of the family, wheat PKABA1 (Anderberg and Walker-Simmons, 1992), is induced at the mRNA level by drought and abscisic acid, but some other members appear to be expressed constitutively. Perhaps the best clue to the function of these kinases has come from a recent genetic study in the green alga *Chlamydomonas reinhardtii* (Davies *et al.*, 1999). In this organism, mutations in the *SAC3* gene lead to a failure to express a high-affinity sulfate transport system in response to depletion of medium sulfate. They also fail to repress fully expression of arylsulfatase (required for metabolism of organic sulfate esters) in the *presence* of medium sulfate. When the gene was cloned, it was found to encode a member of the SnRK2 family. These findings suggest that the Sac3 protein kinase is involved in a signalling pathway that responds to sulfate availability, and that it has both positive (sulfate transporters) and negative (arylsulfatase) effects on sulfate-regulated genes. Whether the SnRK2 kinases in higher plants are also involved in the response to sulfate starvation is unclear, but it is an attractive hypothesis given that the closely related SnRK1 kinases may be involved in the response to the availability of a carbon source.

VI. THE RECEPTOR-LIKE KINASE (RLK) SUBFAMILY

The receptor-like kinases (RLKs) are characterized by an N-terminal hydrophobic signal peptide, and an internal hydrophobic sequence flanked by basic 'stop-transfer' sequences, suggesting that they are type 1 membrane proteins. The C-terminal, intracellular regions are the locations of the protein kinase domains. At the time of writing there were at least 27 RLKs in *Arabidopsis*; by phylogenetic analysis their kinase domain sequences fall into a discrete cluster, although there is a small number of non-transmembrane kinases that also fall into the same cluster (Fig. 2).

The models for this group are the receptor-linked kinases in animal cells, which are also type 1 membrane proteins with an intracellular kinase domain. The majority of animal receptor-linked kinases is protein–tyrosine kinases (a subfamily not yet found in plants) but there is also a small number of protein–serine/threonine kinases. The mechanisms of signal transduction downstream of mammalian receptor-linked kinases may be summarized as follows.

1. The receptor-linked protein–tyrosine kinases usually exist as monomers in the absence of ligand. Ligand binding causes the receptors to form homo- or hetero-oligomers, usually dimers. The cytoplasmic domains phosphorylate each other, often at multiple sites, and this creates docking sites for proteins with domains (e.g. SH2 domains) that form high-affinity binding sites for specific sequence motifs containing phosphotyrosine. By this means, large signalling complexes assemble at the membrane in response to ligand binding. Components of these complexes include enzymes that generate second messengers such as inositol-1,4,5-trisphosphate (IP_3) and

phosphatidylinositol-3,4,5-trisphosphate (PIP_3), factors that promote conversion of Ras family proteins between their active (GTP-bound) and inactive (GDP-bound) states, protein–tyrosine phosphatases, and additional protein–tyrosine kinases [see (2) below].

2. Some protein–tyrosine kinases, e.g. those in the Src family, are not transmembrane proteins. However, they associate with the inner surface of the plasma membrane via their myristoylated N-termini, and contain SH2 domains that cause them to associate with receptor-linked kinases after the latter are activated.
3. Mammalian serine/threonine-kinase receptors include those for transforming growth factor-β, activins, and bone morphogenetic proteins. Each ligand has distinct type I and type II receptors (both of which are protein–serine/threonine kinases), and ligand binding causes formation of heterodimers. The type II receptor phosphorylates the type I receptor, activating the latter which then phosphorylates proteins of the SMAD family. The phosphorylated SMADs associate with other SMADs and translocate to the nucleus, where they trigger changes in gene expression (reviewed in Kretzschmar and Massague, 1998).

The plant receptor-like kinases are thought to be receptors mainly by virtue of their structural similarities to the mammalian receptor kinases, although a ligand has not yet been identified for any of them. However, the idea that they are receptors was strengthened with the cloning of the *BRI1* gene from *Arabidopsis* (Li and Chory, 1997). *Bri1* mutants are resistant to the brassinosteroid brassinolide, and map-based cloning of the *BRI1* gene showed that it encoded an RLK. This is consistent with the idea that BRI1 is a brassinosteroid receptor, although the true ligand may be a complex between the steroid and a binding protein.

Figure 6 shows a phylogenetic tree constructed by analysing the *extracellular* domains of the 27 *Arabidopsis* RLKs. Most of them cluster into five groups depending on the nature of this extracellular domain:

1. *The leucine-rich repeat (LRR) group* (reviewed in Walker, 1994; Braun and Walker, 1996), currently the largest group in *Arabidopsis*. LRRs occur in numerous eukaryotic proteins (Kobe and Deisenhofer, 1994), and are thought to be involved in mediating protein–protein interactions. Intriguingly, LRRs are present in several mammalian receptors for protein/peptide messengers, including some receptor-linked kinases such as the nerve growth factor receptor. LRR proteins contain multiple tandem copies of a $\approx$25 residue repeat with leucines at characteristic spacings. The crystal structure of ribonuclease inhibitor protein shows that its 16 LRRs form a horseshoe-shaped structure with a 16-strand parallel β sheet on the inside and 16 short α helices on the outside. It is proposed that the exposed surface of the β sheet forms the site for protein–protein interactions (Kobe and Deisenhofer, 1993).

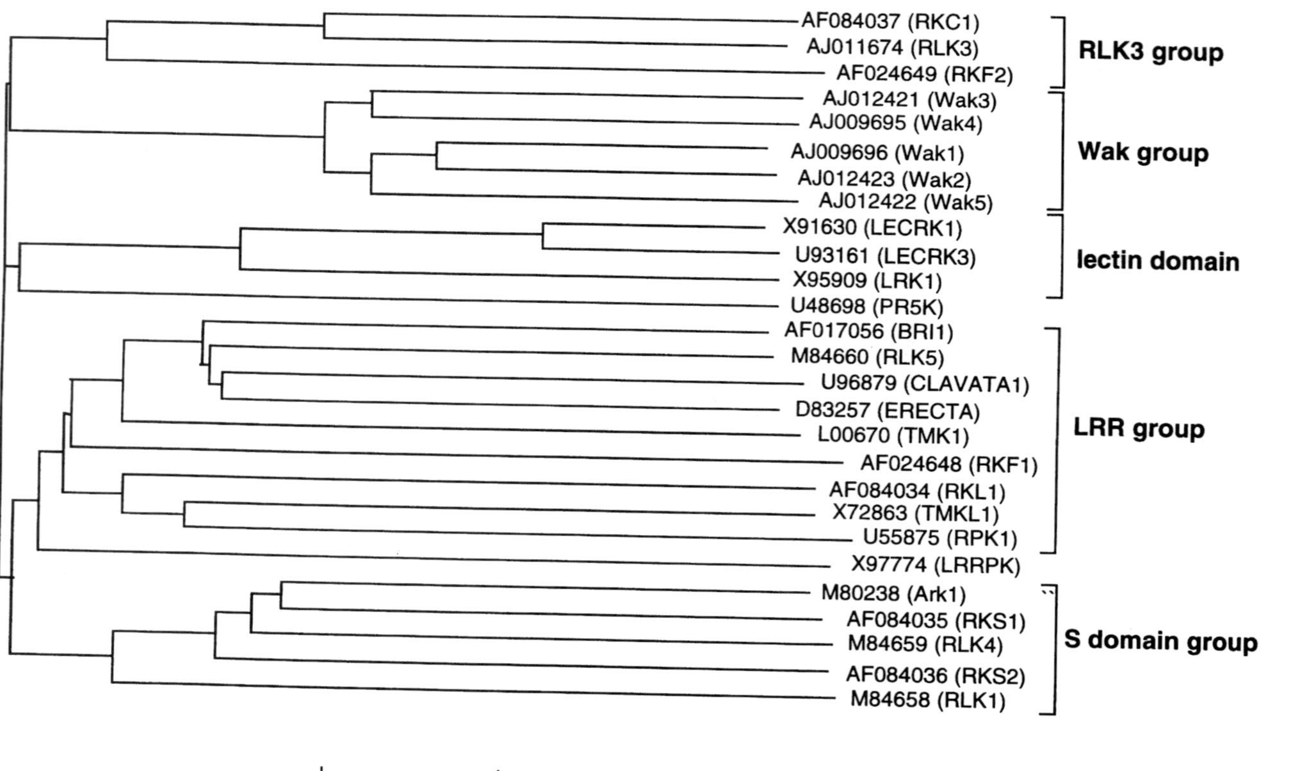

Fig. 6. Phenogram derived from sequences of the *extracellular* domains of *Arabidopsis* receptor-like kinases. The tree was made using the methods described for Fig. 2. The scale bar at the bottom corresponds to 0.1 evolutionary distance units.

2. *The wall-associated kinase (Wak) group*, the original member of this family being Pro25 (Kohorn *et al.*, 1992). Its extracellular domain was found to be tightly associated with the cell wall, so the name was changed to wall-associated kinase 1 (Wak1). Five *Wak* genes (*Wak1–Wak5*) have recently been shown to be present as a tight cluster on chromosome 1 of *Arabidopsis* (He *et al.*, 1999). *Wak4* mRNA was detected only in siliques, while the other isoforms were more widely expressed. The encoded proteins are all transmembrane proteins with the intracellular kinase domains being very closely related. The extracellular regions are less highly conserved, but all have patterns of cysteine residues reminiscent of the mammalian EGF-like repeats, while some have regions with similarity to the mammalian extracellular matrix proteins collagen, neurexin and tenascin. It has been suggested that these kinases provide a signalling pathway between the cell wall and the cell interior: a possible analogy is with the systems which provide signalling pathways between the extracellular matrix and the cytoplasm in mammalian cells, such as the integrin-focal adhesion kinase system (Schlaepfer and Hunter, 1998). A clue to the function of *Wak1* came from observations that its mRNA was induced by salicylate (or its analogue 2,2-dichloroisonicotinic acid, INA), or by infection with the pathogen *Pseudomonas syringae*. Moreover, expression of *Wak1* in antisense showed that normal *Wak1* expression was required for plants to survive treatment with INA, while overexpression of *Wak1* or its kinase domain in sense orientation gave plants that were resistant to high levels of salicylate (He *et al.*, 1998). Thus, Wak1 appears to be involved in the response to pathogen invasion.
3. *The S-domain group*, members of which have extracellular domains related to the S-locus glycoprotein (SLG) of *Brassica* species, involved in the *self-incompatibility* (SI) response (Nasrallah *et al.*, 1994). This response is discussed in more detail below (see also the chapter by Cock), but it is worth noting that the five S-domain RLKs found in *Arabidopsis* to date (Fig. 6) cannot be involved in self-incompatibility because they are expressed in inappropriate locations, and because *Arabidopsis* does not appear to display this phenomenon anyway. An S domain is characterized by a characteristic array of cysteine residues, and other conserved motifs (Walker, 1994).
4. *The lectin domain group*, members of which have an extracellular domain related to lectins. The extracellular domain of LECRK1 is related to legume lectins and to the Lec2 protein of *Arabidopsis* (Herve *et al.*, 1996). These relationships raise the intriguing possibility that these receptors bind oligosaccharides, such as the elicitors derived from breakdown of the cell wall (either of the host or of the pathogen) during fungal infection.
5. *The RLK3 group*, comprising three members including RLK3 itself. Their extracellular domains contain conserved patterns of leucines and cysteines that are reminiscent of the LRR and S-domain class, but they cannot be

assigned to either of those two classes. Although it is expressed under basal conditions, *RLK3* mRNA is strongly induced by salicylic acid and by oxidative stresses such as H_2O_2 and menadione (Czernic *et al.*, 1999).

Other *Arabidopsis* RLKs that cannot readily be placed into any of these groups include PR5K and LRRPK (Fig. 6). PR5K (Wang *et al.*, 1996) has an extracellular domain related to the *Arabidopsis* PR5 protein, which accumulates in the extracellular space in response to infection by various pathogens. Unlike PR5, PR5K is not induced upon infection, but their sequence relationship raises the possibility that PR5K might be a receptor involved in the response to infection. LRRPK (Deeken and Kaldenhoff, 1997) has an extracellular domain with a novel leucine zipper motif. It has the interesting property that levels of its mRNA are repressed by light, but its function remains unclear.

As discussed above, the extracellular domains of many RLKs are related to other proteins, and these relationships can give tantalizing clues as to their possible function. However, much more direct evidence regarding function is available for those cases where the DNA was obtained by cloning of mutant genes. The *BRI1* gene (Li and Chory, 1997), encoding an LRR kinase that may represent a brassinosteroid receptor, has already been mentioned. Two other members of the LRR subfamily in *Arabidopsis*, i.e. *ERECTA* and *CLAVATA1* (*CLV1*), give rise to developmental mutants. *ERECTA* is expressed at and around apical meristems, and mutations lead to plants with compact inflorescences with short, thick peduncles and fruits. It is proposed that the Erecta protein may be involved in cell–cell signalling during meristem development (Torii *et al.*, 1996). *CLV1* is expressed in a patch of cells in the centre of the apical meristem, and mutations lead to greatly enlarged shoot and floral meristems, with large clusters of undifferentiated cells at the centre. These mutations therefore appear to upset the balance between cell proliferation and differentiation. Signalling through Clv1 may inhibit proliferation of stem cells, promote their differentiation into organ primordia, or both (Clark *et al.*, 1997).

A particularly interesting group of plant kinases has been defined by cloning genes responsible for disease resistance. The *Xa21* gene in rice is responsible for resistance to some strains of the bacterial pathogen *Xanthomonas oryzae*, and encodes an RLK in the LRR group (Song *et al.*, 1995). This suggests that it is a receptor for some product present during infection, although the actual ligand is not known. In wheat, the *Lrk10* gene, encoding an RLK with some similarity to the S-domain group, is within the *Lr10* locus conferring wheat rust resistance, and is either the resistance gene itself, or is tightly linked to it (Feuillet *et al.*, 1997). In tomato, the *Pto, Fen* and *Ptil* genes encode protein kinases that are involved in the resistance to pathogens and the insecticide fenthion (see the chapter by Sessa and Martin). Although none of these genes encodes transmembrane proteins and their products are not therefore cell surface receptors, Pto and Fen may form receptors for foreign compounds by forming complexes with the leucine-rich repeat protein Prf (Salmeron *et al.*, 1996).

A final instance where functions of plant RLKs have been studied by genetic approaches is the phenomenon of self-incompatibility in Brassicaceae (see the chapter by Cock). SI prevents pollen grains from genetically similar plants from germinating when they fall on the stigma. In the type found in Brassicaceae, SI is controlled by the *S*-locus complex, a highly polymorphic gene cluster that occurs in a number of distinct haplotypes (Nasrallah *et al.*, 1994). Two of the genes encoded at the *S* locus are the *S*-locus glycoprotein and the *S*-locus receptor-like kinase (SRK), which has an extracellular domain closely related to the SLG present in the same haplotype (Stein *et al.*, 1991). SRK was the original representative of the S-domain RLK subfamily (see above). Both SLG and SRK are co-ordinately expressed in the stigma (Stein *et al.*, 1996). A current hypothesis (Stein *et al.*, 1996) is that SRK (possibly in combination with SLG) forms a receptor for some component of the pollen grain specified by the *S* locus. A match between the pollen component and SLG/SRK causes a response that prevents the pollen grain from germinating.

Overall, these studies on the functions of RLKs would indicate that they are involved in signal transduction in response to interactions with neighbouring cells. These neighbouring cells can either be from the same plant (e.g. Erecta, Clavata1?), or a pathogen (e.g. Xa21, Wak1, PR5K?, Lrk10?), or self-incompatible pollen grains from the same species (e.g. SRK). This suggests an interesting scenario where the response to both friend and foe might have arisen from a common ancestral response system.

VII. THE MAP KINASE (MAPK), MAP KINASE KINASE (MAPKK), AND MAP KINASE KINASE KINASE (MAPKKK) SUBFAMILIES

These are three separate subfamilies, and the MAPK subfamily is separated from the MAPKKs and MAPKKKs in the main phylogenetic tree (Fig. 2), but it is convenient to consider them together. In yeast and mammalian cells they form cascades or 'modules' operating in the direction MAPKKK → MAPKK → MAPK. The *S. cerevisiae* genome encodes six MAPKs, with the two-best characterized MAPK modules acting downstream of mating factor receptors (Ste11 → Ste7 → Fus3) and sensors of high osmolarity (Ssk2/22 → Pbs2 → Hog1) (Madhani and Fink, 1998). In mammalian cells the classical MAPK cascade (Raf1 → MEK1/2 → ERK1/2) lies downstream of growth factor receptors that act via the GTP-binding protein Ras. At least two other MAPK cascades (MEKK1/2 → JNKK1 → JNK, ASK1 → MKK3/

6 → p38) are also activated by cytokines and various cellular stresses (Minden and Karin, 1997).

DNAs encoding at least eight MAPKs, five MAPKKs and eight MAPKKKs have been cloned from *Arabidopsis* (Fig. 7). Representatives have also been found in other plant species including alfalfa, tobacco, oat and petunia (reviewed in Hirt, 1997). In this analysis only the members of the MEKK/Ste11 subfamily (related to mammalian MEKKs and *S. cerevisiae* Ste11) have been counted as MAPKKKs. A more comprehensive recent analysis of MAPKKKs (Jouannic *et al.*, 1999) also included the CTR1/Raf-like subfamily, which is discussed separately below.

All of the MAPKs contain the characteristic MAPK motif Thr–X–Tyr in the 'A loop' or 'activation segment' (Johnson *et al.*, 1996) between conserved kinase motifs VII and VIII. In the yeast and mammalian systems, phosphorylation at both the threonine and the tyrosine, catalysed by a single MAPKK, is necessary to cause activation. The serine or threonine residues on MAPKKs, which are phosphorylated during activation by MAPKKKs, are also conserved in the plant MAPKKs (Shibata *et al.*, 1995; Morris *et al.*, 1997). There have been rather few direct biochemical studies of whether these kinases really exist as cascades in plants. However, two *Arabidopsis* MAPKs are

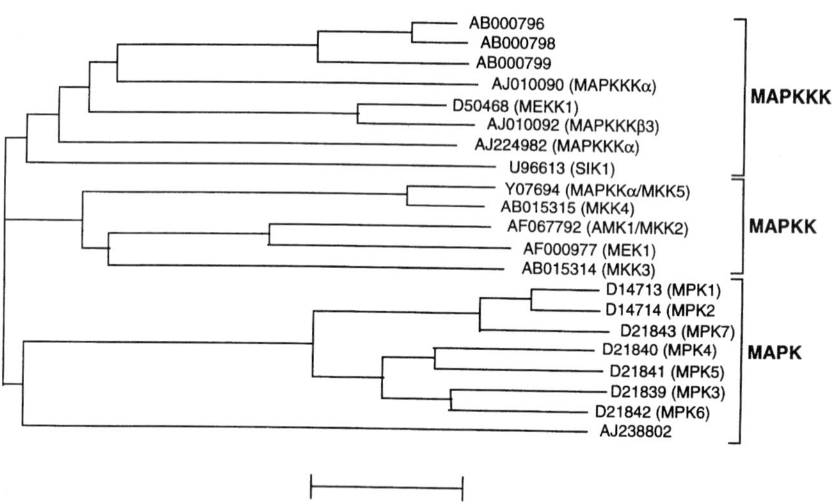

Fig. 7. Phenogram derived from sequences of kinase domains of *Arabidopsis* MAP kinases (MAPKs), MAP kinase kinases (MAPKKs) and MAP kinase kinase kinases (MAPKKKs, MEKK/Ste11 subfamily). The tree was made using the methods described for Fig. 2. The scale bar at the bottom corresponds to 0.1 evolutionary distance units. Note that the MAPKs are more closely related to the CK2, GSK3/shaggy and CDC2 groups than to the MAPKKs and MAPKKKs (see Fig. 2).

phosphorylated and activated by a *Xenopus laevis* MAPKK *in vitro* (Mizoguchi *et al.*, 1994). Overexpression of plant DNAs has also been shown to suppress mutations in specific components of MAP kinase cascades in *S. cerevisiae* (Banno *et al.*, 1993; Jonak *et al.*, 1995b; Mizoguchi *et al.*, 1996; Popping *et al.*, 1996; Nishihama *et al.*, 1997), indicating that these plant kinases can take part in specific MAP kinase cascades at the appropriate level when expressed in yeast. Recently, Mizoguchi *et al.*, 1998) have shown that MPK4 and MEK1, and MEK1 and MEKK1, interact in yeast using the two-hybrid system. In the same paper they demonstrated that co-expression of *Arabidopsis* MEK1 and MPK4 reversed the growth defect of a yeast *mpk1* Δ mutant, whereas expression of MPK4 alone, or MEK1 with any of three other *Arabidopsis* MAPKs, did not. These results suggest that MEKK1 → MEK1 → MPK4 may form a cascade in *Arabidopsis*. The same group showed by similar approaches that MKK2, a kinase detected via a two-hybrid screen with MEKK1 as bait, could replace MEK1 in this cascade (Ichimura *et al.*, 1998).

The functions of plant MAP kinase cascades are considered in more detail in the chapters by Jouannic *et al.* and Ichimura *et al.* Although no single plant MAPK cascade has yet been mapped out in detail, there is abundant evidence that they are rapidly activated by plant hormones and by environmental stimuli such as touch, cold shock, wounding and pathogen infection.

VIII. THE CYCLIN-DEPENDENT KINASE (CDK) SUBFAMILY

The yeast and mammalian cyclin-dependent kinases (CDKs), exemplified by the *cdc2* and *CDC28* gene products from *Schizosaccharomyces pombe* and *S. cerevisiae*, respectively, have catalytic subunits of about 34 kDa that are minimal kinase domains. The CDKs are characterized by being inactive except when bound to a *cyclin* protein by interactions involving the conserved PSTAIRE motif on the CDK (Jeffrey *et al.*, 1995). Some CDKs are critical regulators of entry into key phases of the cell cycle, such as S phase and mitosis. Others, such as Pho85 in *S. cerevisae*, appear to have roles not exclusively connected with the cell cycle (Kaffman *et al.*, 1994). In general, there are more cyclins than there are CDKs, with a single CDK often having multiple roles, depending on the cyclins with which it associates. The activity of CDKs during the cell cycle is regulated by association with cyclins that are synthesized and degraded at different times, by association with CDK inhibitor proteins whose content and/or activity varies during the cycle, and by phosphorylation and dephosphorylation by upstream kinases and phosphatases. An activating phosphorylation event occurs at a conserved threonine within the ‘A loop’, catalysed by CDK-activating kinases, and an inactivating phosphorylation occurs at a tyrosine residue within the sequence GXGXYG (conserved kinase motif I), catalysed in *S. pombe* by the Weel kinase.

Despite differences in the mechanics of cell division between fungi, animals and plants, the basic cell cycle control machinery appears to be conserved. In 1990 Feiler and Jacobs (1990) described cloning of a *cdc2*-like gene, and they also found a Cdc2-like protein kinase activity, in pea. Two *cdc2*-like genes, termed *CDC2a* and *CDC2b*, have been cloned from *Arabidopsis* (Ferreira *et al.*, 1991; Hirayama *et al.*, 1991). A third gene termed *CDC2c* (not represented in Fig. 2) has been found to be expressed only in flowers (Lessard *et al.*, 1999). *CDC2a*, which has an exact match to the PSTAIRE motif, partially complements *cdc2* mutations in *S. pombe*, whereas *CDC2b*, which has a PPTALRE motif instead, does not. In general, PSTAIRE variants are broadly expressed throughout the plant cell cycle, whereas PPTALRE variants are expressed in a narrower window from S phase through to M phase (Fobert *et al.*, 1996; Magyar *et al.*, 1997). It has been proposed that the PPTALRE variants might be involved in formation of the pre-prophase band (Doonan and Fobert, 1997). Recently, a CDK-activating kinase (CAK1) has also been cloned from *Arabidopsis* by complementation of an *S. cerevisiae cak1* mutant. When purified by immunoprecipitation, it phosphorylated and activated a human CDK (Umeda *et al.*, 1998). Although related to mammalian CDK-activating kinases (CDK7), CAK1 is not closely related to *Arabidopsis* CDC2a or CDC2b (Fig. 2). A fourth CDK-like kinase from *Arabidopsis* is MAK homologous kinase (MHK), related to the mammalian male germ cell-associated kinase MAK (Moran and Walker, 1993). The latter is expressed specifically in testicular germ cells and may have a role in meiosis (Jinno *et al.*, 1993), but the functions of MHK and MAK are not well understood. Although originally claimed to be a member of the CDK subfamily, in fact MHK appears to be equidistant between the CDKs and the MAPKs (Fig. 2).

DNAs encoding at least 10 A/B-type mitotic cyclins, plus three D-type G1 cyclins, have been cloned from *Arabidopsis* (Renaudin *et al.*, 1996; Day and Reddy, 1998). Homologues of several important regulatory proteins that interact with CDKs have also been reported in plants. These include a CDK inhibitor related to $p27^{Kip1}$ (Wang *et al.*, 1997), a CDK-binding protein related to *S. pombe* Suc1 (de Veylder *et al.*, 1997), and a homologue of a key target for phosphorylation by CDKs in G1 phase, the retinoblastoma protein (Grafi *et al.*, 1996; Xie *et al.*, 1996).

IX. THE CK1 AND CK2 (CASEIN KINASE) SUBFAMILIES

These two subfamilies will be considered together, although they are not in fact closely related (Fig. 2). The mammalian kinases were originally termed casein kinases because they were detected biochemically using the milk protein casein as substrate, but they are now referred to simply as CK1 and CK2 to reflect the fact that casein is not a true physiological substrate. Both kinases phosphorylate serine or threonine residues in the context of clusters of neighbouring acidic side chains. For CK1, these acidic residues are preferred on the N-terminal side of the serine/threonine, with the P − 3 position being particularly important; for CK2 they are preferred on the C-terminal side, with the P + 3 position being important. An interesting feature of the specificity of both kinases is that phosphoamino acids can substitute for acidic residues, so that other protein kinases can 'prime' phosphorylation by CK1 and CK2. A puzzling feature of both kinases, which means that their cellular functions remain somewhat enigmatic, is that there is little or no evidence for acute regulation *in vivo*. Conceivably, they provide a constitutive, basal kinase activity against which protein phosphatases (and/or kinases phosphorylating neighbouring priming sites) could provide regulation of target protein function.

Mammalian CK1s exist as monomers that are widely expressed and found in both the cytoplasm and nucleus (Cobb, 1995). They phosphorylate many substrate proteins *in vitro*, often without producing any apparent change in function. In *S. cerevisiae*, deletion of a redundant pair of CK1 genes (*YCK1* and *YCK2*, also known as *CKI1/CKI2*) is lethal (Robinson *et al.*, 1992; Wang *et al.*, 1992), while disruption of a third (*HRR25*) leads to hypersensitivity to DNA-damaging agents, suggesting a function in DNA repair (DeMaggio *et al.*, 1992). Figure 2 shows that at the time of writing there were three full-length CK1 sequences from *Arabidopsis* in the EMBL database: two additional CK1 sequences (CKI2, CKI3) have been reported (Klimczak *et al.*, 1995b) but the sequences were incomplete. An interesting feature of ADK1 was that it autophosphorylated on tyrosine residues and could also phosphorylate the exogenous substrate poly (Glu, Tyr) on tyrosine (Ali *et al.*, 1994). A CK1 has also been purified to homogeneity from broccoli (Klimczak and Cashmore, 1993). The biochemical properties of this, and of recombinant *Arabidopsis CKI1* expressed in bacteria (Klimczak *et al.*, 1995b), are similar to those of mammalian CK1s. As yet there are few insights into the functions of CK1s in plant cells.

An unusual feature of mammalian CK2s is that they will utilize GTP as well as ATP, and are relatively insensitive to the general kinase inhibitor staurosporine. They normally exists as $\alpha_2\beta_2$ tetramers, where the α subunits are catalytic. The β subunits are not essential for activity, but stimulate or inhibit activity against certain substrates. CK2s phosphorylate a bewildering array of substrate proteins *in vitro*, and in some cases these targets have been shown to be phosphorylated *in vivo*. Particularly interesting targets include DNA-binding proteins such as p53 (Kapoor and Lozano, 1998) and DNA topoisomerase II (Ishida *et al.*, 1996). Disruption of both of the CK2 genes (*CKA1/CKA2*) in *S. cerevisiae* is lethal, but experiments with temperature-sensitive mutants suggest a role for *CKA1* in the maintenance of cell polarity (Rethinaswamy *et al.*, 1998), and for *CKA2* in cell cycle progression (Hanna *et al.*, 1995). DNAs encoding two isoforms of the CK2 catalytic (α) subunit

(*CKA1* and *CKA2*) have been cloned from *Arabidopsis* (Mizoguchi *et al*., 1993), as have two forms of the β subunit (*CKB1* and *CKB2*; Collinge and Walker, 1994). Active $\alpha_2\beta_2$ tetramers can be reconstituted from recombinant CKA1 and CKB1 (Klimczak *et al*., 1995a). CK2 has also been purified from *Arabidopsis* in two forms, i.e. a monomeric catalytic subunit and a tetrameric $\alpha_2\beta_2$ form (Espunya and Martinez, 1997). A CK2 from maize was localized in both the cytoplasm and the nucleus, depending on the cell type: a probable nuclear localization sequence was identified (Peracchia *et al*., 1999). Plant CK2s have been shown to phosphorylate a wide variety of plant proteins *in vitro*. In many cases it is not clear whether this affects the function of the target protein, but phosphorylation of the transcription factor GBF1 stimulates its binding to DNA containing a 'G box' (Klimczak *et al*., 1995a).

Recently, DNA encoding a third *Arabidopsis* CK2 β subunit (CKB3) was cloned using a two-hybrid screen with the protein CCA1 as bait (Sugano *et al*., 1998). The latter (circadian clock associated-1) is a Myb-related transcription factor involved in regulating genes (e.g. *Lhcb1*3*, encoding a component of the light-harvesting complex) whose expression has a circadian oscillation and is also affected by light via the phytochrome system (Wang and Tobin, 1998). CCA1 also interacted with CKB1, CKB2 and CKB3, and CKA1 and CKA2, and was phosphorylated by CK2 complexes *in vitro*. Phosphorylation did not affect the binding of recombinant CCA1 to DNA, but evidence was presented that a phosphorylation sensitive to CK2 inhibitors was required for binding of endogenous DNA-binding proteins, including CCA1, to the *Lhcb1*3* promoter (Sugano *et al*., 1998).

The catalytic subunit of CK2 from maize occupies a special place in that it recently became the first plant protein kinase for which a crystal structure is available (Niefind *et al*., 1998). The structure confirmed the location of basic residues within the active site which bind the acidic residues around the target serine, and also provided clues as to why CK2s can utilize GTP as well as ATP.

X. THE GSK3/SHAGGY-RELATED PROTEIN KINASE (ASK) SUBFAMILY

Mammalian glycogen synthase kinase-3 (GSK3) was originally described as one of several protein kinases phosphorylating and inactivating rabbit muscle glycogen synthase. When DNA encoding it was cloned, it turned out to be closely related to the *shaggy/zeste-white 3* gene product from *Drosophila melanogaster* (Woodgett, 1991). In mammals, GSK3 occurs in a signal transduction pathway downstream of the insulin and insulin-like growth factor-1 receptors. It is unusual in that it is inactivated (rather than activated) by stimulation of the pathway, via phosphorylation by protein kinase B (also known as Akt) at a serine residue near the N-terminus (Cross *et al*., 1995). Since GSK3 inactivates glycogen synthase this accounts, at least in part, for the observed activation of glycogen synthesis by insulin. In *Drosophila* the *shaggy/zeste-white 3* gene is required for the establishment of segment polarity. It is now known that the GSK3/shaggy protein kinase lies on the pathway downstream of Wingless, a secreted protein messenger involved in the control of development (Hunter, 1997). A related pathway is found in mammalian cells, where the equivalent of Wingless is the Wnt1 protein. Stimulation of mammalian fibroblasts with *Drosophila* Wingless results in inactivation of GSK3 due to phosphorylation; the signal transduction pathway is not known but it is clearly distinct from that utilized by insulin (Cook *et al.*, 1996).

Figure 8 shows that, at the time of writing, there were eight full-length sequences of members of the *Arabidopsis* shaggy-related protein kinase (ASK) subfamily in the EMBL database. Bianchi *et al.* (1994) also reported the existence of two more (ASK-δ and -ϵ) for which full-length sequences are not yet available. As pointed out by Dornelas *et al.* (1999), the full-length sequences cluster into three groups comprising: (i) ASK-ι (iota), -ζ (zeta) and -η (eta); (ii) ASK-β, -τ (tetha) and -κ (kappa, also known as ATK-1); and (iii) ASK-γ and -α. There are also multiple representatives in rice, tobacco, petunia and alfalfa (reviewed in Dornelas *et al.*, 1998).

Because of their sequence similarity with *Drosophila shaggy*, they have been assumed to play roles during plant development. However, although different isoforms do show tissue-specific expression patterns (Pay *et al.*, 1993; Decroocq-Ferrant *et al.*, 1995; Einzenberger *et al.*, 1995; Jonak *et al.*, 1995a; Dornelas *et al.*, 1999), there is currently little evidence available that provides any real insight into the functions of the plant GSK3 family.

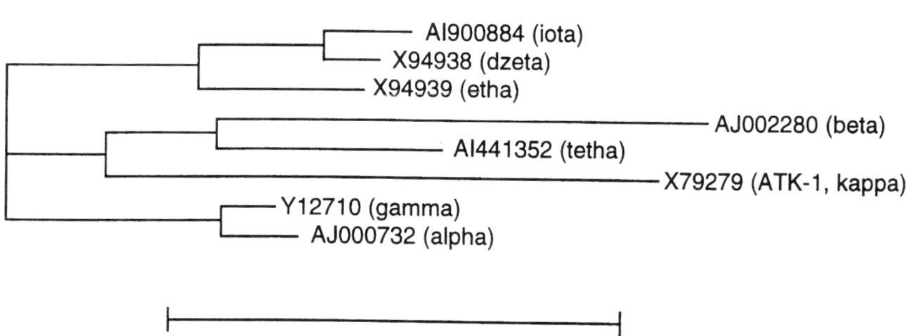

Fig. 8. Phenogram derived from sequences of the kinase domains of *Arabidopsis* GSK3/shaggy-related kinases. The tree was made using the methods described for Fig. 2. The scale bar at the bottom corresponds to 0.1 evolutionary distance units. Note that the scale used for this phenogram is different from that used in most of the others in this chapter.

XI. THE CTR1/Raf-LIKE SUBFAMILY

The *CTR1* (*constitutive triple response-1*) gene was identified via loss-of-function mutations that displayed a 'triple response' in the absence of the hormone ethylene. Cloning of the mutant gene revealed that it encoded a protein kinase related to the mammalian Raf subfamily (Kieber *et al.*, 1993). The archetypal Raf family member, i.e. Raf-1, couples the GTP-binding protein Ras to the MAP kinase (ERK1/ERK2) cascade, and acts at the level of a MAPKKK (although not closely related to MAPKKKs of the MEKK/Ste11 family described above). It was therefore proposed (Kieber *et al.*, 1993) that CTR1 lies upstream of a MAPK cascade, although direct evidence for this is still lacking. The proposed receptors for ethylene, the products of the genes *ETR1* (Chang and Meyerowitz, 1995), *ETR2* (Sakai *et al.*, 1998) and *ERS* (Hua *et al.*, 1995), all of which give rise to ethylene-insensitive mutants, are members of the prokaryotic two-component histidine/aspartate kinase family. Since loss of *CTR1* function gives rise to a constitutive ethylene-like response, the CTR1 pathway must normally switch off ethylene responses, and be *inactivated* by binding of ethylene to its receptor(s). The N-terminal regulatory domain of CTR1 has been shown to interact with the histidine kinase domains of ETR1 and ERS by two-hybrid analysis (Clark *et al.*, 1998). Two protein kinases of unknown function that appear to cluster into the same group as CTR1 according to their kinase domain sequences are ATN1 and ATMRK1 (Fig. 2).

XII. THE LAMMER SUBFAMILY

Bender and Fink (1994) cloned an *Arabidopsis* DNA (*AFC1*) by virtue of its ability to complement a double mutation (*fus3 kss1*) in genes encoding MAPKs that transmit the effect of mating factors to expression of mating-specific genes in yeast. AFC1 was not a MAPK but a member of the *LAMMER* subfamily (Yun *et al.*, 1994), members of which have the sequence *EHLAMMERIL* at kinase conserved motif XI. They also cloned two other *LAMMER* genes, *AFC2* and *AFC3*; unlike *AFC1*, *AFC2* did not complement the *fus3 kss1* mutant. The functions of the *LAMMER* subfamily remain unclear.

XIII. THE PVPK1 SUBFAMILY

PK64 (Mizoguchi *et al.*, 1992), PK5 (Hayashida *et al.*, 1992) and PK7 (Fig. 2) appear to be homologues of PVPK1 from *Phaseolus vulgaris* and G11A from rice. Although the latter two were the first plant protein kinases to be cloned (Lawton *et al.*, 1989), the functions of the members of this subfamily remain unknown. AtPK3, a homologue of PsPK3 from pea (Khanna *et al.*, 1999), also appears to be related to this group (Fig. 2). Along with the S6 kinase homologues and the NPH group (see below), the kinase domains of this subfamily are related to those of mammalian cyclic AMP- and cyclic GMP-dependent protein kinases, and protein kinase C, and thus fall into what is termed the 'AGC group' (Hanks and Hunter, 1995a, b).

XIV. S6 KINASE HOMOLOGUES

DNAs encoding two *Arabidopsis* kinases (AtS6K1 and AtS6K2) related to mammalian p70 ribosomal protein S6 kinase have been cloned (Zhang *et al.*, 1994b; Mizoguchi *et al.*, 1995; Turck *et al.*, 1998). These were induced at the transcriptional level by cold or salt stress (Mizoguchi *et al.*, 1995). The plant kinases contain three of the serine/threonine residues whose phosphorylation in mammalian p70 S6 kinases is required for activity, but lack the N-terminal domain which conveys sensitivity to the macrolide antibiotic, rapamycin. Turck *et al.* (1998) expressed these kinases in mammalian 293 cells, and found that they phosphorylated ribosomal protein S6 from *Arabidopsis* when assayed at 25°C, but not at 37°C, the temperature used in previous attempts to demonstrate S6 kinase activity (Zhang *et al.*, 1994a). The heat lability of these kinases might account for the dephosphorylation of a basic ribosomal protein of 30 kDa (presumed to be S6) in tomato cells by heat shock (Scharf and Nover, 1982). When expressed in 293 cells, AtS6K2 was activated by serum (like the endogenous p70 S6 kinase) and could reverse the inhibition of S6 phosphorylation by rapamycin (Turck *et al.*, 1998). Thus the *Arabidopsis* kinases appear to be true functional homologues of the mammalian p70 S6 kinases.

XV. NAK SUBFAMILY

The original representative of this subfamily was NAK (Moran and Walker, 1993), but it also includes APK1 (Hirayama and Oka, 1992), APK2a (Ito *et al.*, 1997), and ARSK1 (Hwang and Goodman, 1995). NAK was cloned by screening an *Arabidopsis* cDNA library with DNA encoding the catalytic domain of the maize receptor-like kinase ZmPK1 (Moran and Walker, 1993). Although the members of this group do not appear to be transmembrane proteins, their kinase domains do cluster with those of the RLKs in Fig. 2, as does that of EMBL:Z84202. Interesting features of the NAK and APK1 sequences are that they bear some resemblance to mammalian protein–tyrosine as well as protein–serine/threonine kinases. When APK1 was expressed in bacteria it caused bacterial proteins to become reactive with anti-phosphotyrosine antibodies. However, the expressed kinase autophosphorylated, and phosphorylated exogenous proteins such as casein and myosin light chain, on serine and threonine and not tyrosine (Hirayama and Oka, 1992). It therefore does not appear to be a true protein–tyrosine kinase.

XVI. MISCELLANEOUS KINASES

Mutations in the *TOUSLED* gene give rise to plants that exhibit random loss of floral organs, and organ development is impaired. The gene encodes a protein kinase (Roe *et al.*, 1993) with an N-terminal domain that contains a coiled-coil region responsible for oligomerization, and a localization signal that directs the protein to the nucleus (Roe *et al.*, 1997).

Plants with mutations in the *NPH1* gene, which appears to encode a soluble protein kinase, lack all phototropic responses. The N-terminal domain contains two repeated sequences that are related to sequences found in bacterial proteins regulated by redox status (Huala *et al.*, 1997). The sequence of a close relative (NPH2, see Fig. 2) has recently also been deposited in the database.

Deak *et al.* (1999) have cloned DNA encoding a homologue of mammalian 3-phosphoinositide-dependent protein kinase-1 (PDK1). The latter is part of a kinase cascade downstream of the insulin receptor, and binds the second messenger phosphatidylinositol-3,4,5-trisphosphate (PIP_3) via its N-terminal PH domain. Expression of human or *Arabidopsis* (Deak *et al.*, 1999) PDK1 can rescue the lethal phenotype of a yeast strain in which both of the yeast PDK1 genes (*PKH1* and *PKH2*) have been disrupted, demonstrating a functional relationship. *Arabidopsis* PDK1 has a PH domain and binds PIP_3. It can also activate *in vitro* the protein kinase that lies downstream of PDK1 in mammals, i.e. protein kinase B (also known as Akt). However, the significance of these findings remains unclear, because neither PIP_3 nor a homologue of protein kinase B has yet been found in plants. *Arabidopsis* PDK1 does bind other phospholipids such as phosphatidic acid and phosphatidylinositol-3-phosphate, and it is possible that one of these might be the true upstream signal.

Lee *et al.* (1999) cloned an *Arabidopsis* DNA (Dbf2; see Fig. 2) which rendered yeast tolerant to osmotic, heat and cold stress. It encoded a homologue of the yeast Dbf2 protein kinase and complemented a *dbf2* mutation in that organism. Dbf2 is required in yeast for the later stages of mitosis, and is a component of the CCR4 transcription complex. *DBF2* mRNA was induced in both *Arabidopsis* and yeast by osmotic, heat, cold and drought stress. Overexpression of *Arabidopsis* Dbf2 in transgenic plants rendered them more resistant to these stresses, while reduction of expression via an antisense strategy rendered them more sensitive (Lee *et al.*, 1999).

Finally, the *Arabidopsis* protein kinases whose sequences can be found in EMBL entries Y11930 (PPK1), U53501 and U97568 (see Fig. 2) do not fall into any of the subfamilies described above, and nothing appears to be known about their function at present.

XVII. CONCLUSIONS AND PERSPECTIVES

Although the past decade has been a ‘golden age’ for discovery of the sequences of plant protein–serine/threonine kinases, the elucidation of their functions is only just beginning. The success of 'cloning by homology' and genome sequencing has meant that the rate of accumulation of knowledge about the number and sequence of plant kinase genes has greatly outstripped the rate at which we have learned about their physiological roles. Even for those plant kinases identified by cloning of mutant genes, there is still a great deal to be understood. Knowledge of the phenotype of a mutation does not mean that one fully understands the normal function of the encoded protein; the phenotype only reveals one defect produced by the mutation under the particular circumstances of the genetic screen. In most cases, we also know about only one component in any plant signal transduction chain, and we do not understand the full sequence of events. In the next decade, we will need to explore upstream and downstream of the known kinases in order to identify their regulators and target proteins, respectively, and also 'sideways' to find regulatory subunits and other interacting proteins. In my view, this will require greater emphasis on biochemical analysis than has occurred in the past, although the yeast two-hybrid screen, and interaction cloning (Stone *et al.*, 1994) are useful approaches to obtain initial clues as to regulatory partners. Whatever approach is adopted, this field of plant research will remain dynamic and exciting for the foreseeable future, and many surprises and new twists undoubtedly await us.

ACKNOWLEDGEMENTS

Studies on plant protein kinases in this laboratory have been supported by the UK Biotechnology and Biological Sciences Research Council (BBSRC). For phylogenetic trees and other sequence analyses, extensive use was made of the bioinformatics and computing facilities at the European Bioinformatics Institute and the UK MRC HGMP Resource Centre, Hinxton, Cambridge.

REFERENCES

Alderson, A., Sabelli, P. A., Dickinson, J. R., Cole, D., Richardson, M. *et al.* (1991). Complementation of *snf1*, a mutation affecting global regulation of carbon metabolism in yeast, by a plant protein kinase cDNA. *Proceedings of the National Academy of Sciences of the United States of America* **88**, 8602–8605.

Ali, N., Halfter, U. and Chua, N. H. (1994). Cloning and biochemical characterization of a plant protein kinase that phosphorylates serine, threonine and tyrosine. *Journal of Biological Chemistry* **269**, 31626–31629.

Anderberg, R. J. and Walker-Simmons, M. K. (1992). Isolation of a wheat cDNA clone for an abscisic acid-inducible transcript with homology to protein kinases. *Proceedings of the National Academy of Sciences of the United States of America* **89**, 10183–10187.

Ball, K. L., Dale, S., Weekes, J. and Hardie, D. G. (1994). Biochemical characterization of two forms of 3-hydroxy-3-methylglutaryl-CoA reductase kinase from cauliflower *(Brassica oleracea)*. *European Journal of Biochemistry* **219**, 743–750.

Ball, K. L., Barker, J., Halford, N. G. and Hardie, D. G. (1995). Immunological evidence that HMG-CoA reductase kinase-A is the cauliflower homologue of the Rkin1 subfamily of plant protein kinases. *FEBS Letters* **377**, 189–192.

Banno, H., Hirano, K., Nakamura, T., Irie, K., Nomoto, S. *et al.* (1993). *NPK1*, a tobacco gene that encodes a protein with a domain homologous to yeast BCK1, STE11, and Byr2 protein kinases. *Molecular and Cellular Biology* **13**, 4745–4752.

Bateman, A. (1997). The structure of a domain common to archaebacteria and the homocystinuria disease protein. *Trends in Biochemical Science* **22**, 12–13.

Bender, J. and Fink, G. R. (1994). AFC1, a LAMMER kinase from *Arabidopsis thaliana*, activates *STE12*-dependent processes in yeast. *Proceedings of the National Academy of Sciences of the United States of America* **91**, 12105–12109.

Bevan, M., Bancroft, I., Bent, E., Love, K., Goodman, H. *et al.* (1998). Analysis of 1.9 Mb of contiguous sequence from chromosome 4 of *Arabidopsis thaliana*. *Nature* **391**, 485–488.

Bhalerao, R. P., Salchert, K., Bako, L., Okresz, L., Szabados, L. *et al.* (1999). Regulatory interaction of PRL1 WD protein with arabidopsis SNF1-like protein kinases [In Process Citation]. *Proceedings of the National Academy of Sciences of the United States of America* **96**, 5322–5327.

Bianchi, M. W., Guivarch, D., Thomas, M., Woodgett, J. R. and Kreis, M. (1994). *Arabidopsis* homologs of the shaggy and GSK-3 protein kinases–molecular cloning and functional expression in *Escherichia coli*. *Molecular and General Genetics* **242**, 337–345.

Bouly, J. P., Gissot, L., Lessard, P., Kreis, M. and Thomas, M. (1999). *Arabidopsis thaliana* proteins related to the yeast SIP and SNF4 interact with AKIN alpha 1, an SNF1-like protein kinase. *Plant Journal* **18**, 541–550.

Braun, D. M. and Walker, J. C. (1996). Plant transmembrane receptors–new pieces in the signaling puzzle. *Trends in Biochemical Science* **21**, 70–73.

C. elegans sequencing consortium (1998). Genome sequence of the nematode *C. elegans:* a platform for investigating biology. The *C. elegans* Sequencing Consortium [published erratum appears in (1999) *Science* **283**, 35]. *Science* **282**, 2012–2018.

Camoni, L., Fullone, M. R., Marra, M. and Aducci, P. (1998). The plasma membrane H^+-ATPase from maize roots is phosphorylated in the C-terminal domain by a calcium-dependent protein kinase. *Physiologia Plantarum* **104**, 549–555.

Chang, C. and Meyerowitz, E. M. (1995). The ethylene hormone response in *Arabidopsis* – a eukaryotic 2-component signaling system. *Proceedings of the National Academy of Sciences of the United States of America* **92**, 4129–4133.

Clark, K. L., Larsen, P. B., Wang, X. X. and Chang, C. (1998). Association of the *Arabidopsis* CTR1 Raf-like kinase with the ETR1 and ERS ethylene receptors. *Proceedings of the National Academy of Sciences of the United States of America* **95**, 5401–5406.

Clark, S. E., Williams, R. W. and Meyerowitz, E. M. (1997). The *CLAVATA1* gene encodes a putative receptor kinase that controls shoot and floral meristem size in *Arabidopsis*. *Cell* **89**, 575–585.

Cobb, M. H. (1995). Casein kinase 1 (vertebrates). *In* "The Protein Kinase Factsbook" (D. G. Hardie and S. K. Hanks, eds) pp. 347–349. Academic Press, London.

Collinge, M. A. and Walker, J. C. (1994). Isolation of an *Arabidopsis thaliana* casein kinase II β subunit by complementation in *Saccharomyces cerevisiae*. *Plant Molecular Biology* **25**, 649–658.

Cook, D., Fry, M. J., Hughes, K., Sumathipala, R., Woodgett, J. R. and Dale, T. C. (1996). Wingless inactivates glycogen synthase kinase-3 via an intracellular signalling pathway which involves a protein kinase C. *EMBO Journal* **15**, 4526–4536.

Cross, D. A., Alessi, D. R., Cohen, P., Andjelkovich, M. and Hemmings, B. A. (1995). Inhibition of glycogen synthase kinase-3 by insulin mediated by protein kinase B. *Nature* **378**, 785–789.

Czernic, P., Visser, B., Sun, W. N., Savoure, A., Deslandes, L. *et al.* (1999). Characterization of an *Arabidopsis thaliana* receptor-like protein kinase gene activated by oxidative stress and pathogen attack. *Plant Journal* **18**, 321–327.

Dale, S., Arro, M., Becerra, B., Morrice, N. G., Boronat, A. *et al.* (1995a). Bacterial expression of the catalytic domain of 3-hydroxy-3-methylglutaryl-CoA reductase (isoform HMGR1) from *Arabidopsis thaliana*, and its inactivation by phosphorylation at ser577 by *Brassica oleracea* 3-hydroxy-3-methylglutaryl-CoA reductase kinase. *European Journal of Biochemistry* **233**, 506–513.

Dale, S., Wilson, W. A., Edelman, A. M. and Hardie, D. G. (1995b). Similar substrate recognition motifs for mammalian AMP-activated protein kinase, higher plant HMG-CoA reductase kinase-A, yeast SNF1, and mammalian calmodulin-dependent protein kinase-i. *FEBS Letters* **361**, 191–195.

Davies, J. P., Yildiz, F. H. and Grossman, A. R. (1999). Sac3, an Snf1-like serine threonine kinase that positively and negatively regulates the responses of chlamydomonas to sulfur limitation. *Plant Cell* **11**, 1179–1190.

Day, I. S. and Reddy, A. S. N. (1998). Isolation and characterization of two cyclin-like cDNAs from *Arabidopsis*. *Plant Molecular Biology* **36**, 451–461.

de Veylder, L., Segers, G., Glab, N., Casteels, P., Van Montagu, M. and Inze, D. (1997). The *Arabidopsis* Cks1At protein binds the cyclin-dependent kinases Cdc2aAt and Cdc2bAt. *FEBS Letters* **412**, 446–452.

Deak, M., Casamayor, A., Currie, R. A., Downes, C. P. and Alessi, D. R. (1999). Characterisation of a plant 3-phosphoinositide-dependent protein kinase-1 homologue which contains a pleckstrin homology domain. *FEBS Letters* **451**, 220–226.

Decroocq-Ferrant, V., Vanwent, J., Bianchi, M. W., Devries, S. C. and Kreis, M. (1995). *Petunia hybrida* homologs of shaggy/zeste-white-3 expressed in female and male reproductive organs. *Plant Journal* **7**, 897–911.

Deeken, R. and Kaldenhoff, R. (1997). Light-repressible receptor protein kinase: a novel photo-regulated gene from *Arabidopsis thaliana*. *Planta* **202**, 479–486.

DeMaggio, A. J., Lindberg, R. A., Hunter, T. and Hoekstra, M. F. (1992). The budding yeast HRR25 gene product is a casein kinase I isoform. *Proceedings of the National Academy of Sciences of the United States of America* **89**, 7008–7012.

Doonan, J. and Fobert, P. (1997). Conserved and novel regulators of the plant cell cycle. *Current Opinion in Cell Biology* **9**, 824–830.

Dornelas, M. C., Lejeune, B., Dron, M. and Kreis, M. (1998). The *Arabidopsis* SHAGGY-related protein kinase (ASK) gene family: structure, organization and evolution. *Gene* **212**, 249–257.

Dornelas, M. C., Wittich, P., vonRecklinghausen, I., vanLammeren, A. and Kreis, M. (1999). Characterization of three novel members of the *Arabidopsis* SHAGGY-related protein kinase (ASK) multigene family. *Plant Molecular Biology* **39**, 137–147.

Douglas, P., Moorhead, G., Hong, Y., Morrice, N. and MacKintosh, C. (1998). Purification of a nitrate reductase kinase from *Spinacia oleracea* leaves, and its identification as a calmodulin-domain protein kinase. *Planta* **206**, 435–442.

Einzenberger, E., Eller, N., Heberlebors, E. and Vicente, O. (1995). Isolation and expression during pollen development of a tobacco cDNA clone encoding a protein kinase homologous to shaggy glycogen synthase kinase-3. *Biochimica et Biophysica Acta* **1260**, 315–319.

Ellard-Ivey, M., Hopkins, R. B., White, T. J. and Lomax, T. L. (1999). Cloning, expression and N-terminal myristoylation of CpCPK1, a calcium-dependent protein kinase from zucchini (*Cucurbita pepo* L.). *Plant Molecular Biology* **39**, 199–208.

Espunya, M. C. and Martinez, M. C. (1997). Identification of two different molecular forms of *Arabidopsis thaliana* casein kinase II. *Plant Science* **124**, 131–142.

Feiler, H. S. and Jacobs, T. W. (1990). Cell division in higher plants: a *cdc2* gene, its 34kDa product, and histone H1 kinase activity in pea. *Proceedings of the National Academy of Sciences of the United States of America* **87**, 5397–5401.

Ferreira, P. C. G., Hemerly, A. S., Villaroel, R., Vanmontagu, M. and Inze, D. (1991). The *Arabidopsis* functional homolog of the p34cdc2 protein kinase. *Plant Cell* **3**, 531–540.

Feuillet, C., Schachermayr, G. and Keller, B. (1997). Molecular cloning of a new receptor-like kinase gene encoded at the Lr10 disease resistance locus of wheat. *Plant Journal* **11**, 45–52.

Fobert, P. R., Gaudin, V., Lunness, P., Coen, E. S. and Doonan, J. H. (1996). Distinct classes of *cdc2*-related genes are differentially expressed during the cell-division cycle in plants. *Plant Cell* **8**, 1465–1476.

Foretz, M., Carling, D., Guichard, C., Ferré, P. and Foufelle, F. (1998). AMP-activated protein kinase inhibits the glucose-activated expression of fatty acid synthase gene in rat hepatocytes. *Journal of Biological Chemistry* **273**, 14767–14771.

Furumoto, T., Ogawa, N., Hata, S. and Izui, K. (1996). Plant calcium-dependent protein kinase-related kinases (CRKs) do not require calcium for their activities. *FEBS Letters* **396**, 147–151.

Goldberg, J., Nairn, A. C. and Kuriyan, J. (1996). Structural basis for the autoinhibition of calcium/calmodulin-dependent protein kinase I. *Cell* **84**, 875–887.

Grafi, G., Burnett, R. J., Helentjaris, T., Larkins, B. A., Decaprio, J. A. *et al.* (1996). A maize cDNA-encoding a member of the retinoblastoma protein family – involvement in endoreduplication. *Proceedings of the National Academy of Sciences of the United States of America* **93**, 8962–8967.

Halford, N. G. and Hardie, D. G. (1998). SNF1-related protein kinases: global regulators of carbon metabolism in plants? *Plant Molecular Biology* **37**, 735–748.

Hanks, S. K. (1987). Homology probing: identification of cDNA clones encoding members of the protein–serine kinase family. *Proceedings of the National Academy of Sciences of the United States of America* **84**, 388–392.

Hanks, S. K. and Hunter, T. (1995a). The eukaryotic protein kinase superfamily: kinase (catalytic) domain structure and classification. *FASEB Journal* **9**, 576–596.

Hanks, S. K. and Hunter, T. (1995b). The eukaryotic protein kinase superfamily: kinase (catalytic) domain structure and classification based on phylogenetic analysis. *In* "The Protein Kinase Factsbook" (D. G. Hardie and S. K. Hanks, eds) pp. 7–47. Academic Press, London.

Hanks, S. K., Quinn, A. M. and Hunter, T. (1988). The protein kinase family: conserved features and deduced phylogeny of the catalytic domains. *Science* **241**, 42–52.

Hanna, D. E., Rethinaswamy, A. and Glover, C. V. (1995). Casein kinase II is required for cell cycle progression during G1 and G2/M in *Saccharomyces cerevisiae*. *Journal of Biological Chemistry* **270**, 25905–25914.

Hardie, D. G. (1988). Pseudosubstrates turn off protein kinases. *Nature* **335**, 592–593.

Hardie, D. G. (1999). Plant protein serine/threonine kinases: classification and functions. *Annual Review of Plant Physiology and Plant Molecular Biology* **50**, 97–131.

Hardie, D. G. and Carling, D. (1997). The AMP-activated protein kinase – fuel gauge of the mammalian cell? *European Journal of Biochemistry* **246**, 259–273.

Hardie, D. G., Carling, D. and Carlson, M. (1998). The AMP-activated/SNF1 protein kinase subfamily: metabolic sensors of the eukaryotic cell? *Annual Review of Biochemistry* **67**, 821–855.

Hardie, D. G., Salt, I. P., Hawley, S. A. and Davies, S. P. (1999). AMP-activated protein kinase: an ultrasensitive system for monitoring cellular energy charge. *Biochemistry Journal* **338**, 717–722.

Harmon, A. C., Putnam-Evans, C. and Cormier, M. J. (1987). A calcium-dependent but calmodulin-independent protein kinase from soybean. *Plant Physiology* **83**, 830–837.

Harmon, A. C., Yoo, B. C. and McCaffery, C. (1994). Pseudosubstrate inhibition of CDPK, a protein kinase with a calmodulin-like domain. *Biochemistry* **33**, 7278–7287.

Harper, J. F., Sussman, M. R., Schaller, G. E., Putnam-Evans, C., Charbonneau, H. and Harmon, A. C. (1991). A calcium-dependent protein kinase with a regulatory domain similar to calmodulin. *Science* **252**, 951–954.

Harper, J. F., Huang, J. F. and Lloyd, S. J. (1994). Genetic identification of an autoinhibitor in CDPK, a protein kinase with a calmodulin-like domain. *Biochemistry* **33**, 7267–7277.

Hartwell, J., Smith, L. H., Wilkins, M. B., Jenkins, G. I. and Nimmo, H. G. (1996). Higher plant phosphoenolpyruvate carboxylase kinase is regulated at the level of translatable mRNA in response to light or a circadian rhythm. *Plant Journal* **10**, 1071–1078.

Hayashida, N., Mizoguchi, T., Yamaguchi-Shinozaki, K. and Shinozaki, K. (1992). Characterization of a gene that encodes a homolog of protein kinase in *Arabidopsis thaliana*. *Gene* **121**, 325–330.

He, Z. H., He, D. Z. and Kohorn, B. D. (1998). Requirement for the induced expression of a cell wall-associated receptor kinase for survival during the pathogen response. *Plant Journal* **14**, 55–63.

He, Z. H., Cheeseman, I., He, D. Z. and Kohorn, B. D. (1999). A cluster of five cell wall-associated receptor kinase genes, *Wak1–5*, are expressed in specific organs of *Arabidopsis*. *Plant Molecular Biology* **39**, 1189–1196.

Herve, C., Dabos, P., Galaud, J. P., Rouge, P. and Lescure, B. (1996). Characterization of an *Arabidopsis thaliana* gene that defines a new class of putative plant receptor kinases with an extracellular lectin-like domain. *Journal of Molecular Biology* **258**, 778–788.

Hirayama, T., Imajuku, Y., Anai, T., Matsui, M. and Oka, A. (1991). Identification of two cell cycle controlling *cdc2* gene homologs in *Arabidopsis thaliana*. *Gene* **105**, 159–165.

Hirayama, T. and Oka, A. (1992). Novel protein kinase of *Arabidopsis thaliana* (APK1) that phosphorylates tyrosine, serine and threonine. *Plant Molecular Biology* **20**, 653–662.

Hirt, H. (1997). Multiple roles of MAP kinases in plant signal transduction. *Trends in Plant Science* **2**, 11–15.

Hong, Y., Takano, M., Liu, C. M., Gasch, A., Chye, M. L. and Chua, N. H. (1996). Expression of 3 members of the calcium-dependent protein kinase gene family in *Arabidopsis thaliana*. *Plant Molecular Biology* **30**, 1259–1275.

Hu, S. H., Parker, M. W., Lei, J. Y., Wilce, M. C. J., Benian, G. M. and Kemp, B. E. (1994). Insights into autoregulation from the crystal structure of twitchin kinase. *Nature* **369**, 581–584.

Hua, J., Chang, C., Sun, Q. and Meyerowitz, E. M. (1995). Ethylene insensitivity conferred by *Arabidopsis ers gene*. *Science* **269**, 1712–1714.

Huala, E., Oeller, P. W., Liscum, E., Han, I. S., Larsen, E. and Briggs, W. R. (1997). *Arabidopsis* NPH1: a protein kinase with a putative redox-sensing domain. *Science* **278**, 2120–2123.

Huang, J. F., Teyton, L. and Harper, J. F. (1996). Activation of a Ca^{2+}-dependent protein kinase involves intramolecular binding of a calmodulin-like regulatory domain. *Biochemistry* **35**, 13222–13230.

Huber, C., Huber, J. L., Liao, P. C., Gage, D. A., McMichael, R. W. *et al.* (1996). Phosphorylation of serine-15 of maize leaf sucrose synthase – occurrence *in vivo* and possible regulatory significance. *Plant Physiology* **112**, 793–802.

Huber, S. C. and Huber, J. L. (1996). Role and regulation of sucrose-phosphate synthase in higher-plants. *Annual Review of Plant Physiology and Plant Molecular Biology* **47**, 431–444.

Hunter, T. (1997). Oncoprotein networks. *Cell* **88**, 333–346.

Hwang, I. W. and Goodman, H. M. (1995). An *Arabidopsis thaliana* root-specific kinase homolog is induced by dehydration, ABA, and NaCl. *Plant Journal* **8**, 37–43.

Ichimura, K., Mizoguchi, T., Irie, K., Morris, P., Giraudat, J. *et al.* (1998). Isolation of ATMEKK1 (a MAP kinase kinase kinase)–interacting proteins and analysis of a MAP kinase cascade in *Arabidopsis*. *Biochemical and Biophysical Research Communications* **253**, 532–543.

Ishida, R., Iwai, M., Marsh, K. L., Austin, C. A., Yano, T. *et al.* (1996). Threonine 1342 in human topoisomerase IIalpha is phosphorylated throughout the cell cycle. *Journal of Biological Chemistry* **271**, 30077–30082.

Ito, T., Takahashi, N., Shimura, Y. and Okada, K. (1997). A serine/threonine protein kinase gene isolated by an *in vivo* binding procedure using the *Arabidopsis* floral homeotic gene product, AGAMOUS. *Plant Cell Physiology* **38**, 248–258.

Iwata, Y., Kuriyama, M., Nakakita, H., Kojima, H., Ohto, M. and Nakamura, K. (1998). Characterization of a calcium-dependent protein kinase of tobacco leaves that is associated with the plasma membrane and is inducible by sucrose. *Plant Cell Physiology* **39**, 1176–1183.

Jeffrey, P. D., Russo, A. A., Polyak, K., Gibbs, E., Hurwitz, J. *et al.* (1995). Mechanism of CDK activation revealed by the structure of a cyclinA-CDK2 complex. *Nature* **376**, 313–320.

Jinno, A., Tanaka, K., Matsushime, H., Haneji, T. and Shibuya, M. (1993). Testis-specific MAK protein kinase is expressed specifically in the meiotic phase in spermatogenesis and is associated with a 210-kilodalton cellular phosphoprotein. *Molecular and Cellular Biology* **13**, 4146–4156.

Johnson, K. D. and Chrispeels, M. J. (1992). Tonoplast-bound protein kinase phosphorylates tonoplast intrinsic protein. *Plant Physiology* **100**, 1787–1795.

Johnson, L. N., Noble, M. E. M. and Owen, D. J. (1996). Active and inactive protein kinases: structural basis for regulation. *Cell* **85**, 149–158.

Jonak, C., Heberlebors, E. and Hirt, H. (1995a). Inflorescence-specific expression of ATK-1, a novel *Arabidopsis thaliana* homolog of shaggy/glycogen synthase kinase-3. *Plant Molecular Biology* **27**, 217–221.

Jonak, C., Kiegerl, S., Lloyd, C., Chan, J. and Hirt, H. (1995b). Mmk2, a novel alfalfa MAP kinase, specifically complements the yeast mpk1 function. *Molecular and General Genetics* **248**, 686–694.

Jouannic, S., Hamal, A., Leprince, A. S., Tregear, J. W., Kreis, M. and Henry, Y. (1999). Plant MAP kinase kinase kinases structure, classification and evolution. *Gene* **233**, 1–11.

Kaffman, A., Herskowitz, I., Tjian, R. and O'Shea, E. K. (1994). Phosphorylation of the transcription factor PHO4 by a cyclin-CDK complex, PHO80–PHO85. *Science* **263**, 1153–1156.

Kapoor, M. and Lozano, G. (1998). Functional activation of p53 via phosphorylation following DNA damage by UV but not gamma radiation. *Proceedings of the National Academy of Sciences of the United States of America* **95**, 2834–2837.

Khanna, R., Lin, X. and Watson, J. C. (1999). Photoregulated expression of the *PsPK3* and *PsPK5* genes in pea seedlings. *Plant Molecular Biology* **39**, 231–242.

Kieber, J. J., Rothenberg, M., Roman, G., Feldmann, K. A. and Ecker, J. R. (1993). CTR1, a negative regulator of the ethylene response pathway in *Arabidopsis*, encodes a member of the raf family of protein kinases. *Cell* **72**, 427–441.

Klimczak, L. J. and Cashmore, A. R. (1993). Purification and characterization of casein kinase I from broccoli. *Biochemistry Journal* **293**, 283–288.

Klimczak, L. J., Collinge, M. A., Farini, D., Giuliano, G., Walker, J. C. and Cashmore, A. R. (1995a). Reconstitution of *Arabidopsis* casein kinase II from recombinant subunits and phosphorylation of transcription factor Gbf1. *Plant Cell* **7**, 105–115.

Klimczak, L. J., Farini, D., Lin, C., Ponti, D., Cashmore, A. R. and Giuliano, G. (1995b). Multiple isoforms of *Arabidopsis* casein kinase I combine conserved catalytic domains with variable carboxyl-terminal extensions. *Plant Physiology* **109**, 687–696.

Kobe, B. and Deisenhofer, J. (1993). Crystal structure of porcine ribonuclease inhibitor, a protein with leucine-rich repeats. *Nature* **366**, 751–756.

Kobe, B. and Deisenhofer, J. (1994). The leucine-rich repeat: a versatile binding motif. *Trends in Biochemical Science* **19**, 415–421.

Kohorn, B. D., Lane, S. and Smith, T. A. (1992). An *Arabidopsis* serine/threonine kinase homologue with an epidermal growth factor repeat selected in yeast for its specificity for a thylakoid membrane protein. *Proceedings of the National Academy of Sciences of the United States of America* **89**, 10989–10992.

Kretzschmar, M. and Massague, J. (1998). SMADs: mediators and regulators of TGF-beta signaling. *Current Opinion in Genetic Development* **8**, 103–111.

Lakatos, L., Klein, M., Hofgen, R. and Banfalvi, Z. (1999). Potato StubSNF1 interacts with StubGAL83: a plant protein kinase complex with yeast and mammalian counterparts. *Plant Journal* **17**, 569–574.

Lawton, M. A., Yamamoto, R. T., Hanks, S. K. and Lamb, C. J. (1989). Molecular cloning of plant transcripts encoding protein kinase homologs. *Proceedings of the National Academy of Sciences of the United States of America* **86**, 3140–3144.

Leclerc, I., Kahn, A. and Doiron, B. (1998). The 5′-AMP-activated protein kinase inhibits the transcriptional stimulation by glucose in liver cells, acting through the glucose response complex. *FEBS Letters* **431**, 180–184.

Lee, J. H., VanMontagu, M. and Verbruggen, N. (1999). A highly conserved kinase is an essential component for stress tolerance in yeast and plant cells. *Proceedings of the National Academy of Sciences of the United States of America* **96**, 5873–5877.

Lee, J. W., Zhang, Y., Weaver, C. D., Shomer, N. H., Louis, C. F. and Roberts, D. M. (1995). Phosphorylation of nodulin 26 on serine 262 affects its voltage-sensitive channel activity in planar lipid bilayers. *Journal of Biological Chemistry* **270**, 27051–27057.

Lee, J. Y., Yoo, B. C. and Harmon, A. C. (1998). Kinetic and calcium-binding properties of three calcium-dependent protein kinase isoenzymes from soybean. *Biochemistry* **37**, 6801–6809.

Lessard, P., Bouly, J. P., Jouannic, S., Kreis, M. and Thomas, M. (1999). Identification of cdc2cAt: a new cyclin-dependent kinase expressed in *Arabidopsis thaliana* flowers. *Biochimica et Biophysica Acta* **1445**, 351–358.

Li, J., Lee, Y. R. and Assmann, S. M. (1998). Guard cells possess a calcium-dependent protein kinase that phosphorylates the KAT1 potassium channel. *Plant Physiology* **116**, 785–795.

Li, J. M. and Chory, J. (1997). A putative leucine-rich repeat receptor kinase involved in brassinosteroid signal transduction. *Cell* **90**, 929–938.

Lindzen, E. and Choi, J. H. (1995). A carrot cDNA-encoding an atypical protein kinase homologous to plant calcium-dependent protein kinases. *Plant Molecular Biology* **28**, 785–797.

Lino, B., Baizabal Aguirre, V. M. and dela Vara, L. E. G. (1998). The plasma-membrane H^+-ATPase from beetroot is inhibited by a calcium-dependent phosphorylation. *Planta* **204**, 352–359.

Liu, Z. H., Xia, M. and Poovaiah, B. W. (1998). Chimeric calcium/calmodulin-dependent protein kinase in tobacco: differential regulation by calmodulin isoforms. *Plant Molecular Biology* **38**, 889–897.

MacKintosh, R. W., Davies, S. P., Clarke, P. R., Weekes, J., Gillespie, J. G. *et al.* (1992). Evidence for a protein kinase cascade in higher plants: 3-hydroxy-3-methylglutaryl-CoA reductase kinase. *European Journal of Biochemistry* **209**, 923–931.

Madhani, H. D. and Fink, G. R. (1998). The riddle of MAP kinase signaling specificity. *Trends in Genetics* **14**, 151–155.

Maeda, T., Wurglermurphy, S. M. and Saito, H. (1994). A two-component system that regulates an osmosensing MAP kinase cascade in yeast. *Nature* **369**, 242–245.

Magyar, Z., Meszaros, T., Miskolczi, P., Deak, M., Feher, A. *et al.* (1997). Cell cycle phase specificity of putative cyclin-dependent kinase variants in synchronized alfalfa cells. *Plant Cell* **9**, 223–235.

Martin, G. B., Brommonschenkel, S. H., Chunwongse, J., Frary, A., Ganal, M. W. *et al.* (1993). Map-based cloning of a protein kinase gene conferring disease resistance in tomato. *Science* **262**, 1432–1436.

Meshi, T., Moda, I., Minami, M., Okanami, M. and Iwabuchi, M. (1998). Conserved Ser residues in the basic region of the bZIP-type transcription factor HBP-1a(17): importance in DNA binding and possible targets for phosphorylation. *Plant Molecular Biology* **36**, 125–136.

Minden, A. and Karin, M. (1997). Regulation and function of the JNK subgroup of MAP kinases. *Biochimica et Biophysica Acta* **1333**, F85–F104.

Mizoguchi, T., Hayashida, N., Yamaguchishinozaki, K., Harada, H. and Shinozaki, K. (1992). Nucleotide sequence of a cDNA encoding a protein kinase homolog in *Arabidopsis thaliana. Plant Molecular Biology* **18**, 809–812.

Mizoguchi, T., Yamaguchi-Shinozaki, K., Hayashida, N., Kamada, H. and Shinozaki, K. (1993). Cloning and characterization of two cDNAs encoding casein kinase II catalytic subunits in *Arabidopsis thaliana. Plant Molecular Biology* **21**, 279–289.

Mizoguchi, T., Gotoh, Y., Nishida, E., Yamaguchishinozaki, K., Hayashida, N. *et al.* (1994). Characterization of 2 cDNAs that encode MAP kinase homologs in *Arabidopsis thaliana* and analysis of the possible role of auxin in activating such kinase activities in cultured cells. *Plant Journal* **5**, 111–122.

Mizoguchi, T., Hayashida, N., Yamaguchishinozaki, K., Kamada, H. and Shinozaki, K. (1995). Two genes that encode ribosomal protein S6 kinase homologs are induced by cold or salinity stress in *Arabidopsis thaliana. FEBS Letters* **358**, 199–204.

Mizoguchi, T., Irie, K., Hirayama, T., Hayashida, N., Yamaguchishinozaki, K. *et al.* (1996). A gene encoding a mitogen-activated protein kinase kinase kinase is induced simultaneously with genes for a mitogen-activated protein kinase and an S6 ribosomal protein kinase by touch, cold, and water stress in *Arabidopsis thaliana. Proceedings of the National Academy of Sciences of the United States of America* **93**, 765–769.

Mizoguchi, T., Ichimura, K., Irie, K., Morris, P., Giraudat, J. *et al.* (1998). Identification of a possible MAP kinase cascade in *Arabidopsis thaliana* based on pairwise yeast two-hybrid analysis and functional complementation tests of yeast mutants. *FEBS Letters* **437**, 56–60.

Moran, T. V. and Walker, J. C. (1993). Molecular cloning of two novel protein kinase genes from *Arabidopsis thaliana. Biochimica et Biophysica Acta* **1216**, 9–14.

Morris, P. C., Guerrier, D., Leung, J. and Giraudat, J. (1997). Cloning and characterisation of MEK1, an *Arabidopsis* gene encoding a homologue of MAP kinase kinase. *Plant Molecular Biology* **35**, 1057–1064.

Nasrallah, J. B., Stein, J. C., Kandasamy, M. K. and Nasrallah, M. E. (1994). Signaling the arrest of pollen-tube development in self-incompatible plants. *Science* **266**, 1505–1508.

Niefind, K., Guerra, B., Pinna, L. A., Issinger, O. G. and Schomburg, D. (1998). Crystal structure of the catalytic subunit of protein kinase CK2 from *Zea mays* at 2.1 A resolution. *EMBO Journal* **17**, 2451–2462.

Nishihama, R., Banno, H., Kawahara, E., Irie, K. and Machida, Y. (1997). Possible involvement of differential splicing in regulation of the activity of *Arabidopsis* ANP1 that is related to mitogen-activated protein kinase kinase kinases (MAPKKKs). *Plant Journal* **12**, 39–48.

Pandey, S. and Sopory, S. K. (1998). Biochemical evidence for a calmodulin-stimulated calcium-dependent protein kinase in maize. *European Journal of Biochemistry* **255**, 718–726.

Patil, S., Takezawa, D. and Poovaiah, B. W. (1995). Chimeric plant calcium/calmodulin-dependent protein-kinase gene with a neural visinin-like calcium-binding domain. *Proceedings of the National Academy of Sciences of the United States of America* **92**, 4897–4901.

Pay, A., Jonak, C., Bogre, L., Meskiene, I., Mairinger, T. *et al.* (1993). The MSK family of alfalfa protein kinase genes encodes homologs of shaggy/glycogen-synthase kinase-3 and shows differential expression patterns in plant organs and development. *Plant Journal* **3**, 847–856.

Pei, Z. M., Ward, J. M., Harper, J. F. and Schroeder, J. I. (1996). A novel chloride channel in *Vicia faba* guard cell vacuoles activated by the serine/threonine kinase, CDPK. *EMBO Journal* **15**, 6564–6574.

Peracchia, G., Jensen, A. B., CulianezMacia, F. A., Grosset, J., Goday, A. *et al.* (1999). Characterization, subcellular localization and nuclear targeting of casein kinase 2 from *Zea mays. Plant Molecular Biology* **40**, 199–211.

Ponticos, M., Lu, Q. L., Morgan, J. E., Hardie, D. G., Partridge, T. A. and Carling, D. (1998). Dual regulation of the AMP-activated protein kinase provides a novel mechanism for the control of creatine kinase in skeletal muscle. *EMBO Journal* **17**, 1688–1699.

Popping, B., Gibbons, T. and Watson, M. D. (1996). The *Pisum sativum* MAP kinase homologue (PsMAPK) rescues the *Saccharomyces cerevisiae hog1* deletion mutant under conditions of high osmotic stress. *Plant Molecular Biology* **31**, 355–363.

Purcell, P. C., Smith, A. M. and Halford, N. G. (1998). Antisense expression of a sucrose non-fermenting-1-related protein kinase sequence in potato results in decreased expression of sucrose synthase in tubers and loss of sucrose inducibility of sucrose synthase transcripts in leaves. *Plant Journal* **14**, 195–202.

Renaudin, J. P., Doonan, J. H., Freeman, D., Hashimoto, J., Hirt, H. *et al.* (1996). Plant cyclins: a unified nomenclature for plant A-, B- and D-type cyclins based on sequence organization. *Plant Molecular Biology* **32**, 1003–1018.

Rethinaswamy, A., Birnbaum, M. J. and Glover, C. V. (1998). Temperature-sensitive mutations of the *CKA1* gene reveal a role for casein kinase II in maintenance of cell polarity in *Saccharomyces cerevisiae*. *Journal of Biological Chemistry* **273**, 5869–5877.

Robinson, L. C., Hubbard, E. J., Graves, P. R., DePaoli Roach, A. A., Roach, P. J. *et al.* (1992). Yeast casein kinase I homologues: an essential gene pair. *Proceedings of the National Academy of Sciences of the United States of America* **89**, 28–32.

Roe, J. L., Rivin, C. J., Sessions, R. A., Feldmann, K. A. and Zambryski, P. C. (1993). The *Tousled* gene in *A. thaliana* encodes a protein kinase homolog that is required for leaf and flower development. *Cell* **75**, 939–950.

Roe, J. L., Durfee, T., Zupan, J. R., Repetti, P. P., McLean, B. G. and Zambryski, P. C. (1997). TOUSLED is a nuclear serine/threonine protein kinase that requires a coiled-coil region for oligomerization and catalytic activity. *Journal of Biological Chemistry* **272**, 5838–5845.

Sakai, H., Hua, J., Chen, Q. H. G., Chang, C. R., Medrano, L. J. *et al.* (1998). *ETR2* is an *ETR1*-like gene involved in ethylene signaling in *Arabidopsis*. *Proceedings of the National Academy of Sciences of the United States of America* **95**, 5812–5817.

Salmeron, J. M., Oldroyd, G. E. D., Rommens, C. M. T., Scofield, S. R., Kim, H. S. *et al.* (1996). Tomato *Prf* is a member of the leucine-rich repeat class of plant disease resistance genes and lies embedded within the *Pto* kinase gene cluster. *Cell* **86**, 123–133.

Sanders, D., Brownlee, C. and Harper, J. F. (1999). Communicating with calcium. *Plant Cell* **11**, 691–706.

Scharf, K. D. and Nover, L. (1982). Heat shock-induced alterations of ribosomal protein phosphorylation in plant cell cultures. *Cell* **30**, 427–437.

Schlaepfer, D. D. and Hunter, T. (1998). Integrin signalling and tyrosine phosphorylation: just the FAKs? *Trends in Cell Biology* **8**, 151–157.

Sheen, J. (1996). Ca^{2+}-dependent protein kinases and stress signal transduction in plants. *Science* **274**, 1900–1902.

Shibata, W., Banno, H., Ito, Y., Hirano, K., Irie, K. *et al.* (1995). A tobacco protein-kinase, NPK2, has a domain homologous to a domain found in activators of mitogen-activated protein kinases (MAPKKs). *Molecular and General Genetics* **246**, 401–410.

Smertenko, A. P., Jiang, C. J., Simmons, N. J., Weeds, A. G., Davies, D. R. and Hussey, P. J. (1998). Ser6 in the maize actin-depolymerizing factor, ZmADF3, is phosphorylated by a calcium-stimulated protein kinase and is essential for the control of functional activity. *Plant Journal* **14**, 187–193.

Son, M., Gundersen, R. E. and Nelson, D. L. (1993). A second member of the novel Ca^{2+}-dependent protein kinase family from *Paramecium tetraurelia*. Purification and characterization. *Journal of Biological Chemistry* **268**, 594–598.

Song, W. Y., Wang, G. L., Chen, L. L., Kim, H. S., Pi, L. Y. *et al.* (1995). A receptor kinase-like protein encoded by the rice disease resistance gene, *xa21*. *Science* **270**, 1804–1806.

Stein, J. C., Howlett, B., Boyes, D. C., Nasrallah, M. E. and Nasrallah, J. B. (1991). Molecular cloning of a putative receptor protein kinase gene encoded at the self-incompatibility locus of *Brassica oleracea*. *Proceedings of the National Academy of Sciences of the United States of America* **88**, 8816–8820.

Stein, J. C., Dixit, R., Nasrallah, M. E. and Nasrallah, J. B. (1996). SRK, the stigma-specific S-locus receptor kinase of *Brassica*, is targeted to the plasma membrane in transgenic tobacco. *Plant Cell* **8**, 429–445.

Stone, J. M., Collinge, M. A., Smith, R. D., Horn, M. A. and Walker, J. C. (1994). Interaction of a protein phosphatase with an *Arabidopsis* serine–threonine receptor kinase. *Science* **266**, 793–795.

Sugano, S., Andronis, C., Green, R. M., Wang, Z. Y. and Tobin, E. M. (1998). Protein kinase CK2 interacts with and phosphorylates the *Arabidopsis* circadian clock-associated 1 protein. *Proceedings of the National Academy of Sciences of the United States of America* **95**, 11020–11025.

Sugden, C. and Hardie, D. G. (1998). Multiple SNF1-related protein kinases from spinach leaf phosphorylate and inactivate sucrose phosphate synthase. *Plant Physiology*, submitted.

Sugden, C., Crawford, R. M., Halford, N. G. and Hardie, D. G. (1999). Regulation of spinach SNF1-related (SnRK1) kinases by protein kinases and phosphatases is associated with phosphorylation of the T loop and is regulated by 5′-AMP. *Plant Journal* **19**, 1–7.

Takano, M., KajiyaKanegae, H., Funatsuki, H. and Kikuchi, S. (1998). Rice has two distinct classes of protein kinase genes related to SNF1 of *Saccharomyces cerevisiae*, which are differently regulated in early seed development. *Molecular and General Genetics* **260**, 388–394.

Takezawa, D., Ramachandiran, S., Paranjape, V. and Poovaiah, B. W. (1996). Dual regulation of a chimeric plant serine threonine kinase by calcium and calcium-calmodulin. *Journal of Biological Chemistry* **271**, 8126–8132.

Thelen, J. J., Muszynski, M. G., Miernyk, J. A. and Randall, D. D. (1998). Molecular analysis of two pyruvate dehydrogenase kinases from maize. *Journal of Biological Chemistry* **273**, 26618–26623.

Thompson, J. D., Gibson, T. J., Plewniak, F., Jeanmougin, F. and Higgins, D. G. (1997). The CLUSTAL_X windows interface: flexible strategies for multiple sequence alignment aided by quality analysis tools. *Nucleic Acids Research* **25**, 4876–4882.

Torii, K. U., Mitsukawa, N., Oosumi, T., Matsuura, Y., Yokoyama, R. *et al.* (1996). The *Arabidopsis erecta* gene encodes a putative receptor protein kinase with extracellular leucine-rich repeats. *Plant Cell* **8**, 735–746.

Turck, F., Kozma, S. C., Thomas, G. and Nagy, F. (1998). A heat-sensitive *Arabidopsis thaliana* kinase substitutes for human p70(s6k) function *in vivo*. *Molecular and Cellular Biology* **18**, 2038–2044.

Umeda, M., Bhalerao, R. P., Schell, J., Uchimiya, H. and Koncz, C. (1998). A distinct cyclin-dependent kinase-activating kinase of *Arabidopsis thaliana*. *Proceedings of the National Academy of Sciences of the United States of America* **95**, 5021–5026.

Vidal, J. and Chollet, R. (1997). Regulatory phosphorylation of C-4 PEP carboxylase. *Trends in Plant Science* **2**, 230–237.

Walker, J. C. (1994). Structure and function of the receptor-like protein kinases of higher plants. *Plant Molecular Biology* **26**, 1599–1609.

Walker, J. C. and Zhang, R. (1990). Relationship of a putative receptor protein kinase from maize to the S-locus glycoproteins of *Brassica*. *Nature* **345**, 743–746.

Wang, H., Fowke, L. C. and Crosby, W. L. (1997). A plant cyclin-dependent kinase inhibitor gene. *Nature* **386**, 451–452.

Wang, P. C., Vancura, A., Mitcheson, T. G. and Kuret, J. (1992). Two genes in *Saccharomyces cerevisiae* encode a membrane-bound form of casein kinase-1. *Molecular and Cellular Biology* **3**, 275–286.

Wang, X. Q., Zafian, P., Choudhary, M. and Lawton, M. (1996). The PR5K receptor protein kinase from *Arabidopsis thaliana* is structurally related to a family of plant defense proteins. *Proceedings of the National Academy of Sciences of the United States of America* **93**, 2598–2602.

Wang, Z. Y. and Tobin, E. M. (1998). Constitutive expression of the *Circadian clock-associated 1* (*CCA1*) gene disrupts circadian rhythms and suppresses its own expression. *Cell* **93**, 1207–1217.

Watillon, B., Kettmann, R., Boxus, P. and Burny, A. (1993). A calcium/calmodulin-binding serine/threonine protein kinase homologous to the mammalian type II calcium/calmodulin-dependent protein kinase is expressed in plant cells. *Plant Physiology* **101**, 1381–1384.

Wilson, W. A., Hawley, S. A. and Hardie, D. G. (1996). The mechanism of glucose repression/derepression in yeast: SNF1 protein kinase is activated by phosphorylation under derepressing conditions, and this correlates with a high AMP:ATP ratio. *Current Biology* **6**, 1426–1434.

Winder, W. W. and Hardie, D. G. (1999). The AMP-activated protein kinase, a metabolic master switch: possible roles in type 2 diabetes. *American Journal of Physiology* **277**, E1–E10.

Woodgett, J. R. (1991). A common denominator linking glycogen metabolism, nuclear oncogenes and development. *Trends in Biochemical Science* **16**, 177–181.

Woods, A., Munday, M. R., Scott, J., Yang, X., Carlson, M. and Carling, D. (1994). Yeast SNF1 is functionally related to mammalian AMP-activated protein kinase and regulates acetyl-CoA carboxylase *in vivo*. *Journal of Biological Chemistry* **269**, 19509–19515.

Xie, Q., Sanz-Burgos, A. P., Hannon, G. J. and Gutierrez, C. (1996). Plant cells contain a novel member of the retinoblastoma family of growth regulatory proteins. *EMBO Journal* **15**, 4900–4908.

Yahalom, A., Lando, R., Katz, A. and Epel, B. L. (1998). A calcium-dependent protein kinase is associated with maize mesocotyl plasmodesmata. *Journal of Plant Physiology* **153**, 354–362.

Yoo, B. C. and Harmon, A. C. (1996). Intramolecular binding contributes to the activation of CDPK, a protein kinase with a calmodulin-like domain. *Biochemistry* **35**, 12029–12037.

Yoon, G. M., Cho, H. S., Ha, H. J., Liu, J. R. and Lee, H. S. P. (1999). Characterization of *NtCDPK1*, a calcium-dependent protein kinase gene in *Nicotiana tabacum*, and the activity of its encoded protein. *Plant Molecular Biology* **39**, 991–1001.

Yun, B., Farkas, R., Lee, K. and Rabinow, L. (1994). The *Doa* locus encodes a member of a new protein kinase family and is essential for eye and embryonic development in *Drosophila melanogaster*. *Genes Development* **8**, 1160–1173.

Zhang, S. H., Broome, M. A., Lawton, M. A., Hunter, T. and Lamb, C. J. (1994a). *Atpk1*, a novel ribosomal-protein kinase gene from *Arabidopsis*. 2. Functional and biochemical analysis of the encoded protein. *Journal of Biological Chemistry* **269**, 17593–17599.

Zhang, S. H., Lawton, M. A., Hunter, T. and Lamb, C. J. (1994b). *Atpk1*, a novel ribosomal-protein kinase gene from *Arabidopsis*. 1. Isolation, characterization, and expression. *Journal of Biological Chemistry* **269**, 17586–17592.

Zhao, Y., Kappes, B. and Franklin, R. M. (1993). Gene structure and expression of an unusual protein kinase from *Plasmodium falciparum* homologous at its carboxyl terminus with the EF hand calcium-binding proteins. *Journal of Biological Chemistry* **268**, 4347–4354.

Zhou, R. (1998). The Eph family receptors and ligands. *Pharmacology and Therapeutics* **77**, 151–181.

Bioinformatics: Using Phylogenetics and Databases to Investigate Plant Protein Phosphorylation

E. R. INGHAM, T. P. HOLTSFORD and J. C. WALKER

Division of Biological Sciences, University of Missouri–Columbia, MO 65211, USA

I. INTRODUCTION: OVERVIEW

What is bioinformatics? What use is sequencing genomes? What has been and will be done with the growing resource of sequence databases? How is multiple sequence analysis done? These questions will be addressed in this chapter while focusing on plant protein kinases and phosphatases, and an elementary introduction to basic phylogenetic analysis will be provided.

Bioinformatics is the analysis of biological data, particularly sequence and genome information, using computer-based information and statistical methods. Although it is already common practice in the plant molecular community to perform a BLAST (Basic Logical Alignment Search Tool) search with newly cloned sequences, and many papers include some sequence analysis, in-depth use of sequence database information is still rare outside the study of molecular evolution. This is partly due to the fact that until recently

Advances in Botanical Research Vol. 32
incorporating Advances in Plant Pathology
ISBN 0-12-005932-0

relatively few sequences, particularly plant sequences, were available for such investigations. This has changed: the amount of data currently available is large and growing rapidly, and the methods for analyzing it are becoming better and easier to obtain.

Being able to compare sequences and gain information about new genes from previously characterized genes is essential if the information learned from model organisms such as *Arabidopsis* is to be transferred to experimentally less tractable but economically important crop species. To further our understanding of the common and the unique characteristics of the plant kingdom, discoveries need to be extended beyond the specific organism or gene. With the availability of whole genome sequences, and sequences from several different genomes, new approaches in bioinformatics can answer new questions. Although this chapter focuses on using bioinformatics to discover functional relationships between genes and proteins, these are also the techniques used to examine gene and species evolution and are just as powerful in that context. Bioinformatics has come of age and should be used as a common tool by the plant scientist.

II. GENE DISCOVERY AND ANALYSIS

Sequence databases can be used to discover general functional, regulatory, and evolutionary information about a gene of interest. This involves sequence comparison by database searching followed by sequence alignment and the creation of phylogenetic trees. Terms used to describe closely related or similar genes include homologous, paralogous, and orthologous.

Sequences which are similar to one another over a long enough stretch that random convergence is unlikely are thought to be derived from the same ancestral source and are referred to as homologous (Li, 1997). Paralogous and orthologous describe the relationship between homologous genes whose origin via duplication is known. Two genes are paralogous if they arose through duplication within the same organism. Two genes are orthologous if they are the gene of the same origin in two different species; alleles of the same gene are also orthologous. Figure 1 shows an example of orthologous and paralogous relationships. Musα is paralogous to both Musα2 and Humα2 but orthologous to Humα. Humα2 is paralogous to Humα2 and Musα2 and orthologous to Musα2.

A. SEQUENCE COMPARISON

The simplest use of sequence databases is to investigate the function of an uncharacterized gene by searching for homologous (or similar) sequences. An easy way to do this is to use NCBI BLAST (Altschul *et al.*, 1997), available for use online at http://www.ncbi.nlm.nih.gov/BLAST/. The protein or DNA

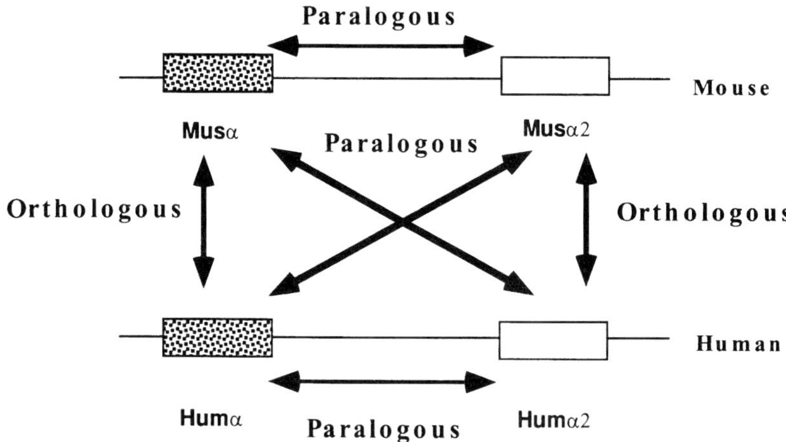

Fig. 1. Paralogous versus orthologous. Genes separated by a duplication event, such as α and $\alpha 2$, are paralogous within a species and between species. Genes which differ by speciation, such as Musα and Humα, are orthologous.

sequences retrieved by this search will be similar to the query sequence and some may be of known function. If the degree of identity or similarity is high this may give a strong indication as to what the new gene product does. High is a relative term; greater than 40% identity is often considered high, but part of the problem in defining a cut-off point is that similarity may be more important over certain areas than others. The amount of sequence similarity required for functional similarity varies depending on the domain and the evolutionary distance between organisms and must be considered on a case-by-case basis. Some information about the statistical relevance of the similarity between retrieved sequences is provided by the BLAST program itself. The results of a BLAST search yield the degree of similarity and identity as well as an alignment between the query sequence and each retrieved sequence, or 'hit'. The statistical likelihood of the match occurring by chance (the E value) is presented along with a score 'bits'. The lower the E value, the lower the probability that the match is spurious or random. The higher the score 'bits' value, the more significant the match is in general. Further details are available in the documentation provided at the BLAST website.

In performing a BLAST search with a gene of interest it must be decided whether this should be done with the amino acid or nucleotide sequence. This decision is based on the question being asked, and how wide a sequence search and comparison is desired. Nucleotide sequences can be informative for closely related genes. For more distantly related genes, amino acid alignments may be more reliable because they have a slower rate of change. Additionally, changes at the third position in a codon of nucleotide sequences are usually (70%)

without functional consequence because they do not change the amino acid in the protein. This means that reversions from one residue to another and then back to the original at the third codon position are functionally meaningless and undetectable (Li, 1997). However, if the genes in question are very closely related, then amino acid differences may not be great enough to provide any information; this is more often the case in evolutionary inquiries. In most instances where protein function is being examined by homology, the query sequence should be amino acid and not nucleotide.

The sequences retrieved by a BLAST search need to be examined over their entire length and at specific areas/domains to uncover all possible homologies which might give functional insight. After an initial investigation using the full-length sequence, specific portions/domains can also be examined in further detail to compare possible sites of regulation and/or substrate specificity. It is also possible that different domains of the protein have different origins, so it may be useful to search with these smaller regions of the protein to uncover multiple sources of a composite gene. Protein kinases often have noncatalytic domains which need this type of separate analysis when identifying potential functions by homology. For instance, the catalytic domain of a protein kinase may be fused to a regulatory domain as in the case of the calcium-dependent protein kinases (CDPKs; Hrabak *et al.*, 1996); see Section III.B.

Examining a gene's possible function using its similarity to other sequences is particularly useful when the type of protein is common and/or has been extensively investigated because there is a large base for comparison. Kinases, particularly serine/threonine/tyrosine protein kinases, fulfill this criteria. There are over 4376 predicted protein kinases which have been sequenced (Protein Kinase Resource http://www.sdsc.edu/kinases, October 5 1998). Many of these also have known biological and biochemical functions. Specific catalytic subdomains of protein kinases have been identified both functionally and structurally, and several subfamilies have been characterized (Hardie and Hanks, 1995). Protein phosphatases have also been classified and extensively studied both functionally and structurally (Barford, 1996). Because of the large amount of sequence and biochemical information available for these protein families, much can be gained from sequence analysis of an uncharacterized protein kinase or phosphatase. Future investigations about plant protein phosphorylation can then make full use of previous work.

Much of what is currently known about the characteristics of protein kinases was brought about by sequence comparison and the use of gene phylogeny. Along with comparing the structural information provided by crystallography, phylogenetic analysis was used to develop the common classification of serine/threonine/tyrosine protein kinases (Hardie and Hanks, 1995). As genome projects advance and researchers finish sequencing the *Arabidopsis* genome, this type of protein family analysis will be available to refine further the classification of novel plant protein kinases/phosphatases and to define other new plant protein families. This is of particular interest since the current protein kinase and phosphatase classifications were done before many of the current plant sequences were available.

The classification of protein kinases and phosphatases, like all phylogenetic analysis, involved more than collecting sequences. Once the sequences have been gathered, they need to be aligned. This is perhaps the most important step and should be done with care. When constructing an alignment a model for substitution rates needs to be chosen. A substitution matrix is necessary because all possible changes between nucleotide or amino acid residues are not of equal probability. Evolutionary constraints also tend to favor conservative amino acid substitutions, such as from serine to threonine, and different organisms can have different rates of nucleotide or amino acid change (Li, 1997). After a substitution model has been chosen, an alignment is made. Phylogenetic trees may then be inferred using a tree-building algorithm. These processes of sequence alignment and analysis are outlined in the following two sections.

1. Substitution Models

Models of substitution rates consist of square matrices, 4×4 for nucleotides, 20×20 for amino acids. These are used when creating an alignment and also when doing database searches with BLAST. Numbers in the matrix estimate the frequency of changing from one residue to the other. For nucleotide data the matrix is based on a model incorporating both rates of substitution and the relative rate of substitution at all different sites. Thus both the rate of overall change and the specific substitution rate are included (Baxevanis and Ouellette, 1998). For amino acids, matrices currently in use only incorporate rates of substitution. Matrices for DNA and protein substitution are provided in alignment programs or user-created matrices can be imported from other files. [For more on substitution models see Durbin *et al.* (1998).]

PAM (point-accepted-mutation; Dayhoff *et al.*, 1978) and BLOSUM (BLOCKS substitution matrix; Henikoff and Henikoff, 1992) are the most commonly used amino acid substitution matrices and are based on empirical differences found in alignments. The PAM matrix values are based on the mutation frequencies between residues in very similar pairs of sequences. The evolutionary relationship between the sequences was examined in a phylogenetic tree made by the maximum parsimony method (see Section II.A.2). The data obtained from the pairs were extrapolated to longer evolutionary distances by estimating the conditional probability for a particular substitution over time. See Durbin *et al.* (1998) for a short explanation of the mathematics involved. In short, pairs of sequences which could be aligned without doubt were used to estimate a PAM distance of one, and this was extrapolated to the distance of 250 based on the phylogenetic tree. The number following a PAM matrix, usually the original 250, refers to the relative distance to which the matrix was extrapolated; values above 200 being most useful because they can estimate the rate of substitution over larger evolutionary distances.

The BLOSUM substitution matrix was based on sequence divergence rather than on sequence similarity. The sequences used for BLOSUM came from the BLOCKS database (Henikoff and Henikoff, 1992) of ungapped alignments of protein families. Proteins were grouped by similarity and the frequency with which a residue occurred in different groups was used to calculate the matrix. The number following a BLOSUM matrix (e.g. BLOSUM90) refers to the maximum identity between the sequences used for its creation. This means that the substitution values in a BLOSUM90 matrix were calculated using sequences not more than 90% identical. High numbered matrices are used for very similar sequences, lower numbered for very divergent. This means that the BLOSUM matrix has been created from a data set comparable in similarity to the experimental set of sequences. BLOSUM62 and 50 are commonly used matrices and are suitable for comparing most protein kinases or phosphatases. For more detailed descriptions and equations see Durbin *et al.* (1998). More sophisticated models (Yang *et al.*, 1998) are being developed and will likely come into use with time as they become incorporated into available programs.

There are several programs available for doing alignments, such as CLUSTAL (Higgins and Sharp, 1988), which is both widely used and easy to obtain as freeware. However, any method has some degree of error and cannot take into account all the information particular to a gene or protein, so it is advisable to refine manually any sequence alignment. For instance, the investigator may know that there is a bias against interrupting certain domains with gaps, and this bias can only be incorporated manually into the alignment. Protein kinase catalytic domains, as an example, usually do not have their subdomains interrupted by gaps or insertions. Of course, as with protein kinases, such deviations from the norm can occur and it is important to keep this in mind when manually adjusting a sequence alignment.

After an alignment is made, to understand further the relationship between the aligned sequences, a phylogenetic gene tree should be made. Looking at a gene tree reveals which of the sequences group together and how they do so in relationship to other groups from the alignment. The closer sequences are on the tree, the more similar and thus the more likely they are to share aspects of function and regulation. Using such gene trees allows for the identification and categorization of orthologous genes and members of gene families. Gene trees also give insights into gene origin, duplication number and timing, and divergence.

There are several basic approaches to making phylogenetic trees, but their outcomes can be used in the same way. Trees provide a means to interpret visually data obtained from applying a given algorithm to sequence information. Pairwise distances among sequences are estimated by using an algorithm, and are put into a matrix which is then used by a tree-building algorithm. The algorithm which calculates the distance matrix is independent of the alignment process and can even use a separate substitution model. Each of the different intermediary algorithms determines the relationship between sequences and their similarity or distance to one another. The soundness of the conclusions made from a tree is based on its quality and the accuracy of the sequence alignment. It is important to be certain of the alignment, as any error will be further compounded in the tree.

2. Constructing Phylogenetic Trees

A quick review of basic phylogenetic concepts is helpful before analyzing sequence sets. For further information see Li (1997), Baxevanis and Ouellette (1998) and Avise (1994). Figure 2 is an outline of the general process of phylogenetic analysis, beginning with a gene of interest, through the tree-building and analysis methods discussed below.

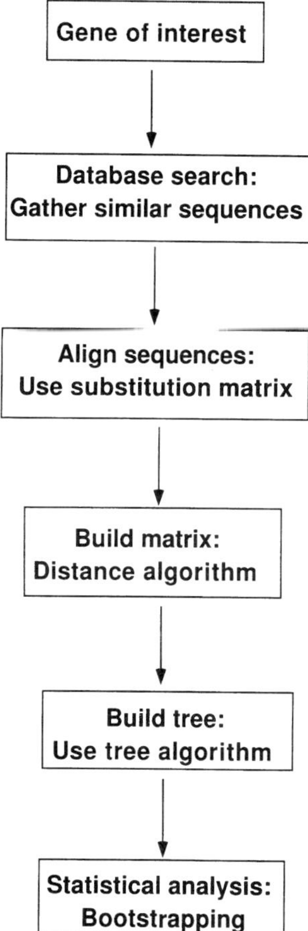

Fig. 2. The general steps for a distance-based phylogenetic sequence analysis. BLAST is commonly used for database searching. CLUSTALX and the PHYLIP package can perform the other steps outlined.

A phylogenetic tree is a graph representing a pattern of descent between sequences or species. The sequences are represented on the tree by nodes. Existing sequences are at external nodes at the ends of branches. Internal nodes or branchings represent points of divergence and/or ancient sequences. The length of the branches themselves may represent the amount of change. However, in some trees, such as those that are printed 'right aligned', branch lengths may be altered for presentation/appearance. It is best to view multiple visual representations of a tree for maximum information. The tree can be either rooted or unrooted. A tree is rooted if there is an outgroup, a particular node, which is outside of the others. The consequence of this is that there is one unique path from the outgroup to any other sequence. This provides a hierarchical, structured relationship between the sequences in the tree. Rooting the tree creates an evolutionary path from the outgroup to the end branches, but requires assumptions to be made about ancestry (Avise, 1994). The outgroup should be homologous to the focal sequences (ingroup), but known from independent evidence to be more distantly related to the ingroup than any of the ingroups' sequences are to each other. For example, a tyrosine kinase may be used in a serine/threonine kinase tree, or a yeast protein kinase in a plant protein kinase gene tree. The decision to root a tree, and how, depends on the question being asked.

Whether making a rooted or an unrooted tree, both an algorithm for distance matrix creation and an algorithm for tree building need to be selected. There are two basic types of algorithms commonly used to infer phylogenies: distance-based and character-state methods. Distance-based matrices are useful tree-building methods for large data sets of more than 50 sequences (especially protein) because of the relative speed of computation. Distance-matrix methods create a matrix of distances between pairs of sequences (or other traits). From this matrix clustering algorithms can be used to infer a phylogenetic tree. The most commonly used clustering algorithm which uses a distance matrix is the neighbor-joining method in which the final tree is formed from an unresolved tree. The unresolved tree starts by having a number of branches equal to the number of sequences radiating from a single point. The tree is then resolved, step by step, by inserting branches between the most similar sequences. The simplest clustering algorithm is the unweighted pair-group method with arithmetic mean (UPGMA; Sokal and Michener, 1958). UPGMA is, however, one of the least accurate phylogenetic methods (Huelsenbeck, 1995), although it is still occasionally used. Other methods using distance matrices are the Fitch–Margoliash and minimum evolution algorithms in which a tree with the shortest branches is selected by minimizing the squared difference between the observed and tree-based distances.

Character-based algorithms assign a character state to each sequence or species. In the case of sequence character states this refers to the residue identity at each position in the sequence (Li, 1997). Commonly used character-based methods include maximum parsimony and maximum likelihood. The difference between these two methods can best be described by defining the maximum likelihood concept. This differs from the other types of tree building (parsimony or distance matrix-based) in that it searches for a model and a tree which best fit the actual data instead of fitting the data to an external model or criterion (Baxevanis and Ouellette, 1998).

Although maximum likelihood is considered the best tree-building algorithm (Huelsenbeck, 1995), it is very computationally intense. Therefore, this method is not used with large data sets of more than 20 species, for long sequences, or for amino acid sequences. For large data sets, distance methods or maximum parsimony are usually employed. Maximum parsimony seeks to build a tree which requires the minimum number of changes to explain the observed character states of sequences (or species). The main problems of parsimony methods are that they cannot correct for multiple substitutions and a phenomenon known as long branch attraction (Huelsenbeck, 1995; Li, 1997). Long branch attraction occurs between sequences which are very different. These sequences may have made some of the same changes by chance since the evolutionary time since their divergence is long. The result of long branch attraction is that two sequences may be grouped together on a tree by a short common branch which is artificial and not representative of the true phylogeny. Despite these drawbacks, maximum parsimony is a valid approach. It is not as computationally intense as maximum likelihood so for medium (20–50 sequences) or larger data sets often both parsimony and distance methods are used and their results compared.

Once a tree has been created, it is important to evaluate it and see how consistent the data are with the tree topology. The most common method for doing this evaluation is to 'bootstrap' the tree. A tree created by any method can be bootstrapped and most tree-building programs support this function whatever algorithm they use. Bootstrapping is a process by which the tree made from the original alignment is checked by random sampling of the data matrix. New pseudoreplicated data matrices are made, usually 500–1000 times, by randomly sampling the sequence. Residues are selected at random with no bias against resampling. For example, for a six amino acid sequence it would be possible to have positions 2,3,2,3,5,5 or 1,1,1,1,1,1, or any other six residue long combination of positions. A tree is then made from this and compared to the original tree. This is repeated as many times as the user selects. The number of times the pseudodata tree and the alignment tree agree at a branch point is the bootstrap value. This is usually represented as a percentage and can be displayed on the tree itself. A value of 70% or greater is considered to support a meaningful branching and has a 95% chance of representing the true phylogeny under conditions where the rate of change is equal and the branches are symmetrical (Hillis and Bull, 1993). Bootstrapping provides a way to measure the consistency of the data set with each branch-point, and is particularly useful for medium and large data sets. Bootstrapping is less accurate with small data sets (Durbin *et al.*, 1998) of very few, short sequences.

Another, less used, statistical evaluation method is the jackknife test. The jackknife test is another resampling technique for looking at trees or other data sets. Jackknifing does not perform replacements like bootstrapping but instead resamples by systematically dropping one or more positions in the data set. Although there are other methods for resampling, bootstrapping and jackknifing are the most widely used and applicable. For those interested, Baxevanis and Ouellette (1998) describe other methods.

Once a tree has been statistically evaluated, it should be viewed. In general, a tree can be viewed radially or branching, scaled or unscaled. Using more than one viewing method is sometimes helpful and should at least be tried. Scaled trees have branch lengths proportional to the number of inferred changes between nodes. This can also provide an aid in determining the accuracy of the tree in later evaluation; short branches can indicate uncertainty, especially if the bootstrap value is low.

B. MAPKs: AN EXAMPLE OF PHYLOGENETIC ANALYSIS

A good example of how to use sequence analysis to understand better and classify proteins is provided by a study of the mitogen-activated protein kinase (MAPK) family (Kültz, 1998). By aligning all known members of this family and creating a phylogenetic tree by the neighbor-joining method, important sequences were identified which divide MAPKs into three subgroups. Within these subgroups, a total of nine subfamilies was found. There was a signature sequence found in all of the MAPKs which distinguished them from other kinase superfamily members as well as sequences characteristic of the nine subfamilies. The structures of two members of each subfamily were modeled on the three-dimensional structure of two known MAPKs. This analysis revealed that the sequence common to all MAPKs is mostly located in the catalytic cleft of the enzyme, whereas the sequences particular to each subfamily were found mostly on the periphery. This implicates the common sequence in catalysis and general MAPK structure maintenance. The subfamily-specific sequences are implicated in specific interactions with other molecules, perhaps substrates or regulatory proteins. Further, mutagenesis targets in subfamily-specific residues were identified by mapping onto the structural model.

What the Kültz paper outlines is how the classification of proteins by sequence analysis can point out novel motifs, provide clues about structure/function relationships, and direct further studies. It also shows how protein kinases are amenable to such studies because of the number of sequences and proteins which have had their three-dimensional structure resolved.

III. FURTHER ANALYSIS AND CONSIDERATIONS

Sequence comparison can be used to identify potential functions for uncharacterized genes, but it is often necessary in such investigations to analyze these relationships further. This is useful for two main things: classifying and discovering evolutionary relationships. Evolutionary relationships can be at the taxonomic or the gene level. As more complete genomes, including the first plant genome *Arabidopsis*, are sequenced, looking at the evolution of genes and genomes becomes even more practical and exciting. This section looks at some of the questions and phenomena that can be addressed by phylogenetic analysis of sequences and gives specific examples of how these have already been applied.

A. GENE DUPLICATION

To understand why sequence databases can be extremely helpful in examining new genes, it is important to understand gene duplication and its consequences. Genes are often duplicated within species. These duplicated genes often have different, but similar, functions. With evolutionary time entire gene families may arise by this method (Li, 1997). These gene families are closely related (usually 50% + amino acid similarity) and much that is learned about one member can be applied to others where they retain this similarity.

Gene duplication can result in isozymes, or proteins which catalyze the same reaction; it can result in families of receptors, or in regulatory cascades. Isozymes are very easy to recognize as they frequently have very similar sequences. Although the sequences of receptor or regulatory families are more divergent, sequence, structure, and functional homologies can be found within them.

One example of a family of homologous receptors is that of the tumor necrosis factor/nerve growth factor (TNF/NGF) family (Orlinick and Chao, 1998). Such proliferation of a receptor type into a class allows for diversification and the recognition of subtly different signals. In the case of TNF receptors their ligands, TNFs, have a 150 amino acid region of similarity. These TNFs perform several functions during immune responses and inflammation, and are necessary for proper regulation of cell proliferation. The complexity of this system in vertebrates is facilitated by the number of subtly different receptors which can sense the cell's environment within the organism. The receptors themselves have a distinctive extracellular region which consists of two to six cysteine-rich domains. Plants also have many receptor candidates, many of them kinases, but they are not as well characterized. In this case phylogeny can play a role in linking structure and function as members of the plant kinase receptor families are characterized.

Another study examining a receptor family was done on response regulators in bacteria by Pao and Saeir (1995). Response regulators are involved in bacterial sensory transduction and have the basic form of a receiver domain covalently attached to an effector domain which can be DNA binding or catalytic in nature. The receiver and effector domains of 49 fully sequenced bacterial response regulators were studied. Areas of sequence similarity and conserved residues were identified. Structural and functional connections were explored when possible and approaches for doing this type of analysis in other studies were outlined. Since plants have homologues of the bacterial two-component system, it is possible to use this information directly for the analysis of plant genes as well.

Not only can families of functional genes arise by gene duplication, but it can also lead to nonfunctional, 'dead' genes called pseudogenes. This is important to keep in mind because most currently available sequence data are based on genome sequencing and not on expressed sequence tags or biochemical investigations. Just because a sequence initially appears to be a functional gene does not necessarily mean that it is expressed and produces a functional protein. Detection of pseudogenes which have become nonfunctional due to deletions, insertions, or frameshifts in the coding sequence is relatively straightforward. Changes that make genes nonfunctional by altering their transcriptional regulation are at this time difficult to detect by sequence alone. Tools to eliminate pseudogenes from functional gene trees based on promoter deficiencies will no doubt be developed in the near future.

B. DOMAIN SWAPPING: GENES WITH MULTIPLE ORIGINS

It is not only genes that can be duplicated, domains can as well. Domain swapping, or shuffling, is the process by which domains are exchanged between genes or by which a domain from one gene is inserted into another coding sequence or even within the same gene. One class of plant protein kinases which may have been formed by shuffling or swapping of domains is the calmodulin-like domain protein kinases (CDPKs; Hrabak *et al.*, 1996). Calmodulin is a protein found in animals, plants, and fungi which binds calcium and activates calmodulin-dependent protein kinases. Plant CDPKs have a calmodulin-like domain at the carboxyl end and can be directly activated by calcium. This type of calmodulin-like domain protein kinase has been found only in plants and protists, not in prokaryotes, animals, or fungi. It is possible that CDPKs simply combine the function of animal/fungi calmodulin and the kinase it regulates. It is also possible that the combination in one protein of regulation and catalytic activity allows for the kinases to respond in a different way or with different kinetics to calcium.

C. FROM FUNCTION OR PATHWAY TO GENE

The function of selected genes can be examined using sequence databases. A pathway or function can also be investigated and additional genes found. To do this it is necessary that other genes having the function of interest have already been identified or that the pathway being investigated is very similar to other known pathways. Again, the more data available, the better the insights to be gained.

One example where a function has been used to identify a new gene based on sequence is the identification of new phosphoglucomutase (PGM) proteins (Whitehouse *et al.*, 1998). In humans three PGM isozymes were biochemically known, but only one had been cloned (PGM1). This was despite trying both molecular and antibody approaches to find other PGMs. To identify potential candidates, a different approach, database searching, was used. The cloned PGM1 amino acid sequence was used for a BLAST search of protein databases. This was followed by a search using the most divergent sequences found in the first search. Forty-seven sequences were found. After building trees by neighbor-joining these fell into six families, three of which contained sequences from only prokaryotic species. It is postulated that the three remaining families containing eukaryotic sequences correspond to the three isozyme forms. This is supported by one group, which does not contain PGM1, and contains a partial human sequence likely, by map position and sequence qualities, to be PGM2. This type of analysis will definitely improve as more sequences are accumulated and is most likely to succeed with a completely sequenced genome.

Sequence databases can also identify new members in signaling or regulatory pathways. This is true if the pathway of interest is similar to well-known pathways or if protein/protein or protein/nucleic acid interaction sites are known. For example, Clarke and Berg (1998) looked for genes that might be regulated by specific zinc finger transcription factors in *Caenorhabditis elegans*. They chose a family of such transcription factors which has a well-known sequence-specific DNA-binding domain. Candidate genes were identified by looking for nearby, upstream sequences matching the zinc-finger binding sites. The search was done by constructing HMMs, or hidden Markov models (a statistical method), for the DNA binding sites using the program HMMER (Eddy: http://genome.wustl.edu/eddy/hmmer.html). HMMs can be constructed for proteins, DNA, or RNA and are useful when interaction sites or particular structural domains are well known.

Using HMMs uncovered more than 200 000 sites. Clearly not all of these were relevant, so further sorting was needed. It was necessary to eliminate hits based on random distribution of nucleotides and to place these sites 5′ of known genes. Doing so reduced the number of candidate genes to five when considering the necessity of three or more DNA sites to bind. Of course, as the authors state, simply counting sites is a somewhat naive approach, but sophistication will improve with time as more tools are developed and more such searches have been performed and followed up with biological data. As more phosphorylation sites on protein kinase substrates are completely determined it may also become possible to use this type of analysis to discover protein kinase substrates and sites of autophosphorylation.

Examining single genes or functions is just one use of sequence analysis. It is also possible to use sequence information to investigate entire pathways. One way to do this is to look at gene pairs. Based on the assumption that proteins which work together will evolve together and either be entirely present or absent from a genome, Pellegrini *et al.* (1999) present a method of looking at correlated evolution. Phylogenetic profiles of proteins using the *Escherichia coli* and 16 other fully sequenced and profiled genomes were constructed. These profiles indicated whether the protein was present or absent in each genome investigated. These data revealed that more than half of the proteins with similar profiles were involved in common complexes or pathways, but that they shared little or no amino acid similarity. The converse was also shown; proteins with very different phylogenetic profiles do not have functional correlations. Reasons for a lack of precision in this analysis include the fact that not all proteins have known function and that both biochemical and signaling pathways are interconnected. More data will only make this type of analysis more powerful. A final test to the accuracy of this procedure will be to test proteins of unknown function and determine whether they interact structurally or functionally with proteins having the same phylogenetic profile. The power of this method will increase as more genomes are sequenced and should be very helpful in transferring known data to new species not as biochemically well studied.

D. EXAMINING DOMAINS

In Section II.A, querying sequence databases with a new gene sequence was discussed. This approach can also provide information at the domain level. An example where this has been the case is the fork-head associated (FHA) domain. This domain can be found in a number of unrelated proteins including the kinase-associated protein phosphate (KAPP) of *Arabidopsis* (Hofmann and Bucher, 1995). It was discovered to occur in a variety of proteins and species by database sequence searching with the domain alone; outside the FHA domain there is little or no homology between the proteins. The function of this FHA domain was unknown and the variety of proteins it was part of puzzling. Now, it appears that the function of the FHA domain may be for a phosphorylation-dependent protein–protein interaction. Both in the budding yeast protein Rad53 (Sun *et al.*, 1998) and in KAPP (Li *et al.*, 1999), this domain is necessary for phosphorylation-dependent binding of the protein's ligand, Rad9 and CLV1/KIK1, respectively. It is likely that the other FHA domains are also important for such protein–protein interactions.

To expand the ability to examine domains by homology, databases of domains and methods for searching using only domains, such as HMMER, have recently been developed. This now allows searches in both directions. That is, one can look at a protein to detect domains, or one can look for proteins containing a particular domain. Such databases include, for example, Pfam at the Sanger Center (http://Pfam.wustl.edu) which is the broadest protein domain database, SMART (Ponting *et al.*, 1999) which contains 250 signaling and extracellular domains, and LIGAND (Goto *et al.*, 1999) which has sections on all known enzymes and their reactions, and on metabolism. Such databases will continue to be formed for additional specific purposes and the information that they can draw on is quickly increasing, adding to their complexity and accuracy.

Protein domains can be classified independently of entire protein function; protein kinase and phosphatase catalytic domains being an example. However, one can do this with even smaller functional sequences such as the sequences that specify membrane protein types. From 2059 known membrane proteins Chou and Elrod (1999) developed an algorithm based on membrane protein type and on the correlation between cellular location and amino acid composition. This model was tested using self-verification by applying it to the test set. The model was then further examined with the jackknife test to explore its power for extrapolation, and then by applying it to an independent data set. The algorithm was found to be approximately 80% accurate for prediction of protein type and 70% for prediction of cellular localization.

Domain homology can also reveal structural information. This kind of structural analysis was done for the C2 domain calcium-binding motif (Nalefski and Falke, 1996). The C2 domain is approximately 130 amino acids long and is found in a wide variety of proteins including calcium-dependent forms of protein kinase C. Looking at 65 C2 sequences revealed that its structure most likely consists of two distinct folds. This prediction is supported by the recent crystallography studies of two C2-containing proteins. Further, with the known structures as a basis, the authors could place the C2 domains into two topological groups and identify candidate calcium co-ordinating residues. This illustrates how primary sequence knowledge can be used to predict tertiary structure when only certain members of a protein family have been crystallized or studied by nuclear magnetic resonance (NMR).

IV. TOOLS

A. PROGRAMS

Programs are helpful and necessary for the analysis of sequences. First, the sequences for the alignment need to be viewed and edited. Then an alignment needs to be created and used to generate trees for phylogenetic analysis at the gene or species level, depending on the question being asked. Several programs are available for DNA and protein sequence viewing, alignment, and phylogenetic tree construction. Below are a few that are easily available either commercially or for free at the Web sites.

A general sequence manipulation package is required to begin. These include such programs as DNA Star (DNAStar Inc., 1228 S. Park St, Madison, WI 53715, USA. http://www.dnastar.com/, 1993) or the Wisconsin Package from GCG (Genetics Computer Group @ http://www.gcg.com), both commercially available. The programs below include freeware packages which are available through the sites listed afterwards.

Clustal (Higgins and Sharp, 1998) is a popular freeware alignment program which builds trees by neighbor-joining. One must be sure to use at least version W, if not X, as this represents a major overhaul and improvement. NJplot is packaged with Clustal and allows the user to view the trees.

FASTA (Pearson and Lipman, 1988) is a sequence-searching program package which provides an alternative to BLAST. It is now contained in commercially available packages which include programs for statistical analysis and alignment. On-line there are several services which will perform free FASTA searches.

HMMER (Eddy, S., University of Washington. http://genome.wustl.edu/eddy/hmmer.html) is a freeware program for constructing protein, DNA or RNA primary structure consensi using the hidden Markov model.

MacClade (Maddison and Maddison) is a commercially available program that allows comparison between and manipulation of trees created using other programs. It has some useful features for looking at heterogeneous rate changes.

PAUP (Swofford) is a widely used and user-friendly commercial package with multiple functions. Version 4 allows maximum parsimony for amino acid and nucleotide sequences but maximum likelihood, invariant and distance methods can be used for nucleotide sequences only.

PHYLIP (Felsenstein, J., Department of Genetics, University of Washington. http://infobiogen.fr/docs/PHYLIPdoc/) is perhaps the most widely used phylogenetic freeware package and contains several programs for use on Macintosh or PC computers. Contained in this package, which is further detailed at the PHYLIP Web site and within the package documentation, are programs for DNA and protein parsimony, neighbor-joining, maximum likelihood tree building as well as other phylogenetic tools. The format is menu-driven line entry, but any confusion that the noncomputer expert has can be cleared up as support for the package is good (see the PHYLIP documentation that comes with the programs).

Sequin (National Center for Biotechnology Information, NCBI, National Library of Medicine, Building 38A, Room 8N805, Bethesda, MD 20894, USA. http://www.ncbi.nlm.nih.gov/) is available for free and puts sequences in the proper form for submission to GenBank, European Molecular Biology Laboratory (EMBL), and DDBJ databases.

Splitstree (Huson, D., FSP Mathematisierung, Universität, 33501 Bielefeld, Germany. http://bioweb.pasteur.fr/docs/doc-gensoft/splitstree/) is a freeware program which uses split decomposition to analyze and view phylogenetic data. It produces a graph with alternate splits which may look like either a tree or a network, depending on the input data.

(p)treevolve (Grassly, N.C. and Rambaut, A., Department of Zoology, University of Oxford, South Parks Road, Oxford OX1 3PS, UK. http://evolve.zoo.ox.ac.uk/Treevolve/treevolve.html) is freeware which simulates DNA or protein evolution using either constant population size or exponential growth and allows the user to set the rate of recombination.

TreeViewPPC (Page, R., Division of Environmental and Evolutionary biology, Institute of Biomedical and Life Sciences, University of Glasgow, Glasgow G12 8QQ, UK. http://taxonomy.zoology.gla.ac.uk/rod/treeview.html) is an easy to use freeware tree viewing/manipulating program which can accept and produce pict, clustal, nexus and other types of files depending on the version.

Other programs available and new ones are currently being written. The BioCatalog Web site (below) has an extensive list of older (1 year+) programs and will no doubt be updated in the near future.

B. SOURCES

1. *Databases*

A molecular biology database list (Burks, 1999) given in *Nucleic Acids Research* contains many database sites, including some newer ones, which have papers concerning them in the same issue.

2. *Web Sites*

Below is a list of some helpful sites which also contain additional links. The list is far from exhaustive and simply provides a starting point. *indicates features of particular interest at the site.

- Australian Biotechnology Association (ABA) – *many links. http://ba-itumacl.lib.unimelb.edu.au/ABA/zPhylogentics/Phylogenetics.html#RTFToC5
- BiBiServ (Bielefeld University Bioinformatics Server) – *programs for use online and to download. http://bibserv.TechFak.Uni-Bielefeld.DE/intor/seqdept.html
- The BioCatalog (European Bioinformatics Institute) – *list of and links to programs. http://www.ebi.ac.uk/biocat/Phylogeny.html
- Cellular and Molecular Pharmacology (University of California, SF) – *good sets of lists. http://www.cmpharm.ucsf.edu/~marrcinj/mytools.html

- Center for Biological Sequence Analysis (Department of Biotechnology, The Technical University of Denmark) – *contains links on different types of search. http://www.cbs.dtu.dk/biolink.html
- Computational Services at EMBL (European Molecular Biology Laboratory) – *alignments, searches, structural modeling and more programs as well as a biocomputing service. http:embl-heidelberg.de/Services/index.html
- HMMER (Sean Eddy's Laboratory at the University of Washington) – *program for construction of HMMs. http://genome.wustl.edu/eddy/hmmer.html
- MOLPHY (Human Genome Mapping Project Resource) – *maximum likelihood tree building. Online access, or programs in C and Perl can be downloaded. Must first register to use. http://sal.kachintech.com/z/2/MOLPHY.html
- NCBI BLAST (National Center for Biotechnology Information) – *online database search tool. http://www.ncbi.nlm.nih.gov/BLAST/
- NIH Bioinformatics (National Institute of Health) – *many different resources, databases, utilities. http://molbio.info.nih.gov/molbio/analysis.html
- Phylogenetic Analysis by Maximum Likelihood (Ziheng Yang at the University College London) – *programs for looking at sequence evolution. Not for building trees, but can be used to analyze and test models. http:/abacus.gene.ucl.ac.uk/ziheng/paml.html#intro
- PHYLIP Package (Joe Felsenstein, Department of Genetics, University of Washington) – *extensive package of freeware widely used for phylogenetic analysis. http://evolution.genetics.washington.edu/phylip.html
- Protein Data Base (PBD) is currently being transferred (as of July 1 1999) to the Research Collaboratory for Structural Bioinformatics (RCSB) whose homepage is: http://www.rcsb.org/ – *database of all known 3D protein structures.
- The Protein Kinase Resource – *this is a database of all known protein kinases as well as tools to analyze sequences. Has kinase-specific modeling tools including ones for three-dimensional structure. http://www.sdsc.edu/kinases
- The Sanger Center Pfam – *Pfam is a large database of multiple alignments. The site has online www access or you can download the database and programs to search it. (Online use is recommended.) http://www.sanger.ac.uk/Pfam

3. *Books and Articles*

A good starting book is: "Bioinformatics: A Practical Guide to the Analysis of Genes and Proteins" (Baxevanis and Ouellette, 1998). This even has examples of how to use particular programs as well as detailed descriptions of many the aspects to consider in sequence alignment. The February 1999 issue of *Cell* contains a review of books by O. K. Pickeral and M. S. Boguski on bioinformatics.

For a more detailed and statistical explanation of biological sequence computation read "Biological Sequence Analysis: Probabilistic Models of Proteins and Nucleic Acids" by Durbin *et al.* (1998).

V. THE FUTURE

The utility of sequence databases will only increase with time as they become more complete, better annotated, and more numerous. Improved tools for identifying genes, searching databases using motifs, and improved algorithms for sequence alignment and evolutionary inferences will also improve the power of the deductions that can be drawn from comparison with characterized sequences.

There is also a great deal of information in noncoding sequences. Noncoding sequences can be used to ask evolutionary questions. They also hold information about how coding sequences are regulated at the level of gene expression and higher level DNA and chromosome architecture. Whole genome analysis of regulation is still nascent but will soon become even more important in understanding biological processes, especially regarding development. Development is only one such area which requires different parts of the genome to be functional in different cells as tissue-specific genes are turned on and other genes are turned off. Genome analysis will allow for pattern searching on a larger scale by using the complete sequences of entire organisms. To summarize, more sequence = more data = more power for searching and the ability to perform more kinds of search.

Primary sequences, DNA or protein, can and have provided much insight. Another area of bioinformatics is also proving useful, structural genomics, or determining and analyzing protein three-dimensional structures (Gaasterland, 1998). Three-dimensional protein databases now exist and with time they may be made tractable to database searching for novel proteins. The collection of data is more time consuming and difficult for protein structure than for sequence data and therefore it lags behind in terms of the amount of information to draw upon. However, international pilot projects to begin providing entire genome data have been started (Gaasterland, 1998). This information will be deposited at the main repository of current protein structure data, the Protein Data Bank (PDB). A test version of a new database which cross-references the PDB to sequence databases is at http:/www.expasy.ch/cgi-bin/swmodel-search-de (Swiss Bioinformatics Institute).

Structural genomics is an interesting area which, along with the rest of bioinformatics, will see much development over the next 10 years as database sizes and numbers increase and programs evolve further. This development will encompass new methods as well as the sequence alignment and phylogenetic tree making outlined in this chapter. More accurate substitution models, more sophisticated matrix and tree-building algorithms, and increased computer processing power will allow for quicker and more accurate inferences about undescribed genes. This and greater structural knowledge will make the application of specific biochemical investigations to groups of homologous proteins more accurate and useful. Such informatic investigations based on sequence comparison are already becoming more common and their importance will no doubt increase.

REFERENCES

Altschul, S. F., Madden, T. L., Schaffer, A. A., Zhang, J., Zhang, Z., Miller, W. and Lipman, D. J. (1997). Gapped BLAST and PSI-BLAST: a new generation of protein database search programs. *Nucleic Acids Research* **25**, 3389–3402.

Avise, J. C. (1994). "Molecular Markers, Natural History and Evolution". Chapman and Hall, New York.

Barford, D. (1996). Molecular mechanism of the serine/threonine phosphatases. *Trends in Biochemical Science* **21**, 407–412.

Baxevanis, A. D. and Ouellette, B. F. F. (1998). "Bioinformatics: A Practical Guide to the Analysis of Genes and Proteins". John Wiley and Sons, New York.

Burks, C. (1999). Molecular biology database list. *Nucleic Acids Research* **27**, 1–9.

Chou, K.-C. and Elrod, D. W. (1999). Prediction of membrane protein types and subcellular location. *Proteins* **34**, 137–153.

Clarke, N. D. and Berg, J. M. (1998). Zinc fingers in *Caenorhabditis elegans*: findings families and probing pathways. *Science* **282**, 2018–2022.

Dayhoff, M. O., Schwartz, R. M. and Orcutt, B. C. (1978). A model of evolutionary change in proteins. *In* "Atlas of Protein Sequences and Structure" (M.O. Dayhoff, ed.) Vol. 5, Suppl. 3, Chapter 22, pp. 345–352). National Biomedical Research Foundation, Washington DC.

Durbin, E., Eddy, S., Krogh, A. and Mitchison, G. (1998). "Biological Sequence Analysis". Cambridge University Press, Cambridge.

Gaasterland, T. (1998). Structural genomics: bioinformatics in the driver's seat. *Nature Biotechnology* **16**, 625–627.

Goto, S., Nishioka, T. and Kanehisa, M. (1999). LIGAND database for enzymes, compounds and reactions. *Nucleic Acids Research* **27**, 377–379.

Hardie, G. and Hanks, S. (1995). "The Protein Kinase Facts Book". Academic Press, London.

Henikoff, S. and Henikoff, J. G. (1992). Amino acid substitution matrices from protein blocks. *Proceedings of the National Academy of Sciences of the United States of America* **89**, 10915–10919.

Higgins, D. G. and Sharp, P. M. (1988). CLUSTAL: a package for performing multiple sequence alignment on a microcomputer. *Gene* **73**, 237–244.

Hillis, D. M. and Bull, J. J. (1993). An empirical test of bootstrapping as a method for assessing confidence in phylogenetic analysis. *Systematic Biology* **42**, 182–192.

Hofmann, K. and Bucher, P. (1995). The FHA domain: a putative nuclear signaling domain found in protein kinases and transcription factors. *Trends in Biochemical Science* **20**, 347–349.

Hrabak, E. M., Dickmann, L. J., Satterlee, J. S. and Sussman, M. R. (1996). Characterization of eight new members of the calmodulin-like domain protein kinase gene family from *Arabidopsis thaliana*. *Plant Molecular Biology* **31**, 405–412.

Huelsenbeck, J. P. (1995). Performance of phylogentic methods in simulation. *Systematic Biology* **44**, 17–48.

Kültz, D. (1998). Phylogenetic and functional classification of mitogen- and stress-activated protein kinases. *Journal of Molecular Evolution* **46**, 571–588.

Li, J., Smith, G. P. and Walker, J. C. (1999). Kinase interaction domain of KAPP, a novel phosphorylation binding domain. *Proceedings of the National Academy of Sciences of the United States of America*, in press.

Li, W.-H. (1997). "Molecular Evolution". Sinauer Associates, Sunderland, WA.

Nalefski, E. A. and Falke, J. J. (1996). The C2 domain calcium-binding motif: structural and functional diversity. *Protein Science* **5**, 2375–2390.

Orlinick, J. R. and Chao, M. V. (1998). TNF-related ligands and their receptors. *Cellular Signaling* **10**, 543–551.

Pao, G. M. and Saier, M. H., Jr (1995). Response regulators of bacterial signal transduction systems: selective domain shuffling during evolution. *Journal of Molecular Evolution* **40**, 136–154.

Pearson, W. R. and Lipman, D. J. (1988). Improved tools for biological sequence comparison. *Proceedings of the National Academy of Sciences of the United States of America* **4**, 2444–2448.

Pellegrini, M., Marcotte, E. D., Thompsons, M. J., Eisenberg, D. and Yeats, T. O. (1999). Assigning protein functions by comparative genome analysis: protein phylogenetic profiles. *Proceedings of the National Academy of Sciences of the United States of America* **96**, 4285–4288.

Ponting, C. P., Schultz J., Milpetz, F. and Bork, P. (1999). SMART: identification and annotation of domains from signaling and extracellular protein sequences. *Nucleic Acids Research* **27**, 229–232.

Sokal, R. R. and Michener, C. D. (1958). A statistical method for evaluating systematic relationships. *University of Kansas Science Bulletin* **28**, 1409–1438.

Sun, Z., Hsiao, J., Fay, D. S. and Stern, D. F. (1998). Rad53 FHA domain associated with phosphorylated Rad9 in the DNA damage checkpoint. *Science* **281**, 272–274.

Whitehouse, D. B., Tomkins, J., Lovegrove, J. U., Hopkinson D. A. and McMillan W. O. (1998). A phylogenetic approach to the identification of phosphoglucomutase genes. *Molecular Biology and Evolution* **15**, 456–462.

Yang, Z., Nielson, R. and Hasegawa, M. (1998). Models of amino acid substitution and applications to mitochondrial protein evolution. *Molecular Biology and Evolution* **15**, 1600–1611.

Protein Phosphatases: Structure, Regulation, and Function

SHENG LUAN

Department of Plant and Microbial Biology, University of California at Berkeley, Berkeley, CA 94720, USA

I. OVERVIEW

Signal transduction regulates cellular processes in response to external/internal stimuli and is crucial for the growth, division, and differentiation of all organisms. It is now recognized that the reversible phosphorylation of proteins is an essential component of almost all signaling pathways in a living cell. Changes in the phosphorylation state of a protein are conducted by two types of enzyme activities: protein kinases that catalyze the covalent attachment of a phosphate group to an amino acid side chain, and protein phosphatases that reverse this process.

Whereas all protein kinases are structurally related to one another (Hunter, 1995; Johnson *et al.*, 1996a), protein phosphatases are defined by at least three distinct families (Stone and Dixon, 1994; Barford, 1995; Cohen, 1997; Neel and Tonks, 1997; Table I). The PPP and PPM families consist of serine (Ser)/

Advances in Botanical Research Vol. 32
incorporating Advances in Plant Pathology
ISBN 0-12-005932-0

TABLE I
Protein phosphatases and their distribution in yeast, plants, and animals

Ser/Thr phosphatases	
PPP family	PP1: yeast, plants, and animals
	PP2A: yeast, plants, and animals
	PP2B: yeast, plants(?), and animals
	Novel: PP4, PP5, PP6, RdgC/PP7 (yeast, plants? and animals)
PPM family	PP2C: yeast, animals, and plants (e.g. ABI, KAPP, MP2C)
Tyrosine phosphatases	
Tyrosine-specific	Receptor-like: animals
	Intracellular: yeast, animals, and plants (e.g. AtPTP1)
Dual-specificity	VH1
	MKPs: yeast, animals, and plants (e.g. AtDsPTP1)
	CDC25: yeast, animals, and plants (?)
	PTEN

threonine (Thr) phosphatases and the protein tyrosine phosphatase (PTPase) family includes both tyrosine-specific and dual-specificity phosphatases (Denu *et al.*, 1996; Tonks and Neel, 1996). Even within the same family, significant structural diversity can be generated by the presence of unique regulatory and targeting domains, or attachment of regulatory subunits to the catalytic subunits. These regulatory domains or subunits may localize the protein complexes to a specific subcellular compartment, modulate the substrate specificity, or alter the catalytic activity.

The attachment to or removal of a phosphate group from a protein often has profound effects on the structure and thereby modifies the functional property of the protein. For example, phosphorylation can regulate an enzyme activity by initiation of allosteric conformational changes, which may directly block the access to the active site (Johnson *et al.*, 1993; Johnson and O'Reilly, 1996). Another important function for phosphorylation is to regulate the interaction among protein partners that must form complexes in order to function (Pawson, 1995). Nearly all aspects of cell function involve reversible phosphorylation. Some examples include metabolism, cell cycle progression, ion transport, developmental control, and stress responses. This diverse spectrum of cellular functions is reflected by a large number of intracellular proteins that is subject to reversible phosphorylation. It may not be surprising that eukaryotic genomes encode approximately 2000 and 1000 protein kinases and phosphatases, respectively, corresponding to 3% of these genomes (Hunter, 1995).

The large number and dynamic nature of protein kinases and phosphatases make it difficult to understand the function of each enzyme. During the past two decades, a number of approaches have been used to define (1) the structure and catalytic mechanism, (2) regulation by regulatory subunits, and (3) cellular processes regulated by protein kinases/phosphatases. While the large number of protein kinases and their functions have been recognized for many years, the number, diversity, and function of protein phosphatases have only been appreciated more recently. Even less is known regarding protein phosphatases in higher plants. In each section of this chapter, major paradigms established in animal (and yeast) systems will first be described, which will be followed by findings on protein phosphatases in higher plants. The comparative approach used here, together with several recent reviews (e.g. Smith and Walker, 1996; Luan 1998), will hopefully bring together the basic concepts and the most recent progress in this field.

II. STRUCTURE AND CATALYSIS

A. Ser/Thr PHOSPHATASES

The earliest biochemical studies in animal systems (reviewed by Cohen, 1989) defined two major types of Ser/Thr phosphatases based on their substrate specificity and pharmacological properties: type 1 and type 2. Type 1 phosphatase (PP1) prefers a β subunit of phosphorylase kinase as the substrate and is inhibited by nanomolar concentrations of two small peptide inhibitors, inhibitors 1 and 2. Type 2 phosphatases (PP2) preferentially dephosphorylate the α subunit of phosphorylase kinase and are insensitive to inhibitors 1 and 2. PP2 enzymes can be further divided into PP2A, -2B, and -2C, simply by their dependence on divalent cations. While PP2B and PP2C require Ca^{2+} and Mg^{2+}, respectively, PP2A, like PP1, is active in the absence of any divalent cations. A group of drugs including okadaic acid, calyculin A, and cantharidin has also been useful in distinguishing members of PP1- and PP2-type enzymes. For example, both okadaic acid and calyculin A potently inhibit the activity of PP1 and PP2A but are not effective on PP2B and PP2C. Cantharidin inhibits only PP2A but not others (Li and Casida, 1992). These reagents have been used as effective tools in defining cellular functions of these enzymes in various systems ranging from mammalian to plant cells (see Section IV).

This simple classification system was used for more than two decades until a large number of genes encoding these phosphatases was identified from eukaryotic organisms. Sequence and structural analyses of these gene products suggest that PP1, PP2A, and PP2B are more closely related and are defined as the PPP family, whereas PP2C, pyruvate dehydrogenase phosphatase, and several other Mg^{2+}-dependent Ser/Thr phosphatases, are more similar to each other and are referred to as the PPM family (Barford, 1996; Cohen, 1997).

1. PPP Family

Although the PPP family enzymes (PP1, PP2A, and PP2B) share a common catalytic region of approximately 280 amino acids, they become divergent when comparing their N- and C-terminal noncatalytic domains. These enzymes are further distinguished by their associated regulatory subunits to form a diverse variety of holoenzymes.

In animals, PP1 is involved in controlling multiple cellular functions including glycogen metabolism, muscle contraction, cell cycle progression, neuronal activities, and splicing of RNA. These processes are regulated by different holoenzymes in which the same catalytic subunit (PP1c) is complexed to distinct regulatory and targeting subunits (Hubbard and Cohen, 1993; Faux and Scott, 1996). Similar to PP1, the diverse functions of PP2A are attributed to the presence of at least 15 B regulatory subunits that individually assemble with each core heterodimer of PP2Ac and a 65-kDa A subunit (Wera and Hemmings, 1995). PP2B is a Ca^{2+}, calmodulin-dependent protein phosphatase (Klee *et al.*, 1988). The holoenzyme contains the catalytic A and regulatory B subunits. The A subunit contains several unique structural domains besides the core catalytic region: the B-interaction, calmodulin-binding, and an auto-inhibitory domain. The B subunit is a Ca^{2+} sensor protein that, like calmodulin, contains four Ca^{2+}-binding EF-hand motifs. When the cytosolic Ca^{2+} concentration reaches micromolar levels, calmodulin assembles into the A–B heterodimer and activates the phosphatase activity (Klee *et al.*, 1998; Figs 1 and 2).

The crystal structure of PP1 and PP2B has been solved and provides models for the structure and catalysis of PPP enzymes (Egloff *et al.*, 1995, 1997; Goldberg *et al.*, 1995; Griffith *et al.*, 1995; Kissinger *et al.*, 1995). Based on the primary sequence, it is not surprising that PP1C and PP2B share a common catalytic domain structure. The catalytic domain of PP1C and PP2B consists of a central 'sandwich' formed by two subdomains of 'helix-sheet' mixture. The crystal structure of the two enzymes also reveals the importance of metal ions in the catalytic reaction of PPP family phosphatases. These metal ions include Mn^{2+} and Fe^{2+}/Fe^{3+} for PP1C (Egloff *et al.*, 1995) and Zn^{2+} and Fe^{2+}/Fe^{3+} for PP2B (Yu *et al.*, 1997), although both enzymes should, based on structure, contain the same metal ions. This point is still debatable. Evidence also indicates regulation of PP1C/PP2B activity by redox potential that changes the oxidative state of the metal center. At least for PP2B, it has been shown that oxidative stress inactivates the phosphatase (Wang *et al.*, 1996).

Regarding the catalytic mechanism, a number of studies suggests that PPPs catalyze dephosphorylation in a single step with a metal-activated water molecule or hydroxide ion (Lohse *et al.*, 1995). The most compelling evidence for this hypothesis is from the studies of purple acid phosphatase that is related to PPPs at the catalytic domain (Klabunde *et al.*, 1995). This phosphatase catalyzes dephosphorylation by inversion of configuration of the oxygen geometry of the phosphate ion. This mechanism of catalysis would not involve

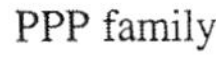

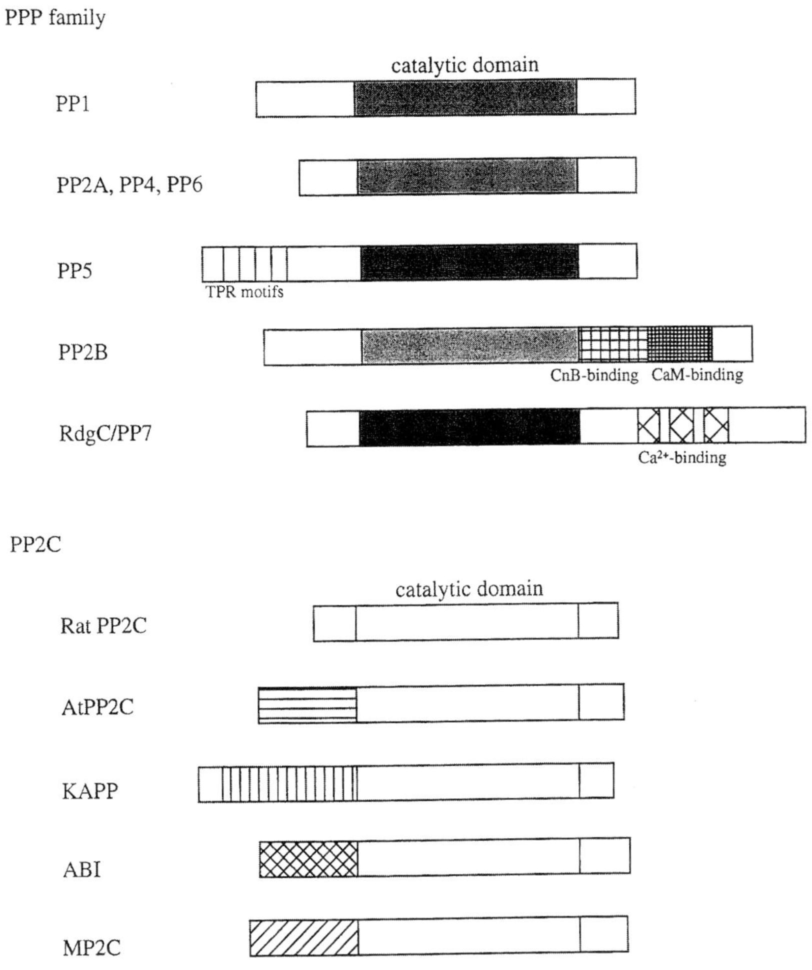

Fig. 1. Structural domains of Ser/Thr phosphatases. The PPP family members all have a catalytic domain of about 280 amino acids. The sequence of this domain is highly conserved among the members (40–60% identity). The PP2C family members have a catalytic domain that is less conserved (20–35% identity). While the PPP family enzymes from plants are similar to animal homologues, the PP2C members are highly diversified.

the formation of a phosphoryl-enzyme intermediate. The two metal-bound water molecules are within the van der Waals distance of the phosphorus atom, and one of them as a water-activated nucleophile that attacks the phosphate group. This mechanism is in sharp contrast with that for tyrosine phosphatases described in later sections.

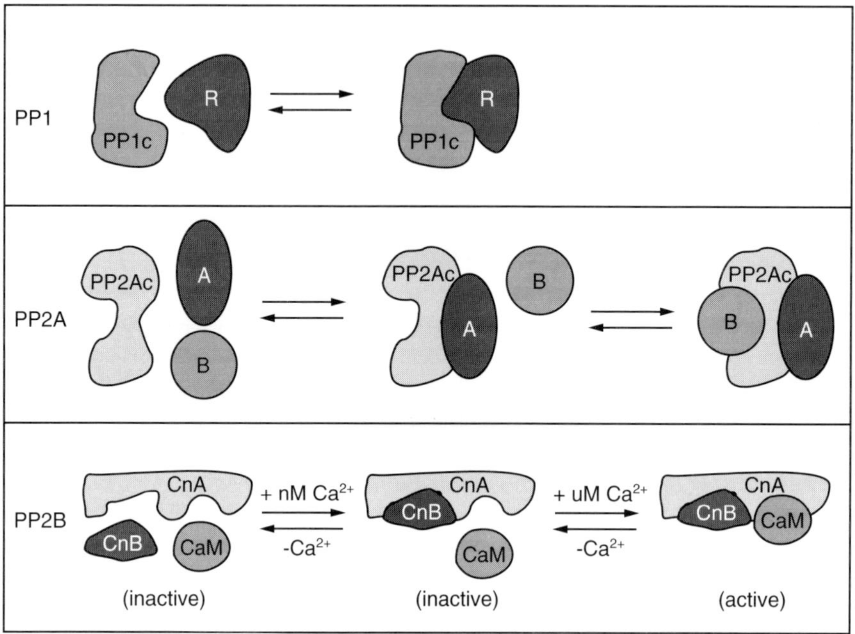

Fig. 2. Subunit composition and regulation of PPP family enzymes. PP1 holoenzymes are heterodimers composed of catalytic PP1C and regulatory R subunit. PP2A holoenzymes exist in either heterodimer or heterotrimer form. PP2B is often present in a heterodimer form with low activity (catalytic CnA and regulatory CnB) under 'resting' levels of calcium. Elevated levels of calcium recruit calmodulin (CaM) into the heterodimer to form a highly active trimeric enzyme.

2. *PPM Family*

PP2C is the representative group of this family that exists in both prokaryotes and eukaryotes. Although the primary sequences of PP2C and other PPM family members share no homology with the PPP enzymes, the structural folds of these two families are strikingly similar (Das *et al.*, 1996). Mammalian PP2C consists of two domains: the N-terminal catalytic region forms a central 'sandwich' structure by two antiparallel sheets; the C-terminal region is a unique antiparallel helix structure remotely attached to the catalytic domain by a loose strand, suggesting a role for defining substrates rather than activity. Also in the catalytic domain are two Mn^{2+} ions in a binuclear metal center coordinated by four invariant Asp and a nonconserved Glu residue. These residues are located on the top of the sandwich channel. Dephosphorylation may be catalyzed by a metal-activated water molecule that serves as a nucleophile, as proposed for the PPP family.

3. Plant Enzymes

Studies on plant protein phosphatases started with biochemical analyses of protein extract from plant tissues (reviewed in Smith and Walker, 1996). Using substrates for mammalian phosphatases and the same pharmacological agents (e.g. okadaic acid), PPP family enzymes such as PP1 and PP2A have been detected in plants such as *Brassica napus* (Mackintosh and Cohen, 1989), pea, carrot and maize (Mackintosh *et al.*, 1991). In *B. napus* seed extract, for example, the predominant protein phosphatase activity (>60% of total activity) prefers β subunit of rabbit phosphorylase kinase and is sensitive to inhibition by inhibitor 1 and 2 from a mammalian source, defining its identity as PP1 (Mackintosh and Cohen, 1989). Further analysis showed that PP1 in rape seed extract is present in a supermolecular complex and is associated primarily with endoplasmic reticulum membranes and other particulate fractions (Mackintosh *et al.*, 1991). In pea leaves and carrot cells, however, PP1 activity is largely cytosolic (Mackintosh *et al.*, 1991). PP1 activity has also been reported to be associated with nucleus and plasma membrane (Sheen, 1993; Vera-Estrella *et al.*, 1994).

Like PP1, PP2A has also been detected in various plant tissues and located in a number of subcellular compartments of plant cells (Mackintosh and Cohen, 1989; Mackintosh *et al.*, 1991; Sheen, 1993; Vera-Estrella *et al.*, 1994). Plant PP2A, like mammalian enzymes, prefers an α subunit of phosphorylase kinase as a substrate and is inhibited by okadaic acid (Mackintosh and Cohen, 1989). In similar studies, PP2C-like activity has also been detected in plant tissues (Mackintosh *et al.*, 1991). Because PP2C prefers a phosphocasein substrate and requires Mg^{2+} for activity, most of the studies use Mg^{2+}-dependent dephosphorylation of phosphocasein as a criterion for PP2C activity. Unlike PP1 and PP2A, the localization of PP2C seems to be more restricted to the cytosolic fraction of plant cells (Mackintosh *et al.*, 1991). More recently, a PP2B-like activity has been reported in the guard cell extract of *Vicia faba* (Luan *et al.*, 1993) and from other plant tissues (C. MacKintosh, personal communications). The subcellular localization of plant PP2B-like activity has not been defined. All of these studies indicate that biochemical properties of PPP and PPM phosphatases are highly conserved among higher eukaryotes ranging from plants to mammals.

It is not surprising that the primary structure of plant phosphatases is also highly similar to the mammalian enzymes. This became obvious when genes encoding members of PP1, PP2A, and PP2C types of phosphatases were cloned from a number of plant species.

At least eight different isogenes encoding PP1 have been identified and more have been recognized in *Arabidopsis* as the genome sequencing projects move forward. The multigene family of PP1 appears to be a common feature of plant PP1. The amino acid sequence of plant PP1 is over 70% similar to that of mammalian enzymes (Smith *et al.*, 1995). Recombinant PP1 from maize is also inhibited by peptide inhibitor 2 and by okadaic acid with a similar sensitivity as measured by IC_{50}. Recombinant maize PP1 requires Mn^{2+} for activity (Smith and Walker, 1991), consistent with the fact that Mn^{2+} is an essential metal ion in the catalytic domain of the PPP family enzymes discussed earlier (Egloff *et al.*, 1995). Although PP1 activity has been shown to be associated with large protein complexes in plant extract (Mackintosh *et al.*, 1991), other proteins (presumably the regulatory subunits) in the complexes and their corresponding genes have not been identified, except for the catalytic subunit (PP1C).

Native PP2A in mammalian cells is found as either a heterodimer consisting of a 36-kDa catalytic subunit and a 65-kDa regulatory A subunit, or as a heterotrimer that is assembled by the heterodimer core and one of the regulatory B subunits (Wera and Hemmings, 1995). Genes for both catalytic and regulatory subunits have been identified from plants. At least five isogenes for the catalytic subunit have been isolated from *Arabidopsis* (Arino *et al.*, 1993; Casamayor *et al.*, 1994; Perez-Callejon *et al.*, 1998). Based on the sequence homology, these genes can be divided into two subgroups, one consisting of isoforms PP2A-1, -2, and -5, and the other consisting of isoforms PP2A-3 and PP2A-4. Catalytic subunit genes have also been identified from other plants such as alfalfa (Pirck *et al.*, 1993) and *B. napus* (Mackintosh *et al.*, 1990). High sequence similarity (over 80%) has been found between plant enzymes and animal homologues. At least three genes for the 65-kDa A regulatory subunit are present in the *Arabidopsis* genome (Slabas *et al.*, 1994). Five genes for B subunits have also been cloned from *Arabidopsis* (Rundle *et al.*, 1995; Latorre *et al.*, 1997). The deduced peptide sequences of plant A and B regulatory subunits share approximately 70% similarity with animal homologues.

Although PP2B-like activity was detected in plant tissues (Luan *et al.*, 1993), no gene has been isolated for a typical catalytic subunit that contains the domains present in the animal PP2B. Two recent reports described identification of calcium-binding proteins related to calcineurin B subunit (Liu and Zhu, 1998; Kudla et al., 1999). These *Arabidopsis* proteins are approximately 30% identical to animal PP2B regulatory subunit and have been shown to bind to rat calcineurin A subunit in a yeast two-hybrid system (Kudla *et al.*, 1999). Further studies to identify the interacting proteins in plants will define the functional identity of these calcineurin B-like proteins.

Higher plants produce a diverse family of PPM enzymes that include four novel PP2C-like enzymes cloned from *Arabidopsis* and alfalfa. Unlike PPP family enzymes, these PP2C members share low homology with animal enzymes (only 20–35% identity at the amino acid level). In addition, the plant enzymes each contain unique N-terminal extensions (Fig. 1). These structural domains are not found in any animal or fungal enzyme. The first *Arabidopsis* PP2C homologue, PP2C-At, was identified by its ability to complement a defect in yeast strains with mutations in cyclic adenosine monophosphate (cAMP)-dependent signaling pathways (Kuromori and Yamamoto, 1994), although its role in *Arabidopsis* remains unknown. PP2C-At contains a C-terminal region with 35% identity to rat PP2C and a unique N-terminal region without significant homology to any proteins in the database. The second *Arabidopsis* PP2C homologue, ABI1, was identified as a component in a signaling pathway in response to plant hormone abscisic acid (ABA) (Leung *et al.*, 1994; Meyer *et al.*, 1994). The C-terminal region of ABI1 protein shares 35% identity with rat PP2C, but the N-terminal region is unique and displays a limited homology to Ca^{2+}-binding protein. The recombinant enzyme is a typical PP2C with Mg^{2+}-dependent phosphatase activity that dephosphorylates phosphocasein (Bertauche *et al.*, 1996). The phosphatase activity of ABI1 is not sensitive to okadaic acid and is not regulated by Ca^{2+}. An ABI1 homologue, ABI2, has been identified recently and is 80% identical to ABI1 (Leung *et al.*, 1997).

The third PP2C from *Arabidopsis* is KAPP, referring to kinase-associated protein phosphatase. This PP2C member was isolated as an interacting partner protein of RLK5, a receptor-like kinase from *Arabidopsis* (Stone *et al.*, 1994). Based on the deduced peptide sequence, KAPP protein contains three domains, a N-terminal membrane anchor, a kinase-interacting domain, and a C-terminal PP2C catalytic domain. Consistent with this structural prediction, recombinant KAPP displays a Mg^{2+}-dependent activity that dephosphorylates phosphocasein (Stone *et al.*, 1994). In addition, the kinase-interacting domain is indeed required and sufficient for interacting with RLK5. Most recently, another plant PP2C isoform, MP2C, has been isolated from alfalfa (Meskiene *et al.*, 1998). In addition to the C-terminal PP2C catalytic domain, MP2C has a unique N-terminal region that show no homology to known proteins. The N-terminal extension of plant PP2Cs may be important for regulation of the protein phosphatases, although this has yet to be demonstrated by further experiments.

In several systems studied, a number of 'novel' protein phosphatases has been found. In particular, the PPP family has several novel variations from the signature phosphatases (Cohen, 1997). For example, PP4 and PP6 are similar to PP2A but have distinct features. PP4 is 65% identical to PP2A (45% identical to PP1) and is located specifically to the centrosome of animal cells, and PP6 (63% identical to PP2A and 53% to PP1) plays a role in cell cycle control (Brewis *et al.*, 1993). PP5 is more similar to PP1 with a N-terminal tetratricopeptide repeat (TPR) structure that has been shown to be present in a number of proteins and is important for protein–protein interactions (Das *et al.*, 1998). Studies show that PP5 interacts with the TPR-interacting domain of HSP90 that forms a heteromultimer complex with the glucocorticoid receptor (Chen *et al.*, 1996).

Another interesting phosphatase, RdgC, has been found in *Drosophila* (Steele *et al.*, 1992). It shares significant homology with the PPP family in the N-terminal catalytic domain but has a C-terminal region that contains multiple EF-hand Ca^{2+}-binding motifs. An RdgC homologue, PP7, has been recently reported in humans (Huang and Honkanen, 1998). Although its sequence is more closely related to PPP family members, the phosphatase activity of recombinant PP7 requires Mg^{2+} and is not sensitive to okadaic acid and other PP1/2A inhibitors. Consistent with the presence of EF-hand motifs, PP7 activity is activated by high micromolar Ca^{2+} but is not sensitive to calmodulin (Huang and Honkanen, 1998). In higher plants, little is known regarding the existence of these 'novel' forms of protein phosphatases. Based on studies in animals and yeast, these enzymes are highly conserved during evolution (Cohen, 1997). It will not be surprising if similar sequences are already in the plant databases. The structure, regulation, and function of these novel protein phosphatases in higher plants await characterization.

B. PROTEIN–TYROSINE PHOSPHATASES

In animals, protein–tyrosine phosphatases (PTPases) consists of hundreds of enzymes that are capable of dephosphorylating tyrosine residues in a protein substrate. Based on phosphoamino acid specificity, PTPases can be further divided into two large groups: tyrosine-specific PTPases and dual-specificity PTPases. Tyrosine-specific PTPases dephosphorylate only tyrosine but not serine/threonine; the dual-specificity PTPases are able to dephosphorylate both tyrosine and serine/threonine residues (Stone and Dixon, 1994). Although the overall protein sequence of tyrosine-specific PTPases and dual-specificity PTPases share little homology, all PTPases contain a signature motif in the catalytic core: (V/I) HCXAGXGR (S/T)G that harbors an essential cysteinyl residue required for the formation of phosphoenzyme intermediate. In addition, the secondary and tertiary structure of all PTPases bears high similarity in the catalytic region. The substrate specificity is determined by sequences outside the catalytic domain (Denu *et al.*, 1996; Tonks and Neel, 1996).

Tyrosine-specific PTPases are a diverse group of proteins. Based on their subcellular localization, these PTPases are classified into two subgroups: receptor-like PTPases and intracellular PTPases (Stone and Dixon, 1994; Neel and Tonks, 1997). The receptor-like PTPases all contain an extracellular domain of variable length and composition, a single transmembrane region, and one or two cytoplasmic PTPase catalytic domains. The intracellular PTPases typically contain a single catalytic domain and various N- and C-terminal extensions that are involved in targeting or other regulatory functions (Mauro and Dixon, 1994).

At least a dozen receptor PTPases have been identified from mammalian systems, each with a distinct extracellular domain. The first receptor-like PTPase to be identified was CD45, which also represents the first PTPase ever found (Charbonneau *et al.*, 1988; Tonks *et al.*, 1988). The extracellular domain of CD45 is a single fibronectin III unit and an extended N-terminal sequence. The intracellular region has two catalytic domains. A subset of the receptor-like PTPases, such as PTPμ and PTPκ, has extracellular domains that contain one immunoglobulin-like domain and four fibronectin III repeats (Gebbink *et al.*, 1993; Jiang *et al.*, 1993). Both molecules have two catalytic domains in the cytoplasmic region. Most of the other subtypes of receptor-like PTPases also have extracellular domains with similarities to cell-adhesion molecules such as fibronectin, suggesting a potential role in cell–cell interaction and recognition (Fig. 3).

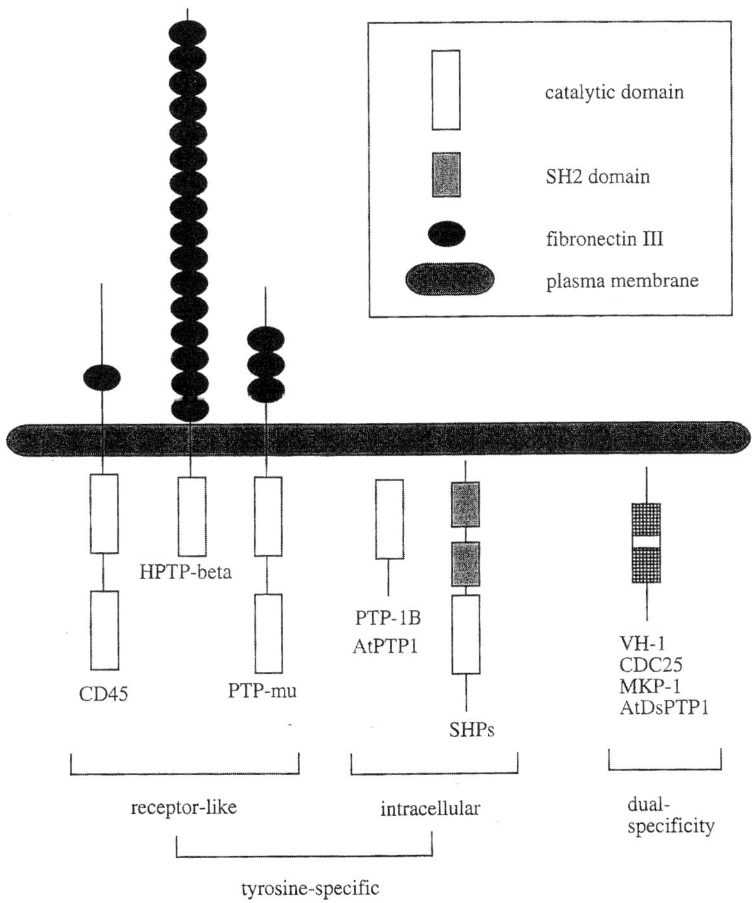

Fig. 3. Structural domains of some representative protein tyrosine phosphatases. Receptor-like PTPases often have two catalytic domains whereas intracellular PTPases have only one. The dual-specificity PTPases are different overall in sequence but always contain the catalytic domain signature motif (as indicated by an open window in the center of the molecule). Plant intracellular (AtPTP1) and dual-specificity PTPase (AtDsPTP1) are listed.

Intracellular PTPases are also distinguished by their noncatalytic domains that serve to localize the enzymes to the membranes, the nucleus, cytoskeleton elements, and tyrosine-phosphorylated signaling proteins (Mauro and Dixon, 1994; Van Vactor *et al.*, 1998). Of particular interest are several PTPases with Src-homology 2 (SH2) domains attached to the catalytic region. These PTPases have been shown to interact with activated receptor tyrosine kinases and other targets (Feng *et al.*, 1993; O'Reilly and Neel, 1998; Timms *et al.*, 1998). The dual-specificity PTPases are also intracellular proteins but unique in their ability to hydrolyze not only phosphotyrosyl but phosphoseryl and threonyl monoesters as well. Dual-specificity PTPases also represent a large number of proteins with a similar catalytic region and distinct noncatalytic regions. As indicated in Table I and Fig. 3, the prototype of this group is VH1, the enzyme encoded by vaccinia virus open reading frame H1 (Guan *et al.*, 1991). The major group of mammalian enzymes is the MAPK phosphatases that dephosphorylate and inactivate mitogen-activated protein kinases (MAPKs). As least half a dozen MKPs have been characterized from mammalian cells; each has a distinct substrate specificity and is involved in the regulation of different MAPK pathways (Keyse, 1995). Another distinct group is CDC25, a cell cycle regulator that is required for activation of CDC2 protein kinase and cell division (Dunphy, 1994). Considering the primary sequence, CDC25 is only distantly related to other dual-specificity PTPases.

Crystal structures of both tyrosine-specific and dual-specificity PTPases have been solved and provide insights into the structural basis for catalysis and substrate specificity (Barford *et al.*, 1994; Stuckey *et al.*, 1994; Yuvaniyama *et al.*, 1996; Hoffman *et al.*, 1997). Despite the sequence divergence between these two subfamilies of PTPases, all crystal structures solved so far reveal essentially the same core structural features: the central four-stranded parallel β sheet surrounded on both sides by one and four α helices. The dual-specificity PTPases appear to be truncated versions of tyrosine-specific PTPases. The signature motif residues are found within a single loop, nestled at the base of a cleft on the surface of the protein. The essential cysteine is in the position for nucleophilic attack on an incoming phosphotyrosyl residue. The remaining residues of the core motif function to increase the nucleophilicity of the catalytic cysteine and to bind to and position the phosphate group. The arginyl residue in the signature motif is particularly important for this function. The depth of the catalytic cleft determines the substrate specificity and is set by an invariant tyrosyl residue in the tyrosine-specific PTPases. Therefore, only phosphotyrosine (not phosphoserine/threonine) is long enough to access the catalytic cysteine (Denu *et al.*, 1996; Tonks and Neel, 1996).

Unlike Ser/Thr phosphatases that have been well conserved among all eukaryotes, PTPases appear to be abundant only in animal, but not in yeast and plant, systems. One reason is that a typical tyrosine kinase has not been identified in yeast and higher plants despite the availability of a complete genome sequence from yeast and approximately half from *Arabidopsis*. There are at least four PTPases characterized from budding yeast, three tyrosine-specific (PTP1, PTP2, and PTP3) and one dual-specificity PTPase (Msg5) (Guan *et al.*, 1992a, b; Zhan *et al.*, 1997). In higher plants, only one tyrosine-specific and one dual-specificity PTPase has been characterized from *Arabidopsis* (Gupta *et al.*, 1998; Xu *et al.*, 1998). At least three more *Arabidopsis* genes have been identified to contain a catalytic core motif of PTPases (S. Luan, unpublished results). Further experiments are needed to determine the enzymatic properties of their protein products. It is conceivable that more genes encoding PTPases are present in plant genomes. The tyrosine-specific PTPase from *Arabidopsis*, AtPTP1, appears to be a member of cytoplasmic PTPases. It contains the catalytic core motif at the C-terminal region and an N-terminal extension of unknown identity. The recombinant AtPTP1 protein specifically dephosphorylates phosphotyrosine but not phosphoserine/threonine in protein substrates. Like other PTPases, the cysteinyl residue in the signature motif is essential for catalytic activity of recombinant AtPTP1. Specific inhibitors for PTPases such as vanadate also inhibit AtPTP1 activity. The dual-specificity PTPase from *Arabidopsis*, AtDsPTP1, is similar to the animal MPKs and is capable of dephosphorylating both phosphotyrosine and phosphoserine/threonine. Besides the signature motif in the catalytic core (of about 10 amino acids), AtPTP1 and AtDsPTP1 share no sequence homology. This is also true with other PTPase-like genes that we have recently identified from the *Arabidopsis* genome (S. Luan, unpublished results). Identification of these PTPases provides a stepping-stone to further understanding the functional significance of tyrosine phosphorylation in higher plants.

III. REGULATION OF PROTEIN PHOSPHATASES

A critical feature of all signal transduction molecules is that they must have the structural basis for sensitive regulation by extracellular and intracellular signals. Protein kinases and phosphatases are pivotal regulators of almost all signaling pathways in a eukaryotic cell. The various mechanisms by which protein phosphatases are regulated in the cell are discussed in this next section. These include regulation at the level of expression, localization, substrate specificity, and activity of these enzymes.

A. REGULATED EXPRESSION

Most of serine/threonine protein phosphatases in animals are ubiquitously expressed. Exceptions are those enzymes that were summarized earlier in the 'novel' category. For example, the *Drosophila* RdgC and human PP7 have been shown to be expressed only in retina (Steele *et al.*, 1992; Huang and Honkanen, 1998). In higher plants, expression of some Ser/Thr phosphatases appears to be highly regulated by spatial and temporal cues, although only a few studies have been dedicated to this topic. For example, genes for PP2A catalytic subunits are co-expressed in most tissues but the expression levels are developmentally regulated (Arino *et al.*, 1993; Casamayor *et al.*, 1994; Perez-Callejon *et al.*, 1998). Expression of a regulatory subunit of PP2A is strongly induced under heat shock (Latorre *et al.*, 1997). The transcript level of a PP2C member, MP2C, is normally not detectable, but increases dramatically upon wounding (Meskiene *et al.*, 1998). A calcineurin B-like protein, AtCBL1, is also highly responsive to stress signals including drought, cold, and wounding (Kudla *et al.*, 1999). In the case of MP2C and AtCBL1, the stress-induced expression is transient, consistent with a possible role of these gene products in early steps of stress signaling pathways.

Unlike Ser/Thr phosphatases, many PTPases are expressed in a tissue- or developmental stage-specific manner and their expression levels are often modulated by environmental conditions. These expression patterns are closely related to their specific functions. For example, CD45 has been shown to be expressed preferentially in the lymphocytic tissues in mammalian systems and is important in signal transduction in T and B cells (Charbonneau *et al.*, 1988; Ledbetter *et al.*, 1988). Several receptor-like PTPases are selectively expressed in the neuronal tissues in *Drosophila* and are involved in neural cell adhesion (Desai *et al.*, 1994). An cytoplasmic SH2-containing PTPase, SHP-1, is expressed at a high level in the hematopoietic cells and plays a critical role in lymphoid cell differentiation (Neel and Tonks, 1997). Dual-specificity PTPases play a role in regulation of MAPK pathways and their expression is often induced by mitogenic and stress signals (Keyse and Emslie, 1992; Sun and Tonks, 1994). This is also true in yeast where tyrosine-specific PTPases are involved in MAPK regulation and are responsive to stress signals (Wurgler-Murphy *et al.*, 1997). In addition, Msg5, a dual-specificity PTPase involved in pheromone signal transduction, is also regulated by pheromone application (Doi *et al.*, 1994). AtPTP1, the only plant tyrosine-specific PTPase identified so far, is regulated by stress factors (Xu *et al.*, 1998). The *Arabidopsis* dual-specificity PTPase (AtDsPTp1) is not regulated by stress signals (Gupta *et al.*, 1998).

Further studies are required to examine the detailed expression pattern of both AtPTP1 and AtDsPTP1 in *Arabidopsis* at various developmental stages and under other stress conditions, such as pathogen infection, which also activates MAPK pathways (Ligterink *et al.*, 1997; Zhang and Klessig, 1997).

B. POST-TRANSLATIONAL MODIFICATION

Phosphorylation is a common post-translational modification of many protein molecules. As enzymes involved in reversible phosphorylation, protein phosphatases are also regulated by phosphorylation. For example, CDK2 phosphorylates PP1C C-terminal Thr residue in a cell cycle-dependent manner and inhibits the PP1 activity in animal cells. PP1 inactivation by CDK2 prevents the dephosphorylation of retinoblastoma protein, which must be phosphorylated for cells to enter the S phase from G1 (Berndt *et al.*, 1997). PP2A catalytic subunits from mammalian cells have been shown to be phosphorylated at a tyrosine residue close to the C-terminus and this modification leads to inactivation (Chen *et al.*, 1992). In addition, threonine phosphorylation by a novel protein kinase inactivates PP2A enzyme (Guo and Damuni, 1993). However, these regulation mechanisms for PP2A have yet to be generalized to PP2A from other eukaryotes including plants. Another novel form of covalent modification of PP2A is carboxymethylation of a Leu residue in the C-terminal region, which increases the activity of PP2A (Favre *et al.*, 1994; Turowski *et al.*, 1995; Lee *et al.*, 1996). Several PTPases have been shown to interact with tyrosine kinases and are phosphorylated by these kinases, although the functional significance of these phosphorylation events is not understood (Lorenz *et al.*, 1994; Shifrin *et al.*, 1997; Van Vactor *et al.*, 1998).

The noncovalent form of modification is exemplified by redox control of phosphatase activity. The central metal cluster in the PPP family can be oxidized leading to inactivation of the catalytic core. This has been demonstrated for the PP2B catalytic subunit (Wang *et al.*, 1996). In addition, all PTPases contain a catalytic cysteine that must be in a reduced form for phosphatase activity. Oxidation of cysteine under oxidative stress reversibly inactivates the PTPase (Denu and Tanner, 1998). To date, little is known about the regulation of plant protein phosphatases by covalent or noncovalent modification of the protein.

C. TARGETING DOMAINS

Although the catalytic region of the same family of protein phosphatases is highly homologous, these members always contain distinct noncatalytic regions that represent functional domains involved in the regulation of various aspects of a specific phosphatase. Perhaps the most common function is to direct the localization of the catalytic activity in a cell. The receptor-like PTPases all contain a transmembrane domain for plasma membrane targeting (Stone and Dixon, 1994; Neel and Tonks, 1997). Some dual-specificity PTPases such as MKP1 and MKP2 are strictly localized to the nucleus (Rohan *et al.*, 1993; Guan and Butch, 1995). Some cytoplasmic PTPases contain SH2 domain for interacting with receptor tyrosine kinases in the plasma membrane (Feng *et al.*, 1993). A novel PPP family enzyme, PP5, contains TPR domain for interaction with cytoplasmic anchor chaperone HSP90 (Chen *et al.*, 1996; Das *et al.*, 1998). PP2B catalytic subunit contains a region for interacting with the regulatory subunit and calmodulin, respectively, which in turn can direct the enzyme to the plasma membrane through myristoylation or other mechanisms.

The noncatalytic regions also have other functions including the regulation of the catalytic activity. For example, the RgdC and human PP7 contain multiple Ca^{2+}-binding motifs for activation by high levels of Ca^{2+} in the cell (Huang and Honkanen, 1998). PP2B has a calmodulin-binding domain for activation by interaction with calmodulin (Klee *et al.*, 1998). The C-terminal region of PP2B is an autoinhibitory domain that inhibits the phosphatase activity in the absence of calmodulin binding (Klee *et al.*, 1998). All receptor-like PTPases are regulated by ligand binding to the extracellular domains. A recent hypothesis suggests that ligand binding to the receptor-like PTPases may inactivate the PTPase activity (Neel and Tonks, 1997). In the plant enzymes, such a regulation mechanism also exists. KAPP, a PP2C homologue, contains a membrane anchor and a domain for interacting with receptor-like kinases. Both these domains would facilitate KAPP localization to the plasma membrane (Stone *et al.*, 1994). The N-terminal extension of MP2C phosphatase has been shown to be essential for substrate specificity (H. Hirt, personal communications).

D. REGULATORY SUBUNITS

Interaction with regulatory proteins may be the most common mechanism for regulation of protein phosphatases (Fig. 2). In mammalian systems, the only catalytic subunit of PP1 has been shown to dephosphorylate a number of physiological substrates. This seemingly paradoxical situation was resolved by the discovery of many regulatory subunits that target the same catalytic subunit to various subcellular locations and therefore to distinct substrates. In addition, these regulatory subunits often change the specific activity of a protein phosphatase towards different substrates. For example, the M110 subunit, responsible for the association of PP1C with the myofibrils of the muscle cells, enhances the activity of PP1C against the myosin light chain and suppresses activity towards glycogen phosphorylase (Johnson *et al.*, 1996b, 1997). Besides defining substrate specificity, some regulatory subunits allow the catalytic subunit to be modified by specific events. For instance, the GM regulatory subunit of PP1 is phosphorylated by PKA, resulting in dissociation of the PP1C and G_M and hence the substrate glycogen (Dent *et al.*, 1992). Because the binding of distinct regulatory subunits to PP1C is mutually exclusive, the binding sites must be the same or overlapping the region of the PP1C protein. Indeed, structural analysis has defined a region of PP1C and similar motifs in different regulatory subunits to be important for interaction (Egloff *et al.*, 1997). To date, more than 15 regulatory subunits of PP1 have been characterized. Approximately 100 more novel PP1C-binding proteins have recently been reported (Damer *et al.*, 1998). In contrast to this vigorous research effort, study on PP1 regulatory subunits in higher plants has yet to be initiated.

Much like PP1, PP2A is also highly regulated by its regulatory subunits. Although PP2Ac is anchored by a 65-kDa A subunit, PP2Ac activity and substrate specificity is largely controlled by a number of regulatory B subunits. At least 20 B regulatory subunits (54–130 kDa in molecular weight) have been identified from mammalian systems. These B subunits are further divided into three groups, the 55-kDa B, 54–74-kDa B′, and 72–130-kDa B′ subunits (McCright and Virshup, 1995; Wera and Hemmings, 1995). The best known function of these subunits is to target the PP2Ac to distinct subcellular locations. For example, the 55-kDa B subunit targets PP2Ac to the microtubule (Sontag *et al.*, 1995), and B′ subunits facilitate nuclear location of PP2Ac (McCright *et al.*, 1996). As discussed earlier, several studies showed the presence of PP2A regulatory subunits in higher plants (Rundle *et al.*, 1995; Latorre *et al.*, 1997). However, the specific function of these subunits remains unknown.

PP2B (or calcineurin) is normally found in the cell as a heterodimer of a 58–64-kDa calmodulin-binding catalytic A and a Ca^{2+}-binding regulatory B subunit, calcineurin B. This dimeric structure, unique among the protein phosphatases, is highly conserved from yeast to human. *In vitro* studies have shown that the B subunit interacts with the A subunit and increases the stability and catalytic activity of the catalytic subunit (reviewed in Klee *et al.*, 1998). The other regulatory protein for PP2B is calmodulin, another Ca^{2+}-binding protein. As the cellular Ca^{2+} level is elevated by an extracellular signal, calmodulin interacts with the A/B heterodimer and stimulates the PP2B activity many fold (Fig. 2). Further analysis indicates that only one of the four EF-hand motifs in the calcineurin B subunit displays a high affinity to Ca^{2+} ($K_d < 0.1\ \mu M$) and the other three EF-hands show low affinity (K_d around 0.5–1 μM). Ca^{2+} binding to the high affinity site is required and is sufficient for the interaction with the A subunit, but the enzyme is inactive, consistent with the fact that A/B are tightly bound together even in 'resting' cells. The occupancy of the three low-affinity sites on calcineurin B only slightly increases the phosphatase activity. Only when calmodulin is recruited into the complex under high Ca^{2+} levels is the PP2B activity activated to more than 20-fold of the basal levels.

Despite the accumulating evidence of calcineurin function in higher plants (Luan *et al.*, 1993; Allen and Sanders, 1995; Bethke and Jones, 1997; Pardo *et al.*, 1998), the biochemical nature of plant calcineurin remains elusive. Two recent studies report the identification of a calcineurin B-like protein from *Arabidopsis* (Liu and Zhu, 1998; Kudla *et al.*, 1999). Although the report by Kudla *et al.* (1999) confirmed that these plant proteins are indeed capable of binding Ca^{2+} and interacting with the animal catalytic subunit, their endogenous protein partner has yet to be identified. The regulatory function of these calcineurin B-like proteins can only be addressed after their interacting subunit(s) are found.

E. ENDOGENOUS AND EXOGENOUS INHIBITORS AND ACTIVATORS

Protein phosphatases, like many other signaling molecules, can be inhibited or activated by small molecules naturally occurring in the cells. The discovery of endogenous inhibitors for PP1 established this paradigm (Cohen, 1989). Two heat-stable proteins, inhibitor 1 and 2 (I-1 and I-2), were first purified from liver and muscle extract for their ability to inhibit PP1 activity. It is noteworthy that the function of these peptide inhibitors is highly regulated by phosphorylation. The I-1 (18 kDa) is only effective after it is phosphorylated at a specific threonine by PKA (Endo *et al.*, 1996). While I-2 (22 kDa) does not require phosphorylation to be an active inhibitor, the potency of inhibition increases 10^6-fold after phosphorylation (Desdouits *et al.*, 1995). Secondary structure prediction provides a model for the mechanism of PP1C inhibition by I-2. The N-terminal region of I-1 forms a β strand and an α helix, respectively, positioning the pThr-35 at the catalytic site. A similar secondary structural element is found in I-2 (Goldberg *et al.*, 1995), although the two peptide inhibitors share little homology in the primary structure. PP1 is also potently inhibited by a natural toxin, microcystin LR, a complex cyclic heptapeptide. The crystal structure of PP1C complexed with microcystin LR reveals several contacting sites on the PP1C surface (Goldberg *et al.*, 1995). One involves a carboxylate and carbonyl group of the toxin that interacts with two of the metal-bound water molecules, thus directly blocking substrate binding. Other sites involve the formation of covalent bonds between PP1C and a toxin side chain.

PP2A holoenzymes have been shown to be activated by ceramide, a lipid second messenger in mammalian systems (Dobrowsky *et al.*, 1993; Galadari *et al.*, 1998). Ceramide is produced from the sphingomycin cycle and specifically activates two of the regulatory subunit-containing holoenzymes of PP2A. This indicates that these PP2A holoenzymes may be modulated by the extracellular signals that trigger sphingosine hydrolysis.

An important breakthrough in the study of PP2B (calcineurin) regulation and function is the finding that calcineurin serves as a functional target for the immunosuppressants cyclosporin A and FK506. Cyclosporin A and FK506 bind to their intracellular receptors, cyclophilin and FKBP, respectively, forming ligand–receptor complexes that interact with the heterodimer or heterotrimer formed by the calcineurin A/B subunits or calcineurin A/B/calmodulin. The formation of these supermolecular complexes renders calcineurin inactive (Liu *et al.*, 1991, 1992). The mechanism of inhibition by the FKBP12/FK506 complex has been solved by structural analyses of the calcineurin–FKBP12–FK506 complex (Griffith *et al.*, 1995; Kissinger *et al.*,

1995). Neither FK506 nor its receptor FKBP12 individually is capable of interacting with calcineurin; the FK506–FKBP12 composite surface is required for interaction with calcineurin. On the other side, both A and B subunits of calcineurin are required for the interaction with FK506–FKBP12 complex (Liu *et al.*, 1991, 1992). The major contact site on calcineurin is the base of the B subunit-binding helix of the A subunit together with the B subunit, and a minor contact of the catalytic domain (Griffith *et al.*, 1995; Kissinger *et al.*, 1995). FKBP12 sterically blocks access of large molecules to the catalytic site, whereas FK506 does not participate directly in phosphatase inhibition. The tertiary structure of the FKBP12–FK506 complex does not change significantly, yet FK506 binding is required for the FKBP12 to take a proper conformation for the interaction with calcineurin. Based on these studies, FK506 and cyclosporin A have been widely used as specific inhibitors of calcineurin (PP2B) in systems such as mammalian, yeast, and plants (Heitman *et al.*, 1993; Luan *et al.*, 1993; Rao, 1997).

A ubiquitous protein factor, 14-3-3 protein, has been recently identified for its ability to regulate a number of signaling molecules including the cell cycle regulator phosphatase, CDC25. When phosphorylated by a protein kinase Chk1, a DNA replication checkpoint kinase (Furnari *et al.*, 1997; Sanchez *et al.*, 1997), the CDC25 protein interacts with 14-3-3 and cannot dephosphorylate CDC2, thereby arresting the cell cycle at the G2 phase.

Small molecule pharmacological agents have been crucial in the study of cellular function of specific phosphatases. Okadaic acid, calyculin A, and microcystin LR are potent inhibitors of both PP1 and PP2A. Although *in vitro* studies indicate that okadaic acid inhibits PP2A with higher potency as compared to its potency to PP1, *in vivo* application of okadaic acid to distinguish PP1 and PP2A has proved to be difficult (Shenolikar, 1994). More specific inhibitors include peptide inhibitors I-1 and I-2 for PP1 and cantharidin for PP2A (Li and Casida, 1992). Because PP2C and PP2B require divalent cations Mg^{2+} and Ca^{2+}, respectively, it has been successful to use ethylenediaminetetraacetic acid (EDTA) and EGTA to inhibit these phosphatase by chelating the divalent cations. The most common inhibitor for PTPases are phosphate analogues such as vanadate and tungstate that specifically inhibit all PTPases but not Ser/Thr phosphatases at high micromolar concentrations. In order to distinguish effectively one phosphatase from the others, it is often necessary to use a combination of these inhibitors.

IV. FUNCTION OF PROTEIN PHOSPHATASES

As reversible phosphorylation plays a role in almost every aspect of cellular life, functions of protein phosphatases encompass the whole spectrum of cell and developmental biology. The well-studied functions of protein phosphatases in other systems such as animal and yeast will now be introduced, and then findings on the function of plant enzymes will be collated. As will be shown, the study of the plant phosphatases has lagged far behind in most of the cases discussed below.

A. Ser/Thr PROTEIN PHOSPHATASES

1. PP1/2A

In animals, Ser/Thr phosphorylation and tyrosine phosphorylation play an equally important part in regulation of cellular activities. It is generally true that cellular localization of catalytic activity, regulated by the targeting domains or regulatory subunit, controls the function of a specific phosphatase. This is best illustrated by the function of PP1 in animal cells. Glycogen metabolism may be the first pathway in which a regulatory role of a protein phosphatase was recognized (reviewed in Cohen, 1989). The neuronal and hormonal control of glycogen metabolism is mediated by changes in the phosphorylation state of glycogen phosphorylase, phosphorylase kinase, and glycogen synthase. The glycogen synthase is activated, and enzymes for glycogen hydrolysis are inhibited, by phosphatases. PP1 is the major phosphatase in the regulation of glycogen metabolic enzymes. For this purpose, a specific regulatory subunit, RG1, targets the catalytic subunit PP1C to the glycogen particles and stimulates the activity of PP1C towards glycogen-bound enzymes such as glycogen synthase and glycogen phosphorylase, increasing glycogen accumulation. The M110 subunit recruits PP1C to the myofibrils and increases the activity of PP1 to dephosphorylate myosin light chain, inducing muscle relaxation (Johnson *et al.*, 1996a).

Another putative regulatory subunit, PNUTS (phosphatase 1 nuclear targeting subunit), is implicated in targeting PP1C into the nucleus where PP1 dephosphorylates a number of substrates including histone H1 (Allen *et al.*, 1998), retinoblastoma protein (Nelson and Ludlow, 1997; Nelson *et al.*, 1997), and nuclear lamin (Thompson *et al.*, 1997), which all affect cell cycle progression. Nuclear localization of PP1 may also regulate RNA splicing because PP1 regulates the distribution of pre-messenger RNA (mRNA) splicing factors (Misteli and Spector, 1996). In addition, PP1 plays a critical role in neuronal activity by regulating ion-channel activities (Yan and Surmeier, 1997).

Like PP1, PP2A catalytic subunits are also located to various compartments in a cell and play diverse functions ranging from metabolism and cell cycle control, to telomerase activity. A specific isoform of PP2Ac has been shown to be essential for early embryonic development (Goetz *et al.*, 1998). A mutation in the 65-kDa A subunit of PP2A is found in numerous cancers, implicating PP2A as a tumor suppressor (Wang *et al.*, 1998). In fission yeast, two catalytic subunit genes encode PP2A activity. Deletion of both genes is lethal and one of the genes is responsible for negative regulation of mitosis (Kinoshita *et al.*,

1993). Further evidence for this notion came from studies in *Xenopus* oocytes where PP2A inhibits histone H1 kinase and hence cell cycle progression (Lee *et al.*, 1994). There is at least one example that shows inhibition of telomerase activity by PP2A. Incubation of cell nuclear telomerase extracts with protein phosphatase 2A (PP2A) abolished the telomerase activity. In contrast, cytoplasmic telomerase activity was unaffected, and protein phosphatases 1 and 2B were ineffective (Li *et al.*, 1997a). Several studies have shown that PP2A regulates a variety of ion channels in animal cells (Groschner *et al.*, 1996; Zhou *et al.*, 1996; Sansom *et al.*, 1997).

Our knowledge about the function of PP1 and PP2A in higher plants comes mostly from the studies using pharmacological approaches. Application of specific inhibitors such as okadaic acid and others in a specific cellular or developmental process provides much of the information regarding the importance of a subclass of phosphatases in plants. Because it is difficult to distinguish PP1 from PP2A by using okadaic acid, most studies defined the okadaic acid-sensitive process as involving PP1/PP2A. For example, PP1/2A is required for activation of metabolic enzymes, including nitrate reductase (NR) and sucrose phosphate synthase (SPS), that play a key role in carbon and nitrogen metabolism. The current view on the mechanism and significance of reversible phosphorylation in regulation of these enzymes is described by Toroser and Huber in this volume (Toroser and Huber, 2000). Inhibitor analysis also indicates that PP1/2A is involved in ion-channel regulation. For instance, okadaic acid inhibits inward K^+ channel activity in the guard cells but activates the outward K^+ channels in mesophyll cells of *Vicia faba*, suggesting that PP1/2A plays a distinct role in the regulation of different K^+ channels in various cell types (Li *et al.*, 1994). Okadaic acid also activates anion channels in guard cells of *Vicia faba* and the *Arabidopsis* plant (Schmidt *et al.*, 1995; Pei *et al.*, 1997), implicating PP1/2A in the regulation of anion channels. In addition, signal transduction pathways triggered by light, hormones, and pathogens all involve action of PP1/2A. Using okadaic acid and calyculin A, a study showed that PP1/2A is required for the light-inducible gene expression in maize mesophyll cells (Sheen, 1993).

In contrast, ethylene-induced gene expression can be mimicked by application of okadaic acid, implicating PP1/2A as a negative regulator of ethylene signaling processes (Raz and Fluhr, 1993). Similarly, PP1/2A also play a negative role in pathogen response as indicated by the observation that okadaic acid can mimic the action of pathogen elicitors in inducing gene expression and oxygen burst (Mackintosh *et al.*, 1994; Chandra and Low, 1995). If PP1/2A serves as a negative regulator of pathogen response, protein kinases may positively regulate the same pathway. Indeed, MAPK isoforms have been shown to be activated by the pathogen elicitors, and okadaic acid blocks the inactivation of MAPK, suggesting that PP1/2A may be involved in inactivation of MAPK (Suzuki and Shinshi, 1995). An example of PP1/2A involvement in plant development is the finding of okadaic acid inhibition of root cell growth in *Arabidopsis* (Smith *et al.*, 1994; Baskin and Wilson, 1997). A more detailed description of such studies appears in previous reviews (Smith and Walker, 1996; Luan, 1998).

A role of the PP1/2A function in root growth is further indicated by a genetic analysis of *Arabidopsis rcn1* mutant. *RCN1* gene encodes a 65-kDa A subunit of PP2A (Garbers *et al.*, 1996). Mutation in the *RCN1* gene causes a defect in the *Arabidopsis* root curling and hypocotyl hook formation. In addition, auxin transport in the *rcn1* mutant exhibits more sensitivity to the polar transport inhibitor naphthylphthalamic acid. Biochemical analysis has linked the mutation in the *RCN1* gene with a reduction in PP2A activity in the protein extract of the *rcn1* mutant (J. Deruere and D. Söll, personal communications).

Another approach to understanding the function of plant PP1 is yeast complementation. For example, *Arabidopsis* TOPP2 but not TOPP1 can rescue the lethality caused by the interruption of PP1, *glc7*, in budding yeast (Smith *et al.*, 1995). However, TOPP2 cannot complement the glycogen deficiency phenotype of the *glc7* mutant. This study suggests that *Arabidopsis* PP1 may serve a functional role in mitosis but not in glycogen metabolism in yeast, consistent with the fact that mitosis, but not glycogen metabolism, is highly conserved between plant and fungal systems. A cold-sensitive phenotype caused by a *PP1* mutation in fission yeast is suppressed by *Arabidopsis* TOPP1 but not by TOPP2, again indicating a distinct function of the two PP1 catalytic subunits (Nitschke *et al.*, 1992). The function of TOPP1/2 in *Arabidopsis* has not been characterized.

2. *PP2B*

PP2B is also referred to as calcineurin because it was first identified as a calmodulin-binding protein from brain tissues. Although calcineurin constitutes almost 1% of the total protein from the brain, its function in the brain is not yet clear. Studies on calcineurin function have been revolutionized by the discovery of immunosuppressants cyclosporin A and FK506 as specific inhibitors. Studies using these chemical probes have shown that PP2B plays a crucial role in a number of signal transduction pathways. In particular, calcineurin is involved in signaling pathways leading to T-cell activation (Rao, 1997). In brief, a critical transcriptional factor in T cells, NFAT, requires dephosphorylation by calcineurin in order to be translocated into the nucleus and to activate expression of a number of genes required for T-cell proliferation. Calcineurin action in hippocampal neurons induces desensitization of the postsynaptic *N*-methyl-D-aspartate (NMDA) receptor-coupled Ca^{2+} channels (Lieberman and Mody, 1994). In addition, calcineurin has been shown to regulate the Ca^{2+}-releasing activity of ryanodine receptors and IP3 receptors in mammalian cells (Snyder *et al.*, 1998).

In yeast, like in mammalian cells, calcineurin plays an important role in activation of gene expression in response to calcium signals (Heitman *et al.*, 1993). Two major pathways for calcineurin action include pheromone response

(Cyert and Thorner, 1992) and adaptation to salt stress (Nakamura *et al.*, 1993). Other processes in fungal systems such as mating, sporalation, cytokinesis, and Ca^{2+} sequestration are also regulated by calcineurin activity (Means, 1994; Rasmussen *et al.*, 1994; Yoshida *et al.*, 1994).

Perhaps the first evidence for the presence of a PP2B-like activity came from a study using immunosuppressants as molecular probes (Luan *et al.*, 1993). In this study, a Ca^{2+}-dependent pathway for K^+ channel inhibition in guard cells was used as a model system. Application of cyclosporin A and FK506–FKBP12 complex specifically blocked Ca^{2+}-dependent K^+ channel regulation, implicating PP2B in Ca^{2+} signaling in guard cells. This notion is further supported by the observation that a constitutively active form of PP2B from an animal source mimicked the function of high levels of Ca^{2+} in the inactivation of the K^+ channel in the guard cells (Luan *et al.*, 1993). Several other studies also suggest that PP2B may regulate other ion-channel activities in plant cells (Allen and Sanders, 1995; Bethke and Jones, 1997). In addition, overexpression of yeast PP2B in transgenic tobacco increased the salt tolerance of the plants (Pardo *et al.*, 1998), consistent with a role of PP2B in salt adaptation in yeast. Recent studies on a group of calcineurin B-like (CBL) proteins may provide important information regarding the function of a PP2B-like protein. One of the CBL proteins, SOS3, is required for salt tolerance in *Arabidopsis* (Liu and Zhu, 1998), consistent with the fact that PP2B plays a role in salt adaptation in yeast. Regulation of AtCBL1 expression by stress signals including cold, drought, and wounding also implicates AtCBL1 in stress signal transduction (Kudla *et al.*, 1999). Further identification of protein partners for these calcineurin B-like proteins will be crucial for defining the functional identity of these proteins.

3. *PP2C*

The function of PP2C is the least well understood among all Ser/Thr phosphatases. Several studies in yeast and mammalian systems suggest that PP2C is involved in the stress signal transduction but the mechanisms are different. In mammalian cells, PP2C regulates stress signaling by modulating p38 and the JNK MAPK pathway (Hanada *et al.*, 1998; Takekawa *et al.*, 1998), whereas in fission yeast, PP2C regulates the stress response independently of the stress-activated kinase cascade (Shiozaki *et al.*, 1994; Shiozaki and Russell, 1994, 1995b; Gaits *et al.*, 1997). PP2C has also been shown to play a role in the regulation of the cystic fibrosis transmembrane conductance regulator, a chloride channel (Luo *et al.*, 1998). A novel form of PP2C has been recently isolated from *Dictyostelium* and is essential for cell-type differentiation (Aubry and Firtel, 1998). One PP2C isoform from *C. elegans* has been shown to have a unique N-terminal extension required for a sex determination process (Hansen and Pilgrim, 1998).

Plants contain a more diverse family of PP2C than do animal and yeast systems. One PP2C subfamily, ABI1 and ABI2, was identified as a component in the ABA signal transduction pathway (Leung *et al.*, 1994, 1997; Meyer *et al.*, 1994). The identified mutations in *ABI* genes result in dramatically decreased sensitivity of the mutant plants to ABA in several aspects, including stomatal regulation. The *abi* mutants display enlarged stomatal apertures in the presence of ABA that normally causes stomatal closure. Because stomatal movement is controlled by ion channels in the plasma membrane and tonoplast of guard cells (Assmann 1993; Ward *et al.*, 1995; Blatt and Grabov, 1997), studies were directed to ion-channel regulation by ABA in both wild-type and *abi* mutants. An important channel activity for initiating stomatal closure is a guard cell anion current responsible for anion efflux (Ward *et al.*, 1995). In the wild-type *Arabidopsis* plants this anion channel is activated by ABAs whereas in mutant plants this anion channel becomes insensitive to ABA, thereby blocking ABA-induced stomatal closure (Pei *et al.*, 1997). In the tobacco plant, however, expression of the same *abi* mutant protein has a different effect on ion channels in guard cells: the anion channel is not affected by the *abi* mutant protein but both outward and inward K^+ channels become less sensitive to ABA application (Armstrong *et al.*, 1995). These studies suggest that different plants may utilize distinct molecular components to respond to ABA. Although *ABI1* and *ABI2* both encode a PP2C, they appear to function differently in guard cells (Pei *et al.*, 1997). It is worth noting that both *abi* mutations are dominant, suggesting that they may be gain-of-function mutations. As a result, it has been difficult to define the mechanism of ABI action in ABA signaling. In a transient system, ABI appears to serve as a negative regulator of ABA-induced gene expression in maize mesophyll cells (Sheen, 1998). It is not known whether mesophyll cells and guard cells respond to ABA by the same or distinct mechanisms. Loss of function mutants may facilitate further understanding of ABI action in plants.

Another PP2C member, KAPP, was identified as a partner protein for a receptor-like protein kinase (Stone *et al.*, 1994). Recent studies suggest that KAPP regulates the function of some RLKs. For example, KAPP interacts with CLV1, a RLK required for normal development of shoot meristem in *Arabidopsis* (Clark *et al.*, 1997). Biochemical studies show that KAPP can dephosphorylate autophosphorylated CLV1 (Williams *et al.*, 1997; Stone *et al.*, 1998). In addition, overexpression of KAPP in transgenic *Arabidopsis* mimics the phenotype of the *clv1* mutant (Williams *et al.*, 1997). Reduction of KAPP levels suppressed the *clv1* phenotype (Stone *et al.*, 1998). Both studies suggest that KAPP serves as a negative regulator of the CLV1 signaling pathway, possibly by acting directly on the receptor-like kinase. To date, several dozens of receptor kinases have been identified from higher plants. KAPP and KAPP-like protein phosphatases may be an important component of receptor kinase-mediated signaling pathways.

As described earlier, PP2C has been shown to play a role in MAPK regulation in animals. Using the yeast strain with defects in the MAPK pathways for pheromone response, an alfalfa complementary DNA (cDNA) clone was identified by its ability to suppress the MAPK pathway (Meskiene *et al.*, 1998). The gene encodes a PP2C member that acts at the level of MAPK kinase kinase (Ste11) in yeast. This PP2C is able to inactivate the stress-activated MAPK pathway (SAMK) in plants. The PP2C mRNA levels are induced by wounding, which also activates the SAMK pathway (Meskiene *et al.*, 1998). While MAPKs are mostly activated at the post-translational level by their kinases, negative regulators of MAPK (including tyrosine-specific PTPase, dual-specificity PTPases, and PP2C) are often regulated at the level of mRNA accumulation (Keyse, 1995).

B. PROTEIN–TYROSINE PHOSPHATASES

A diverse array of functions has been identified with PTPases in animal cells (Stone and Dixon, 1994; Neel and Tonks, 1997). The receptor-like PTPases play crucial roles in cell adhesion, consistent with the fact that their extracellular domains all contain regions with high similarity to cell adhesion molecules. Several studies using biochemical and genetic analyses all support this notion. For example, expression of receptor-like PTPases (RPTP) in insect cells makes them aggregate. Deletion analysis indicates that the extracellular domains mediate the cell–cell interaction (Gebbink *et al.*, 1993b; Sap *et al.*, 1994). The extracellular domain alone of PTPμ was also capable of self-association (homophilic interaction). Cells expressing different types of RPTP were mixed and found to have homophilic interactions between the same types of cells. Four out of five genes encoding receptor-like PTPases with cell adhesion molecule homology are expressed in the nerve system. Mutant embryos lacking expression of some of these PTPases were generated (Desai *et al.*, 1996; Krueger *et al.*, 1996). Some mutants show defects in their growth cones that fail to recognize muscle targets, or follow pathways that bypass these targets altogether. Manipulation of some adhesion molecules had a similar phenotype, suggesting the overlapping function of receptor-like PTPases and cell adhesion molecules (Lin and Goodman, 1994).

The SH2-containing PTPases (SHPs) have been identified from all animal systems studied including mammals, *Xenopus*, and *Drosophila*. Two SHPs are present in vertebrates: SHP-1 is expressed in the highest levels in hematopoietic cells and SHP-2 is ubiquitously expressed. The presence of the SH2 domain clearly indicates the possibility for these molecules to interact with receptor-like tyrosine kinases. Both biochemical and genetic approaches have validated this hypothesis. SHP1 has been shown to regulate negatively multiple hematopoietic signaling pathways, including those that are downstream of cytokine and immune-recognition receptors (reviewed in Neel and Tonks, 1997). For example, activation of signaling through the B-cell antigen receptor is abrogated upon cross-linking of the inhibitory Fc receptor that interacts with SHP-1. Interaction analysis demonstrated that SHP-1 also interacts with and downregulates the receptor tyrosine kinase c-kit *in vivo*. A SHP-like PTPase from *Drosophilia*, named Corkscrew, is required for proper embryonic head–tail development, which is regulated by a receptor tyrosine kinase (Perkins *et al.*, 1992, 1996). Corkscrew may also regulate the pathway mediated by *daughter of sevenless* gene product, another receptor tyrosine kinase (Allard *et al.*, 1996).

A family of dual-specificity PTPases has been shown to play a critical role in the regulation of MAPK pathways. MAPK family members are essential components in many signaling pathways including the growth factor receptor signaling pathways and the stress response pathways. For activation, MAPK requires phosphorylation, by a dual-specificity kinase, of both a threonine and a tyrosine residue in the kinase activation domains (Ahn *et al.*, 1992; Guan, 1994). A significant body of literature implicates dual-specificity PTPases in the dephosphorylation of MAPKs (Guan, 1994; Sun and Tonks, 1994; Keyse, 1995). Studies in yeast indicate that both tyrosine-specific and dual-specificity PTPases play a role in regulating MAPKs (Wurgler-Murphy and Saito, 1997; Wurgler-Murphy *et al.*, 1997; Zhan *et al.*, 1997). In mammalian cells, a variety of dual-specificity PTPases exhibits activity towards activated forms of MAPK both *in vivo* and *in vitro*. In each case, dephosphorylation of a MAPK isoform by a dual-specificity PTPase results in a loss of MAPK activity. It was initially puzzling that multiple isoforms of dual-specificity PTPases could be found in a single cell type, as can multiple forms of MAPKs. Studies show that dual-specificity PTPases have rather strict substrate specificity towards different isoforms of MAPKs (Keyse, 1995). Therefore various combinations of MAPK and dual-specificity PTPase isoforms may constitute modules in distinct signaling pathways in a single cell type (Keyse, 1995). Studies in both budding and fission yeast demonstrate that tyrosine-specific PTPases are also important for MAPK regulation. For example, the osmosensing pathway in budding yeast is initiated by a two-component histidine kinase Sln1 that activates a kinase cascade involving a MAPK member, Hog1. Both biochemical and genetic analyses indicate that Hog1 is inactivated by PTP2 (Wurgler-Murphy and Saito, 1997; Wurgler-Murphy *et al.*, 1997). A similar MAPK-mediated pathway has been shown to link the environmental changes to the onset of mitosis in fission yeast. In this pathway, Spc/Sty1 MAPK is inactivated by typical PTPases Pyp1 and Pyp2 (Shiozaki and Russell, 1995a).

An important subclass of dual-specificity PTPase is CDC25 protein phosphatase that is required in cell cycle progression. As its name indicates, CDC25 was initially identified as a *cdc* gene in which a loss of function mutation halts the cell cycle in yeast. Homologues of CDC25 have since been found in all animal systems examined. Biochemical and genetic analyses have defined a role for CDC25 in activation of CDC2 protein kinase (reviewed by Dunphy, 1994). Inactivation of CDC2 involves phosphorylation of a specific tyrosine residue in the protein. The phosphatase activity of CDC25 is responsible for dephosphorylating the inhibitory tyrosine residue and activating CDC2 kinase.

Most recently, a subgroup of dual-specificity PTPases has been identified for its function as a tumor suppressor. This subclass, represented by PTEN/MMC1, contains the catalytic core of PTPases and has a structural domain with high similarity to a cytoskeleton protein, tensin (Li and Sun, 1997; Li *et al.*, 1997b; Steck *et al.*, 1997). Biochemical studies show that PTEN has a low dual-specificity phosphatase activity towards protein substrates but has a high phosphatase activity against phosphotidylinositol 3,4,5-triphosphate (PIP3) (Myers *et al.*, 1997, 1998; Maehama and Dixon, 1998). The phosphatase activity specifically dephosphorylates the 3′ phosphate group in PIP3 (Maehama and Dixon, 1998). Therefore, the finding of PTEN as a tumor suppressor is consistent with the fact that production of PIP3 by PI3 kinase (PI3K) is a key step for cell transformation into tumor (Fruman *et al.*, 1998). Further studies have confirmed that PTEN functions in the PI3K pathway for balancing the activity of PI3K (Li and Sun, 1998; Stambolic *et al.*, 1998; Wu *et al.*, 1998).

In higher plants, protein tyrosine kinase has not been identified. The importance of tyrosine phosphorylation in plant cells has been controversial until the finding of MAPK pathways in numerous signal transduction pathways in higher plants (Mizoguchi *et al.*, 1997). For example, a specific MAPK activity is rapidly activated by cold, drought, mechanical stimuli, and wounding (Seo *et al.*, 1995; Usami *et al.*, 1995; Jonak *et al.*, 1996). MAPK activation is an early step in the wounding response because systemin, but not jasmonic acid, can mimic the effect of wounding by inducing MAPK activity (Stratmann and Ryan, 1997). In both tobacco and parsley cells, a MAPK is rapidly activated in response to pathogen elicitors (Zhang and Klessig, 1997). Salicylic acid, a signaling molecule in the pathogen response, also transiently activates a MAPK in tobacco cells (Wilson *et al.*, 1997). Besides stress and pathogen responses, MAPKs also participate in developmental processes such as pollen development (Wilson *et al.*, 1997). In the ethylene signaling pathway, both histidine kinase and Raf-like serine/threonine protein kinase have been shown to play a role, although the MAPK component of the pathway has yet to be identified (Chang and Meyerowitz, 1995). Taken together, these findings demonstrate that MAPK activation is a common, and an early step, in the response to stress, hormonal, and developmental signals.

A number of MAPK isoforms may be involved in pathways initiated by different signals (Hare *et al.*, 1997; Seo *et al.*, 1997; Meskiene *et al.*, 1998). Because tyrosine phosphorylation of MAPK protein is required for activation of the kinase, dephosphorylation of the tyrosine residue must occur to reset the enzyme to the inactive form. Therefore either a tyrosine-specific or a dual-specificity PTPase must exist in higher plants. To date, one tyrosine-specific PTPase (AtPTP1) has been identified from *Arabidopsis* and its expression seems to be regulated by stress factors including high salt and cold stress (Xu *et al.*, 1998). Further biochemical studies demonstrated that AtPTP1 dephosphorylates a MAPK from *Arabidopsis in vitro* and inactivates the kinase (Huang, Y., Gupta, R., Luan, S. and Kieber, J.J., unpublished results). A dual-specificity PTPase (AtDsPTP1) has also been identified from *Arabidopsis* and shown to dephosphorylate and inactivate a MAPK from the same plant (Gupta *et al.*, 1998). These studies suggest that plant PTPases may function to regulate MAPK pathways in response to a number of extracellular signals. Further studies are being conducted to identify the *in vivo* targets of these phosphatases and their function in plant cellular and developmental processes (S. Luan, unpublished results).

In organisms ranging from yeast to mammals, regulation of CDC2 by CDC25 phosphatase is highly conserved (Dunphy, 1994). At least one study (Zhang *et al.*, 1996) shows that cell cycle progression in tobacco suspension culture is regulated by PTPases. In addition, this regulation by PTPase may be a component of the cytokinin signaling pathway responsible for G2-M transition in the cell cycle. It is reasonable to suggest that a plant CDC25-like protein phosphatase may exist. However, such a CDC25 homologue has yet to be identified from a higher plant.

V. FUTURE PERSPECTIVES

The biochemical properties of protein phosphatases have been highly conserved among eukaryotic organisms. Future effort will certainly be directed to the functional analysis of these enzymes in a context of plant cell and developmental biology. As genome sequencing projects move along, all of the genes for protein phosphatases in model plants (e.g. *Arabidopsis*) will be identified. The question is how to make the connection from the sequence information to the functional significance.

Several approaches have already been taken in the study of plant protein phosphatases and will continue to shine light on the subject. For example, a genetic screening identified ABI phosphatases in a pathway related to ABA signaling. Because of a gain-of-function mutation, it has been difficult from the initial genetic analysis to understand the role of ABI in ABA signal transduction. A transient assay procedure (Sheen, 1998) complements the genetics approach and shows a possible role for ABI as a negative regulator of the Ca^{2+}-dependent pathway in ABA signaling. However, it is still unresolved as to whether ABI functions in a similar or distinct manner in different cell types and in different plant species (Armstrong *et al.*, 1995; Pei *et al.*, 1997). A biochemical approach is now under way to isolate protein partners for ABI with the hope of identifying the physiological substrate of ABI phosphatases (J. Sheen, personal communication). Meanwhile, more genetic analyses are being conducted to identify suppressors of *abi* mutants (J. Giraudat, personal communications).

Another interesting phosphatase, KAPP, has been identified by a biochemical approach (Stone *et al.*, 1994). Its function has recently been approached by molecular genetic studies using genetic mutants and transgenic plants (Williams *et al.*, 1997; Stone *et al.*, 1998). The first tyrosine-specific PTPase and dual-specificity PTPase have been identified by a molecular and genomic approach, respectively (Gupta *et al.*, 1998; Xu *et al.*, 1998). The essential biochemical characterization of the enzymes has been completed and functional analysis is in progress. A critical approach has been 'reverse genetics' using transgenic plants and insertional mutagenesis, which is successful for the *AtPTP1* gene. A phenotypic analysis suggests that this tyrosine-specific phosphatase plays a role in normal plant development (S. Luan, unpublished results). As more genes for protein phosphatases are identified by genomic sequencing, the reverse genetics and functional genomics approaches will be increasingly important in dissecting the role of these signaling enzymes. Despite the structural similarities in all eukaryotes, protein phosphatases certainly play distinct functions in various organisms with different developmental programs. It is these unique functions in plant growth and development that make plant biologists interested in protein phosphatases.

ACKNOWLEDGEMENTS

The author wishes to thank all colleagues who provided unpublished results and apologize to those whose research was not discussed due to subject focus and page limitation. The research in the author's laboratory is supported by the National Institute of Health and US Department of Agriculture.

REFERENCES

Ahn, N. G., Seger, R. and Krebs, E. G. (1992). The mitogen-activated protein kinase activator. *Current Opinion in Cell Biology* **4**, 992–999.

Allard, J. D., Chang, H. C., Herbst, R., McNeill, H. and Simon, M. A. (1996). The SH2-containing tyrosine phosphatase corkscrew is required during signaling by sevenless, Ras1 and Raf. *Development* **122**, 1137–1146.

Allen, G. J. and Sanders, D. (1995). Calcineurin, a type 2B protein phosphatase, modulates the Ca^{2+}-permeable slow vacuolar ion channel of stomatal guard cells. *Plant Cell* **7**, 1473–1483.

Allen, P. B., Kwon, Y. G., Nairn, A. C. and Greengard, P. (1998). Isolation and characterization of PNUTS, a putative protein phosphatase 1 nuclear targeting subunit. *Journal of Biological Chemistry* **273**, 4089–4095.

Arino, J., Perez-Callejon, E., Cunillera, N., Camps, M., Posas, F. and Ferrer, A. (1993). Protein phosphatases in higher plants multiplicity of type 2a phosphatases in *Arabidopsis thaliana*. *Plant Molecular Biology* **21**, 475–485.

Armstrong, F., Leung, J., Grabov, A., Brearley, J., Giraudat, J. and Blatt, M. R. (1995). Sensitivity to abscisic acid of guard-cell K^+ channels is suppressed by *abil-1*, a mutant *Arabidopsis* gene encoding a putative protein phosphatase. *Proceedings of the National Academy of Sciences of the United States of America* **92**, 9520–9524.

Assmann, S. M. (1993). Signal transduction in guard cells. *Annual Review of Cell Biology* **9**, 345–375.

Aubry, L. and Firtel, R. A. (1998). Spalten, a protein containing G-alpha-protein-like and PP2C domains, is essential for cell-type differentiation in *Dictyostelium. Genes and Development* **12**, 1525–1538.

Barford, D. (1995). Protein phosphatases. *Current Opinion in Structural Biology* **5**, 728–734.

Barford, D. (1996). Molecular mechanisms of the protein serine-threonine phosphatases. *Trends in Biochemical Science* **21**, 407–412.

Barford, D., Flint, A. J. and Tonks, N. K. (1994). Crystal structure of human protein tyrosine phosphatase 1B. *Science* **263**, 1397–1403.

Baskin, T. I. and Wilson, J. E. (1997). Inhibitors of protein kinase and phosphatase alter root morphology and disorganize cortical microtubules. *Plant Physiology* **113**, 493–502.

Berndt, N., Dohadwala, M. and Liu, C. W. Y. (1997). Constitutively active protein phosphatase 1-alpha causes Rb-dependent G1 arrest in human cancer cells. *Current Biology* **7**, 375–386.

Bertauche, N., Leung, J. and Giraudat, J. (1996). Protein phosphatase activity of abscisic acid insensitive 1 (ABI1) protein from *Arabidopsis thaliana. European Journal of Biochemistry* **241**, 193–200.

Bethke, P. C. and Jones, R. L. (1997). Reversible protein phosphorylation regulates the activity of the slow-vacuolar ion channel. *Plant Journal* **11**, 1227–1235.

Blatt, M. R. and Grabov, A. (1997). Signal redundancy, gates and integration in the control of ion channels for stomatal movement. *Journal of Experimental Biology* **48**, 529–537.

Brewis, N. D., Street, A. J., Prescott, A. R. and Cohen, P. T. W. (1993). PPX a novel protein serine-threonine phosphatase localized to centrosomes. *EMBO Journal* **12**, 987–996.

Casamayor, A., Perez-Callejon, E., Pujol, G., Arino, J. and Ferrer, A. (1994). Molecular characterization of a fourth isoform of the catalytic subunit of protein phosphatase 2A from *Arabidopsis thaliana. Plant Molecular Biology* **26**, 523–528.

Chandra, S. and Low, P. S. (1995). Role of phosphorylation in elicitation of the oxidative burst in cultured soybean cells. *Proceedings of the National Academy of Sciences of the United States of America* **92**, 4120–4123.

Chang, C. and Meyerowitz, E. M. (1995). The ethylene hormone response in *Arabidopsis*: a eukaryotic two-component signaling system. *Proceedings of the National Academy of Sciences of the United States of America* **92**, 4129–4133.

Charbonneau, H., Tonks, N. K., Walsh, K. A. and Fischer, E. H. (1988). The leukocyte common antigen (CD45): a putative receptor-linked protein tyrosine phosphatase. *Proceedings of the National Academy of Sciences of the United States of America* **85**, 7182–7186.

Chen, J., Martin, B. L. and Brautigan, D. L. (1992). Regulation of protein serine-threonine phosphatase type-2A by tyrosine phosphorylation. *Science* **257**, 1261–1264.

Chen, M. S., Silverstein, A. M., Pratt, W. B. and Chinkers, M. (1996). The tetratricopeptide repeat domain of protein phosphatase 5 mediates binding to glucocorticoid receptor heterocomplexes and acts as a dominant negative mutant. *Journal of Biological Chemistry* **271**, 32315–32320.

Clark, S. E., Williams, R. W. and Meyerowitz, E. M. (1997). The *CLAVATA1* gene encodes a putative receptor kinase that controls shoot and floral meristem size in *Arabidopsis. Cell* **89**, 575–585.

Cohen, P. (1989). Structure and regulation of protein phosphatases. *Annual Review of Biochemistry* **58**, 453–508.

Cohen, P. T. W. (1997). Novel protein serine-threonine phosphatases: variety is the spice of life. *Trends in Biochemical Science* **22**, 245–251.

Cyert, M. S. and Thorner, J. (1992). Regulatory subunit (*CNB1* gene product) of yeast calcium/calmodulin-dependent phosphoprotein phosphatases is required for adaptation to pheromone. *Molecular and Cellular Biology* **12**, 3460–3469.

Damer, C. K., Partridge, J., Pearson, W. R. and Haystead, T. A. J. (1998). Rapid identification of protein phosphatase 1-binding proteins by mixed peptide sequencing and database searching. Characterization of a novel holoenzymic form of protein phosphatase 1. *Journal of Biological Chemistry* **273**, 24396–24405.

Das, A. K., Helps, N. R., Cohen, P. T. W. and Barford, D. (1996). Crystal structure of the protein serine-threonine phosphatase 2C at 2.0 A resolution. *EMBO Journal* **15**, 6798–6809.

Das, A. K., Cohen, P. T. W. and Barford, D. (1998). The structure of the tetratricopeptide repeats of protein phosphatase 5: implications for TPR-mediated protein–protein interactions. *EMBO Journal* **17**, 1192–1199.

Dent, P., Macdougall, L. K., Mackintosh, C., Campbell, D. G. and Cohen, P. (1992). A myofibrillar protein phosphatase from rabbit skeletal muscle contains the beta isoform of protein phosphatase-1 complexed to a regulatory subunit which greatly enhances the dephosphorylation of myosin. *European Journal of Biochemistry* **210**, 1037–1044.

Denu, J. M. and Tanner, K. G. (1998). Specific and reversible inactivation of protein tyrosine phosphatase by hydrogen peroxide: evidence for a sulfenic acid intermediate and implications for redox regulation. *Biochemistry* **37**, 5633–5642.

Denu, J. M., Stuckey, J. A., Saper, M. A. and Dixon, J. E. (1996). Form and function in protein dephosphorylation. *Cell* **87**, 361–364.

Desai, C. J., Popova, E. and Zinn, K. (1994). A *Drosophila* receptor tyrosine phosphatase expressed in the embryonic CNS and larval optic lobes is a member of the set of proteins bearing the 'HRP' carbohydrate epitope. *Journal of Neuroscience* **14**, 7272–7283.

Desai, C. J., Gindhart, J. G. Jr, Goldstein, L. S. B. and Zinn, K. (1996). Receptor tyrosine phosphatases are required for motor axon guidance in the *Drosophila* embryo. *Cell* **84**, 599–609.

Desdouits, F., Cheetham, J. J., Huang, H.-B., Kwon, Y.-G., Da Cruz, E. Silva, E. F., Denefle, P., Ehrlich, M. E., Nairn, A. C., Greengard, P. and Girault, J.-A. (1995). Mechanism of inhibition of protein phosphatase 1 by DARPP-32: studies with recombinant DARPP-32 and synthetic peptides. *Biochemical and Biophysical Research Communications* **206**, 652–658.

Dobrowsky, R. T., Kamibayashi, C., Mumby, M. C. and Hannun, Y. A. (1993). Ceramide activates heterotrimeric protein phosphatase 2a. *Journal of Biological Chemistry* **268**, 15523–15530.

Doi, K., Gartner, A., Ammerer, G., Errede, B., Shinkawa, H., Sugimoto, K. and Matsumoto, K. (1994). MSG5, a novel protein phosphatase promotes adaptation to pheromone response in *S. cerevisiae*. *EMBO Journal* **13**, 61–70.

Dunphy, W. G. (1994). The decision to enter mitosis. *Trends in Cell Biology* **4**, 202–207.

Egloff, M. P., Cohen, P. T. W., Reinemer, P. and Barford, D. (1995). Crystal structure of the catalytic subunit of human protein phosphatase 1 and its complex with tungstate. *Journal of Molecular Biology* **254**, 942–959.

Egloff, M. P., Johnson, D. F., Moorhead, G., Cohen, P. T. W., Cohen, P. and Barford, D. (1997). Structural basis for the recognition of regulatory subunits by the catalytic subunit of protein phosphatase 1. *EMBO Journal* **16**, 1876–1887.

Endo, S., Zhou, X., Connor, J., Wang, B. and Shenolikar, S. (1996). Multiple structural elements define the specificity of recombinant human inhibitor-1 as a protein phosphatase-1 inhibitor. *Biochemistry* **35**, 5220–5228.

Faux, M. C. and Scott, J. D. (1996). More on target with protein phosphorylation: conferring specificity by location. *Trends in Biochemical Sciences* **8**, 312–315.

Favre, B., Zolnierowicz, S., Turowski, P. and Hemmings, B. A. (1994). The catalytic subunit of protein phosphatase 2A is carboxyl-methylated *in vivo*. *Journal of Biological Chemistry* **269**, 16311–16317.

Feng, G. S., Hui, C. C. and Pawson, T. (1993). Sh2-containing phosphotyrosine phosphatase as a target of protein tyrosine kinases. *Science* **259**, 1607–1611.

Fruman, D. A., Meyers, R. E. and Cantley, L. C. (1998). Phosphoinositide kinases. *Annual Review of Biochemistry* **67**, 481–507.

Furnari, B., Rhind, N. and Russell, P. (1997). CDC25 mitotic inducer targeted by Chk1 DNA damage checkpoint kinase. *Science* **277**, 1495–1497.

Gaits, F., Shiozak, K. and Russell, P. (1997). Protein phosphatase 2C acts independently of stress-activated kinase cascade to regulate the stress response in fission yeast. *Journal of Biological Chemistry* **272**, 17873–17879.

Galadari, S., Kishikawa, K., Kamibayashi, C., Mumby, M. C. and Hannun, Y. A. (1998). Purification and characterization of ceramide-activated protein phosphatases. *Biochemistry* **37**, 11232–11238.

Garbers, C., Delong, A., Deruere, J., Bernasconi, P. and Soll, D. (1996). A mutation in protein phosphatase 2A regulatory subunit A affects auxin transport in *Arabidopsis*. *EMBO Journal* **15**, 2115–2124.

Gebbink, M. F. B. G., Verheijen, M. H. G., Zondag, G. C. M., Van Etten, I. and Moolenaar, W. H. (1993a). Purification and characterization of the cytoplasmic domain of human receptor-like protein tyrosine phosphatase RPTP-mu. *Biochemistry* **32**, 13516–13522.

Gebbink, M. F. B. G., Zondag, G. C. M., Wubbolts, R. W., Beijersbergen, R. L., Van Etten, I. and Moolenaar, W. H. (1993b). Cell–cell adhesion mediated by a receptor-like protein tyrosine phosphatase. *Journal of Biological Chemistry* **268**, 16101–16104.

Goetz, J., Probst, A., Ehler, E., Hemmings, B. and Kues, W. (1998). Delayed embryonic lethality in mice lacking protein phosphatase 2A catalytic subunit C alpha. *Proceedings of the National Academy of Sciences of the United States of America* **95**, 12370–12375.

Goldberg, J., Huang, H. B., Kwon, Y. G., Greengard, P., Nairn, A. C. and Kuriyan, J. (1995). Three-dimensional structure of the catalytic subunit of protein serine-threonine phosphatase-1. *Nature* **376**, 745–753.

Griffith, J. P., Kim, J. L., Kim, E. E., Sintchak, M. D., Thomson, J. A., Fitzgibbon, M. J., Fleming, M. A., Caron, P. R., Hsiao, K. and Navia, M. A. (1995). X-ray structure of calcineurin inhibited by the immunophilin-immunosuppressant FKBP12–FK506 complex. *Cell* **82**, 507–522.

Groschner, K., Schuhmann, K., Mieskes, G., Baumgartner, W. and Romanin, C. (1996). A type 2A phosphatase-sensitive phosphorylation site controls modal gating of L-type Ca^{2+} channels in human vascular smooth-muscle cells. *Biochemistry Journal* **318**, 513–517.

Guan, K., Broyles, S. S. and Dixon, J. E. (1991). A tyrosine serine protein phosphatase encoded by vaccinia virus. *Nature* **350**, 359–362.

Guan, K., Deschenes, R. J. and Dixon, J. E. (1992a). Isolation and characterization of a second protein tyrosine phosphatase gene, PTP2, from *Saccharomyces cerevisiae*. *Journal of Biological Chemistry* **267**, 10024–10030.

Guan, K., Hakes, D. J., Wang, Y., Park, H. D., Cooper, T. G. and Dixon, J. E. (1992b). A yeast protein phosphatase related to the vaccinia virus Vh1 phosphatase is induced by nitrogen starvation. *Proceedings of the National Academy of Sciences of the United States of America* **89**, 12175–12179.

Guan, K. L. (1994). The mitogen activated protein kinase signal transduction pathway: from the cell surface to the nucleus. *Cellular Signalling* **6**, 581–589.

Guan, K. L. and Butch, E. (1995). Isolation and characterization of a novel dual specific phosphatase, HVH2, which selectively dephosphorylates the mitogen-activated protein kinase. *Journal of Biological Chemistry* **270**, 7197–7203.

Guo, H. and Damuni, Z. (1993). Autophosphorylation-activated protein kinase phosphorylates and inactivates protein phosphatase 2a. *Proceedings of the National Academy of Sciences of the United States of America* **90**, 2500–2504.

Gupta, R., Huang, Y., Kieber, J. J. and Luan, S. (1998). Identification of a dual-specificity protein phosphatase that inactivates MAP kinase from *Arabidopsis*. *Plant Journal* **16**, 581–590.

Hanada, M., Kobayashi, T., Ohnishi, M., Ikeda, S., Wang, H., Katsura, K., Yanagawa, Y., Hiraga, A., Kanamaru, R. and Tamura, S. (1998). Selective suppression of stress-activated protein kinase pathway by protein phosphatase 2C in mammalian cells. *FEBS Letters* **43**, 172–176.

Hansen, D. and Pilgrim, D. (1998). Molecular evolution of a sex determination protein: FEM-2 (PP2C) in *Caenorhabditis*. *Genetics* **149**, 1353–1362.

Hare, P. D., Cress, W. A. and Van Staden, J. (1997). The involvement of cytokinins in plant responses to environmental stress. *Plant Growth Regulation* **23**, 1–2.

Heitman, J., Koller, A., Cardenas, M. E. and Hall, M. N. (1993). Identification of immunosuppressive drug targets in yeast. *Methods* **5**, 176–187.

Hoffmann, K. M. V., Tonks, N. K. and Barford, D. (1997). The crystal structure of domain 1 of receptor protein–tyrosine phosphatase mu. *Journal of Biological Chemistry* **272**, 27505–27508.

Huang, X. and Honkanen, R. E. (1998). Molecular cloning, expression, and characterization of a novel human serine-threonine protein phosphatase, PP7, that is homologous to *Drosophila* retinal degeneration C gene product (rdgC). *Journal of Biological Chemistry* **273**, 1462–1468.

Hubbard, M. J. and Cohen, P. (1993). On target with a new mechanism for the regulation of protein phosphorylation. *Trends in Biochemical Science* **18**, 172–177.

Hunter, T. (1995). Protein kinases and phosphatases: the yin and yang of protein phosphorylation and signaling. *Cell* **80**, 225–236.

Jiang, Y. P., Wang, H., D'Eustachio, P., Musacchio, J. M., Schlesinger, J. and Sap, J. (1993). Cloning and characterization of R-PTP-KAPPA, a new member of the receptor protein tyrosine phosphatase family with a proteolytically cleaved cellular adhesion molecule-like extracellular region. *Molecular and Cellular Biology* **13**, 2942–2951.

Johnson, L. N. and O'Reilly, M. (1996). Control by phosphorylation. *Current Opinion in Structural Biology* **6**, 762–769.

Johnson, L. N., Snape, P., Martin, J. L., Acharya, K. R., Barford, D. and Oikonomakos, N. G. (1993). Crystallographic binding studies on the allosteric inhibitor glucose-6-phosphate to T state glycogen phosphorylase B. *Journal of Molecular Biology* **232**, 253–267.

Johnson, L. N., Noble, M. E. M. and Owen, D. J. (1996a). Active and inactive protein kinases: structural basis for regulation. *Cell* **85**, 149–158.

Johnson, D. F., Moorhead, G., Caudwell, F. B., Cohen, P., Chen, Y. H., Chen, M. X. and Cohen, P. T. W. (1996b). Identification of protein-phosphatase-1-binding domains on the glycogen and myofibrillar targetting subunits. *European Journal of Biochemistry* **239**, 317–325.

Johnson, D., Cohen, P., Chen, M. X., Chen, Y. H. and Cohen, P. T. W. (1997). Identification of the regions on the M-110 subunit of protein phosphatase 1M that interact with the M-21 subunit and with myosin. *European Journal of Biochemistry* **244**, 931–939.

Jonak, C., Kiegerl, S., Ligterink, W., Barker, P. J., Huskisson, N. S. and Hirt, H. (1996). Stress signaling in plants: a mitogen-activated protein kinase pathway is activated by cold and drought. *Proceedings of the National Academy of Sciences of the United States of America* **93**, 11274–11279.

Keyse, S. M. (1995). An emerging family of dual specificity MAP kinase phosphatases. *Biochimica et Biophysica Acta* **1265**, 2–3.

Keyse, S. M. and Emslie, E. A. (1992). Oxidative stress and heat shock induce a human gene encoding a protein–tyrosine phosphatase. *Nature* **359**, 644–647.

Kinoshita, N., Yamano, H., Niwa, H., Yoshida, T. and Yanagida, M. (1993). Negative regulation of mitosis by the fission yeast protein phosphatase ppa2. *Genes and Development* **7**, 1059–1071.

Kissinger, C. R., Parge, H. E., Knighton, D. R., Lewis, C. T., Pelletier, L. A., Tempczyk, A., Kalish, V. J., Tucker, K. D., Showalter, R. E., Moomaw, E. W., Gastinel, L. N., Habuka, N., Chen, X., Maldonado, F., Barker, J. E., Bacquet, R. and Villafranca, J. E. (1995). Crystal structures of human calcineurin and the human FKBP12–FK506–calcineurin complex. *Nature* **378**, 641–644.

Klabunde, T., Straeter, N., Krebs, B. and Witzel, H. (1995). Structural relationship between the mammalian Fe (III)-Fe (II) and the Fe (III) -Zn (II) plant purple acid phosphatases. *FEBS Letters* **367**, 56–60.

Klee, C. B., Ren, H. and Wang, X. (1998). Regulation of the calmodulin-stimulated protein phosphatase, calcineurin. *Journal of Biological Chemistry* **273**, 13367–13370.

Krueger, N. X., Van Vactor, D., Wan, H. I., Gelbart, W. M., Goodman, C. S. and Saito, H. (1996). The transmembrane tyrosine phosphatase DLAR controls motor axon guidance in *Drosophila*. *Cell* **84**, 611–622.

Kudla, J., Xu, Q., Harter, K., Gruissem, W. and Luan, S. (1999). Molecular characterization of a calcineurin B-like calcium sensor protein encoded by a stress-responsive gene in *Arabidopsis*. *Proceedings of the National Academy of Sciences of the United States of America* **96**, 4718–4723.

Kuromori, T. and Yamamoto, M. (1994). Cloning of cDNAs from *Arabidopsis thaliana* that encode putative protein phosphatase 2C and a human Dr1-like protein by transformation of a fission yeast mutant. *Nucleic Acids Research* **22**, 5296–5301.

Latorre, K. A., Harris, D. M. and Rundle, S. J. (1997). Differential expression of three *Arabidopsis* genes encoding the B′ regulatory subunit of protein phosphatase 2A. *European Journal of Biochemistry* **245**, 156–163.

Ledbetter, J. A., Tonks, N. K., Fischer, E. H. and Clark, E. A. (1988). CD45 regulates signal transduction and lymphocyte activation by specific association with receptor molecules on T or B cells. *Proceedings of the National Academy of Sciences of the United States of America* **85**, 8628–8632.

Lee, J., Chen, Y., Tolstykh, T. and Stock, J. (1996). A specific protein carboxyl methylesterase that demethylates phosphoprotein phosphatase 2A in bovine brain. *Proceedings of the National Academy of Sciences of the United States of America* **93**, 6043–6047.

Lee, T. H., Turck, C. and Kirschner, M. W. (1994). Inhibition of cdc2 activation by INH-PP2A. *Molecular and Cellular Biology* **5**, 323–338.

Leung, J., Bouvier-Durand, M., Morris, P. C., Guerrier, D., Chefdor, F. and Giraudat, J. (1994). *Arabidopsis* ABA response gene *ABI1*: features of a calcium-modulated protein phosphatase. *Science* **264**, 1448–1452.

Leung, J., Merlot, S. and Giraudat, J. (1997). The *Arabidopsis ABSCISIC ACID-INSENSITIVE2* (*ABI2*) and *ABL1* genes encode homologous protein phosphatases 2C involved in abscisic acid signal transduction. *Plant Cell* **9**, 759–771.

Li, D.-M. and Sun, H. (1997). TEP1, encoded by a candidate tumor suppressor locus, is a novel protein tyrosine phosphatase regulated by transforming growth factor beta. *Cancer Research* **57**, 2124–2129.

Li, D. M. and Sun, H. (1998). Expression of PTEN down-regulates phosphorylation of protein kinase B/Akt and arrests cells in G1 phase in human glioblastoma cell line. *Molecular and Cellular Biology* **9**, 15406–15411.

Li, H., Zhao, L. L., Funder, J. W. and Liu, J. P. (1997a). Protein phosphatase 2A inhibits nuclear telomerase activity in human breast cancer cells. *Journal of Biological Chemistry* **272**, 16729–16732.

Li, J., Yen, C., Liaw, D., Podsypanina, K., Bose, S., Wang, S. I., Puc, J., Miliaresis, C., Rodgers, L., McCombie, R., Bigner, S. H., Giovanella, B. C., Ittmann, M., Tycko, B., Hibshoosh, H., Wigler, M. H. and Parsons, R. (1997b). PTEN, a putative protein tyrosine phosphatase gene mutated in human brain, breast, and prostate cancer. *Science* **275**, 1943–1947.

Li, W., Luan, S., Schreiber, S. L. and Assmann, S. M. (1994). Evidence for protein phosphatase 1 and 2A regulation of K^+ channels in two types of leaf cells. *Plant Physiology* **106**, 963–970.

Li, Y. M. and Casida, J. E. (1992). Cantharidin-binding protein identification as protein phosphatase 2a. *Proceedings of the National Academy of Sciences of the United States of America* **89**, 11867–11870.

Lieberman, D. N. and Mody, I. (1994). Regulation of NMDA channel function by endogenous Ca^{2+}-dependent phosphatase. *Nature* **369**, 235–239.

Ligterink, W., Kroj, T., Nieden, U. Z., Hirt, H. and Scheel, D. (1997). Receptor-mediated activation of a MAP kinase in pathogen defense of plants. *Science* **276**, 2054–2057.

Lin, D. M. and Goodman, C. S. (1994) Ectopic and increase expression of fasiculin III alters motorneuron growth cone guidance. *Neuron* **13**, 507–523.

Liu, J. and Zhu, J. K. (1998). A calcium sensor homolog required for plant salt tolerance. *Science* **280**, 1943–1945.

Liu, J., Farmer, J. D. Jr, Lane, W. S., Friedman, J., Weissman, I. and Schreiber, S. L. (1991). Calcineurin is a common target of cyclophilin–cyclosporin A and FKBP–FK506 complexes. *Cell* **66**, 806–816.

Liu, J., Albers, M. W., Wandless, T. J., Luan, S., Alberg, D. G., Belshaw, P. J., Cohen, P., Mackintosh, C., Klee, C. B. and Schreiber, S. L. (1992). Inhibition of T cell signaling by immunophilin-ligand complexes correlates with loss of calcineurin phosphatase activity. *Biochemistry* **31**, 3896–3901.

Lohse, D. L., Denu, J. M. and Dixon, J. E. (1995). Insights derived from the structures of the Ser-Thr phosphatases calcineurin and protein phosphatase 1. *Structure* **3**, 987–990.

Lorenz, U., Ravichandran, K. S., Pei, D., Walsh, C. T., Burakoff, S. J. and Neel, B. G. (1994). Lck-dependent tyrosyl phosphorylation of the phosphotyrosine phosphatase SH-PTP1 in murine T cells. *Molecular and Cellular Biology* **14**, 1824–1834.

Luan, S. (1998) Protein phosphatases and signaling cascades in higher plants. *Trends in Plant Science* **3**, 271–275.

Luan, S., Li, W., Rusnak, F., Assmann, S. M. and Schreiber, S. L. (1993). Immunosuppressants implicate protein phosphatase regulation of potassium channels in guard cells. *Proceedings of the National Academy of Sciences of the United States of America* **90**, 2202–2206.

Luo, J., Pato, M. D., Riordan, J. R. and Hanrahan, J. W. (1998). Differential regulation of single CFTR channels by PP2C, PP2A, and other phosphatases. *American Journal of Physiology* **274**, C1397–C1410.

Mackintosh, C. and Cohen, P. (1989). Identification of high levels of type 1 and type 2A protein phosphatases in higher plants. *Biochemistry Journal* **262**, 335–340.

Mackintosh, C., Coggins, J. and Cohen, P. (1991). Plant protein phosphatases: subcellular distribution, detection of protein phosphatase 2C and identification of protein phosphatases 2A as the major quinate dehydrogenase phosphatase. *Biochemistry Journal* **273**, 733–738.

Mackintosh, R. W., Haycox, G., Hardie, D. G. and Cohen, P. T. W. (1990). Identification by molecular cloning of two complementary DNA sequences from the plant brassica-napus which are very similar to mammalian protein phosphatase-1 and protein phosphatase-2a. *FEBS Letters* **276**, 1–2.

Mackintosh, C., Lyon, G. D. and Mackintosh, R. W. (1994). Protein phosphatase inhibitors activate anti-fungal defence responses of soy bean cotyledons and cell cultures. *Plant Journal* **5**, 137–147.

Maehama, T. and Dixon, J. E. (1998). The tumor suppressor, PTEN-MMAC1, dephosphorylates the lipid second messenger, phosphatidylinositol 3,4,5-triphosphate. *Journal of Biological Chemistry* **273**, 13375–13378.

Mauro, L. J. and Dixon, J. E. (1994). 'Zip codes' direct intracellular protein tyrosine phosphatases to the correct cellular 'address'. *Trends in Biochemical Science* **19**, 151–155.

McCright, B. and Virshup, D. M. (1995). Identification of a new family of protein phosphatase 2A regulatory subunits. *Journal of Biological Chemistry* **270**, 26123–26128.

McCright, B., Rivers, A. M., Audlin, S. and Virshup, D. M. (1996). The B56 family of protein phosphatase 2A (PP2A) regulatory subunits encodes differentiation-induced phosphoproteins that target PP2A to both nucleus and cytoplasm. *Journal of Biological Chemistry* **271**, 22081–22089.

Means, A. R. (1994). Calcium, calmodulin and cell cycle regulation. *FEBS Letters* **347**, 1–4.

Meskiene, I., Bogre, L., Glaser, W., Balog, J., Brandstotter, M., Zwerger, K., Ammerer, G. and Hirt, H. (1998a). MP2C, a plant protein phosphatase 2C, functions as a negative regulator of mitogen-activated protein kinase pathways in yeast and plants. *Proceedings of the National Academy of Sciences of the United States of America* **95**, 1938–1943.

Meskiene, I., Ligterink, W., Bogre, L., Jonak, C., Kiegerl, S., Balog, J., Eklof, S., Ammerer, G. and Hirt, H. (1998b). The SAM kinase pathway: an integrated circuit for stress signaling in plants. *Journal of Plant Research* **111**, 339–344.

Meyer, K., Leube, M. P. and Grill, E. (1994). A protein phosphatase 2C involved in ABA signal transduction in *Arabidopsis thaliana*. *Science* **264**, 1452–1455.

Misteli, T. and Spector, D. L. (1996). Serine-threonine phosphatase 1 modulates the subnuclear distribution of pre-mRNA splicing factors. *Molecular and Cellular Biology* **7**, 1559–1572.

Mizoguchi, T., Ichimura, K. and Shinozaki, K. (1997). Environmental stress response in plants: the role of mitogen-activated protein kinases. *Trends in Biotechnology* **15**, 15–19.

Myers, M. P., Stolarov, J. P., Eng, C., Li, J., Wang, S. I., Wigler, M. H., Parsons, R. and Tonks, N. K. (1997). P-TEN, the tumor suppressor from human chromosome 10q23, is a dual-specificity phosphatase. *Proceedings of the National Academy of Sciences of the United States of America* **94**, 9052–9057.

Myers, M. P., Pass, I., Batty, I. H., Van Der Kaay, J., Stolarov, J. P., Hemmings, B. A., Wigler, M. H., Downes, C. P. and Tonks, N. K. (1998). The lipid phosphatase activity of PTEN is critical for its tumor suppressor function. *Proceedings of the National Academy of Sciences of the United States of America* **95**, 13513 13518.

Nakamura, T., Liu, Y., Hirata, D., Namba, H., Harada, S. I., Hirokawa, T. and Miyakawa, T. (1993). Protein phosphatase type 2B (calcineurin)-mediated, FK506-sensitive regulation of intracellular ions in yeast is an important determinant for adaptation to high salt stress conditions. *EMBO Journal* **12**, 4063–4071.

Neel, B. G. and Tonks, N. K. (1997). Protein tyrosine phosphatases in signal transduction. *Current Opinion in Cell Biology* **9**, 193–204.

Nelson, D. A. and Ludlow, J. W. (1997). Characterization of the mitotic phase pRb-directed protein phosphatase activity. *Oncogene* **14**, 2407–2415.

Nelson, D. A., Krucher, N. A. and Ludlow, J. W. (1997). High molecular weight protein phosphatase type 1 dephosphorylates the retinoblastoma protein. *Journal of Biological Chemistry* **272**, 4528–4535.

Nitschke, K., Fleig, U., Schell, J. and Palme, K. (1992). Complementation of the cs dis2-11 cell cycle mutant of *Schizosaccharomyces pombe* by a protein phosphatase from *Arabidopsis thaliana*. *EMBO Journal* **11**, 1327–1333.

O'Reilly, A. and Neel, B. G. (1998). Structural determinants of SHP-2 function and specificity in *Xenopus* mesoderm induction. *Molecular and Cellular Biology* **18**, 161–177.

Pardo, J. M., Reddy, M. P., Yang, S., Maggio, A., Huh, G.-H., Matsumoto, T., Coca, M. A., Paino-D'Urzo, M., Koiwa, H., Yun, D.-J., Watad, A. A., Bressan, R. A. and Hasegawa, P. M. (1998). Stress signaling through Ca^{2+}–calmodulin dependent protein phosphatase calcineurin mediates salt adaptation in plants. *Proceedings of the National Academy of Sciences of the United States of America* **95**, 9681–9686.

Pawson, T. (1995). Protein modules and signalling networks. *Nature* **373**, 573–580.

Pei, Z. M., Kuchitsu, K., Ward, J. M., Schwarz, M. and Schroeder, J. I. (1997). Differential abscisic acid regulation of guard cell slow anion channels in *Arabidopsis* wild-type and *abi1* and *abi2* mutants. *Plant Cell* **9**, 409–423.

Perez-Callejon, E., Casamayor, A., Pujol, G., Camps, M., Ferrer, A. and Arino, J. (1998). Molecular cloning and characterization of two phosphatase 2A catalytic subunit genes from *Arabidopsis thaliana*. *Gene* **209**, 1–2.

Perkins, L. A., Larsen, I. and Perrimon, N. (1992). Corkscrew encodes a putative protein tyrosine phosphatase that functions to transduce the terminal signal from the receptor tyrosine kinase torso. *Cell* **70**, 225–236.

Perkins, L. A., Johnson, M. R., Melnick, M. B. and Perrimon, N. (1996). The nonreceptor protein tyrosine phosphatase corkscrew functions in multiple receptor tyrosine kinase pathways in *Drosophila*. *Developmental Biology* **180**, 63–81.

Pirck, M., Pay, A., Heberle-Bors, E. and Hirt, H. (1993). Isolation and characterization of a phosphoprotein phosphatase type 2A gene from alfalfa. *Molecular and General Genetics* **240**, 126–131.

Rao, A. (1997). Transcription factors of the NFAT family: regulation and function. *Annual Review of Immunology* **15**, 707–747.

Rasmussen, C., Garen, C., Brining, S., Kincaid, R. L., Means, R. L. and Means, A. R. (1994). The calmodulin-dependent protein phosphatase catalytic subunit (calcineurin A) is an essential gene in *Aspergillus nidulans*. *EMBO Journal* **13**, 2545–2552.

Raz, V. and Fluhr, R. (1993). Ethylene signal is transduced via protein phosphorylation events in plants. *Plant Cell* **5**, 523–530.

Rohan, P. J., Davis, P., Moskaluk, C. A., Kearns, M., Krutzsch, H., Siebenlist, U. and Kelly, K. (1993). Pac-1 a mitogen-induced nuclear protein tyrosine phosphatase. *Science* **259**, 1763–1766.

Rundle, S. J., Hartung, A. J., Corumi, J. W. and O'Neill, M. (1995). Characterization of a cDNA encoding of the 55 kDa B regulatory subunit of *Arabidopsis* protein phosphatase 2A. *Plant Molecular Biology* **28**, 257–266.

Sanchez, Y., Wong, C., Thoma, R. S., Richman, R., Wu, Z., Piwnica-Worms, H. and Elledge, S. J. (1997). Conservation of the Chk1 checkpoint pathway in mammals: linkage of DNA damage to Cdk regulation through Cdc25. *Science* **277**, 1497–1501.

Sansom, S. C., Stockand, J. D., Hall, D. and Williams, B. (1997). Regulation of large calcium-activated potassium channels by protein phosphatase 2A. *Journal of Biological Chemistry* **272**, 9902–9906.

Sap, J., Jiang, Y. P., Friedlander, D., Grumet, M. and Schlessinger, J. (1994). Receptor tyrosine phosphatase R-PTP-kappa mediates homophilic binding. *Molecular and Cellular Biology* **14**, 1–9.

Schmidt, C., Schelle, I., Liao, Y. J. and Schroeder, J. I. (1995). Strong regulation of slow anion channels and abscisic acid signaling in guard cells by phosphorylation and dephosphorylation events. *Proceedings of the National Academy of Sciences of the United States of America* **92**, 9535–9539.

Seo, S., Okamoto, M., Seto, H., Ishizuka, K., Sano, H. and Ohashi, Y. (1995). Tobacco MAP kinase: a possible mediator in wound signal transduction pathways. *Science* **270**, 1988–1992.

Seo, S., Sano, H. and Ohashi, Y. (1997). Jasmonic acid in wound signal transduction pathways. *Physiologia Plantarum* **101**, 740–745.

Sheen, J. (1993). Protein phosphatase activity is required for light-inducible gene expression in maize. *EMBO Journal* **12**, 3497–3505.

Sheen, J. (1998). Mutational analysis of protein phosphatase 2C involved in abscisic acid signal transduction in higher plants. *Proceedings of the National Academy of Sciences of the United States of America* **95**, 975–980.

Shenolikar, S. (1994). Protein serine-threonine phosphatases: new avenues for cell regulation. *Annual Review of Biochemistry* **63**, 55–86.

Shifrin, V. I., Davis, R. J. and Neel, B. G. (1997). Phosphorylation of protein–tyrosine phosphatase PTP-1B on identical sites suggests activation of a common signaling pathway during mitosis and stress response in mammalian cells. *Journal of Biological Chemistry* **272**, 2957–2962.

Shiozaki, K. and Russell, P. (1994). Cellular function of protein phosphatase 2C in yeast. *Cellular and Molecular Biology Research* **40**, 241–243.

Shiozaki, K. and Russell, P. (1995a). Cell-cycle control linked to extracellular environment by MAP kinase pathway in fission yeast. *Nature* **378**, 739–743.

Shiozaki, K. and Russell, P. (1995b). Counteractive roles of protein phosphatase 2C (PP2C) and a MAP kinase kinase homolog in the osmoregulation fission yeast. *EMBO Journal* **14**, 492–502.

Shiozaki, K., Akhavan-Niaki, H., McGowan, C. H. and Russell, P. (1994). Protein phosphatase 2C, encoded by ptc1+, is important in the heat shock response of *Schizosaccharomyces pombe*. *Molecular and Cellular Biology* **14**, 3742–3751.

Slabas, A. R., Fordham-Skelton, A. P., Fletcher, D., Martinez-Rivas, J. M., Swinhoe, R., Croy, R. R. D. and Evans, I. M. (1994). Characterisation of cDNA and genomic clones encoding homologues of the 65-kDa regulatory subunit of protein phosphatase 2A in *Arabidopsis thaliana*. *Plant Molecular Biology* **26**, 1125–1138.

Smith, R. D. and Walker, J. C. (1991). Isolation and expression of a maize type 1 protein phosphatase. *Plant Physiology* **97**, 677–683.

Smith, R. D. and Walker, J. C. (1996). Plant protein phosphatases. *Annual Review of Plant Physiology and Plant Molecular Biology* **47**, 101–125.

Smith, R. D., Wilson, J. E., Walker, J. C. and Gerbers, T. I. (1994). Protein-phosphatase inhibitors block root hair growth and alter cortical cell shape of *Arabidopsis* roots. *Planta* **194**, 516–524.

Smith, R. D., Lin, Q., Cannon, J. F. and Walker, J. C. (1995). Type-1 and type-2C protein phosphatases of higher plants. *Advances in Protein Phosphatases* **9**, 105–120.

Snyder, S. H., Lai, M. M. and Burnett, P. E. (1998). Immunophilins in the nervous system. *Neuron* **21**, 283–294.

Sontag, E., Nunbhakdi-Craig, V., Bloom, G. S. and Mumby, M. C. (1995). A novel pool of protein phosphatase 2A is associated with microtubules and is regulated during the cell cycle. *Journal of Cellular Biology* **128**, 1131–1144.

Stambolic, V., Suzuki, A., De La Pompa, J. L., Brothers, G. M., Mirtsos, C., Sasaki, T., Ruland, J., Sasaki, T., Rutland, J., Penninger, J. M., Siderovski, D. P. and Mak, T. W. (1998). Negative regulation of PKB/Akt-dependent cell survival by the tumor suppressor PTEN. *Cell* **95**, 29–39.

Steck, P. A., Pershouse, M. A., Jasser, S. A., Yung, W. K. A., Lin, H., Ligon, A. H., Langford, L. A., Baumgard, M. L., Hattier, T., Davis, T., Frye, C., Hu, R., Swedlund, B., Teng, D. H. F. and Tavtigian, S. V. (1997). Identification of a candidate tumour suppressor gene, MMAC1, at chromosome 10q23.3 that is mutated in multiple advanced cancer. *Nature Genetics* **15**, 356–362.

Steele, F. R., Washburn, T., Rieger, R. and O'Tousa, J. E. (1992). *Drosophila* retinal degeneration C (rdgC) encodes a novel serine/threonine protein phosphatase. *Cell* **69**, 669–676.

Stone, R. L. and Dixon, J. E. (1994). Protein-tyrosine phosphatases. *Journal of Biological Chemistry* **269**, 31323–31326.

Stone, J. M., Collinge, M. A., Smith, R. D., Horn, M. A. and Walker, J. C. (1994). Interaction of a protein phosphatase with an *Arabidopsis* serine-threonine receptor kinase. *Science* **266**, 793–795.

Stone, J. M., Trotochaud, A. E., Walker, J. C. and Clark, S. E. (1998). Control of meristem development by CLAVATA1 receptor kinase and kinase-associated protein phosphatase interactions. *Plant Physiology* **117**, 1217–1225.

Stratmann, J. W. and Ryan, C. A. (1997). Myelin basic protein kinase activity in tomato leaves is induced systemically by wounding and increases in response to systemin and oligosaccharide elicitors. *Proceedings of the National Academy of Sciences of the United States of America* **94**, 11085–11089.

Stuckey, J. A., Schubert, H. L., Fauman, E. B., Zhang, Z. Y., Dixon, J. E. and Saper, M. A. (1994). Crystal structure of *Yersinia* protein tyrosine phosphatase at 2.5 ANG and the complex with tungstate. *Nature* **370**, 571–575.

Sun, H. and Tonks, N. K. (1994). The coordinated action of protein tyrosine phosphatases and kinases in cell signaling. *Trends in Biochemical Science* **19**, 480–485.

Suzuki, K. and Shinshi, H. (1995). Transient activation and tyrosine phosphorylation of a protein kinase in tobacco cells treated with a fungal elicitor. *Plant Cell* **7**, 639–647.

Takekawa, M., Maeda, T. and Saito, H. (1998). Protein phosphatase 2C-alpha inhibits the human stress-responsive p38 and JNK MAPK pathways. *EMBO Journal* **17**, 4744–4752.

Thompson, L. J., Bollen, M. and Fields, A. P. (1997). Identification of protein phosphatase 1 as a mitotic lamin phosphatase. *Journal of Biological Chemistry* **272**, 29693–29697.

Timms, J. F., Carlberg, K., Gu, H., Chen, H., Kamatkar, S., Nadler, M. J. S., Rohrschneider, L. R. and Neel, B. G. (1998). Identification of major binding proteins and substrates for the SH2-containing protein tyrosine phosphatase SHP-1 in macrophages. *Molecular and Cellular Biology* **18**, 3838–3850.

Tonks, N. K. and Neel, B. G. (1996). From form to function: signaling by protein tyrosine phosphatases. *Cell* **87**, 365–368.

Tonks, N. K., Charbonneau, H., Diltz, C. D., Fischer, E. H. and Walsh, K. A. (1988). Demonstration that the leukocyte common antigen CD 45 is a protein tyrosine phosphatase. *Biochemistry* **27**, 8695–8701.

Toroser, D. and Huber, S. C. (2000). Carbon and nitrogen metabolism and reversible protein phosphorylation. *In* "Advances in Botanical Research" (J. A. Callow, ed.) pp. 437–460. Academic Press, London.

Turowski, P., Fernandez, A., Favre, B., Lamb, N. J. C. and Hemmings, B. A. (1995). Differential methylation and altered conformation of cytoplasmic and nuclear forms of protein phosphatase 2A during cell cycle progression. *Journal of Cellular Biology* **129**, 397–410.

Usami, S., Banno, H., Ito, Y., Nishihama, R. and Machida, Y. (1995). Cutting activates a 46-kilodalton protein kinase in plants. *Proceedings of the National Academy of Sciences of the United States of America* **92**, 8660–8664.

Van Vactor, D., OReilly, A. M. and Neel, B. G. (1998). Genetic analysis of protein tyrosine phosphatases. *Current Opinion in Genetics and Development* **8**, 112–126.

Vera-Estrella, R., Barkla, B. J., Higgins, V. J. and Blumwald, E. (1994). Plant defense response to fungal pathogens: activation of host-plasma membrane H^+-ATPase by elicitor-induced enzyme dephosphorylation. *Plant Physiology* **104**, 209–215.

Wang, S. S., Esplin, E. D., Li, J. L., Huang, L., Gazdar, A., Minna, J. and Evans, G. A. (1998). Alterations of the *PPP2R1B* gene in human lung and colon cancer. *Science* **282**, 284–287.

Wang, X., Culotta, V. C. and Klee, C. B. (1996). Superoxide dismutase protects calcineurin from inactivation. *Nature* **383**, 434–437.

Ward, J. M., Pei, Z. M. and Schroeder, J. I. (1995). Roles of ion channels in initiation of signal transduction in higher plants. *Plant Cell* **7**, 833–844.

Wera, S. and Hemmings, B. A. (1995). Serine-threonine protein phosphatases. *Biochemistry Journal* **311**, 17–29.

Williams, R. W., Wilson, J. M. and Meyerowitz, E. M. (1997). A possible role for kinase-associated protein phosphatase in the *Arabidopsis* CLAVATA1 signaling pathway. *Proceedings of the National Academy of Sciences of the United States of America* **94**, 10467–10472.

Wilson, C., Voronin, V., Touraev, A., Vicente, O. and Heberle-Bors, E. (1997). A developmentally regulated MAP kinase activated by hydration in tobacco pollen. *Plant Cell* **9**, 2093–2100.

Wu, X., Senechal, K., Nechat, M. S., Whang, Y. E. and Sawyers, C. S. (1998). The PTEN/MMC1 tumor suppressor phosphatase functions as a negative regulator of the phosphoinositide 3-kinase/Akt pathway. *Proceedings of the National Academy of Sciences of the United States of America* **95**, 15587–15591.

Wurgler-Murphy, S. M. and Saito, H. (1997). Two-component signal transducers and MAPK cascades. *Trends in Biochemical Science* **22**, 172–176.

Wurgler-Murphy, S. M., Maeda, T., Witten, E. A. and Saito, H. (1997). Regulation of the *Saccharomyces cerevisiae* HOG1 mitogen-activated protein kinase by the PTP2 and PTP3 protein tyrosine phosphatases. *Molecular and Cellular Biology* **17**, 1289–1297.

Xu, Q., Fu, H. H., Gupta, R. and Luan, S. (1998). Molecular characterization of a tyrosine-specific protein phosphatase encoded by a stress-responsive gene in *Arabidopsis*. *Plant Cell* **10**, 849–857.

Yan, Z. and Surmeier, D. J. (1997). D-5 dopamine receptors enhance Zn^{2+}-sensitive GABA-A currents in striatal cholinergic interneurons through a PKA-PP1 cascade. *Neuron* **19**, 1115–1126.

Yoshida, T., Toda, T. and Yanagida, M. (1994). A calcineurin-like gene *ppb1*+ in fission yeast: mutant defects in cytokinesis, cell polarity, mating and spindle pole body positioning. *Journal of Cellular Science* **107**, 1725–1735.

Yu, L., Golbeck, J., Yao, J. and Rusnak, F. (1997). Spectroscopic and enzymatic characterization of the active site dinuclear metal center of calcineurin: implications for a mechanistic role. *Biochemistry* **36**, 10727–10734.

Yuvaniyama, J., Denu, J. N., Dixon, J. E. and Saper, M. A. (1996). Crystal structure of the dual specificity protein phosphatase VHR. *Science* **272**, 1328–1331.

Zhan, X. L., Deschenes, R. J. and Guan, K. L. (1997). Differential regulation of FUS3 MAP kinase by tyrosine-specific phosphatases PTP2-PTP3 and dual-specificity phosphatase MSG5 in *Saccharomyces cerevisiae*. *Genes and Development* **11**, 1690–1702.

Zhang, K., Letham, D. S. and John, P. C. L. (1996). Cytokinin controls the cell cycle at mitosis by stimulating the tyrosine dephosphorylation and activation of p34-cdc2-like H1 histone kinase. *Planta* **200**, 2–12.

Zhang, S. and Klessig, D. F. (1997). Salicylic acid activates a 48-kD MAP kinase in tobacco. *Plant Cell* **9**, 809–824.

Zhou, X. B., Ruth, P., Schlossmann, J., Hofmann, F. and Korth, M. (1996). Protein phosphatase 2A is essential for the activation of Ca^{2+}-activated K^{+} currents by cGMP-dependent protein kinase in tracheal smooth muscle and Chinese hamster ovary cells. *Journal of Biological Chemistry* **271**, 19760–19767.

Histidine Kinases and the Role of Two-Component Systems in Plants

G. ERIC SCHALLER

Department of Biochemistry and Molecular Biology, University of New Hampshire, Durham, NH 03824, USA

I. INTRODUCTION

Until recently, the thought of devoting an entire review, much less a sentence, to plant histidine kinases would have been met with an incredulous stare. If pressed, a plant scientist might have remarked, remembering something from a college microbiology course, 'Aren't those something found in bacteria?' That picture of plants and plant signaling changed in 1993 with the publication of a

Advances in Botanical Research Vol. 32
incorporating Advances in Plant Pathology
ISBN 0-12-005932-0

paper entitled '*Arabidopsis* ethylene-response gene *ETR1*: similarity of product to two-component regulators' (Chang *et al.*, 1993). This paper can be considered emblematic of our time in two ways. First, it points to the important role that genetic studies in the model plant *Arabidopsis* have played in unraveling the mysteries of plant growth and development; the authors describe a gene involved in sensing the plant hormone ethylene. Second, it points to the role of databases in our analysis of genes, whereby we can make predictions of function based upon sequence similarities; the authors show homology between the *ETR1* gene product and proteins previously known only in bacteria, namely histidine kinases and response regulators. This paper also presaged the finding of additional histidine kinases and response regulators in plants, and set a precedent for these proteins being instrumental in plant signal transduction.

Histidine kinases and response regulators are collectively known as 'two-component systems'. They confer upon bacteria the ability to sense and respond to environmental stimuli (Parkinson, 1993; Swanson *et al.*, 1994; Stock *et al.*, 1995). The first component is a histidine kinase that autophosphorylates a conserved histidine residue in response to an environmental stimulus. This phosphate is then transferred to an aspartic acid residue on the second component referred to as the response regulator. Sometimes signaling involves a phosphate relay between a series of proteins, and here a third component is involved; this component is also capable of being phosphorylated on a conserved histidine residue and is known as a histidine-containing phosphotransfer protein (Appleby *et al.*, 1996).

In plants, proteins with homology to two-component systems have been implicated in signaling by the hormones ethylene (Bleecker and Schaller, 1996) and cytokinin (Kakimoto, 1998), in sensing changes in osmotic pressure (Urao *et al.*, 1997, 1999), in responses to light (Quail, 1997b), and in regulation of pyruvate dehydrogenase (Thelen *et al.*, 1998). Genes encoding histidine kinases (Chang *et al.*, 1993; Kakimoto, 1996), response regulators (Brandstatter and Kieber, 1998; Imamura *et al.*, 1998; Sakai *et al.*, 1998a; Urao *et al.*, 1998), and phosphotransfer proteins (Miyata *et al.*, 1998; Suzuki *et al.*, 1998) have all been identified in *Arabidopsis*. Thus all components needed for establishing a phosphorelay are represented in plants. In addition to two-component proteins implicated in specific signaling pathways, plants contain a number of two-component proteins whose functions in the plant are currently unknown. These may provide 'missing links' in some of the pathways just described. Of particular note is that some of the plant two-component proteins have diverged from the bacterial model. This is apparent with phytochromes and pyruvate dehydrogenase kinases, both related to histidine kinases, but which appear to have evolved new enzymatic activities (Thelen *et al.*, 1998; Yeh and Lagarias, 1998).

In this review, the main focus is on members of two-component systems that have been identified in *Arabidopsis*. *Arabidopsis* gives the most complete picture currently available on how these signaling systems are represented and function in plants. Results from other plant systems are brought in where they add to the picture. Among the two-component proteins, there will be a focus in detail on the ETR1 ethylene receptor. There exists a wealth of genetic, molecular, and biochemical information on ETR1 owing to it having been the first two-component protein identified in plants.

II. CONCEPTUAL FRAMEWORK

Much of what is known about two-component systems arises from studies in bacteria. These studies provide a wealth of information and serve as the basis for understanding eukaryotic two-component systems. However, it is to be expected that a system that evolved in prokaryotes will have undergone further evolutionary changes in its adaptation to eukaryotic signaling pathways. For this reason it is also worthwhile examining the best characterized eukaryotic two-component system: the osmosensing system of yeast.

A. BACTERIAL TWO-COMPONENT SYSTEMS

The two-component systems of bacteria confer the ability to sense and respond to environmental stimuli (Parkinson, 1993; Swanson *et al.*, 1994; Stock *et al.*, 1995). They are involved in such diverse processes as the 'swimming' of bacteria towards attractants and away from repellents (chemotaxis), osmotic sensing, oxygen sensing, and host recognition. The proteins involved in these systems vary considerably, which is to be expected given the different stimuli that are recognized. However, all these sensory systems contain two conserved elements, termed the histidine kinase and the response regulator, and for this reason are called two-component systems. Histidine kinase and response regulator domains can be incorporated into proteins in a variety of ways. A common strategy is to have one protein containing a sensing or receptor domain, as well as the histidine kinase domain (Fig. 1). This allows direct regulation of the histidine kinase by the stimulus. The response regulator is often present as a separate protein, and regulates downstream events in the signaling pathway.

Because they are involved in sensing environmental stimuli, the histidine kinase sensors are typically found in the cytoplasmic membrane of the bacterium (Parkinson, 1993). This places them in direct contact with the extracellular environment: the sensing domain on the extracellular side of the membrane, the histidine kinase domain on the intracellular side of the membrane. The histidine kinases function as dimers, and in response to the environmental stimulus, each histidine kinase phosphorylates its partner on a conserved histidine residue using adenosine triphosphate (ATP) as a phosphodonor. This phosphate is then transferred to a conserved aspartic

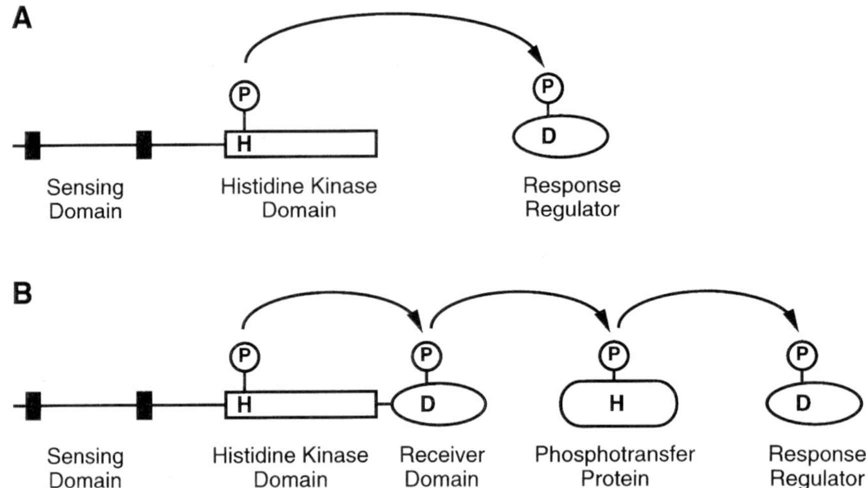

Fig. 1. Examples of two-component systems. Vertical black bars indicate transmembrane domains. The sites of phosphorylation (P) at histidine (H) and aspartate (D) residues are indicated. (A) A simple two-component system that incorporates a histidine kinase and a response regulator. (B) A multi-step phosphorelay that incorporates a hybrid kinase with both histidine kinase and receiver domains, a histidine-containing phosphotransfer protein, and a response regulator.

acid residue within the receiver domain of the response regulator, the second component in the two-component system (Fig. 1). Phosphorylation of the response regulator modulates its ability to mediate downstream signaling in the pathway. In bacteria many of the response regulators are transcription factors (Stock *et al.*, 1995). Thus, by use of two proteins one is able to create a signaling circuit, capable of converting an external stimulus into a change in transcription.

Some histidine kinases possess a phosphatase activity in addition to their histidine kinase activity (Igo *et al.*, 1989; Parkinson, 1993). The histidine kinase domain is not simply on or off, but rather switches back and forth between kinase and phosphatase activities, the enzymatic mode of the protein being dictated by binding of the signaling ligand. The phosphatase activity is targeted against the response regulator, and is responsible for shortening the lifetime of its phosphorylated form from hours to minutes. In this manner the two-component system is able to respond very rapidly to changes in concentration of the signaling ligand.

There are many permutations of the two-component system. These permutations give credence to the concept that the signaling motifs can function as modules to be mixed and matched as needed for a particular signaling pathway. Of particular relevance to the plant two-component systems are the 'hybrid' kinases (Swanson *et al.*, 1994). These contain both histidine kinase and receiver domains in one protein (Fig. 1). In these cases a separate response regulator operates downstream as well. The fused receiver domain has two possible roles: it may operate to modulate activity of the histidine kinase, or it may serve as one step in a phosphorelay signaling sequence (Appleby *et al.*, 1996). In such multi-step phosphorelays the phosphate is transferred from histidine kinase to receiver domain within the hybrid kinase, then from there to a histidine-containing phosphotransfer (HPt) protein, then finally to a response regulator. The phosphate is moved from amino acid to amino acid in the sequence His → Asp → His → Asp.

Members of two-component systems all have conserved sequence motifs, with different motifs being diagnostic for histidine kinases, response regulators, and phosphotransfer proteins. Histidine kinases contain conserved sequences around the histidine that serves as the site of autophosphorylation (termed the H-box), and within the region implicated in binding ATP (N, G1, F, and G2 boxes) (Parkinson, 1993). Structures of the histidine kinase domains from the bacterial proteins EnvZ and CheA have been reported, and these provide information as to how the conserved domains function in ATP binding and catalysis (Bilwes *et al.*, 1998; Tanaka *et al.*, 1998). Response regulators are all homologous within their receiver domain, with a core sequence of conserved regions and invariant amino acids (Volz, 1995). The structures for many response regulators have been determined (Volz, 1995). Although phosphotransfer proteins show the least overall similarity, they too have a conserved domain around the histidine that is phosphorylated (Appleby *et al.*, 1996). The conserved motifs found within elements of two-component systems serve both as diagnostics for function and as sequences to use in searching genome databases. The structures of histidine kinases will serve as the basis for rational drug design; some of these drugs will be useful as diagnostic agents to study plant histidine kinases and could potentially function as herbicides.

The bacterial systems have been the basis for defining the terminology used for two-component systems. However, this terminology has been modified and adapted based on our increasing awareness of the complexities possible with two-component systems. For instance, the term 'two-component system' originates from a time when only two components were understood to function in these signaling systems. Now with the numerous examples of multi-step phosphorelays, we have a third type of component present in the phosphotransfer protein. For this reason, the term 'histidyl-aspartyl phosphorelay' is sometimes used because this term emphasizes the enzymatic nature rather than the number of components. The term two-component system has been retained for this review because, with plants, proteins derived from two-component systems may no longer participate in histidyl-aspartyl phosphorelays. The presence of multi-step phosphorelays has also led to greater use of the term 'receiver'. The 'receiver' domain is the core structure found in both hybrid kinases and response regulators that participate in the aspartyl phosphorylation. The 'response regulator' is the protein that contains a receiver domain and functions in the last step of a histidyl-aspartyl phosphorelay.

B. THE YEAST OSMOSENSING SYSTEM

The yeast *Saccharomyces cerevisiae* contains a single histidine kinase of the two-component model (Ota and Varshavsky, 1993), something that is known with certainty following the complete sequencing of the yeast genome. This histidine kinase operates as part of a multi-step phosphorelay involving the well-established elements of two-component systems. Several features make the yeast system of particular interest. First, the two-component system feeds into and regulates a mitogen-activated protein kinase (MAP kinase) cascade (Maeda *et al.*, 1994; Posas *et al.*, 1996; Posas and Saito, 1998). Thus, two distinct signaling pathways, one associated with prokaryotes and one associated with eukaryotes, have been integrated together. Evidence exists that a similar signaling mechanism may operate in plants. Second, the multi-step phosphorelay bifurcates, thereby regulating multiple signaling pathways (Ketela *et al.*, 1998; Li *et al.*, 1998); this complexity may be commonplace in eukaryotic signaling systems.

Two osmosensing mechanisms have been identified in yeast, one activated by low osmolarity, the second activated by high osmolarity (Fig. 2). The low osmolarity sensing system makes use of a 'two-component system' and initiates at the SLN1 osmosensor (Ota and Varshavsky, 1993; Maeda *et al.*, 1994, 1995). SLN1 is a hybrid kinase containing both a histidine kinase domain and a receiver domain. SLN1 has two transmembrane domains. The high osmolarity sensing system initiates at another membrane protein SHO1 (Maeda *et al.*, 1995), which has four transmembrane domains and a C-terminus with an SH3 domain. SHO1 has no sequence related to proteins of the two-component system. The osmolarity response in yeast is therefore maintained at a delicate balance, the level of response dictated by the relative activity of two opposing sensory systems.

In response to low osmolarity, the kinase domain of SLN1 autophosphorylates on its conserved histidine residue; from here the phosphate is passed on to the receiver domain of SLN1 (Fig. 2) (Posas *et al.*, 1996). The phosphate is transferred from SLN1 to YPD1, a histidine-containing phosphotransfer protein (Posas *et al.*, 1996). The phosphate is next passed on to a response regulator, SSK1, the last element in this two-component phosphorelay (Posas *et al.*, 1996). At this point the two-component system intersects with and regulates the HOG1 MAP-kinase pathway, which serves to elicit the osmolarity responses (Maeda *et al.*, 1994; Posas and Saito, 1998). Phosphorylated SSK1 is not able to activate the SSK2/SSK22 MAP kinase kinase kinase (MAPKKK) (Posas and Saito, 1998). Thus, at low osmolarity,

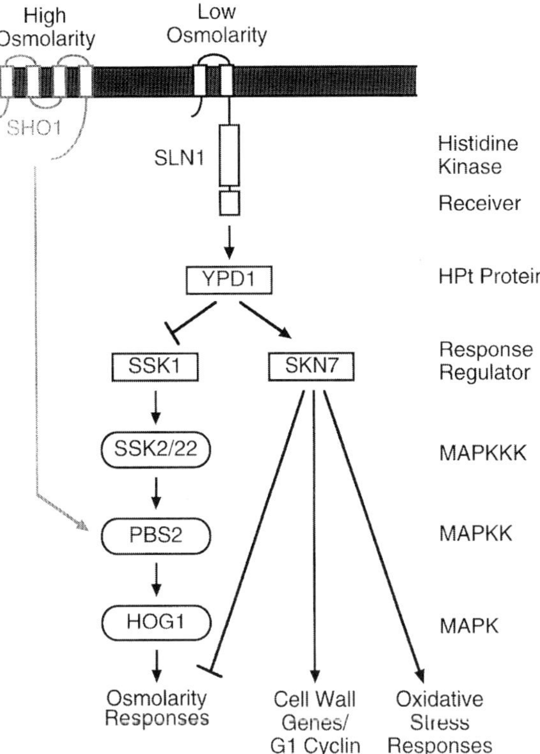

Fig. 2. The yeast osmosensing system. The low osmolarity signal is sensed by SLN1, then transduced through a multi-step phosphorelay. Signaling through the phosphorelay bifurcates at the YPD1 protein to regulate two response regulators (SSK1 and SKN7), thereby allowing the initial signal to affect several independent downstream pathways including the HOG1 MAP kinase pathway. The high osmolarity signal is transduced through the SHO1 protein that is not related to two-component systems.

the SSK1 response regulator is phosphorylated and activity of the HOG1 MAP kinase pathway is reduced.

The mechanism by which the yeast SSK2 MAPKKK is activated by the SSK1 response regulator has been examined (Posas and Saito, 1998). These proteins directly interact, and the regions of interaction have been defined by two-hybrid analysis. The receiver domain of the SSK1 response regulator interacts with a site near the N-terminus of SSK2. This interaction site on SSK2 is separate from the C-terminal kinase domain, a location consistent with the model for activation of MAPKKKs. In this model, the N-terminal noncatalytic domains of MAPKKKs are autoinhibitory for kinase activity. Kinase activity of MAPKKKs can be stimulated by deleting the N-terminal domain or by binding of regulators such as SSK1 to the domain. A two-step model is proposed for the activation of SSK2 by SSK1 (Posas and Saito, 1998).

Initially, binding of the unphosphorylated SSK1 response regulator induces a conformational change in SSK2, thereby activating kinase activity of SSK2. Following autophosphorylation on serine/threonine residues of SSK2, kinase activity of SSK2 becomes independent of SSK1.

The HOG1 pathway is not the only pathway regulated through the SLN1 osmosensor. Following phosphotransfer from SLN1 to the YPD1 histidine-containing phosphotransfer protein, the phosphorelay bifurcates (Fig. 2). Phosphotransfer to the SSK1 response regulator serves to reduce activity of the HOG1 pathway (Posas and Saito, 1998). Phosphotransfer to the SKN7 response regulator serves to stimulate expression of genes for cell wall biosynthesis and G1 cyclin (Ketela *et al.*, 1998; Li*et al.*, 1998). Unlike the SSK1 response regulator that interacts with a MAPKKK, SKN7 contains a domain that may mediate DNA interactions, thereby allowing for a direct regulation of gene expression. SKN7 can also regulate the HOG1 pathway for osmolarity responses in a manner that is dependent on its phosphorylation (Ketela *et al.*, 1998). Interestingly, SKN7 also appears to activate the expression of oxidative stress genes in a manner that is independent of two-component systems (Li *et al.*, 1998). Therefore, even though SKN7 is an active response regulator, it has both two-component dependent and independent roles in regulating gene expression.

III. HISTIDINE KINASES IN PLANTS

A variety of histidine kinases has been identified in plants. Some have been identified through mutant screens in *Arabidopsis* designed to uncover elements of signaling pathways. The strength of this genetic approach is that any histidine kinase identified is immediately implicated as performing a role within a particular pathway. Thus, through mutant screens, histidine kinases have been implicated in both ethylene and cytokinin responses. Other histidine kinases have been identified using sequence-based approaches that take advantage of the conserved domains found in all histidine kinases. For example, histidine kinases have been identified by direct examination of DNA sequence resulting from the genome sequencing projects and also isolated by the polymerase chain reaction (PCR) using primers to the conserved domains. The challenge now is to assign roles to these kinases.

What is apparent from an examination of the plant sequences is that while some of the histidine kinase domains contain all the conserved residues indicative of histidine kinase activity, others lack residues considered essential for activity. This situation has been found even within the single family of proteins involved in ethylene perception, which contains bona fide histidine kinases as well as diverged histidine kinase-like porteins. Phytochromes and pyruvate dehydrogenase kinases offer additional examples of how diverged histidine kinase-like proteins can be incorporated into regulatory pathways.

Phylogenetic analysis supports the functional differences among the plant histidine kinase family (Fig. 3), because proteins predicted to function as histidine kinases are more closely related to each other than to the diverged histidine kinase-like proteins.

A. INVOLVEMENT IN ETHYLENE RESPONSES

Ethylene is a simple two-carbon gas important at many stages of the plant's life, including seed germination, seedling growth, leaf and petal abscission, organ senescence, and pathogen responses (Abeles *et al.*, 1992). A family of at least five proteins (ETR1, ETR2, ERS1, ERS2, and EIN4) functions in ethylene perception by *Arabidopsis* (Solano and Ecker, 1998). ETR1 and ERS1 have been shown to act as ethylene receptors based on their ability to bind ethylene (Schaller and Bleecker, 1995; Hall *et al.*, 2000). The structural features of these proteins are depicted in Fig. 3. All five contain a histidine kinase-like domain. In two of these proteins (ETR1 and ERS1) there is complete conservation of the residues considered essential for histidine kinase activity (Chang *et al.*, 1993; Hua *et al.*, 1995); in the others (ETR2, ERS2, and EIN4) the histidine kinase domains lack residues considered essential for activity (Hua *et al.*, 1998; Sakai *et al.*, 1998b). The histidine kinase domains of ETR1 and ERS1 are more closely related to other histidine kinases than to the analogous domains of ETR2, ERS2, and EIN4 (Fig. 3). Several of the ethylene receptors are hybrid kinases, containing a receiver domain in addition to the histidine kinase-like domain. Similar proteins have been implicated in ethylene signaling by other plant species (Wilkinson *et al.*, 1995; Zhou *et al.*, 1996; Sato-Nara *et al.*, 1999; Tieman and Klee, 1999), indicating that the mechanism of ethylene perception is conserved in plants.

Of all the histidine kinases in plants, the ETR1 ethylene receptor of *Arabidopsis* has been characterized in the most detail, owing to it having been the first two-component protein identified in plants (Chang *et al.*, 1993). ETR1 contains transmembrane domains and an ethylene-binding site in the N-terminal region. ETR1 contains regions with homology to histidine kinases and response regulators in the C-terminal region (Fig. 3), and histidine kinase activity has been demonstrated for the enzyme (Gamble *et al.*, 1998). ETR1 is thus structurally and biochemically similar to its bacterial counterparts in that it is membrane localized, contains a sensory domain near the N-terminus, and has an active histidine-kinase output domain. Based on this similarity in design, binding of ethylene or other signaling factors may regulate activity of the histidine kinase domain.

1. Transmembrane Domain

The ethylene receptor-like proteins have hydrophobic N-terminal regions. Both ETR1 (Chang *et al.*, 1993) and ERS1 (Hua *et al.*, 1995) are predicted to have

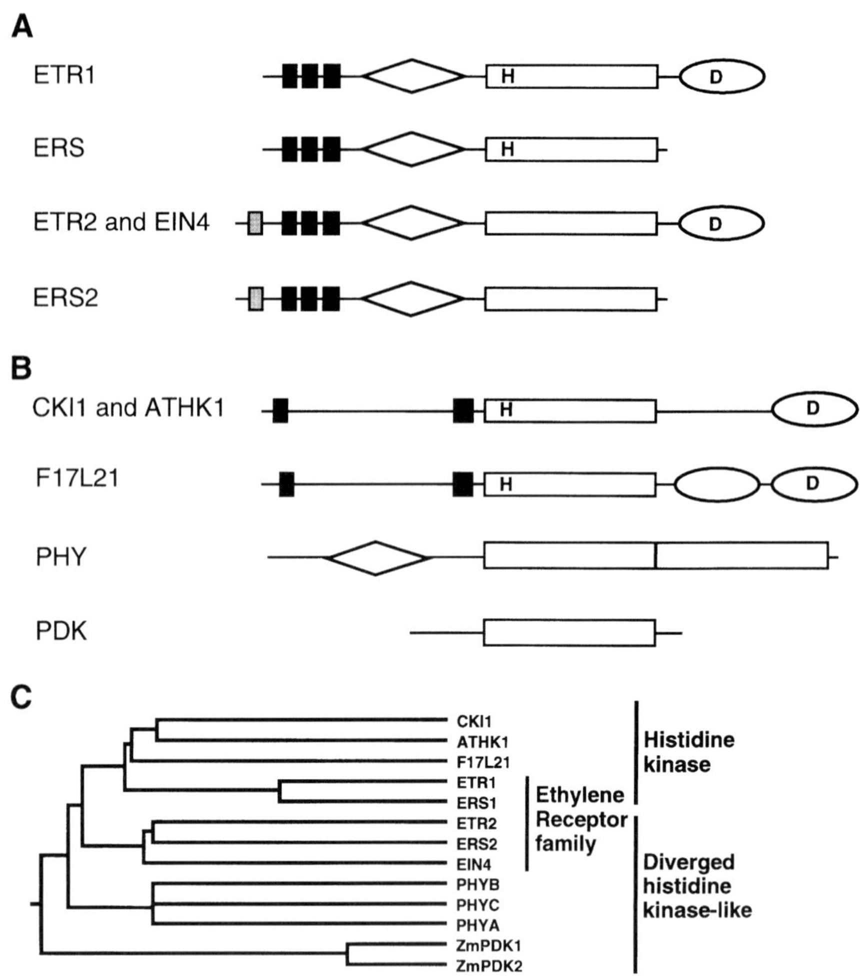

Fig. 3. Histidine kinases and related proteins of plants. Gray bars indicate predicted signal sequences. Black bars indicate predicted transmembrane domains. Triangles indicate GAF domains. Rectangles indicate histidine kinase domains. An 'H' in the rectangle indicates that the domain has all the residues predicted to encode an active histidine kinase, including the histidine predicted to act as the site of phosphorylation; lack of an 'H' indicates that the domain is diverged from the canonical histidine kinase. Ovals indicate receiver domains. A 'D' in the oval indicates that the domain has all the residues required for the receiver to be active, including the aspartate residue that serves as the site of phosphorylation. (A) Structural features of the ethylene receptor family from *Arabidopsis*. (B) Structural features of other histidine-kinase-like proteins of plants. (C) Phylogenetic analysis of the plant histidine kinase family. The dendogram was generated from the histidine kinase domains using MegAlign (DNAStar) with Clustal. All sequences are from *Arabidopsis* with the exception of PDK1 and PDK2 from maize (Zm).

three membrane-spanning segments. Algorithms for topological prediction place the N-terminus of these proteins on the extra-cytosolic side of the membrane (Schaller *et al.*, 1995), followed by the three transmembrane domains, leaving the bulk of the protein exposed on the cytosolic side of the membrane. This topology would place the histidine kinase domain in the correct context for interaction with cytosolic signaling factors.

ETR2, ERS2, and EIN4 have three predicted transmembrane domains that correspond to those in ETR1 and ERS1 based on relative position and sequence conservation (Hua *et al.*, 1998; Sakai *et al.*, 1998b). However, ETR2, ERS2, and EIN4 also have an extra hydrophobic stretch of amino acids preceding the predicted transmembrane domains. This extra hydrophobic stretch shows little sequence conservation between the different isoforms and is predicted to be a cleaved signal sequence, indicating that these proteins may be targeted to the secretory pathway.

Consistent with predictions from hydropathy analysis, ETR1 has been shown to be a membrane protein (Schaller *et al.*, 1995). Antibodies generated against ETR1 recognize a 79-kDa polypeptide in the membrane fraction of *Arabidopsis*, in good agreement with the 83 kDa predicted from the amino acid sequence. The specific membrane localization of the ethylene receptors is not yet known. Ethylene is readily diffusable through both aqueous and lipid environments, so there is no a priori necessity for the receptor to be localized to the plasma membrane. However, the signal sequences present in some of the ethylene receptor-like proteins suggest that they enter the secretory pathway, in which case transport to the plasma membrane may be the default if no retention signals are present in the protein sequence. All the ethylene receptor-like proteins could be localized to the same membrane system, even though signal sequences are present in only a subset of the ethylene receptor-like proteins; cleavable signal sequences are not necessarily required for transit though the secretory system (Gennis, 1989).

2. *Ethylene-Binding Domain*

The site of ethylene binding by ETR1 has been localized to the hydrophobic N-terminal region (Schaller and Bleecker, 1995; Rodriguez *et al.*, 1999). Ethylene binding by ETR1 was determined using a transgenic yeast system. Expression of ETR1 in yeast resulted in the creation of high-affinity binding sites for [^{14}C]ethylene; control yeast showed no saturable binding sites for ethylene. Kinetic analysis of ethylene binding by ETR1 in transgenic yeast yielded a dissociation constant (K_d) for ethylene of 0.04 μl l^{-1} of ethylene (gas phase), a value consistent with concentrations of ethylene known to induce growth responses in plants (Schaller and Bleecker, 1995). Expression of truncated versions of ETR1 in yeast has been used to localize the ethylene-binding site, which lies in the first 128 amino acids of the polypeptide (Schaller and Bleecker, 1995; Rodriguez *et al.*, 1999).

It has long been hypothesized that an ethylene receptor would require a metal co-factor, copper being considered the most likely candidate (Burg and Burg, 1967). The strength of an ethylene–metal complex is such that a stable ethylene-binding site could be formed from which the ethylene molecule could also dissociate, thereby allowing for reversible binding. The presence of a copper co-factor is supported by experiments with the yeast-expressed ETR1 protein; ethylene-binding capability of ETR1 is lost during purification from yeast, but can be reconstituted by the addition of copper (Rodriguez *et al.*, 1999). Silver ions can also reconstitute ethylene binding, but to a lesser extent (Rodriguez *et al.*, 1999). This last result is of interest because silver is a known inhibitor of ethylene responses in the plant. Silver may be able to replace copper at the binding site, still interact with ethylene, but fail to induce the conformational change required for signaling. *In vivo* evidence for a copper requirement by the receptors comes from the genetic identification of a copper-transporting ATPase (RAN1) required for ethylene signaling (Hirayama *et al.*, 1999); RAN1 may deliver the necessary copper co-factor to the ethylene receptors. Two amino acids of ETR1 have been directly implicated in co-ordinating the metal ligand: Cys65 and His69 (Schaller and Bleecker, 1995; Rodriguez *et al.*, 1999). These lie on the second proposed transmembrane helix, one helical turn apart. This would place them on the same face of an α helix, adjacent to each other, and would allow both to interact with the same copper molecule. These two amino acids are conserved in all the ethylene-binding proteins belonging to this receptor family.

Some mutations that affect ethylene signal transduction in plants result from single amino-acid changes within the ethylene-binding domain (Chang *et al.*, 1993). These mutations are dominant and lead to ethylene insensitivity in the plant. Such mutations could theoretically either prevent the binding of ethylene or, if ethylene binding does occur, interfere with signaling. Evidence for the first possibility has been found (Schaller and Bleecker, 1995). The *etr1-1* mutation results in a conversion of Cys65 to tyrosine (Tyr). As previously mentioned, Cys65 has been implicated in co-ordinating a metal ligand involved in ethylene binding. Analysis of the etr1–1 protein demonstrates that it has indeed lost the ability to chelate copper (Rodriguez *et al.*, 1999), and is also no longer capable of binding ethylene (Schaller and Bleecker, 1995; Rodriguez *et al.*, 1999).

The ethylene-binding domain may have had its evolutionary origin with the cyanobacteria that gave rise to present-day chloroplasts. The complete genome of the cyanobacterium *Synechocystis* (strain 6803) has been sequenced and found to contain a gene (*slr1212*) that encodes a protein with sequence and structural characteristics in common with ETR1 (Rodriguez *et al.*, 1999). Sequence conservation is particularly evident with the ethylene sensor domain of ETR1. The slr1212 protein confers the ability to bind ethylene upon *Synechocystis* (Rodriguez *et al.*, 1999), but it is not clear whether that is its normal function, as *Synechocystis* is not known to have ethylene responses. It may be that the ethylene-binding domain of slr1212 normally functions as a copper sensor. A fortuitous capability for binding ethylene could have been exploited during the evolution of plants.

3. Histidine Kinase Domain

ETR1 and ERS1 both contain histidine kinase domains in which all the amino acids diagnostic for activity are conserved (Chang *et al.*, 1993; Hua *et al.*, 1995). Histidine kinase activity has been demonstrated for ETR1 (Gamble *et al.*, 1998). This was accomplished by transgenic expression and purification of the histidine kinase domain from yeast. The purified kinase domain autophosphorylated upon incubation with radiolabeled ATP. The incorporated phosphate was resistant to alkali treatment but was sensitive to acid treatment, characteristics consistent with the phosphorylation of histidine. Autophosphorylation was abolished by a site-directed mutation of the conserved histidine that serves as the presumptive site of phosphorylation. Autophosphorylation was also abolished by a site-directed mutation within the G1 box of the catalytic domain. An examination of cation requirements indicated that ETR1 requires Mn^{2+} for autophosphorylation (Gamble *et al.*, 1998); Mg^{2+} more commonly fulfills the requirement of histidine kinases for a divalent cation, although there are examples in bacteria where Mn^{2+} is preferred (Hess *et al.*, 1991). These results demonstrate that higher plants contain proteins with histidine kinase activity. These results also indicate that aspects of ethylene signaling may be regulated by changes in histidine kinase activity of the receptor.

ETR2, ERS2, and EIN4 all contain diverged histidine kinase-like domains that lack residues considered essential for kinase activity (Hua *et al.*, 1998; Sakai *et al.*, 1998b). Their involvement in ethylene signaling suggests that histidine kinase activity may only be involved in a subset of responses to ethylene, with that subset of responses being mediated by ETR1 and ERS1. Alternatively, the diverged receptors, although lacking histidine kinase activity themselves, may rely upon the activity of ETR1 and ERS1 for proper signal transduction.

4. GAF Domain

All five ethylene receptor-like proteins of *Arabidopsis* contain a GAF domain located between the ethylene-sensing domain and the histidine kinase domain (Aravind and Ponting, 1997). GAF domains are signaling motifs found in a variety of proteins of eubacteria and eukaryotes, where they have been implicated in binding of ligands and small molecules (Aravind and Ponting, 1997). One well-characterized function of GAF domains is that of cGMP binding. In addition, they are present at the chromophore attachment site of phytochromes and a conserved cysteine residue required for the chromophore attachment. Interestingly, the slr1212 protein of *Synechocystis* contains a GAF domain with the conserved residue for chromophore attachment (Rodriguez *et al*., 1999), indicating that it may be able to perceive both ethylene and light. The GAF domains of plant ethylene receptor-like proteins lack the cysteine used in chromophore attachment. The presence of GAF domains in the ethylene receptor-like proteins suggests that other agents besides ethylene may regulate signaling.

5. *Dimerization*

Bacterial histidine kinases function as dimers, with the kinase of one monomer phosphorylating the conserved histidine of the other monomer *in trans* (Parkinson, 1993). The dimeric form appears to be in place prior to signal perception; binding of ligand induces a conformational change in the dimer that affects the ability to trans-phosphorylate. Biochemical characterization of the ETR1 protein of *Arabidopsis* indicates that it too forms a dimer (Schaller *et al*., 1995). When extracts from Arabidopsis were analyzed by sodium dodecyl sulfate–polyacrylamide gel electrophoresis (SDS–PAGE) and Western blotting, immunodetectable ETR1 protein migrated at 79 kDa (monomer) in the presence of a reducing agent, but at 147 kDa (dimer) in the absence of a reducing agent. The sensitivity to reducing agent indicates that the ETR1 dimer contains disulfide linkages. The ETR1 protein behaved similarly when expressed in transgenic yeast, and this system was used to define further the requirements for dimerization (Schaller *et al*., 1995). Expression of truncated forms of ETR1 indicated that the N-terminal domain was necessary and sufficient for dimerization. Site-directed mutagenesis indicated that a pair of cysteine residues (Cys4 and Cys6) is required for formation of the disulfide-linked dimer. These cysteines are conserved in the other plant ethylene receptor-like proteins. The finding that ETR1 forms a dimer indicates that ETR1 may employ a similar signaling mechanism to that of bacterial histidine kinases. Noncovalent associations maintain dimers of bacterial histidine kinases (Pan *et al*., 1993; Parkinson, 1993) and these could also play a part in formation of the ETR1 dimer, with additional stabilization potentially provided by the disulfide linkage (Schaller *et al*., 1995).

6. *Interaction of Ethylene Receptors with CTR1*

CTR1 belongs to the Raf family of protein serine/threonine kinases, which would make CTR1 a MAPKKK (Kieber *et al*., 1993). CTR1 contains a large N-terminal domain (amino acids 1–550) that may function as a regulatory domain, followed by the C-terminal kinase domain (amino acids 551–821). Based on genetic analysis, CTR1 has been implicated in ethylene responses (Kieber *et al*., 1993). Null mutants of the gene result in constitutive ethylene responses. Double mutant analysis positions CTR1 downstream of the ethylene receptors in the pathway for ethylene signal transduction. Thus, the pathway for ethylene signal transduction involves elements of two-component systems and elements of MAP kinase pathways.

A physical interaction of ethylene receptors with CTR1 has been demonstrated by two-hybrid analysis as well as by *in vitro* protein association assays (Clark *et al.*, 1998). Both ETR1 and ERS1 were found to interact with CTR1. Truncated proteins were used to localize the sites of interaction. The histidine kinase domains of ETR1 and ERS1, as well as the receiver domain of ETR1, were capable of interacting with CTR1. The site of CTR1 interaction was localized to between amino acids 171 and 521, and thus falls within the N-terminal region that is hypothesized to function as a regulatory domain for MAPKKKs (Clark *et al.*, 1998). These results demonstrate that ethylene receptors not only are capable of forming self-associations, as has been found in ETR1 dimerization (Schaller *et al.*, 1995), but can also associate with downstream signaling components. As a result of this direct interaction, signaling could involve conformational changes propagated between elements of the receptor complex.

7. *Loss-of-Function Analysis*

To assess clearly the role of genes in a physiological response, loss-of-function mutations are required. By using several complementary methods, loss-of-function mutations have been isolated in four (ETR1, ETR2, EIN4, and ERS2) of the five ethylene receptor-like proteins (Hua and Meyerowitz, 1998). The effects of the mutations upon ethylene signal transduction have been characterized individually and in combination.

Single loss-of-function mutants had little or no effect upon ethylene signal transduction and all displayed similar ethylene sensitivity when compared to wild-type plants (Hua and Meyerowitz, 1998). Only the *etr1* loss-of-function mutants displayed a slight growth effect, characterized by an overall slight reduction in cell elongation that could be discriminated in seedlings and plants.

In combination with an *etr1* loss-of-function mutant, the other loss-of-function mutants began to show more substantial effects upon ethylene responses (Hua and Meyerowitz, 1998). The ethylene responses activated in the mutant plants include reduced leaf expansion, reduced root elongation, increased root hair formation, and upregulation of the ethylene-induced gene *chitinase B* (*CHIB*). Double loss-of-function mutants displayed a weak ethylene response phenotype in air. Triple loss-of-function mutants displayed strong ethylene response phenotypes in air. Quadruple loss-of-function mutants displayed a constitutive ethylene response phenotype in air, and died before bolting. These results indicate that there is functional redundancy among the receptor family members. In addition, these results indicate that the receptors serve as negative regulators of the ethylene response pathway, for elimination of receptors activates the pathway.

8. *Expression Analysis*

Dominant mutations in any of the *Arabidopsis* ethylene receptor-like genes affect ethylene responses throughout the plant, without any apparent tissue specificity. This result suggests ubiquitous expression of the genes. Northern and *in situ* analysis confirms this general pattern of expression throughout the plant, with some members of the family showing higher expression in certain cell types (Hua *et al.*, 1998; Sakai *et al.*, 1998b). Given the role of these genes in ethylene signal transduction, the effects of ethylene on expression have also been tested (Hua *et al.*, 1998). Northern analysis performed on leaf tissue indicates that *ETR2, ERS1*, and *ERS2* are all induced about six-fold after 12 h treatment with ethylene. *ETR1* and *EIN4* did not show significant induction by ethylene at this time point.

Expression of ethylene receptor-like genes has also been examined in tomato (Wilkinson *et al.*, 1995; Payton *et al.*, 1996; Zhou *et al.*, 1996; Lashbrook *et al.*, 1998; Tieman and Klee, 1999), a system that has advantages over *Arabidopsis* for the physiological characterization of ethylene responses. Not only is the plant larger than *Arabidopsis*, thereby facilitating tissue dissection, but tomato fruit ripening is dependent upon ethylene as a hormonal trigger. The *NR* (*LeETR3*) gene of tomato (similar to the *Arabidopsis ERS1* gene) is of particular interest as it shows considerable variability in expression (Wilkinson *et al.*, 1995; Lashbrook *et al.*, 1998). Expression of the *NR* gene is induced by ethylene, but the extent of this induction varies with both tissue and developmental stage of the plant. The most striking change in *NR* expression occurs in ripening tomato fruit. Prior to ripening, transcripts for various ethylene receptor-like genes (*NR, LeETR1*, and *LeETR2*) are present at low but relatively constant levels (Lashbrook *et al.*, 1998). However, upon exposure of mature green tomato fruit to either endogenous or exogenous ethylene, the *NR* transcript is specifically induced, reaching maximum abundance during the light red stage of the fruit (Wilkinson *et al.*, 1995; Lashbrook *et al.*, 1998). *NR* messenger RNA (mRNA) was induced at least 25-fold, and its abundance closely paralleled the level of ethylene production by the tomato. What is interesting is that whereas *NR* mRNA is strongly induced by ethylene during fruit ripening, increases in endogenous ethylene production had much less effect upon *NR* expression during seed hydration and shoot elongation (Lashbrook *et al.*, 1998). These results demonstrate that regulation of *NR* expression is under tight developmental control.

Increased expression of ethylene receptors could be significant to ethylene signal transduction in several ways. The increased expression of specific family members could allow for regulation of a particular subset of ethylene signaling pathways in the plant. For example, *NR* might have a specific role in the tomato fruit that cannot be substituted for by other ethylene receptors. In addition, because the receptors negatively regulate the ethylene responses (Hua and Meyerowitz, 1998), an increase in the number of ethylene receptors could result in desensitization of the cell to ethylene; thus, ethylene would set up a feedback loop whereby the cells become less sensitive to ethylene. In a more general context, these studies indicate that two-component signaling systems may be regulated at the expression level by the signal that they transduce.

B. INVOLVEMENT IN CYTOKININ RESPONSES

Cytokinins play important roles throughout growth and development. Cytokinins regulate metabolism, stimulate branching and chlorophyll synthesis, induce flowering, and delay senescence (Binns, 1994). CKI1, a protein with homology to histidine kinases, has been implicated in cytokinin responses (Kakimoto, 1996). In addition, proteins with homology to response regulators are induced by cytokinins, suggesting that they may also function in cytokinin signal transduction (Kakimoto, 1998). The CKI1 histidine kinase is addressed here and the possible role of response regulators in cytokinin signaling is covered in Section IV.

The *CKI1* gene was identified based on its ability, when overexpressed, to yield plants that showed constitutive cytokinin responses (Kakimoto, 1996). Like ETR1, the CKI1 protein has homology to histidine kinases and the receiver domains of response regulators (Fig. 3). The N-terminal half of the CKI1 protein contains two hydrophobic domains, each capable of spanning a membrane. The histidine kinase and receiver domains of CKI1 are in the C-terminal half of the protein, and are predicted to be on the cytoplasmic side of the membrane. Both histidine kinase and receiver domains contain all amino acids known to be necessary for activity, including the conserved sites for phosphorylation. Based on these characteristics, CKI1 is a hybrid kinase (Kakimoto, 1996).

The exact role that CKI1 plays in cytokinin signaling is not yet known, nor has its expression been examined in wild-type plants, owing to its low abundance. Given the similarity of CKI1 to the bacterial sensing proteins and the precedent set by the ETR1 ethylene receptor, CKI1 may represent a cytokinin receptor (Kakimoto, 1996). The mutant phenotype in plant cells that overexpress CKI1 could then be explained in the following manner: high levels of a cytokinin receptor would potentially allow plant cells to respond to low concentrations of cytokinins, so even the endogenous levels of cytokinin could induce shoot formation. However, it is also possible that CKI1 does not normally function in cytokinin signaling, but rather in another pathway. The mutant phenotype observed in plants overexpressing CKI1 could be due to aberrantly high levels of the kinase activating a pathway in which it is normally not involved. This would represent an example of 'cross-talk', a phenomenon known to occur with histidine kinases (Stock *et al.*, 1989). Regardless of its exact role, it should be emphasized that the work with CKI1 demonstrates that two-component systems perform an important function in cytokinin signal transduction.

C. INVOLVEMENT IN OSMOSENSING

Plants respond and adapt to the varying levels of water in their environment, a necessity given that water is essential to all life and that plants are immobile. Dehydration and water stress can arise from drought, high salinity, or low-temperature conditions. Changes in the osmotic environment of the cell are perceived and then propagated through abscisic acid-dependent and independent pathways to effect changes in gene expression. Some of the induced gene products are directly involved in stress tolerance and encode such proteins as water channels and enzymes required for biosynthesis of osmoprotectants. Other genes encode regulatory proteins such as protein kinases and transcription factors (Shinozaki and Yamaguchi-Shinozaki, 1997).

Osmosensing pathways in bacteria (Pratt and Silhavy, 1995) and yeast (Maeda *et al.*, 1994; Posas *et al.*, 1996) are regulated by two-component systems, and this has prompted a search for a similar system in plants. Using primers based on conserved domains of histidine kinases, an *Arabidopsis* histidine–kinase homologue was isolated by PCR. This gene (*ATHK1*) contains two putative membrane-spanning domains flanking a predicted extracytoplasmic domain, a histidine kinase domain, and a receiver domain (Fig. 3). Thus ATHK1 has the features of a hybrid kinase (Shinozaki and Yamaguchi-Shinozaki, 1997; Urao *et al.*, 1999). ATHK1 has been implicated in osmosensing based on two lines of evidence. First, expression analysis by Northern blot indicates that it is expressed predominantly in the roots, and that its expression is induced by dehydration. Second, overexpression of ATHK1 can complement a yeast mutant lacking its osmosensing histidine kinase (Shinozaki and Yamaguchi-Shinozaki, 1997; Urao *et al.*, 1999).

In yeast, osmosensing involves elements of two-component systems and a MAP kinase pathway. MAP kinase cascades have also been implicated in plant stress responses. In *Arabidopsis*, genes for a MAPKKK (*ATMEKK1*) and a MAPK (*ATMPK3*) are induced by drought, high salinity, low temperature, and touch (Mizoguchi *et al.*, 1996). An alfalfa MAPK (MMK4), is regulated at both the transcriptional and post-translational levels by drought and cold (Jonak *et al.*, 1996). These results suggest that plant osmosensing may link a two-component signal transduction system to a MAP kinase cascade, perhaps in a similar fashion to that observed with the yeast osmosensing system (Fig. 2).

D. PYRUVATE DEHYDROGENASE KINASES

Nearly all biochemistry courses mention the mitochondrial pyruvate dehydrogenase complex when covering metabolism, usually shortly after glycolysis and before the Krebs cycle. In eukaryotes, the mitochondrial pyruvate dehydrogenase complex (PDH complex) is responsible for the oxidative decarboxylation of pyruvate. The products of this reaction, acetyl-coenzyme A (acetyl-CoA) and NADH, then enter the Krebs cycle and oxidative phosphorylation. The activity of the PDH complex is regulated by reversible phosphorylation mediated by the pyruvate dehydrogenase kinase (PDH kinase), with phosphorylation of the PDH complex on serine residues serving to inhibit the enzyme (Popov *et al.*, 1993, 1997). Regulation of this metabolic step allows cells to respond to changing energy requirements, as well as to alternative sources for acetyl-CoA and NADH. PDH kinase, first cloned from rat, is a soluble protein localized to the mitochondrial matrix where it associates with the PDH complex (Popov *et al.*, 1993, 1997). Although the kinase was known to phosphorylate the PDH complex on serine residues, the kinase lacked all the diagnostic motifs for a eukaryotic serine/threonine kinase. Instead it contains the diagnostic sequences for histidine kinases. The PDH kinase even has the conserved histidine that serves as a site for autophosphorylation in histidine kinases. This information established PDH kinase as a new type of serine/threonine kinase.

In plants there are two separate PDH complexes, one located in the mitochondrial matrix as in animals, and a second located in the plastid stroma (Thelen *et al.*, 1998). Although catalyzing the same reaction, both complexes have their own unique features. Only the mitochondrial PDH complex is targeted for regulation by the PDH kinase; phosphorylation (inactivation) of the PDH complex occurs in the light. Two plant PDH kinases, named PDK1 and PDK2, have been characterized from maize (Fig. 3) (Thelen *et al.*, 1998). The two isoforms have different expression patterns: PDK1 is expressed at similar levels in etiolated shoots, roots, leaves, husks, and ear shoots, while PDK2 is expressed predominantly in leaves. High expression of PDK2 in leaves may be related to light regulation of the PDH complex. Like their animal counterparts, both PDK1 and PDK2 have the conserved motifs indicative of histidine kinases (Thelen *et al.*, 1998). They also have several other regions of conservation within their kinase domain particular to the PDH-kinases; some of these sequences may confer serine/threonine phosphorylation capabilities upon the enzyme. Even though the maize PDK enzymes have all the residues considered essential for histidine kinase activity, it is clear from phylogenetic analysis (Fig. 3) that they have diverged from the histidine kinases predicted to participate in histidyl-aspartyl phosphorelays.

The PDH kinases represent an example in which histidine kinases have been adapted to new enzymatic capabilities. Adaptation to serine/threonine phosphorylation may not be that great an evolutionary stretch. Histidine kinases belong to a superfamily of ATPases, with the ATP binding portion of histidine kinases being similar to that of DNA gyrase B (Bilwes *et al.*, 1998; Tanaka *et al.*, 1998). These enzymes differ in that GyrB hydrolyzes ATP, while histidine kinases must also be capable of phosphate transfer. GyrB contains residues required for nucleophilic attack upon the ATP γ-phosphate within the ATP-binding site. With histidine kinases, it is the conserved histidine that participates in nucleophilic attack and that becomes phosphorylated. Serine or threonine residues could conceivably be placed into the correct context to be phosphorylated by a similar mechanism. The branched chain α-keto acid dehydrogenase kinase has similar sequence and enzymatic features to the PDH kinase (Popov *et al.*, 1992, 1997), establishing these enzymes as a new family of eukaryotic protein kinases which resemble histidine kinases but function as serine/threonine kinases.

E. PHYTOCHROMES

The phytochrome family of photoreceptors allows plants to regulate development in response to light (Quail, 1997b). Phytochromes monitor relative levels of red and far-red wavelengths of light, which serve as indicators for the time of day, the season, and the amount of shade. Phytochromes exist in two photoconvertible forms: Pr, the red light absorbing form; and Pfr, the far-red light absorbing form. Upon absorbing red light, the Pr form converts to the Pfr form; upon absorbing far-red light, the Pfr form converts to the Pr form. The Pfr form is generally considered to be the active form and is implicated in such diverse plant responses as seed germination, de-etiolation, shade avoidance, and flowering.

In *Arabidopsis*, there are five members of the phytochrome family (PhyA through to PhyE), all with similar structural features (Fig. 3) (Quail, 1997a). The phytochromes are soluble proteins of about 125 kDa, with two major domains connected by a short linker. Within the N-terminal domain, phytochrome contains a covalently attached tetrapyrrole chromophore; the site of attachment for the chromophore is a conserved cysteine residue within a GAF sequence (Aravind and Ponting, 1997). Within the C-terminal domain, phytochromes contain sequences related to histidine kinases. However, in none of the plant phytochromes is there a histidine kinase-like sequence that contains all the amino acid residues diagnostic for histidine kinase activity (Quail, 1997b). In fact, analysis indicates that plant phytochromes contain not one but two diverged histidine kinase-like sequences in a tandem arrangement with each other (Yeh and Lagarias, 1998). Phytochromes are dimers, and the determinants for dimerization have been localized to a region near the C-terminus of the phytochrome polypeptide (Quail, 1997a).

The biochemical mechanism of action for phytochromes has been hotly debated. Sensory perception by phytochromes is reasonably well characterized in how wavelength, intensity, and duration of light elicit the photoconversion between the Pr and Pfr forms (Quail, 1997a). The chief question is the nature of the output signal. In other words, what is the mechanism by which phytochromes regulate activity of proteins downstream in the signal transduction pathway? It was postulated that light might regulate an enzymatic activity of phytochrome, and serine/threonine kinase activity has been found associated with highly purified preparations of phytochrome (R. W. McMichael and Lagarias, 1990). The finding that the proposed output domain of phytochrome was related to histidine kinases gave further credence to the kinase hypothesis (Schneider-Poetsch *et al.*, 1991). However, it was unclear how the diverged histidine kinase domain would operate. Significantly, site-directed mutations of conserved histidine–kinase residues present in oat PhyA did not affect biological activity of the phytochrome when transgenically expressed in *Arabidopsis* (Boylan and Quail, 1996; Quail, 1997a).

The question of enzymatic activity of plant phytochrome was re-addressed following complete genome sequencing of several cyanobacteria. In particular, the cyanobacterium *Synechocystis* contains a functional phytochrome histidine kinase (Cph1) along with a cognate response regulator (Rcp1) (Hughes *et al.*, 1997; Yeh *et al.*, 1997). Histidine kinase activity of Cph1 is light regulated: the Pr form of Cph1 autophosphorylates on the conserved histidine residue, and this phosphate is then transferred to the aspartate residue in the Rcp1 response regulator (Yeh *et al.*, 1997).

Plant phytochrome thus has structural features similar to those of many histidine kinases: an input domain for a 'ligand' (light); an output domain with similarity to the histidine kinases; and a dimeric unit of signaling. The finding that the ancestral form of plant phytochrome is a functioning histidine kinase has strengthened the protein kinase hypothesis for phytochrome action (Yeh *et al.*, 1997). As a further test of the enzymatic hypothesis, oat and green algal phytochromes were expressed in and purified from yeast. Both recombinant phytochromes autophosphorylated in a light-dependent manner, greater autophosphorylation being observed with the Pfr form. Based on acid/base stability of the incorporated phosphate, phosphorylation was on serine/threonine residue(s) (Yeh and Lagarias, 1998). These data support the hypothesis that kinase activity of plant phytochromes has diverged from the histidine kinase ancestor, with the new domain allowing for serine/threonine phosphorylation.

The identification of kinase activity associated with plant phytochromes has prompted a search for substrates. Two substrates of potential physiological relevance have been identified: cryptochrome (CRY) and PKS1. The CRY1 and CRY2 cryptochromes are blue-light photoreceptors in *Arabidopsis*, and both are substrates for phytochrome A-associated kinase activity (Ahmad *et al.*, 1998). PKS1 is a phytochrome-binding protein that acts as a negative regulator of phytochrome signaling (Fankhauser *et al.*, 1999). PKS1 is phosphorylated by recombinant PhyA *in vitro* (Frankhauser *et al.*, 1999). Significantly, the phosphorylation of PKS1 is light dependent. Red light stimulated phosphorylation of PKS1 about two-fold over control reactions run in the dark or far-red light. The light dependence for phosphorylation could be eliminated by a single site-directed mutation (*Ser599Lys*) within PhyA. Ser599 is preferentially phosphorylated *in vivo* when phytochrome is in the Pfr form; thus this amino acid may play a role in light regulation of phytochrome kinase activity.

The example of PDH kinase indicates that histidine kinase domains can have serine/threonine specificity (Popov *et al.*, 1993). However, PDH kinase has retained substantially more histidine kinase-like sequences than plant phytochromes. The duplicated histidine kinase-like domains of phytochromes

(Yeh and Lagarias, 1998) are also quite unlike the structure of PDH kinase. Thus, the mechanism by which phytochromes would bind ATP and transfer the phosphate group to substrates is novel, and may involve quite different residues than those utilized by other members of the histidine kinase family.

F. OTHER HISTIDINE KINASES

The previous examples give some idea of the breadth of histidine kinase-like proteins present in plants, in terms of both the signaling pathways involved and the enzymatic possibilities inherent within this signaling motif. This list of proteins is not definitive. More histidine kinase-like proteins are being uncovered through plant genome projects, and these are likely to play equally important regulatory roles in plant signal transduction.

One interesting example (product F17L21.11 from accession number AC004557) present in the *Arabidopsis* database appears to represent a hybrid kinase, but with a novel grouping of signaling elements. Whereas the standard hybrid kinase has a histidine kinase domain and one receiver domain, this histidine kinase has two receiver-like domains (Fig. 3). The histidine kinase domain contains all the residues diagnostic for function, as does the second of the receiver domains. The first receiver domain has diverged from the canonical sequence; in particular, it has a Glu instead of an Asp at the normal site of phosphorylation. How two receiver-like domains might participate in signal transduction is not known. There is precedent for multiple receiver domains following a histidine kinase within gene products of the cyanobacterium *Synechocystis* (Kaneto *et al.*, 1996). In addition, the PleD response regulator, which is involved in stalk-cell differentiation of *Caulobacter crescentus*, has two receiver domains, one canonical and one diverged (Hecht and Newton, 1995).

IV. RESPONSE REGULATORS IN PLANTS

Response regulators represent discrete proteins that participate in the phosphorelay initiated at the histidine kinase. They contain a region of conservation referred to as the receiver domain, along with other domains that allow them to interact with and regulate additional proteins or nucleic acid sequences. Response regulators contain a conserved aspartic acid residue within their receiver domain that becomes phosphorylated (Parkinson, 1993; Stock *et al.*, 1995). In the standard two-component system, the source of the phosphate is the phosphohistidine found on the histidine kinase. Phosphorylation of the aspartic acid is an active process and is mediated by the response regulator; the histidine kinase does not phosphorylate the response regulator, rather the response regulator removes the phosphate from the histidine kinase. In a simple two-component system, only one receiver domain is involved in signal transduction. However, in a multi-step phosphorelay, two receiver domains are involved, one acquiring its phosphate from the histidine kinase domain, the other acquiring its phosphate from the histidine-containing phosphotransfer protein (Fig. 1).

In *Arabidopsis*, several different families of response regulators have been identified (Fig. 4). The receiver domains of these proteins have the same diagnostic sequences and structural features as those found in bacterial systems. Although the focus of this section is on response regulator proteins, one should also keep in mind that many of the plant histidine kinases are hybrid kinases and have fused receiver domains (Fig. 3). Thus phosphotransfer from histidine to aspartic acid can presumably take place both with response regulators as well as with fused receiver domains. Although response regulators

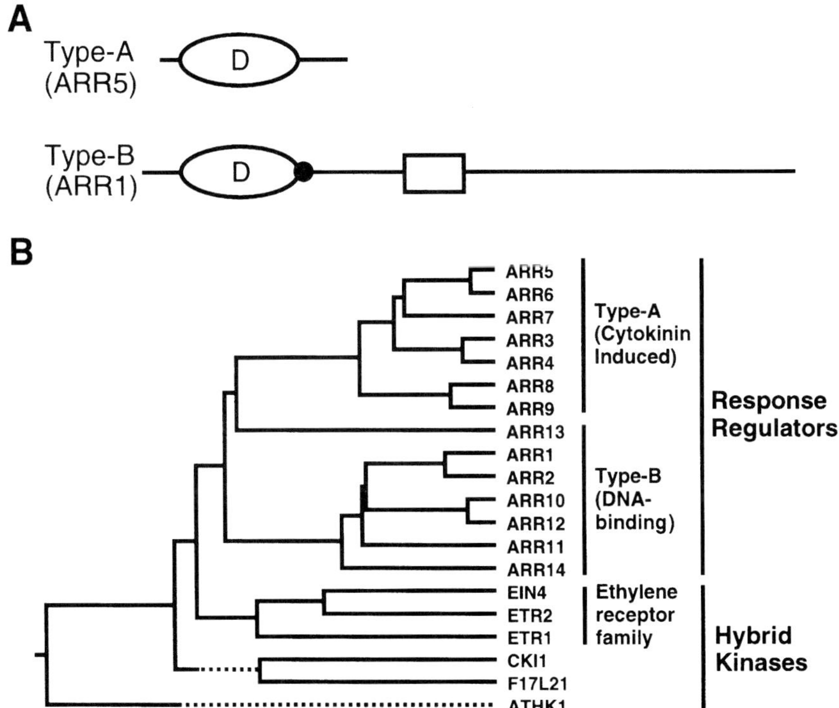

Fig. 4. Response regulators of *Arabidopsis*. (A) Structural features of representative *Arabidopsis* response regulators from the type-A and type-B families. The core sequence of the receiver domain is shown by an oval. In ARR1, the black circle indicates a nuclear-localization signal and the rectangle indicates the Myb-like domain (B-motif). (B) Phylogenetic analysis of the receiver domains from response regulators and hybrid kinases. The dendogram was generated from receiver domains using MegAlign (DNAStar) with Clustal.

and hybrid kinases both contain receiver domains, phylogenetic analysis indicates that the receiver domains of response regulators and the receiver domains of hybrid kinases are found in separate groupings (Fig. 4). A functional difference between the receiver domains could arise due to their actions in different signaling pathways. A functional difference could also arise due to their positions in a multi-step phosphorelay: the receiver domain of a hybrid kinase would interact with the histidine kinase domain, whereas the receiver domain of a response regulator would interact with a phosphotransfer protein.

A. TYPE-A RESPONSE REGULATORS

Seven similar response regulators have been identified in *Arabidopsis* that are referred to as the type-A response regulator family. The type-A response regulators are similar in that they are relatively small, containing a receiver domain along with short N- and C-terminal extensions. Members of the type-A family are all induced to some extent by the plant hormone cytokinin. The type-A response regulators are called ARR (for *Arabidopsis* response regulator) (Imamura *et al.*, 1998), ATRR (for *Arabidopsis thaliana* response regulator) (Urao *et al.*, 1998), or IBC (for induced by cytokinin) (Brandstatter and Kieber, 1998). Some of the response regulators have all three designations as they were characterized independently. These seven genes are *ARR3*, *ARR4* (=*IBC7*/*ATRR1*), *ARR5* (=*IBC6*/*ATRR2*), *ARR6*, *ARR7*, *ARR8* (=*ATRR3*), and *ARR9* (=*ATRR4*) (Fig. 4). All seven genes are similar in sequence within the receiver domain. There is, however, divergence within the short N- and C-terminal extensions.

The functionality of the response regulators was demonstrated by both *in vitro* and *in vivo* experiments performed in bacteria (Imamura *et al.*, 1998) with ARR3, ARR4, and ARR6. For the *in vitro* experiment, a portion of the bacterial ArcB histidine kinase ($ArcB^c$) was used as a phosphodonor. The *Arabidopsis* response regulators were expressed in bacteria, affinity purified, then incubated with phosphorylated $ArcB^c$. The radiolabel on phosphorylated ArcB was rapidly lost, and the *Arabidopsis* response regulators became labeled. These results are consistent with phosphotransfer between the two proteins. For the *in vivo* experiment, the investigators asked whether the *Arabidopsis* response regulators could participate in an *Escherichia coli* His to Asp phosphotransfer. They found that expression of the *Arabidopsis* response regulators could compete with phosphotransfer between ArcB and its conjugate response regulator OmpR. This competition was dependent upon the *Arabidopsis* response regulator having the conserved aspartic acid that serves as the phosphoacceptor. These results indicate that the *Arabidopsis* response regulators can function in a His–Asp phosphotransfer. It should also be noted that in these experiments plant response regulators were capable of interacting bacterial two-component systems. This emphasizes one aspect of response regulators that may make identification of their natural partners in plants difficult, namely their promiscuity.

Expression of the *Arabidopsis* response regulator genes has been analyzed by Northern blot. At the tissue level, there are some discrepancies between results reported by different groups. Two groups report that the response regulators are expressed predominantly in the roots, with lower but significant levels found in seedlings, leaves, stems, flowers, and siliques (Imamura *et al.*, 1998; Urao *et al.*, 1998). In contrast, another group finds little difference between expression levels in the various plant tissues of two of these same response regulators (ARR4 and ARR5), in fact reporting that root has lower levels than other tissues (Brandstatter and Kieber, 1998).

Some of the discrepancies in the tissue expression analysis may be explained by the finding that these response regulator transcripts can be induced by various conditions. First, cytokinin has been shown to induce expression for all seven of the response regulators (Brandstatter and Kieber, 1998; Tanaguchi *et al.*, 1998; Kiba *et al.*, 1999). Induction by cytokinin can be extremely rapid, with a significant increase in transcript level apparent within 15 min for ARR4 and ARR5 (Brandstatter and Kieber, 1998). The final induction varies from three- to 11-fold depending on the gene. The cytokinin effect has been examined in detail for two of the response regulator genes (*ARR4* and *ARR5*), including time course for induction and specificity of the hormone effect. Of the plant hormones tested, only cytokinin induced expression for these two response regulators; no induction was apparent for abscisic acid, auxin, gibberellin, or ethylene (Brandstatter and Kieber, 1998). Second, nitrogen starvation induced expression of the type-A response regulators in a manner similar to that of cytokinin (Tanaguchi *et al.*, 1998); nitrogen signal transduction is thought to be mediated by cytokinin. Third, expression of *ARR4* and *ARR5* has been shown to be slightly induced by dehydration and salt and substantially induced by cold temperature (Urao *et al.*, 1998), suggesting that these same response regulators may have a role in stress responses. A gene encoding a similar response regulator (*ZmCip1*), sharing 60% amino acid identity with ARR5 over the receiver domain, has been identified in maize (Sakakibara *et al.*, 1998); expression of this gene has also been shown to be regulated by cytokinin and nitrogen starvation. Owing to the similarities between these response regulators in sequence and their ability to be induced by cytokinin, it has been suggested that they might all function downstream of the same input signal (Kakimoto, 1998).

Although the type-A response regulators can all be induced by cytokinin, they are not all responsive to the same environmental factors. In particular, *ARR8* and *ARR9* have been contrasted with *ARR4* and *ARR5* based upon how stress affects their expression. Salt, dehydration, and cold temperature had no significant effect upon the expression of *ARR8* and *ARR9* (Urao *et al.*, 1998). ARR8 and ARR9 also form a separate subgroup of response regulators within the type-A family when examined by phylogenetic analysis (Fig. 4), consistent with their having a somewhat different role in plants from the rest of the family.

B. TYPE-B RESPONSE REGULATORS

In the simplest model of the bacterial two-component system, only two proteins are required to transduce an extracellular signal to the transcriptional complex, thereby leading to a change in gene expression (Parkinson, 1993). Such a system includes a sensor protein with histidine kinase activity and a separate response regulator that functions as a transcriptional activator. The response regulators previously described have a receiver domain for phosphorylation on aspartic acid, but do not contain additional domains with known function. Type-B response regulators, in contrast, contain receiver domains along with extensive C-terminal extensions that contain sequences suggestive of a role in transcriptional regulation (Fig. 4) (Sakai *et al.*, 1998a; Imamura *et al.*, 1999). Their presence in *Arabidopsis* indicates that signals may move from a histidine kinase to the nucleus solely through elements of a two-component system.

The type-B response regulator family in *Arabidopsis* is composed of the proteins ARR1, ARR2, and ARR7 through to ARR14 (Sakai *et al.*, 1998a; Imamura *et al.*, 1999). Some of these response regulators have alternative designations, having been independently identified by several research groups: ARR2 = ARP5 (Buchholz *et al.*, GenBank accession no. ATAJ5196); ARR8 = ATRR3 and ARR9 = ATRR4 (Urao *et al.*, 1998); ARR10 = ARP4 and ARR11 = ARP3 (Buchholz *et al.*, GenBank accession nos ATAJ5195 and ATAJ5194). Complementary DNA clones (cDNAs) exist for all the type-B response regulators except ARR12 through to ARR14 which are predicted from the genome sequencing project (Imamura *et al.*, 1999). The type-B receiver domains contain all the conserved residues indicative of activity. The proposed role in transcriptional regulation is based on identification of several motifs within the C-terminal extension (Sakai *et al.*, 1998a; Imamura *et al.*, 1999). First, they contain potential nuclear localization signals scattered throughout their C-terminal extensions. Second, all the proteins contain a conserved motif (the B-motif) with similarity to the DNA-binding protein Myb. Third, the C-terminal extensions are rich in glutamine and proline, characteristics commonly observed in the transcriptional activation domain of eukaryotes. Direct demonstration that members of this family are nuclear localized comes from the analysis of GFP-fusions following transient expression in parsley protoplasts (Lohrmann *et al.*, 1999). A related response regulator was also identified in the rice expressed sequence tags (EST) library, indicating that this family of response regulators is present in other plant species (Sakai *et al.*, 1998a).

Expression analysis of *ARR1* and *ARR2* was carried out using RNA from roots, rosette leaves, cauline leaves, stems, flowers, and siliques (Sakai *et al.*, 1998a). The expression pattern was similar for both genes, with mRNA being present in all tissues examined, but showing relatively high expression levels in the roots. *ARR10* was also found to be present in all tissues examined (Imamura *et al.*, 1999). In contrast to the type-A response regulators, the expression of the type-B response regulators ARR1, ARR2, and ARR10 was not induced by cytokinins (Imamura *et al.*, 1999; Kiba *et al.*, 1999). Neither were they induced by ethylene (Kiba *et al.*, 1999).

C. OTHER RESPONSE REGULATOR-LIKE PROTEINS

It is clear from the above discussion that a variety of response regulators is functioning in *Arabidopsis*, and that they fall into different families. At least one family appears to be cytosolic (the type-A response regulators), while another group is localized to the nucleus (type-B response regulators). An examination of the *Arabidopsis* databases also indicates the presence of additional response regulator-like proteins that do not fall into any of the families described in the literature. Of particular interest are response regulator-like proteins that, while they contain sequence characteristics indicative of receiver domains, nevertheless lack the conserved aspartic acid required for phosphorylation. Instead of the aspartic acid residue, these proteins have a glutamic acid residue. The presence of Glu may be significant in terms of activity. Site-directed mutation of Asp to Glu with the yeast SKN7 response regulator forces the protein into an active conformation (Krems *et al.*, 1996; Li *et al.*, 1998). Given that the phospho-aspartic acid linkage is readily hydrolyzed, the Glu substitution may allow for greater stability of the active form. In such a case, other methods of regulation are likely to come into play to co-ordinate interaction of the response regulator with downstream signaling elements; for example, localization, turnover, or phosphorylation could mediate regulation by serine/threonine protein kinases.

V. HISTIDINE-CONTAINING PHOSPHOTRANSFER PROTEINS IN PLANTS

Histidine-containing phosphotransfer (HPt) proteins contain a conserved histidine residue and perform an intermediary role in signal transduction that utilizes a multi-step phosphorelay (Fig. 1) (Appleby *et al.*, 1996). Three genes encoding histidine-containing phosphotransfer proteins have been characterized in *Arabidopsis* (Miyata *et al.*, 1998; Suzuki *et al.*, 1998) and are called either *AHP* or *ATHP*. These genes are designated *AHP1/ATHP3*, *AHP2/ATHP1*, and *AHP3/ATHP2* (Fig. 5). Additional genes (designated *AHP4* and

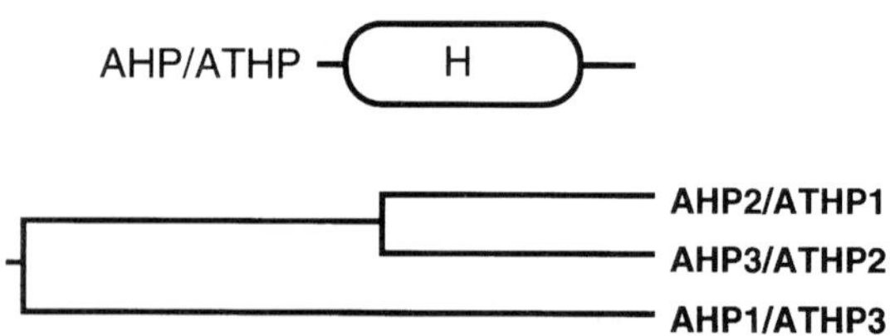

Fig. 5. Histidine-containing phosphotransfer proteins of *Arabidopsis*. Structure of an AHP/ATHP protein is shown with the site of phosphorylation indicated by an 'H'. The dendogram for the three *Arabidopsis* HPt proteins was generated from the core regions using MegAlign (DNAStar) with Clustal.

AHP5) for putative HPt proteins have also been identified in the *Arabidopsis* database (Suzuki *et al.*, 1998).

Genes encoding proteins with HPt domains were initially identified by searching the *Arabidopsis* EST database with sequences for known HPt domains (e.g. the HPt domain from the yeast YPD1 protein). In this manner, three separate sequences were identified and full-length cDNA clones subsequently obtained (Miyata *et al.*, 1998; Suzuki *et al.*, 1998). Of these, AHP2/ATHP1 and AHP3/ATHP2 are the most similar in sequence and have 81% amino-acid identity; in contrast, AHP2/ATHP1 has only 48% identity with AHP1/ATHP3. Like the yeast YPD1 protein, the *Arabidopsis* phosphotransfer proteins are short, ranging from 154 to 156 amino acids in length. The AHP/ATHP proteins have a low level of sequence similarity to yeast YPD1 throughout their sequence, but most significantly they all have a short region of conservation around the histidine that serves as the site of phosphorylation.

To assess whether the AHP/ATHP proteins were functional phosphotransfer proteins, both *in vivo* and *in vitro* experiments were performed (Miyata *et al.*, 1998; Suzuki *et al.*, 1998). For an *in vivo* experiment, the ability of *AHP/ATHP* genes to complement a deletion in the yeast *YPD1* gene was tested. Deletions of *YPD1* are lethal, but the lethality can be suppressed by adding back a functional *YPD1* gene. Each of the three *AHP/ATHP* genes of *Arabidopsis* was capable of suppressing the lethality of the *YPD1* deletion (Miyata *et al.*, 1998; Suzuki *et al.*, 1998). A site-directed mutation demonstrated that the ability of AHP1 to suppress lethality was dependent upon the conserved histidine that serves as a site of phosphorylation (Suzuki *et al.*, 1998). For *in vitro* experiments, the abilities of AHP/ATHP proteins to become phosphorylated and to donate the phosphate to a response regulator protein were tested. AHP1/ATHP3 and AHP2/ATHP1 could be phosphorylated by incubation with *E. coli* membranes; the incorporated phosphate was acid labile but alkali stable, indicative of phosphohistidine. In addition, mutation of the conserved histidine residue of AHP1 rendered the protein incapable of being phosphorylated, consistent with this residue being the site of phosphorylation. AHP1/ATHP3 was also found capable of phosphotransfer when incubated with either the ARR3 or ARR4 type-A response regulators of *Arabidopsis* (Suzuki *et al.*, 1998). Interestingly, AHP2 was unable to transfer its phosphate to the type-B response regulator ARR10, although the two were able to interact based on yeast two-hybrid analysis (Imamura *et al.*, 1999).

Expression of the *AHP*/*ATHP* genes was characterized by Northern blot using RNA preparations from seedlings, roots, leaves, stems, flowers, and siliques (Miyata *et al.*, 1998; Suzuki *et al.*, 1998). The *AHP1*/*ATHP3* transcript was expressed predominantly in the roots. The other transcripts were detected in all the plant tissues examined.

VI. START MAKING SENSE

The previous sections demonstrate that *Arabidopsis* contains all the elements required to establish a histidyl-aspartyl phosphorelay. These elements include the histidine kinases and response regulators that make up the standard 'two-component' system, as well as the histidine-containing phosphotransfer proteins associated with a multi-step phosphorelay. The functionality of the elements has been demonstrated by building on the vast body of work established through study of bacterial systems. This has allowed the design of suitable *in vitro* systems to examine phosphorylation, and the use of *in vivo* systems whereby mutations in members of a phosphorelay can be complemented by overexpression of the relevant plant gene. The big question now is how do these individual elements participate within the overall context of plant signal transduction?

A. BUILDING A PHOSPHORELAY

The very nature of the phosphorelay indicates a certain linearity of information transfer (Parkinson, 1995). The elements of a phosphorelay function in defined order with the phosphate transferred sequentially from histidine to aspartate amino-acid residues. From this standpoint, one can place histidine kinases at the initial step of information transfer, and response regulators as the final step in the system. Furthermore, the majority of histidine kinases identified in plants are hybrid kinases, containing both histidine kinase and receiver domains. Hybrid kinases typically function in multi-step phosphorelays (Appleby *et al.*, 1996; Loomis *et al.*, 1997), and their prevalence in *Arabidopsis* would suggest that the majority of plant systems operate through multi-step phosphorelays. One can then postulate that the typical phosphorelay in plants utilizes the following proteins in the order given: hybrid kinase → histidine-containing phosphotransfer protein → response regulator.

Given this hypothetical order of components and the identification of many of these components in *Arabidopsis*, it might seem a relatively simple task to determine which components form a physiologically relevant phosphorelay. However, even with the aid of exquisitely sensitive methods for analyzing protein–protein interactions such as the two-hybrid system and its permutations (Colas and Brent, 1998; Vidal and Legrain, 1999), there are still significant problems with finding the appropriate partner for any individual component. Chief among these problems is promiscuity; members from one two-component pathway can interact with members from a separate two-component pathway, a phenomenon termed 'cross-talk' (Stock *et al.*, 1989). The ability of two-component proteins to participate in biologically irrelevant pathways can be readily seen from *in vivo* experiments in which plant response regulators and phosphotransfer proteins functioned in bacterial and yeast two-component pathways (Imamura *et al.*, 1998; Suzuki *et al.*, 1998). Response regulators, in particular, are noted for their abilities to use a variety of proteins and compounds as phosphodonors (Stock *et al.*, 1989; Lukat *et al.*, 1992). Given the potential for promiscuous phosphotransfer between components, biologically relevant interactions are probably greatly dependent upon concentration and localization at the cellular and subcellular levels. Data on expression and localization can serve as clues as to which plant components might participate within a defined phosphorelay. Genetic experiments, whereby particular components are eliminated and their effects on signaling examined, will be the ultimate test of biological relevance. However, as with the case of the ethylene receptors, loss-of-function mutations in multiple family members may be required to see a mutant phenotype (Hua and Meyerowitz, 1998).

With osmosensing and cytokinin responses, there are some hints as to possible members in a phosphorelay. ATHK1, the hybrid kinase implicated in osmosensing, is highly expressed in roots and is induced by dehydration and low temperature (Shinozaki and Yamaguchi-Shinozaki, 1997; Urao *et al.*, 1997, 1999). Downstream elements in the two-component system for osmosensing might show similar expression characteristics. Of the three phosphotransfer proteins characterized, AHP1/ATHP3 alone shows root specificity (Miyata *et al.*, 1998; Suzuki *et al.*, 1998). Most of the response regulators show elevated expression levels in roots, with both ARR4 and ARR5 being induced by dehydration and low temperature (Urao *et al.*, 1998). Thus a pathway for osmosensing may proceed from ATHK1 to AHP1/ATHP3 to ARR4 and ARR5. The CKI1 hybrid kinase has been implicated in cytokinin signaling based on overexpression leading to constitutive cytokinin effects in *Arabidopsis* (Kakimoto, 1996). The finding that expression of several response regulators (ARR4 and ARR5) is highly induced by cytokinin lends further support to the involvement of a two-component system in the regulation of cytokinin responses (Brandstatter and Kieber, 1998). CKI1 along with the response regulators ARR4 and ARR5 could participate in a histidyl–aspartyl phosphorelay that regulates cytokinin responses.

When considering information transfer involving a two-component system, one can also consider the steps that follow the two-component signaling system, and how information is transmitted to these downstream elements. With the yeast osmosensing system, one has the case of a two-component system linked to a MAP kinase cascade (Maeda *et al.*, 1994; Posas *et al.*, 1996; Posas and Saito, 1998). Both the ethylene and osmosensing pathways of plants may operate in a similar manner. In ethylene signaling, the histidine kinase-like proteins that function in ethylene perception (ETR1, ETR2, ERS1, ERS2, and EIN4) genetically and/or physically interact with a MAPKKK-like protein (CTR1) (Solano and Ecker, 1998). In osmosensing, a histidine kinase (ATHK1), a MAPKKK (ATMEKK1), and a MAPK (ATMPK3) have all been implicated in signal transduction (Shinozaki and Yamaguchi-Shinozaki, 1997).

B. INTEGRATION OF SIGNALING PATHWAYS

Although phosphotransfer among signaling components occurs in a linear manner, this does not preclude two possibilities that add complexity to the signaling pathways. First, it is possible that several signals may converge at a particular element in the pathway. Second, it is possible for pathways to branch. Thus, one need not have discrete two-component pathways operating in parallel to each other, but potentially a network of interactions with intersections among the pathways. Such a system would allow for integration of multiple stimuli that affect plant growth and development.

The histidine kinases represent one possible point at which multiple inputs can converge, each regulating the level of histidine kinase activity. Examples of this are seen in bacteria. The VirA histidine kinase of Agrobacterium is able to sense three different signals: (1) a phenolic inducer, acetosyringone; (2) glucose, which potentiates the affects of acetosyringone; and (3) pH. A similar situation exists with the bacterial chemoreceptor Tsr, which senses the amino acids serine and leucine, as well as pH; here, perception is tightly linked to regulation of the associated histidine kinase CheA. In these cases, the cell needs to elicit a similar response to divergent stimuli (i.e. the same output for different inputs). Thus, a single histidine kinase output domain is hooked up to multiple input domains. Such a signaling mechanism is facilitated by the modular nature of proteins which allows for several input domains to be present on the same polypeptide. The same situation may hold true for some of the plant histidine kinases. This possibility is most apparent with the ethylene receptors, each of which contains an ethylene binding domain and a GAF domain; the GAF domain potentially allows for a second signal to regulate output from the receptor.

Phosphotransfer proteins and response regulators represent another point at which multiple input signals can be integrated. These proteins are considered promiscuous in their ability to interact with various members of the two-component pathways. This promiscuity may be exploited within the organism to allow for integration of multiple signaling pathways. Thus, a response regulator may not function in one single defined signal transduction pathway, but rather be targeted by several pathways. The potential involvement of the response regulators ARR4 and ARR5 in both osmosensing and cytokinin signaling could represent such a situation. An overlap between the osmosensing and cytokinin signaling pathways is supported by physiological and molecular experiments (Thomas *et al.*, 1995). For example, cytokinins and salt stress both induce an accumulation of proline and osmotin in tobacco plants (Thomas *et al.*, 1995).

It is also possible for pathways to branch, such that one phosphotransfer protein might interact with several response regulators. This situation is found in yeast, where the phosphotransfer protein YPD1 can interact with either the SSK1 or SKN7 response regulators. These response regulators function in separate pathways, wherein SSK1 interacts with a MAPKKK to control the HOG1 MAP kinase pathway, and SKN7 potentially functions as a transcription factor to control expression of genes for cell wall biosynthesis and G1 cyclin. Given that this complexity exists in yeast, which only has one histidine kinase, it seems likely that similar if not more complex branching of pathways occurs in plants, which have multiple histidine kinases.

The increased capacity for branching and regulation may help to explain the preponderance of multi-step phosphorelays in eukaryotic two-component systems. The multi-step phosphorelay is known for yeast and other fungi, and is presumed in plants due to the presence of hybrid kinases and phosphotransfer proteins. In contrast to serine/threonine kinase cascades, two-component systems confer no amplification of signal during phosphotransfer, so the addition of more components does not increase the signaling strength of the system. The amplitude of the signal mediated by a multi-step phosphorelay is no greater than that of a simple two-component system. However, the addition of more components can offer more points of control. Thus the informational content of the system can be modified more readily. The possibility exists that several different pathways can regulate the same phosphotransfer proteins and response regulators. Such a possibility is particularly tempting when considering plant growth regulators which, although they elicit some responses particular to the growth regulator, also overlap for many responses. For example, ethylene and light both regulate cell expansion and thus could share downstream components. Ethylene and cytokinin can have either opposing or shared responses depending on the growth response being evaluated (e.g. senescence or hypocotyl inhibition in dark-grown seedlings). The use of shared and pathway-specific downstream elements could help to mediate these effects.

C. NEW METHODS FROM OLD MODULES

The two-component systems of plants are the evolutionary descendants of signaling systems that originally arose in bacteria (Parkinson, 1993; Swanson *et al.*, 1994). Signaling systems that function in a multicellular eukaryote must deal with quite a different environment than the signaling systems present in a single-celled prokaryote. For this reason, it is not surprising that the plant two-component systems differ in some respects from the prokaryotic systems. These differences can be relatively minor, such as the presence of new input domains coupled to the histidine kinase domains. Such systems allow for sensing of plant-specific factors but with signal transduction still proceeding through a histidyl–aspartyl phosphorelay. However, the differences between bacteria and plants can also be more pronounced, involving a significant conceptual shift from the bacterial paradigm. The presence of diverged two-component elements in plants represents such a shift, with the examples of PDH kinase and phytochromes indicating that these elements can take on new enzymatic activities.

Given that the two-component systems were adopted from prokaryotes, one question to consider is at what point in evolution did plants acquire two-component elements. Pyruvate dehydrogenase kinase is found in plants, yeast, and animals (Popov *et al.*, 1997; Thelen *et al.*, 1998), thus the progenitor of PDH kinase must have been acquired early in evolution. Interestingly, pyruvate dehydrogenase kinase is localized to the mitochondria (Popov *et al.*, 1997; Thelen *et al.*, 1998), an organelle shared by all eukaryotes. It is possible that PDH kinase is descended from a histidine kinase present within the original endosymbiote that gave rise to the mitochondria. Plants also contain a plethora of histidine kinase-like proteins not found in other eukaryotes. Examination of the genome from the cyanobacterium *Synechocystis* has been particularly edifying in terms of understanding the plant two-component systems. *Synechocystis* is thought to be similar to the endosymbiote that gave rise to plant chloroplasts. Proteins related to plant phytochromes (Hughes *et al.*, 1997; Yeh *et al.*, 1997) and ethylene receptors (Rodriguez *et al.*, 1999) have been found in *Synechocystis*. This would suggest that the precursor chloroplast may have served as the evolutionary source for many of the plant two-component systems. Thus the precursor chloroplast may have had a role in conferring on plants, not just the ability to harvest light energy, but also a significant portion of their signal transduction machinery.

One of the most visible differences between eukaryotic and prokaryotic cells is the wealth of membrane-bound compartments within the eukaryotic cell. In bacterial cells, sensors with histidine kinase activity are typically localized to the cytoplasmic membrane (Parkinson, 1993). By direct analogy to bacteria, one could make the prediction that the plant histidine kinases would be present on the plasma membrane, thus placing the protein at the juncture between the cell and its environment. However, an alternative possibility would be targeting of the histidine kinase to a bacteria-derived organelle, such as mitochondria or plastid, in which case the evolutionary positioning would be absolutely conserved between prokaryote and eukaryote. Although plant histidine kinases have hydrophobic domains for membrane localization, most do not have obvious sequences defining their target membrane. Thus, the membrane localization of the individual plant histidine kinases remains to be elucidated.

As mentioned, one of the most obvious differences between the histidine kinases of plants and bacteria is the number of diverged histidine kinase-like proteins found in plants. We are in the relatively strange situation that not all histidine kinases in plants function as histidine kinases. The situation does not end there, as diverged response regulators are also found in plants. From this it is apparent that not only have plants adopted a signaling mechanism from bacteria, but also they have significantly adapted the mechanism to new functions. These novel forms may possess new enzymatic activities; both pyruvate dehydrogenase kinases and phytochromes appear to have serine/threonine kinase activity (Ahmad *et al.*, 1998; Thelen *et al.*, 1998; Yeh and Lagarias, 1998; Fankhauser *et al.*, 1999). The development of serine/threonine phosphorylation may be favored as a replacement for histidine and aspartate phosphorylation owing to the greater stability of the phosphorylated amino acid within the eukaryotic cell. Two-component systems may also function through nonenzymatic mechanisms in eukaryotic cells (e.g. reversible protein–protein interactions). The SKN7 response regulator of yeast, for example, has a role in signal transduction that is independent of two-component systems (Ketela *et al.*, 1998). In plants, the ETR1 ethylene receptor interacts with the CTR1 Raf-like kinase (Clark *et al.*, 1998); this interaction could serve to localize CTR1 to the correct cellular location for signal transduction. Nonenzymatic mechanisms would provide one means by which the diverged histidine kinases and response regulators of *Arabidopsis* could participate in signal transduction.

VII. PERSPECTIVES

This chapter will finish with an analogy that Tom Davis (Department of Plant Biology, University of New Hampshire) shared with the author in conversation. According to the Davis analogy, signal transduction can be thought of as being like a soccer game. The proteins within a signal transduction pathway are the players on one team. In signal transduction, the object is to transfer information from the plasma membrane to the nucleus, resulting in a change in transcription. In soccer, the object of the game is to transfer the ball (information) from one end of the field (plasma membrane) to the other (nucleus), resulting in a goal (change in transcription). In the case of a two-component system the ball can be considered as being a phosphate that is relayed from one player to another. In addition, just as there are several forwards, half-backs, and full-backs in a soccer game, there are often several proteins with redundant function (e.g. the ethylene receptor family) within eukaryotic signal transduction pathways. There are also other signal transduction pathways (the opposing team) that impinge upon the pathway in question, potentially redirecting or halting the transfer of information. Now, the reader should put themselves in the position of an alien trying to decipher the rules of soccer. Let us say that the alien has decided to focus on the actions of one player. Just to make the situation a little more fun, let us say that instead of the typical soccer game played between two teams, the alien is watching a new variant that involves 50 teams all on the same field at once. Now this gives a situation similar to that of the scientist trying to understand signal transduction.

The situation described by the Davis analogy can seem daunting at first. Indeed, in the case of plant two-component systems, one might say that little more has been done than identify a few players and suggest on what team they are playing. However, this was all accomplished in less than a decade. Now, with the plant genome sequencing projects, there will finally be keys to all the players in the game (Meinke *et al.*, 1998). From previous studies in bacteria, a great deal is known about the capabilities of the players and the rules for passing the ball between players (Parkinson, 1993; Stock *et al.*, 1995). With the advent of DNA chip technology, there will be the ability to take snapshots of the game at any particular moment, being able to see which player is called into action and which player is sent to the bench (Schena *et al.*, 1995; Gerhold *et al.*, 1999). Improved techniques for the generation of loss-of-function mutations will allow the removal of one player, several players, or a whole team from the field and the evaluation of their effect upon the game (Krysan *et al.*, 1996; Hua and Meyerowitz, 1998). Thus, in the near future, there should be the possibility of building quite accurate models of signal transduction. The two-component systems of *Arabidopsis* are especially attractive subjects for analysis given their limited number and their involvement in critical signal transduction pathways.

ACKNOWLEDGEMENTS

The author wishes to thank Estelle Hrabak and Rebekah Gamble for critical readings of the manuscript. He also thanks K. Harter, J. Kieber, and K. Shinozaki for sharing information prior to publication. He gratefully acknowledges the financial support from the National Science Foundation and the United States Department of Agriculture for research conducted in his laboratory.

REFERENCES

Abeles, F. B., Morgan, P. W. and Saltveit, M. E., Jr (1992). "Ethylene in Plant Biology", 2nd edition. Academic Press, San Diego.

Ahmad, M., Jarillo, J. A., Smirnova, O. and Cashmore, A. R. (1998). The CRY1 blue light photoreceptor of *Arabidopsis* interacts with phytochrome A *in vitro*. *Molecular Cell* **1**, 939–948.

Appleby, J. L., Parkinson, J. S. and Bourret, R. B. (1996). Signal transduction via the multi-step phosphorelay: not necessarily a road less traveled. *Cell* **86**, 845–848.

Aravind, L. and Ponting, C. P. (1997). The GAF domain: an evolutionary link between diverse phototransducing proteins. *Trends in Biochemical Science* **22**, 458–459.

Bilwes, A. M., Alex, L. A., Crane, B. R. and Simon, M. I. (1998). Structure of CheA, a signal-transducing histidine kinase. *Cell* **96**, 131–141.

Binns, A. N. (1994). Cytokinin accumulation and action: biochemical, genetic, and molecular approaches. *Annual Review of Plant Physiology and Plant Molecular Biology* **45**, 173–196.

Bleecker, A. B. and Schaller, G. E. (1996). The mechanism of ethylene perception. *Plant Physiology* **111**, 653–660.

Boylan, M. T. and Quail, P. H. (1996). Are the phytochromes protein kinases? *Protoplasma* **195**, 59–67.

Brandstatter, I. and Kieber, J. J. (1998). Two genes with similarity to bacterial response regulators are rapidly and specifically induced by cytokinin in *Arabidopsis*. *Plant Cell* **10**, 1009–1019.

Burg, S. P. and Burg, E. A. (1967). Molecular requirements for the biological activity of ethylene. *Plant Physiology* **42**, 144–152.

Chang, C., Kwok, S. F., Bleecker, A. B. and Meyerowitz, E. M. (1993). *Arabidopsis* ethylene response gene *ETR1*: similarity of product to two-component regulators. *Science* **262**, 539–544.

Clark, K. L., Larsen, P. B., Wang, X. and Chang, C. (1998). Association of the *Arabidopsis* CTR1 Raf-like kinase with the ETR1 and ERS1 ethylene receptors. *Proceedings of the National Academy of Sciences of the United States of America* **95**, 5401–5406.

Colas, P. and Brent, R. (1998). The impact of two-hybrid and related methods on biotechnology. *Trends in Biotechnology* **16**, 355–363.

Fankhauser, C., Yeh, K.-C., Lagarias, J. C., Zhang, H., Elich, T. D. and Chory, J. (1999). PKS1, a substrate phosphorylated by phytochrome that modulates light signaling in *Arabidopsis*. *Science* **284**, 1539–1541.

Gamble, R. L., Coonfield, M. L. and Schaller, G. E. (1998). Histidine kinase activity of the ETR1 ethylene receptor from *Arabidopsis*. *Proceedings of the National Academy of Sciences of the United States of America* **95**, 7825–7829.

Gennis, R. B. (1989). "Biomembranes: Molecular Structure and Function". Springer, New York.

Gerhold, D., Rushmore, T. and Caskey, C. T. (1999). DNA chips: promising toys have become useful tools. *Trends in Biochemical Science* **24**, 168–173.

Hall, A. E., Findell, J. L., Schaller, G. E., Sister, E. C. and Bleecker, A. B. (2000). Ethylene perception by the ERS1 protein in *Arabidopsis*. *Plant Physiology*, in press.

Hecht, G. B. and Newton, A. (1995). Identification of a novel response regulator required for the swarmer-to-stalked-cell transition in *Caulobacter crescentus*. *Journal of Bacteriology* **177**, 6223–6229.

Hess, J. F., Bourret, R. B. and Simon, M. I. (1991). Phosphorylation assays for proteins of the two-component regulatory system controlling chemotaxis in *E. coli*. *Methods in Enzymology* **200**, 188–204.

Hirayama, T., Kieber, J. J., Hirayama, N., Kogan, M., Guzman, P., Nourizadeh, S., Alonso, J. M., Dailey, W. P., Dancis, A. and Ecker, J. R. (1999). RESPONSIVE-

TO-ANTAGONIST1, a Menkes/Wilson disease-related copper transporter, is required for ethylene signaling in *Arabidopsis*. *Cell* **97**, 383–393.

Hua, J. and Meyerowitz, E. M. (1998). Ethylene responses are negatively regulated by a receptor gene family in *Arabidopsis thaliana*. *Cell* **94**, 261–271.

Hua, J., Chang, C., Sun, Q. and Meyerowitz, E. M. (1995). Ethylene sensitivity conferred by *Arabidopsis ERS* gene. *Science* **269**, 1712–1714.

Hua, J., Sakai, H., Nourizadeh, S., Chen, Q. G., Bleecker, A. B., Ecker, J. R. and Meyerowitz, E. M. (1998). Ein4 and ERS2 are members of the putative ethylene receptor family in *Arabidopsis*. *Plant Cell* **10**, 1321–1332.

Hughes, J., Lamparter, T., Mittmann, F., Hartmann, E., Gartner, W., Wilde, A. and Borner, T. (1997). A prokaryotic phytochrome. *Nature* **386**, 663–664.

Igo, M. M., Ninfa, A. J., Stock, J. B. and Silhavy, T. J. (1989). Phosphorylation and dephosphorylation of a bacterial transcriptional activator by a transmembrane receptor. *Genes and Development* **3**, 1725–1734.

Imamura, A., Hanaki, N., Umeda, H., Nakamura, A., Suzuki, T., Ueguchi, C. and Mizuno, T. (1998). Response regulators implicated in His-to-Asp phosphotransfer signaling in *Arabidopsis*. *Proceedings of the National Academy of Sciences of the United States of America* **95**, 2691–2696.

Imamura, A., Hanaki, N., Nakamura, A., Suzuki, T., Taniguchi, M., Kiba, T., Ueguchi, C., Sugiyama, T. and Mizuno, T. (1999). Compilation and characterization of *Arabidopsis thaliana* response regulators implicated in His-Asp phosphorelay signal transduction. *Plant Cellular Physiology* **40**, 733–742.

Jonak, C., Kiegerl, M., Ligterink, W., Barker, P. J., Huskisson, N. S. and Hirt, H. (1996). Stress signaling in plants: a MAP kinase pathway is activated by cold and drought. *Proceedings of the National Academy of Sciences of the United States of America* **93**, 11274–11279.

Kakimoto, T. (1996). CKI1, a histidine kinase homologue involved in cytokinin signal transduction. *Science* **274**, 982–985.

Kakimoto, T. (1998). Cytokinin signaling. *Current Opinion in Plant Biology* **1**, 399–403.

Kaneto, T. *et al.* (1996). Sequence analysis of the genome of the unicellular cyanobacterium *Synechocystis* sp. strain PCC6803. II. Sequence determination of the entire genome and assignment of petential protein-coding regions. *DNA Research* **3**, 109–136.

Ketela, T., Brown, J. L., Stewart, R. C. and Bussey, H. (1998). Yeast Skn7p activity is modulated by the Sln1p-Ypd1p osmosensor and contributes to regulation of the HOG pathway. *Molecular and General Genetics* **259**, 372–378.

Kiba, T., Taniguchi, M., Imamura, A., Ueguchi, C., Mizuno, T. and Sugiyama, T. (1999). Differential expression of genes for response regulators in response to cytokinins and nitrate in *Arabidopsis thaliana*. *Plant Cellular Physiology* **40**, 767–771.

Kieber, J. J., Rothenberg, M., Roman, G., Feldman, K. A. and Ecker, J. R. (1993). *CTR1*, a negative regulator of the ethylene response pathway in *Arabidopsis*, encodes a member of the Raf family of protein kinases. *Cell* **72**, 427–441.

Krems, B., Charizanis, C. and Entian, K. D. (1996). The response regulator-like protein Pos9/Skn7 of *Saccharomyces cerevisiae* is involved in oxidative stress resistance. *Current Genetics* **29**, 327–334.

Krysan, P. J., Young, J. C., Tax, F. and Sussman, M. R. (1996). Identification of transferred DNA insertions within *Arabidopsis* genes involved in signal transduction and ion transport. *Proceedings of the National Academy of Sciences of the United States of America* **93**, 8145–8150.

Lashbrook, C. C., Tieman, D. M. and Klee, H. J. (1998). Differential regulation of the tomato *ETR* gene family throughout plant development. *Plant Journal* **15**, 243–252.

Li, S., Ault, A., Malone, C. L., Raitt, D., Dean, S., Johnston, L. H., Deschenes, R. J. and Fassler, J. S. (1998). The yeast histidine protein kinase, Sln1p, mediates phosphotransfer to two response regulators, Ssk1p and Skn7p. *EMBO Journal* **17**, 6952–6962.

Lohrmann, J., Buchholz, G., Keitel, C., Sweere, U., Kircher, S., Baurle, I., Kudla, J., Schafer, E. and Harter, K. (1999). Differential expression and nuclear localization of response regulator-like proteins from *Arabidopsis thaliana. Plant Biology* **1**, 495–505.

Loomis, W. F., Shaulsky, G. and Wang, N. (1997). Histidine kinases in signal transduction pathways of eukaryotes. *Journal of Cellular Science* **110**, 1141–1145.

Lukat, G. S., McCleary, W. R., Stock, A. M. and Stock, J. B. (1992). Phosphorylation of bacterial response regulator proteins by low molecular weight phospho-donors. *Proceedings of the National Academy of Sciences of the United States of America* **89**, 718–722.

Maeda, T., Wurgler-Murphy, S. M. and Saito, H. (1994). A two-component system that regulates an osmosensing MAP kinase cascade in yeast. *Nature* **369**, 242–245.

Maeda, T., Takekawa, M. and Saito, H. (1995). Activation of yeast PBS2 MAPKK by MAPKKKs or by binding of an SH3-containing osmosensor. *Science* **269**, 554–558.

McMichael, R. W. Jr and Lagarias, J. C. (1990). Polycation-stimulated phytochrome phosphorylation: is phytochrome a protein kinase? *Current Topics in Plant Biochemistry and Physiology* **9**, 259–270.

Meinke, D. W., Cherry, J. M., Dean, C., Rounsley, S. D. and Koornneef, M. (1998). *Arabidopsis thaliana*: a model plant for genome analysis. *Science* **282**, 662–682.

Miyata, S.-I., Urao, T., Yamaguchi-Shinozaki, K. and Shinozaki, K. (1998). Characterization of genes for two-component phosphorelay mediators with a single HPt domain in *Arabidopsis thaliana. FEBS Letters* **437**, 11–14.

Mizoguchi, T., Irie, K., Hirayama, T., Hayashida, N., Yamaguchi-Shinozaki, K., Matsumoto, K. and Shinozaki, K. (1996). A gene encoding a MAP kinase kinase kinase is induced simultaneously with genes for a MAP kinase and an S6 kinase by touch, cold, and water stress in *Arabidopsis thaliana. Proceedings of the National Academy of Sciences of the United States of America* **93**, 765–769.

Ota, I. M. and Varshavsky, A. (1993). A yeast protein similar to bacterial two-component regulators. *Science* **262**, 566–569.

Pan, S. Q., Charles, T., Jin, S., Wu, Z.-L. and Nester, E. W. (1993). Preformed dimeric state of the sensor protein VirA is involved in plant-*Agrobacterium* signal transduction. *Proceedings of the National Academy of Sciences of the United States of America* **90**, 9939–9943.

Parkinson, J. S. (1993). Signal transduction schemes of bacteria. *Cell* **73**, 857–871.

Parkinson, J. S. (1995). Genetic approaches for signaling pathways and proteins. *In* "Two Component Signal Transduction" (J. A. Hoch and T. J. Silhavy, eds) pp. 9–24. American Society for Microbiology, Washington, DC.

Payton, S., Fray, R. G., Brown, S. and Grierson, D. (1996). Ethylene receptor expression is regulated during fruit ripening, flower senescence and abscission. *Plant Molecular Biology* **31**, 1227–1231.

Popov, K. M., Zhao, Y., Shimomura, Y., Kuntz, M. J. and Harris, R. A. (1992). Branched-chain alpha-ketoacid dehydrogenase kinase: molecular cloning, expression, and sequence similarity with histidine protein kinases. *Journal of Biological Chemistry* **267**, 13127–13130.

Popov, K. M., Kedishvili, N. Y., Zhao, Y., Shimomura, Y., Crabb, D. W. and Harris, R. A. (1993). Primary structure of pyruvate dehydrogenase kinase establishes a new family of eukaryotic protein kinases. *Journal of Biological Chemistry* **268**, 26602–26606.

Popov, K. M., Hawes, J. W. and Harris, R. A. (1997). Mitochondrial a-ketoacid dehydrogenase kinases: a new family of protein kinases. *Advances in Second Messenger and Phosphoprotein Research* **31**, 105–111.

Posas, F. and Saito, H. (1998). Activation of the yeast SSK2 MAP kinase kinase kinase by the SSK1 two-component response regulator. *EMBO Journal* **17**, 1385–1394.

Posas, F., Wurgler-Murphy, S. M., Maeda, T., Witten, E. A., Thai, T. C. and Saito, H. (1996). Yeast HOG1 MAP kinase cascade is regulated by a multistep phosphorelay mechanism in the SLN1-YPD1-SSK1 'two component' osmosensor. *Cell* **86**, 865–875.

Pratt, L. A. and Silhavy, T. J. (1995). Porin regulation of *Escherichia coli*. *In* "Two-Component Signal Transduction" (J. A. Hoch and T. J. Silhavy, eds) pp. 105–128. American Society for Microbiology, Washington, DC.

Quail, P. H. (1997a). An emerging molecular map of phytochromes. *Plant Cell and Environment* **20**, 657–665.

Quail, P. H. (1997b). The phytochromes: a biochemical mechanism of signaling in sight? *BioEssays* **19**, 571–579.

Rodriguez, F. I., Esch, J.J., Hall, A. E., Binder, B. M., Schaller, G. E. and Bleecker, A. B. (1999). A copper cofactor for the ethylene receptor ETR1 from *Arabidopsis*. *Science* **283**, 996–998.

Sakai, H., Aoyama, T., Bono, H. and Oka, A. (1998a). Two-component response regulators from *Arabidopsis thaliana* contain a putative DNA-binding motif. *Plant Cellular Physiology* **39**, 1232–1239.

Sakai, H., Hua, J., Chen, Q. G., Chang, C., Medrano, L. J., Bleecker, A. B. and Meyerowitz, E. M. (1998b). *ETR2* is an *ETR1*-like gene involved in ethylene signaling in *Arabidopsis*. *Proceedings of the National Academy of Sciences of the United States of America* **95**, 5812–5817.

Sakakibara, H., Suzuki, M., Takei, K., Deji, A., Taniguchi, M. and Sugiyama, T. (1998). A response-regulator homologue possibly involved in nitrogen signal transduction mediated by cytokinin in maize. *Plant Journal* **14**, 337–344.

Sato-Nara, K., Yuhashi, K.-I., Higashi, K., Hosoya, K., Kubota, M. and Ezura, H. (1999). Stage- and tissue-specific expression on ethylene receptor homolog genes during fruit development in muskmelon. *Plant Physiology* **119**, 321–329.

Schaller, G. E. and Bleecker, A. B. (1995). Ethylene-binding sites generated in yeast expressing the *Arabidopsis ETR1* gene. *Science* **270**, 1809–1811.

Schaller, G. E., Ladd, A. N., Lanahan, M. B., Spanbauer, J. M. and Bleecker, A. B. (1995). The ethylene response mediator ETR1 from *Arabidopsis* forms a disulfide-linked dimer. *Journal of Biological Chemistry* **270**, 12526–12530.

Schena, M., Shalon, D., Davis, R. W. and Brown, P. O. (1995). Quantitative monitoring of gene expression patterns with a complementary DNA microarray. *Science* **270**, 467–470.

Schneider-Poetsch, H. A. W., Braun, B., Marx, S. and Schaumburg, A. (1991). Phytochromes and bacterial sensor proteins are related by structural and functional homologies. *FEBS Letters* **281**, 245–249.

Shinozaki, K. and Yamaguchi-Shinozaki, K. (1997). Gene expression and signal transduction in water-stress response. *Plant Physiology* **115**, 327–334.

Solano, R. and Ecker, J. R. (1998). Ethylene gas: perception, signaling, and response. *Current Opinion in Plant Biology* **1**, 393–398.

Stock, J. B., Ninfa, A. J. and Stock, A. M. (1989). Protein phosphorylation and regulation of adaptive responses in bacteria. *Microbiology Reviews* **53**, 450–490.

Stock, J. B., Surette, M. G., Levit, M. and Park, P. (1995). Two-component signal transduction systems: structure–function relationships and mechanisms of catalysis. *In* "Two-Component Signal Transduction." (J. A. Hoch and T. J. Silhavy, eds) pp. 25–51. American Society for Microbiology, Washington, DC.

Suzuki, T., Imamura, A., Ueguchi, C. and Mizuno, T. (1998). Histidine-containing phosphotransfer (HPt) signal transducers implicated in His-to-Asp phosphorelay in *Arabidopsis*. *Plant Cellular Physiology* **39**, 1258–1268.

Swanson, R. V., Alex, L. A. and Simon, M. I. (1994). Histidine and aspartate phosphorylation: two-component systems and the limits of homology. *Trends in Biochemistry* **19**, 485–490.

Tanaguchi, M., Kiba, T., Sakakibara, H., Ueguchi, C., Mizuno, T. and Sugiyama, T. (1998). Expression of *Arabidopsis* response regulator homologs is induced by cytokinins and nitrate. *FEBS Letters* **429**, 259–262.

Tanaka, T. *et al.* (1998). NMR structure of the histidine kinase domain of the *E. coli* osmosensor EnvZ. *Nature* **396**, 88–92.

Thelen, J. T., Muszynski, M. G., Miernyk, J. A. and Randall, D. D. (1998). Molecular analysis of two pyruvate dehydrogenase kinases from maize. *Journal of Biological Chemistry* **273**, 26618–26623.

Thomas, J. C., Smigocki, A. C. and Bohnert, H. J. (1995). Light-induced expression of *ipt* from *Agrobacterium tumefaciens* results in cytokinin accumulation and osmotic stress symptoms in transgenic tobacco. *Plant Molecular Biology* **27**, 225–235.

Tieman, D. M. and Klee, H. J. (1999). Differential expression of two novel members of the tomato ethylene-receptor family. *Plant Physiology* **120**, 165–172.

Urao, T., Yakubov, B., Yamaguchi-Shinozaki, K. and Shinazaki, K. (1998). Stress-responsive expression of genes for two-component response regulator-like proteins in *Arabidopsis thaliana*. *FEBS Letters* **427**, 175–178.

Urao, T., Yakubov, B., Satoh, R., Yamaguchi-Shinozaki, K., Seki, M., Hirayama, T. and Shinozaki, K. (1999). A transmembrane hybrid-type histidine kinase in *Arabidopsis* functions as an osmosensor. *Plant Cell* **11**, 1743–1754.

Vidal, M. and Legrain, P. (1999). Yeast forward and reverse 'n'-hybrid systems. *Nucleic Acids Research* **27**, 919–929.

Volz, K. (1995). Structural and functional conservation in response regulators. *In* "Two-Component Signal Transduction." (J. A. Hoch and T. J. Silhavy, eds) pp. 53–64. American Society for Microbiology, Washington, DC.

Wilkinson, J. Q., Lanahan, M. B., Yen, H.-C., Giovannoni, J. J. and Klee, H. J. (1995). An ethylene-inducible component of signal transduction encoded by *Never-ripe*. *Science* **270**, 1807–1809.

Yeh, K.-C. and Lagarias, J. C. (1998). Eukaryotic phytochromes: Light-regulated serine/threonine protein kinases with histidine kinase ancestry. *Proceedings of the National Academy of Sciences of the United States of America* **95**, 13976–13981.

Yeh, K. C., Wu, S. H., Murphy, J. T. and Lagarias, J. C. (1997). A cyanobacterial phytochrome two-component light sensory system. *Science* **277**, 1505–1508.

Zhou, D., Mattoo, A. K. and Tucker, M. L. (1996). The mRNA for an *ETR1* homologue in tomato is constitutively expressed in vegetative and reproductive tissues. *Plant Molecular Biology* **30**, 1331–1338.

Light and Protein Kinases

JOHN C. WATSON

Department of Biology, Indiana University–Purdue University Indianapolis, 723 West Michigan Street, Indianapolis, IN 46202-5132, USA

I. INTRODUCTION

Several environmental factors influence the growth and development of green plants, but among them light is perhaps the most important. Aside from its role as the energy source for photosynthesis, light provides essential environmental cues that control virtually every stage of the plant life cycle, from seed germination, through stem and leaf development, to flowering. The elegant photosensory systems that have evolved in plants allow them to respond to the intensity, wavelength, duration, periodicity, and directionality of light. Perhaps at no time are the effects of light on plant development more apparent then when seedlings are grown in its absence. Etiolation, the response to growth in darkness, results in greatly elongated stems, a substantial reduction or arrest in leaf development, and no chlorophyll accumulation. Complex biochemical and

Advances in Botanical Research Vol. 32
incorporating Advances in Plant Pathology
ISBN 0-12-005932-0

morphological changes occur when etiolated seedlings are exposed to light. One major consequence of de-etiolation (greening) is the assembly of the photosynthetic apparatus, requiring major changes to the patterns of gene expression and protein synthesis (Watson, 1989). The ability of an etiolated seedling to sense and respond to the light environment, and so trigger the transition to photosynthetic growth, is pivotal to its establishment and survival. This is particularly true for species whose seeds germinate deep within the soil, as with many crop plants. Thus, de-etiolation has important ecological and agricultural consequences.

During photomorphogenesis, distinct photoreceptor families allow light over a broad range of wavelengths to cue plant responses. The phytochromes are the primary sensors for red (R) and far-red (FR) light (Quail, 1994; Quail *et al.*, 1995). There is a multiplicity of photoreceptors for the blue (B)/ultraviolet (UV)-A portion of the spectrum, including the cryptochromes (Senger and Schmidt, 1994; Ahmad and Cashmore, 1996) and phototropin, and a UV-B photoreceptor(s) (Senger and Schmidt, 1994). That protein phosphorylation participates in a fundamental way during the signal transduction that occurs during photomorphogenesis has been an important concept over the years (Singh and Song, 1990; Roux, 1994; Short and Briggs, 1994).

Of the photoreceptors active in the visible portion of the spectrum, the phytochromes are the best understood. Phytochrome is a photointerconvertible pigment: it is synthesized in the dark as a R-absorbing form, called Pr, and upon illumination with R is converted to a FR-absorbing form, called Pfr. FR drives Pfr back to the Pr form, thus providing the mechanism for the classic R/FR reversibility of many phytochrome-mediated responses. In *Arabidopsis*, there are five genes encoding phytochromes, *PHYA–E* (reviewed in Clack *et al.*, 1994; Quail, 1994). *PHYA* encodes the PHYA apoprotein that binds its chromophore to form the dark-abundant, light-labile phyA holoprotein. *PHYB* encodes the PHYB aprotein that yields the low-abundance, light-stable phyB holoprotein. Analysis of mutants in these phytochromes has greatly enhanced our understanding of phytochrome action and photomorphogenesis. In many responses, phyA and phyB have discreet functions (Furuya, 1993; Quail *et al.*, 1995; Smith, 1995). For example, phyA mediates the very low fluence response, a nonphotoreversible response saturable at exceedingly low levels of Pfr (Smith, 1995; Shinomura *et al.*, 1996). phyB mediates most (Smith, 1995; Shinomura *et al.*, 1996), but not all (Reed *et al.*, 1994), of the low fluence responses (the classical R/FR reversible responses) in seedlings. Perhaps the clearest distinction between the action of phyA and phyB is with hypocotyl elongation in *Arabidopsis*. The response to continuous R light (Rc) known as the R high irradiance response (R-HIR) is mediated by phyB, whereas phyA mediates the FR-HIR triggered by continuous FR (FRc) (Furuya, 1993; Quail *et al.*, 1995; Smith, 1995; Chory *et al.*, 1996).

Based on photobiological data, it has been clear for many years that several photoreceptors in the B/UV region of the spectrum were likely to exist (Senger and Schmidt, 1994). Thus far, the UV-B photoreceptor has not been isolated. In the B/UV-A, there are two known classes of photoreceptors: the cryptochromes (Ahmad and Cashmore, 1996; Batschauer, 1998; Khurana *et al.*, 1998; Cashmore *et al.*, 1999) and phototropin (Short and Briggs, 1994; Batschauer, 1998; Briggs and Huala, 1999). In *Arabidopsis*, there are two cryptochrome genes, *CRY1* and *CRY2*, encoding proteins with two chromophore-binding regions. Each apoprotein binds a flavin and a pterin to form the holoproteins. The cryptochromes regulate a variety of B responses in plants, such as hypocotyl elongation, flowering time, and circadian rhythms. A finding of major importance in recent years is that cryptochromes exist in animals where they entrain circadian rhythms (Cashmore *et al.*, 1999). Although they may modulate the response, the cryptochromes are not the primary B light receptors for phototropism, one of the classic B responses in plants. A newly discovered B light receptor, phototropin, is a protein that binds flavin and triggers curvature in response to unilateral light (Briggs and Huala, 1999).

The major focus of this review is on the role that protein kinases play in the perception and transduction of light signals. As discussed below, there is a substantial body of evidence to show that light causes rapid changes in protein phosphorylation and in the expression of protein kinase genes. These types of data are consistent with the notion that protein kinases play a role in transducing light signals. However, this role has become both more intriguing and integral with the recent revelations that phytochrome and phototropin are protein kinases. These exciting developments are also discussed here. Certain aspects of phytochrome structure and signaling are not discussed, so the reader is referred to the many excellent reviews on phytochrome structure, roles, and the downstream signaling proteins that have been implicated by mutation (Quail *et al.*, 1995; Chory *et al.*, 1996; von Arnim and Deng, 1996; Fankhauser and Chory, 1997; Quail, 1997, 1998; Batschauer, 1998; Khurana *et al.*, 1998; Whitelam *et al.*, 1998). Likewise, the reader is directed to excellent reviews on the combined pharmacological, photobiological, and genetic work that has shown that a G protein, cyclic GMP, Ca^{2+}, and calmodulin are central players in phytochrome signaling (Bowler, 1997; Mustilli and Bowler, 1997; Khurana *et al.*, 1998).

II. EFFECTS OF LIGHT ON PROTEIN PHOSPHORYLATION

Protein kinases are often mentioned as potential components of signaling pathways activated by the photoreceptors (Singh and Song, 1990; Roux, 1994; Short and Briggs, 1994). In part, this is because there are rapid changes in the phosphorylation status of particular polypeptides in response to light. Several different approaches and species have been used, as outlined in Table I. Most of these experiments evaluated the effects of the light treatment on phosphorylation by sodium dodecyl sulfate–polyacrylamide gel electrophoresis (SDS–

TABLE I

Systems used to demonstrate light-regulated protein phosphorylation

Species	Material irradiated	Material labeled	Label	Effective light treatment	Reference
Pea	Nuclei	Nuclei	γ-^{32}P-ATP	R/FR	Datta *et al.* (1985)
Oat	Coleoptile tips	Coleoptile tips	^{32}Pi	R/FR	Otto and Schäfer (1988)
Pea	Seedlings	Microsomes	γ-^{32}P-ATP	B	Gallagher *et al.* (1988)
Oat	Protoplasts	Protoplasts	^{32}Pi	R/FR	Park and Chae (1989)
Oat	Nuclei	Nuclei	γ-^{32}P-ATP	R/FR	Romero *et al.* (1991)
Sorghum	Seedlings	Soluble extracts	γ-^{32}P-ATP	R/FR	Doshi *et al.* (1992)
Wheat	Protoplasts	Protoplasts	^{32}Pi	R/FR	Fallon *et al.* (1993)
Wheat	Protoplasts	Blotted proteins	γ-^{32}P-ATP	WL, R/FR, B	Fallon and Trewavas (1994)
Parsley	Cytosol	Cytosol	γ-^{32}P-ATP	R/FR, WL	Harter *et al.* (1994)
Pea	Seedlings	Membranes	γ-^{32}P-ATP	R/FR	Hamada *et al.* (1996)
Pea	Nuclei	Blotted proteins	Anti-PTyr	R/FR	Tong *et al.* (1996)
Oat	Extracts	Immunoprecipitates	Anti-PTyr	R/FR	Sommer *et al.* (1996)
Wheat	Seedlings	Soluble extracts	γ-^{32}P-ATP	WL	Sharma *et al.* (1997b)

PAGE) and autoradiography. Several demonstrated phytochrome regulation by virtue of R/FR reversibility. Despite this, the best understood of the effects summarized in Table I is the B light-dependent phosphorylation of a 120-kDa plasma membrane protein from pea stem segments (Gallagher *et al.*, 1988), because this was the original observation that led to the discovery of phototropin. The features of phototropin are discussed in detail in a later section.

In two other cases it has been possible to establish the identity of the protein whose phosphorylation status is changed by light. One of these is the phytochrome effect on labeling of nucleoside diphosphosphate kinase in peas (Hamada *et al.*, 1996; Tanaka *et al.*, 1998). The other is noteworthy because it utilizes purified cytosol obtained from evacuolated protoplasts from parsley (Harter *et al.*, 1994). Harter *et al.* (1994) found exceedingly rapid labeling of several polypeptides upon irradiation of the cytosol. One of the polypeptides whose labeling increased in a R/FR-dependent manner was approximately 44 kDa. Wellmer *et al.* (1999) noted that this labeled band was about the same size as CPRF2, a bZIP transcription factor from parsley. They immunoprecipitated CPRF2 from ^{32}Pi-labeled, evacuolated protoplasts and showed that its labeling was strongly R/FR-dependent. Moreover, recombinant CPRF2 becomes extensively labeled in the cytosol. The likely CPRF2 kinase is approximately 40 kDa and is part of a high molecular weight complex containing CPRF2 itself. Phosphorylation is known to affect the DNA binding of some transcription factors, but Wellmer *et al.* (1999) found no apparent change in CPRF2 binding that was phosphorylation dependent. If the rapid, phytochrome-triggered phosphorylation of CPRF2 is not involved in regulating DNA recognition, then what is its role? An important clue is provided by Kircher *et al.* (1999). CPRF2 is localized to the cytoplasm of dark-adapted parsley cells and is imported to the nucleus only in response to phytochrome activation (Kircher *et al.*, 1999b). Thus, it seems reasonable to speculate that the phosphorylation of CPRF2 is involved in nuclear import – possibly its existence in a large cytosolic complex might anchor it there until phytochrome triggers its phosphorylation (Wellmer *et al.*, 1999).

III. PHOTOREGULATED PROTEIN KINASE GENES

With the idea of identifying protein kinases that might transduce light signals, kinase genes, whose expression is regulated by light, were identified. The underlying hypothesis was that light could conceivably exert an effect on kinase signaling over very short times by modulating its activity and over slightly longer times by changing the amount of enzyme protein by altering the gene expression. Using reverse transcriptase polymerase chain reactions (PCR), a suite of five partial complementary DNA (cDNA) clones, which encode distinct forms of protein–serine/threonine kinase homologs from the garden pea, were obtained (Lin *et al.*, 1991). These partial cDNAs, called *PsPK1* through to *PsPK5*, correspond to rare class messenger RNAs (mRNAs) that are encoded by single copy nuclear genes. The PsPK3 sequence predicts that it is localized to the nucleus (Khanna and Watson, 1997). On the other hand, PsPK4 (Mitra and Watson, unpublished) and PsPK5 (Lin and Watson, 1992) are close relatives of *Arabidopsis* phototropin (Huala *et al.*, 1997). The five *PsPK* genes are differentially expressed in apical buds during de-etiolation in continuous white light (WLc). Exposure of etiolated pea seedlings to 24 h of WLc causes *PsPK3* and *PsPK5* mRNAs to decline to their minimum levels within apical buds. In contrast, *PsPK4* RNA increases gradually over a period of days in light, while *PsPK1* and *PsPK2* transcripts show little change (Lin *et al.*, 1991). The *Arabidopsis PK3At* gene encodes a protein that is quite closely related to *PsPK3* (Ma *et al.*, 1998). As with *PsPK3*, the *PK3At* transcript declines when dark-grown seedlings are transferred to WLc (Khanna *et al.*, unpublished results). To determine whether the *PsPK3* and *PsPK5* mRNAs could reaccumulate after reaching their minimum levels, seedlings were given 24 h of WLc and then returned to darkness for an additional 24 h. Unlike *PsPK3, PsPK5* mRNA does reaccumulate several-fold during the dark period, indicating that light does not permanently switch off *PsPK5* expression (Lin, 1993). To gain insight into the rapidity with which the expression changes, 6-day-old, dark-grown pea seedlings were transferred to WLc, and *PsPK3* and *PsPK5* RNA levels monitored over the next 24 h. The *PsPK3* and *PsPK5* mRNAs decline rapidly to their minimum levels while transcripts from the *RBCS* gene family increase. *PsPK3* mRNA declines four-fold in ~8 h, whereas *PsPK5* mRNA declines 10-fold in ~2 h (Khanna *et al.*, 1999).

A considerable body of evidence indicated that inductive single pulses of light given to dark-grown seedlings triggered virtually all known phytochrome- and B-light photoreceptor-mediated gene responses (Watson, 1989; Thompson and White, 1991). Hence, single pulses of light were used to elucidate which photoreceptor triggers the negative regulation of *PsPK3* and *PsPK5* gene expression (Khanna *et al.*, 1999). To assess phytochrome involvement, etiolated seedlings were treated with single pulses of R, R followed by FR, or FR light alone. *RBCS* induction by a R pulse is reversed by a subsequent FR light pulse, clearly showing that phytochrome mediates its induction. Similarly, *RBCS* expression is induced with a single pulse of B or a dichromatic pulse of R + B light. However, none of these pulses triggers the *PsPK3* and *PsPK5* mRNA levels to decline in apical buds. Taken together, these results show that the *PsPK3* and *PsPK5* genes do not exhibit the classic responses of etiolated seedlings to light pulses: the very low fluence and low-fluence responses are driven by phytochrome (Mancinelli, 1994), or the low-fluence B-light response is mediated by one or more of the blue light photoreceptors (Kaufman, 1993; Short and Briggs, 1994).

Given the lack of effectiveness of light pulses, etiolated seedlings were transferred to continuous light of three different qualities to determine the spectral sensitivity of *PsPK3* and *PsPK5* gene expression (Khanna *et al.*, 1999).

Exposure to Rc, FRc, or Bc causes the *PsPK3* and *PsPK5* mRNAs to decline and transcripts from the *RBCS* and *CAB* gene families to increase in apical buds. The most likely explanation is that phytochrome A mediates the responses of these genes to FRc. The effectiveness of Rc and Bc in triggering the reduction in *PsPK3* and *PsPK5* mRNA levels and the increase in *RBCS* and *CAB* mRNAs may imply the participation of additional phytochromes and/or B-light photoreceptors. Thus, the *PsPK3* and *PsPK5* genes exhibit responsiveness to continuous light, but a lack of responsiveness to single light pulses, which is unusual, and perhaps unique, among light-regulated genes.

To gain insight into expression during seedling development and in different organs, apical buds and plumular hooks were examined on a daily basis from 3–7-day-old pea seedlings that were either grown in complete darkness or exposed to WLc for 1 day prior to harvest (Khanna, 1998). Both the *PsPK3* and *PsPK5* transcripts are at their highest levels in 3-day-old seedlings, and exhibit a small decline during seedling development in darkness. Even in young, 3-day-old seedlings in both buds and hooks, transcript accumulation from both genes is light responsive. However, the extent of photoresponsiveness increases during seedling development in both organ systems. Interestingly, transcript levels from both *PsPK3* and *PsPK5* are more light responsive in hooks than in buds at later stages of seedling development. This trend is especially pronounced with *PsPK5* mRNA in 7-day-old seedlings, which has a dark-to-light ratio of approximately 50 in hooks and 10 in buds.

The rapid decline of *PsPK3* and *PsPK5* mRNAs must mean that the mRNAs turn over rapidly in WLc. One simple model for *PsPK* expression would be that the mRNAs are inherently unstable because they contain instability determinants. In this model, exposure of etiolated seedlings to light rapidly results in a downregulation of *PsPK3* and *PsPK5* gene transcription. This, coupled with light-independent, inherent instability of the mRNAs, would lead to rapid declines in transcript levels. This basic mechanism accounts for the downregulation of *PHYA* mRNA in oat (Seeley *et al.*, 1992; Quail, 1994) and *PorA* mRNA in barley (Holtorf *et al.*, 1995; Holtorf and Apel, 1996). It is tempting to speculate that the *PsPK3* and *PsPK5* mRNAs are unstable because they contain sequence motifs within their 3′ untranslated regions (UTRs) that are reminiscent of known instability determinants in plant mRNAs. One of these is the $A(U)_3A$ motif (Johnson *et al.*, 1998): the *PsPK5* 3′ UTR contains two copies of the $A(U)_3A$ motif while *PsPK3* contains one. The second *cis*-acting instability element is the DST element originally found in the *SAUR* genes (Johnson *et al.*, 1998): the *PsPK3* 3′ UTR contains a reasonable match to the DST consensus while *PsPK5* does not. A potential DST element resides in the 3′ UTR of barley *PorA* mRNA (Holtorf and Apel, 1996). An alternative model is that light regulates the stability of the *PsPK* mRNAs. Light-dependent alteration of mRNA stability has been implicated in regulation of the ferredoxin gene in pea and *Arabidopsis*, the *RBCS* genes in several species (Abler and Green, 1996), and more recently of the pea *Lhcb1*4* gene (Anderson *et al.*, 1999). For many of the light-regulated genes where changes in mRNA stability occur, there are increases in transcription as well (Terzaghi and Cashmore, 1995; Abler and Green, 1996). Thus, a third possibility for photoregulation of the *PsPK3* and *PsPK5* genes is that both transcription and mRNA stability are subject to control.

To ascertain whether light affects the abundance of the polypeptides encoded by *PsPK3* and *PsPK5*, antibodies were prepared against their catalytic domains expressed in *Escherichia coli*, and used in immunodetections on protein gel blots of pea seedling protein extracts (Khanna, 1998). The levels of the polypeptides detected by the PsPK3 antiserum showed little if any decrease when 6-day-old seedlings were exposed to 6 h of light, but were significantly lower in both buds and hooks following exposure to 1 day of WLc. For the polypeptide detected with the PsPK5 antiserum, exposure to 6 h of WLc caused a slight decline in polypeptide in hooks, but no apparent change in buds. Its abundance decreased slightly in buds, but more substantially in the hooks, following 1 day of WLc. The results suggest that light can negatively regulate accumulation of the polypeptides encoded by the *PsPK3* and *PsPK5* mRNAs.

Besides *PsPK3* and *PsPK5*, several other protein kinase genes respond to light at the mRNA level. When etiolated wheat seedlings are exposed to WLc, the *wpk4* transcript increases rapidly (Sano and Youssefian, 1994). The *wpk4* gene encodes a SNF-like kinase that also responds to hormonal and nutritional signals. The light-repressible receptor protein kinase (*llrpk*) gene from *Arabidopsis* encodes a receptor-like kinase whose transcripts decline in abundance when etiolated seedlings are irradiated with a pulse of UV, blue, red, or far-red light (Deeken and Kaldenhoff, 1997). The wheat *TaPK3* and *CAB* transcripts appear with similar kinetics in seedlings growing in WLc, suggesting that *TaPK3* may be photoregulated (Holappa and Walker-Simmons, 1997). *TaPK3* is closedly related to the ABA-regulated *PKABA1* gene. Transcripts of the zucchini *CpCPK1* gene, encoding a calcium-dependent protein kinase (CDPK) are more abundant in the hooks and hypocotyls of dark-grown seedlings than in light-grown seedlings (Ellard-Ivey *et al.*, 1999). Lastly, the cucumber *CsPK3* gene encodes a protein that is closely related to *PsPK3* (Chono *et al.*, 1998). In etiolated hypocotyls, *CsPK3* mRNA accumulation is strongly induced by auxin, and less so by gibberellin. Transfer of etiolated seedlings from darkness to WLc triggers a rapid and substantial decline in *CsPK3* mRNA level. Since it is responsive to both auxin and light, it will be extremely interesting to determine whether CsPK3 is involved in stem elongation and the suppression of elongation that occurs when dark-grown seedlings are illuminated.

IV. CKII AND LIGHT-REGULATED GENES

CKII, formerly known as casein kinase II, is a serine/threonine kinase composed of two regulatory α subunits and two catalytic β subunits. CKII is known to be involved in a wide variety of cellular functions, but has attracted considerable attention because among its substrates are transcription factors (Schwechheimer *et al.*, 1998). In several cases, modification of DNA binding properties results from phosphorylation of a transcription factor by CKII (Schwechheimer *et al.*, 1998). Two recent reports link CKII with photo-regulated gene expression in plants. Lee *et al.* (1999) used transgenic *Arabidopsis* plants expressing an antisense construct for the α subunit. Etiolated seedlings from the antisense lines expressed higher levels of *CHS* mRNA, whereas the response to R for the *RBCS* and *CAB* genes was increased. One attractive interpretation of theses results is that CKII may be a negative regulator of photoregulated gene expression (Lee *et al.*, 1999). It will be extremely interesting to determine whether the alteration in light-responsive gene expression is via aberrant transcription factor phosphorylation.

The second system not only links CKII and light-regulated gene expression but also involves circadian rhythms. The *Arabidopsis* gene called *CIRCADIAN CLOCK ASSOCIATED 1 (CCA1)* encodes a myb-type transcription factor that binds to the regions of a *Lhcb* (also known as *CAB*) promoter that is sufficient to confer circadian rhythmicity (Wang *et al.*, 1997). Indeed, *CCA1* mRNA and protein levels themselves oscillate with a circadian rhythm. To test the hypothesis that CCA1 may be closely associated with the central oscillator of the circadian clock, the *CCA1* gene was overexpressed in *Arabidopsis*. Consistent with this idea, the rhythm of several clock-regulated genes was abolished in the overexpressing lines (Wang and Tobin, 1998). A similar effect was seen when a close relative of *CCA1*, called *LATE ELONGATED HYPOCOTYL (LHY)*, was overexpressed (Shaffer *et al.*, 1998). Using the yeast two-hybrid screen for CCA1 interaction partners identified a novel β subunit for CKII called CKB3 from *Arabidopsis* (Sugano *et al.*, 1998). Recombinant CKII phosphorylates CCA1 *in vitro*. Significantly, CKB3 associates with CCA1 and promotes DNA binding. Moreover, *Arabidopsis* extracts contain a protein that phosphorylates CCA1 and affects complex formation. The tantalizing suggestion is that CKII plays a role in the function of CCA1 *in vivo* (Sugano *et al.*, 1998). Because this seems likely, and because the central oscillator of the clock is closely linked to CCA1, it is tempting to speculate that CKII plays an important role in circadian rhythms.

V. PHOTORECEPTOR KINASES AND THEIR RELATIVES

The structures of the photoreceptors that either have been shown or are very likely to possess protein kinase activity are shown diagrammatically in Fig. 1.

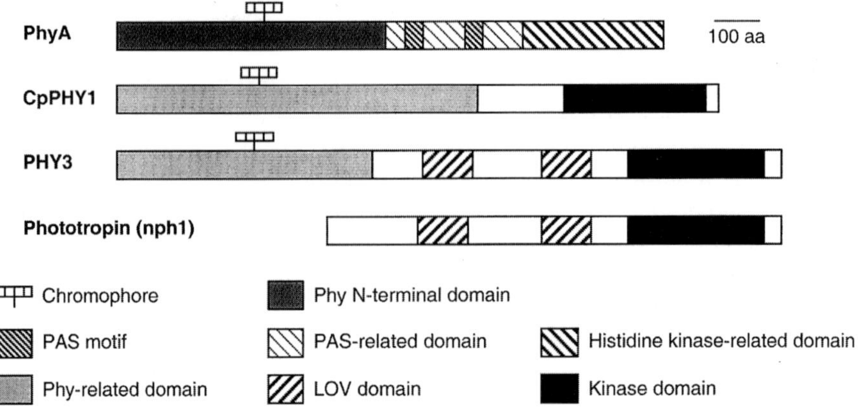

Fig. 1. Structures of photoreceptor kinases in plants.

At the present time, it is clear that photoreceptors active in both the R and B regions of the spectrum act as protein kinases. These are the phytochromes and phototropin, respectively. Their general structure will be discussed here, while additional information concerning their kinase activity will be considered in more detail in following sections.

The phytochromes, which form dimers via the C-terminal region, have an amino-terminal chromophore binding domain (Quail, 1997). In a rather fascinating process, the tetrapyrrole chromochore phytochromobilin autocatalytically attaches at a specific cysteine of the apoprotein, forming the spectrally active holoprotein (Lagarias and Lagarias, 1989; Murphy and Lagarias, 1997). The C-terminal half of phytochrome is occupied by a PAS-related domain (PRD) and a histidine kinase-related domain (HKRD) (Quail, 1997; Yeh *et al.*, 1997; Yeh and Lararias, 1998). The PRD contains two PAS motifs, which, if they behave as other known PAS domains, may be involved in mediating protein–protein interactions (Quail, 1997; Yeh *et al.*, 1997; Yeh and Lararias, 1998). Interestingly, the 'core region' where the majority of missense mutations that disrupt phytochrome function cluster is the area in and between the two PAS domains (Quail, 1997; Yeh and Lararias, 1998). Also of note is the recent realization that the PRD exhibits sequence relatedness to prokaryotic histidine kinases (Yeh *et al.*, 1997; Yeh and Lararias, 1998). The *CpPHY1* gene of the moss *Ceratodon purpureus* encodes a phytochrome that has a kinase catalytic domain at its C-terminus (Thümmler *et al.*, 1992). The kinase domain of CpPHY1 exhibits sequence relatedness to *Dictyostelium* tyrosine kinases, the Raf family, and to the HKRD of phytochrome (Thümmler *et al.*, 1992, 1995a). However, when transiently overexpressed in fibroblasts CpPHY1 phosphorylated serine and threonine (Thümmler *et al.*, 1995b). Interestingly, *Ceratodon* has a second phytochrome gene, *CpPHY2*, that encodes a conventional phytochrome (Pasentsis *et al.*, 1998).

The photoreceptor kinase at the B end of the spectrum is phototropin, the recently named B-light photoreceptor for phototropism (Christie *et al.*, 1999). Phototropin, encoded by the *NPH1* gene of *Arabidopsis*, has a long N-terminal region containing two LOV domains (Huala *et al.*, 1997). The LOV domains, which resemble PAS domains, act as the binding site for flavin chromophores and might mediate protein–protein interactions (Christie *et al.*, 1998, 1999). At the C-terminus of phototropin exists a serine/threonine kinase catalytic domain. The fern *Adiantum capillus-veneris* possesses a rather amazing 'chimera' between phytochrome and phototropin called PHY3 (Nozue *et al.*, 1998). *PHY3* was isolated using the chromophore domain probe from one of the conventional *Adiantum* phytochromes. At the N-terminus of PHY3 is a phytochrome chromophore domain, but the bulk of the protein exhibits high sequence relatedness to, and has the same general architecture as, NPH1. A chromophore adduct of recombinant PHY3 has an absorption difference spectrum similar to other reconstituted, recombinant phytochromes (Nozue *et al.*, 1998). Moreover, the LOV domains of PHY3 bind flavin (Christie *et al.*, 1999). In the future, it will be extremely interesting to understand the photosensory functions of PHY3 including triggering responses to both B and R.

In their classification scheme for the catalytic domains of eukaryotic protein kinases, Hanks and Hunter (1995) divided the serine/threonine kinases into three groups: the AGC group [named for cyclic adenosine monophosphate (cAMP)-dependent protein kinase (PKA), cGMP-dependent protein kinase (PKG), and protein kinase C (PKC)], the CaMK group (calcium/calmodulin regulated kinases), and the C-M-G-C group [named for cyclin-dependent kinases (CDKs), mitogen-activated protein (MAP) kinases, glycogen synthase kinase 3 (GSK3), and cyclin-like kinases (Clk)]. The NPH1 catalytic domain is a member of the AGC group, as are several other plant kinases (see Fig. 2). More specifically, the NPH1 kinase domain belongs within the AGC-VIII group of the Hanks and Hunter classification (Hanks and Hunter, 1995). There are two notable differences between the AGC-VIII kinases and the vast majority of other catalytic domains. The first of these is within subdomain VII, which most often contains the highly conserved DFG motif. In the AGC-VIII kinases, the DFG motif is replaced by DFD. This is interesting because the aspartate of the DFG motif is involved in chelating the activating Mg^{2+} ions that bridge the β and γ phosphates of ATP. There are nearly 50 kinase sequences in the database that have the DFD motif at subdomain VII, and of these all but three are from plants. Those that are not are all fungal (nrc-2 of *Neurospora crassa*, KIN82 from budding yeast and its homolog from fission yeast). The second notable difference is the presence of an insertion between subdomains VII and VIII when compared to the same region of PKA. The length and sequence of this insert varies from kinase to kinase, and can be thought of as a signature motif for each kinase. The length difference between

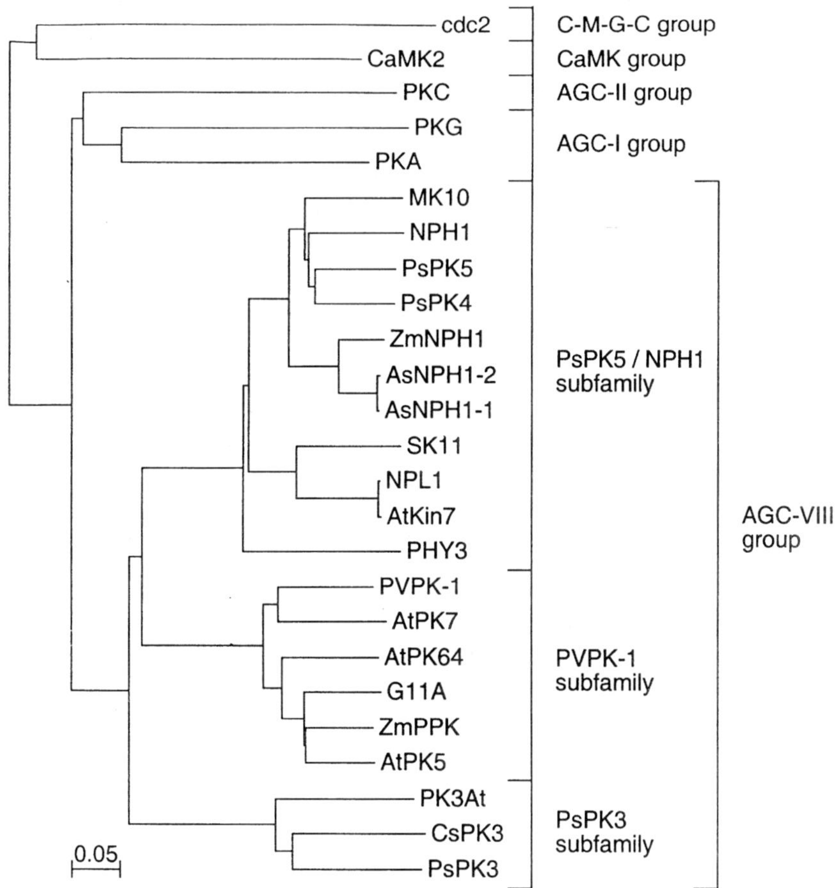

Fig. 2. Sequence relationships among the catalytic domains of protein kinases of the AGC-VIII group. Catalytic domains were first defined by pairwise alignment with the kinase domain of bovine PKA (Hanks *et al.*, 1988; Knighton *et al.*, 1991; Hanks and Hunter, 1995), then multiple alignments were generated with CLUSTAL W version 1.7 (Thompson *et al.*, 1994), and the tree was visualized with NJplot. CaMK and cdc2 were included as an outgroup of other serine/threonine kinases. Sequences used with organism and accession numbers in parentheses: AsNPH1-1 (Oat, AF033096), AsNPH1-2 (Oat, AF033097), AtKin7 (*Arabidopsis*, U79744), AtPK5 (*Arabidopsis*, D10909), AtPK7 (*Arabidopsis*, D10910), AtPK64 (*Arabidopsis*, D10937), CaMK2 (Bovine, J02942), cdc2 (*S. pombe*, M12912), CsPK3 (Cucumber, D85618), G11A (Rice, J04556), MK10 (Iceplant, Z30333), NPH1 (*Arabidopsis*, AF030864), NPL1 (*Arabidopsis*, AF053941), PHY3 (*Adiantum*, AB012082), PKA (Bovine, X67154), PKC (Bovine, M13947), PKG (Bovine, X16086), PK3At (*Arabidopsis*, AF082391), PsPK3 (Pea, U11553), PsPK4 (Pea, U83281), PsPK5 (Pea, M92989), PVPK-1 (Bean, J04555), SK11 (Spinach, Z30332), ZmNPH1 (Maize, AF033263), ZmPPK (Maize, M62985).

the inserts is the primary determinant for the differences in catalytic domain lengths within the AGC-VIII group and in comparison to PKA. Because of the role of subdomain VII in Mg^{2+} chelation and subdomain VIII in recognition of peptide substrates (Knighton *et al.*, 1991; Hanks and Hunter, 1995), it is tempting to speculate that these inserts affect catalytic activity in ways that are yet to be discovered.

Hanks and Hunter (1995) also called the AGC-VIII group the PVPK-1 family since PVPK-1 was the first kinase of this type to be cloned (Lawton *et al.*, 1989). The original members of the AGC-VIII group were PVPK-1, rice G11A, AtPK5, AtPK7, AtPK64, ZmPPK, and PsPK5 (Hanks and Hunter, 1995). All of these catalytic domains are included in the phylogeny shown in Fig. 2, as well as relatives of PsPK5 and PsPK3 that have been reported more recently. The 20 plant sequences compared define three distinct branches within the AGC-VIII group, so it is likely that the group comprises (at least) three subfamilies: the PVPK-1 subfamily, the PsPK5/NPH1 subfamily, and the PsPK3 subfamily. Several additional members of the AGC-VIII group are present in the database, but they all cluster with the PVPK-1 subfamily and for simplicity are not included in Fig. 2. The first branch within the PsPK5/NPH1 subfamily leads to *Adiantum* PHY3, and then another branch leads either to the monocot and dicot NPH1s or to a group containing NPL1 (Jarillo *et al.*, 1998). The predicted NPL1 polypeptide is slightly smaller at the N-terminus than NPH1, contains two LOV domains, and has a C-terminal kinase domain. What role, if any, is played by NPL1 in B sensing remains to be elucidated

The closest relative of the PsPK4 and PsPK5 catalytic domains is that from NPH1. The PsPK4 and PsPK5 catalytic domains are 82% identical to the NPH1 catalytic domain, and 84% identical to each other. Moreover, conceptual translation of the open reading frames upstream of the first methionine codons in the *PsPK4* and *PsPK5* cDNAs extends well into the LOV2 domain, and is over 90% identical to NPH1. This overall level of identity makes it extremely likely that PsPK4, PsPK5, and NPH1 are functionally related. Under conditions where the cDNA clones do not cross-hybridize, Northern blots reveal *PsPK4* and *PsPK5* transcripts of sufficient length to encode NPH1 in addition to shorter transcripts (Khanna *et al.*, 1999; Lin *et al.*, 1991). It seems reasonable to suspect that these longer transcripts encode polypeptides containing the amino-terminal domain of NPH1 that is 'missing' from the existing *PsPK4* and *PsPK5* cDNA sequences. Given that *NPH1* has 19 introns, perhaps the shorter *PsPK4* and *PsPK5* transcripts observed in pea arise by alternate splicing. Work is underway to test the hypothesis that the *PsPK4* and *PsPK5* genes represent a small multigene family from peas that encodes NPH1-like B-light photoreceptors.

The catalytic domains of PsPK3, PK3At, and CsPK3 also belong to the AGC-VIII group. The terminal domains of PsPK3, PK3At, and CsPK3 have striking features: (1) the N-terminal domains are rich in serines and threonines, and thus contain several potential phosphorylation sites; and (2) the C-terminal domains are basic, being rich in lysines and arginines (as in many nucleic acid binding proteins). The basic domains of PsPK3 and PK3At each contain a potential, bipartite nuclear localization signal. The latter observation leads to the hypothesis that PsPK3 is localized to the nucleus. The potential nuclear location of *PsPK3* is interesting because there are phytochrome-mediated, R/FR reversible changes in the phosphorylation of nuclear proteins (Datta *et al.*, 1985; Romero *et al.*, 1991), and the recent realizations that phytochrome can reside in the nucleus (Sakamoto and Nagatani, 1996; Kircher et al., 1999b) and interact with potential transcription factors (Neuhaus *et al.*, 1993; Ni *et al.*, 1998, 1999).

VI. PHOTOTROPIN (NPH1)

Phototropism, a response in which a plant redirects its growth in relation to the direction of light, has been intensively investigated to understand differential growth in plants and to elucidate the nature of B-light photoreceptors. In the laboratory, unilateral light is used to elicit the differential growth response, which is measured as curvature. The fluence dependence of phototropism is decidedly complex (Firn, 1994; Poff *et al.*, 1994). A typical fluence response curve for B-induced curvature is shown diagrammatically in Fig. 3. The first positive phototropic curvature is induced with brief pulses of light and obeys the law of reciprocity. First positive curvature increases to a maximum and then decreases, exhibiting a bell-shaped fluence response curve, followed by a zone of indifference where there is basically no curvature. With still higher

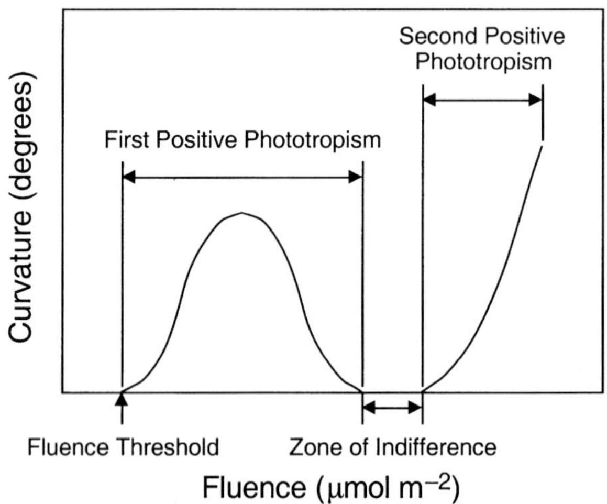

Fig. 3. Idealized fluence response curve for phototropism. Adapted from Firn (1994) and Poff *et al.* (1994).

fluences, the second positive phototropic curvature is triggered, which usually requires prolonged exposures and is time dependent, meaning that it does not obey the reciprocity law. Because of this reciprocity failure, most work has centered on first positive curvature.

A major contributing factor to the complexity of the fluence response for phototropism is photosensory adaptation (Poff *et al.*, 1994). Simply put, once a plant is illuminated, adaptation is a change in the plant's sensitivity and/or responsiveness to a subsequent illumination. In *Arabidopsis*, adaptation in phototropism has distinct components (Poff *et al.*, 1994). There is a rapid desensitization brought on by B during which time there is no response to a second irradiation. After a refractory period, there is a slow recovery of the response to unilateral B, followed even more slowly by an enhancement of curvature. The enhancement caused by preirradiation is mediated by multiple phytochromes in *Arabidopsis* (Janoudi *et al.*, 1997). Phototropism is usually thought of as purely a B-light response because most action spectra show peaks in the B/UV-A region of the spectrum (Firn, 1994; Poff *et al.*, 1994). However, yet another level of complexity is that there are clear instances where phytochrome is implicated at the primary photoreceptor for phototropism, for example in dark-adapted maize (Iino *et al.*, 1984) and etiolated pea seedlings (Parker *et al.*, 1989).

A. BLUE-LIGHT-INDUCED PHOSPHORYLATION

The identification of phototropin began with the observation that when microsomal membranes from pea stem segments were incubated with γ-^{32}P-adenosine triphosphate (ATP), one of the major polypeptides that became labeled was 120 kDa (Gallagher *et al.*, 1988). Significantly, *in vitro* labeling of the 120-kDa protein was affected by prior irradiation of the stem segments with B but not R (Gallagher *et al.*, 1988). Thus it was hypothesized that the phosphorylation was an early step in a signaling chain triggered by a B-light photoreceptor. This led to several careful studies of the B-light-dependent phosphorylation of the 120-kDa protein by Briggs and colleagues establishing strong linkage between the phosphorylation and first positive phototropic curvature (reviewed by Short and Briggs, 1994).

Not only could the B-dependent labeling of the 120-kDa protein be detected by incubating membranes with γ-^{32}P-ATP, but it became labeled in response to B when stem segments were fed ^{32}Pi (Short *et al.*, 1992). It was found that labeling of the 120-kDa protein could be altered by treating either stem segments or isolated membranes with B (Short and Briggs, 1990), thus greatly facilitating further analyses. The phosphoprotein is found in all plant species examined but varies slightly in size between species (Reymond *et al.*, 1992a). For convenience, this protein will simply be referred to here as the 120-kDa phosphoprotein. The size difference between species proved useful because

Reymond *et al.* (1992) were able to show that B-treated membranes were able to cause labeling of the 120-kDa protein in unirradiated membranes. Thus the 120-kDa phosphoprotein is itself a protein kinase (Reymond *et al.*, 1992a). It becomes labeled primarily on serine, with only traces of phosphothreonine (Palmer *et al.*, 1993b; Short *et al.*, 1994). Its activity is enriched in plasma membrane fractions (Gallagher *et al.*, 1988; Hager and Brich, 1993a; Palmer *et al.*, 1993b; Short *et al.*, 1993; Salomon *et al.*, 1996; Sharma *et al.*, 1997a) and is highest in the tissues that exhibit phototropic sensitivity (Short and Briggs, 1990; Hager and Brich, 1993a; Palmer *et al.*, 1993a; Salomon *et al.*, 1996; Sharma *et al.*, 1997a). B light triggers a detectable increase in labeling within seconds that reaches a maximum within 1–2 min (Short *et al.*, 1992; Hager and Brich, 1993a; Palmer *et al.*, 1993b; Short and Briggs, 1994). The spectral sensitivity between 340 and 500 nm for the phosphorylation reaction and phototropism is quite similar (Hager and Brich, 1993a; Palmer *et al.*, 1993a). Fluences of B that trigger the phosphorylation reaction also elicit phototropism (Short and Briggs, 1990; Short *et al.*, 1992; Palmer *et al.*, 1993a; Salomon *et al.*, 1996, 1997a, b). Significantly, the fluence dependence (quantum efficiency) of phosphorylation is the same before and after solubilization of the membranes with triton, suggesting a very tight association of all the requisite functions (Short *et al.*, 1993). The kinetics of recovery in darkness (the time to return to ground state after the light pulse) are very similar for B-induced phosphorylation and phototropism (Short and Briggs, 1990; Hager and Brich, 1993a; Palmer *et al.*, 1993a). Flavin antagonists (Short *et al.*, 1992) as well as redox (Hager and Brich, 1993b) and thiol reagents (Rüdiger and Briggs, 1995) affect the phosphorylation reaction. Taken together, this substantial body of correlative evidence is consistent with the notion that, at the least, the 120-kDa protein is an autophosphorylating protein kinase that acts at a very early stage in the perception of phototropic stimuli. Alternatively, the 120-kDa protein could be the B-light photoreceptor itself.

It seems reasonable to hypothesize that the light gradient across the stem produced by unilateral light (of proper fluence) should establish a difference in the amount of activated photoreceptor between the illuminated and shaded sides. Using an extremely sensitive micromethod allowed Salomon *et al.* (1997a, b) to measure the capacity of the B-light-dependent phosphorylation in sections of oat coleoptiles. Their results provide important insight into the complexity of the fluence dependence of phototropism. First, they established that the amount of the B-induced phosphorylation of the 120-kDa protein decreases exponentially from tip to base in the coleoptile. This correlates well with phototropic sensitivity, which also falls off sharply with distance from the tip. Next, they subjected coleoptiles to unilateral illumination and measured phosphorylation capacity on the lit and shaded sides of the coleoptile tips and basal segments. In the basal segments, there was little difference except at the highest fluence used. In contrast, there were significant fluence-dependent differences between the illuminated and shaded sides of the coleoptile tip.

Importantly, when the difference in phosphorylation of the 120-kDa protein between the illuminated and shaded sides was plotted against fluence, a bell-shaped curve was obtained. Although shaped just like the fluence response curve for first positive phototropic curvature, the phosphorylation curve had a fluence threshold approximately 100-fold higher than curvature. A higher threshold for the phosphorylation reaction probably results from the existence of nonresponsive cells in the tissue harvested for assay or from phosphorylation on multiple sites *in vitro* where phosphorylation of only a few sites *in vivo* may trigger the response (Palmer *et al.*, 1993a; Salomon *et al.*, 1997a). Nevertheless, we at last have insight into why the fluence response curve for phototropism is bell shaped (Fig. 3). On the ascending arm of the curve, as fluence increases so does the difference in phosphorylation between the illuminated and shaded sides up to the point where phosphorylation on the illuminated side is saturated. Then on the descending arm of the curve at even higher fluences, phosphorylation increases on the shaded side up to saturation, resulting again in a net difference of zero between the lit and shaded sides (Salomon *et al.*, 1997a).

B. STRUCTURE AND FUNCTION OF PHOTOTROPIN

Reymond *et al.* (1992) took advantage of two different mutants of *Arabidopsis* that were defective in phototropism. One of the mutants that they analyzed, the *JK224* mutant, was thought to be a photoreceptor mutant and was 20 to 30 times less sensitive to B in triggering first positive phototropism than wild type. The other mutant, *JK218*, did not respond to up to 2 h of unilateral light and was in a different complementation group than *JK224*. B-induced phosphorylation of the 120-kDa protein in *JK224* was greatly diminished compared to wild type or the *JK218* mutant (Reymond *et al.*, 1992b). This genetic evidence strongly supported the proposal that the light-induced phosphorylation plays a role early on during the transduction of the phototropic signal.

Liscum and Briggs (1995) identified and characterized several mutants of *Arabidopsis* that either completely lacked or were substantially impaired in second positive phototropic curvature. The mutants fell into four complementation groups, called *nph1* through to *4*, for nonphototropic hypocotyl. They found that the *JK224* mutation was allelic to *nph1*, and since *JK224* was thought to be a photoreceptor mutant, initial characterization centered on the *nph1* allelic series. Strong alleles of *nph1* failed to exhibit second positive curvature in response to UV-A, green, and B. This was an important observation because previous analyses predicted that at least two pigments, one more active in the green wavelengths and another more active in the B/UV-A region of the spectrum, controlled phototropism. One possible explanation was that the *NPH1* gene product bound more than one chromophore. Of extreme significance is that the *nph1* mutants not only are defective in phototropism, but lack the B-dependent phosphorylation of the 120-kDa protein and the corresponding stained polypeptide on the gels. Light-induced phosphorylation was normal in the other three *nph* mutants. Thus it seemed likely that the *NPH1* gene encoded the apoprotein that bound two or more chromophores to form the photoreceptor holoprotein for phototropism. Moreover, a mutant at *hy4* (a.k.a. *cry1*) exhibited normal phototropism and phosphorylation (Liscum and Briggs, 1995). Since *hy4* mutants are defective in the suppression of hypocotyl elongation under B (Ahmad and Cashmore, 1993, 1996), it was clear that the photocontrol of hypocotyl elongation and phototropism were genetically separable, and that the CRY1 and NPH1 proteins were not functionally redundant.

To facilitate positional cloning, several of the *nph1* null alleles were generated by fast neutron irradiation (Liscum and Briggs, 1995; Huala *et al.*, 1997). The wild-type genomic clone, when transformed into the *nph1–5* mutant background, fully restores curvature in response to unilateral light (Huala *et al.*, 1997). Three mutant alleles were also sequenced, and each contained a lesion within the gene. Thus, the gene cloned is *NPH1*. The predicted NPH1 polypeptide is 996 amino acids with a predicted size of 112 kDa, and has a serine/threonine kinase catalytic domain located at the C-terminus (see Fig. 1). Within the large N-terminal domain, NPH1 contains two repeats of 107 amino acids, called LOV1 and LOV2 (see Fig. 1). The LOV repeats are evolutionarily conserved in certain proteins from prokaryotes to eukaryotes; in those cases where a function is known, the proteins are regulated by light, oxygen, or voltage (hence the name LOV). In two of these cases, redox sensing seems to occur through a flavin prosthetic group (Huala *et al.*, 1997). As for NPH1, it was tempting to hypothesize that the LOV domains bind flavin-type chromophores and modulate activity of the protein kinase domain via B-induced redox changes.

Christie *et al.* (1998) used a baculovirus vector to express NPH1 in insect cells (BacNPH1). Most of the BacNPH1 was insoluble, the remainder being soluble as opposed to being membrane associated, as is the case with the plant-derived protein. Nevertheless, when isolated from dark-grown cells, BacNPH1 became phosphorylated in response to B. Moreover, the fluence dependence and time course of autophosphorylation of BacNPH1 closely matched those of the 120-kDa protein in *Arabidopsis* membranes. Clearly, the NPH1 from insect cells had acquired chromophore(s) and its kinase activity was activated by B. The next step was to identify the chromophore. The insoluble BacNPH1 contained a noncovalently bound pigment that when separated from the protein had fluorescence excitation and emission spectra, as well as an absorption spectrum, typical of a flavin. Thin-layer chromatography revealed that the co-factor is flavin mononucleotide (FMN). Further, the BacNPH1 (containing FMN) has a fluorescence excitation spectrum that is astonishingly similar to the action spectrum for phototropism (Christie *et al.*, 1998). These findings strongly support the idea that *Arabidopsis* NPH1 is a flavoprotein photoreceptor kinase that mediates phototropism.

The next challenge was to test the hypothesis that the LOV domains bind the FMN chromophore. For this, Christie *et al.* (1999) tested the LOV domains from oat and *Arabidopsis* NPH1 or *Adantium* PHY3 (see Fig. 1). The two LOV domains were expressed either singly or in combination in *E. coli*. Both LOV1 and LOV2 from oat NPH1 bind FMN, so both domains are functional in chromophore binding. Quantitation of the amount of FMN bound to the various protein constructs showed that FMN and protein were present in equimolar amounts. All the FMN–protein complexes had similar absorption and fluorescence excitation spectra. Of particular interest was that the absorption spectra of full-length NPH1 and LOV containing FMN exhibit fine structure in the B region similar to that seen in the action spectrum for phototropism, but not in the absorption spectrum of free flavins. Christie *et al.* (1999) concluded that NPH1 represents a novel class of flavoprotein photoreceptor that is distinct from the cryptochromes. This new type of photoreceptor that triggers phototropism was named phototropin – the nph1 holoprotein.

C. DOWNSTREAM OF PHOTOTROPIN

Since phototropin is the B light receptor for phototropism, the remaining *nph* mutations (*nph2, nph3*, and *nph4*) are likely to act downstream of NPH1. Analyses of these genes in Liscum's laboratory are providing substantial insight into the phototropin signaling pathway. NPH2 and NPH3 appear to participate in a phototropism-specific pathway because their gravitropic responses are normal, and so they are not defective in differential growth per se. In contrast, *nph4* mutants are affected in gravitropism and are thus likely to participate directly in differential growth responses downstream of B perception. Indeed, NPH4 is required for a number of differential growth responses in addition to phototropism (Stowe-Evans *et al.*, 1998). Moreover, the *nph4* mutants are resistant to auxin and exhibit major impairment of auxin-dependent gene expression. Taken together, these data show that NPH4 plays an important role in auxin-dependent differential growth (Stowe-Evans *et al.*, 1998).

Motchoulski and Liscum (1999) used positional cloning to obtain the *NPH3* gene. The predicted structure of NPH3 is provocative; it is a novel protein of 795 amino acids, although related sequences exist in the database from plants (but not other taxa). There are four highly conserved regions, one of which encompasses a potential tyrosine phosphorylation site, the tyrosine of which is deleted in the *nph3–2* null mutant. Several other potential serine/threonine phosphorylation sites exist as well. Of special note is the presence of two potential protein–protein interaction domains. The one near the N-terminus is a BTB/POZ domain, while near the C-terminus is a coiled-coil domain. NPH3 is enriched in plasma membrane fractions (as is phototropin, see above). The electrophoretic mobility of NPH3 is altered after treatment of dark-grown seedlings with B, but not R, and in the *nph1–5* background. Motchoulski and Liscum (1999) speculate that perhaps phototropin and NPH3 interact in darkness, preventing NPH3 modification, but B-triggered changes in phototropin, possibly by autophosphorylation, or just the absence of *nph1*, allow NPH3 to be modified. It will be interesting to elucidate the nature of the modification. The yeast two-hybrid system was used to show that the N-terminal domain of NPH1 interacts with the coiled-coil region of NPH3. The interaction *in vitro* is FMN dependent. Based on their data, Motchoulski and Liscum (1999) hypothesize that NPH3 represents a scaffold/adapter protein with which phototropin, and perhaps other signaling proteins, associate to form a phosphorelay system. The importance of scaffold proteins in signaling has become apparent in MAP kinase pathways in yeast and mammalian cells, where the scaffold cannot only facilitate activation of a suite of kinases but also insulate them from stimuli that are irrelevant (Whitmarsh and Davis, 1998). It will be interesting to know whether NPH3 is a substrate for phototropin, and whether NPH3 is a scaffold protein. Understanding the structure and function of a kinase module assembled on its scaffold will provide important insights into plant signal transduction.

Might phototropin govern responses other than phototropism? Thus far, the analyses of *nph1* mutants suggest that lack of phototropism is the only phenotype associated with the defective phototropin. However, there is reason to suspect additional functions, albeit not the primary function. The presence of the B-induced protein kinase activity in parts of the oat coleoptile that do not bend (Salomon *et al.*, 1997a), and in the primary leaf of etiolated wheat seedlings (Sharma *et al.*, 1997a) implies that phototropin might have some functionality in addition to phototropism. If the speculation that PsPK5 is the functional equivalent of nph1 is correct (Khanna *et al.*, 1999), then the presence of PsPK5 in pea buds may also imply a function there. Indeed, one such alternative function has recently been identified. Kaufman's laboratory discovered and characterized two types of responses to B light in seedlings (Kaufman, 1993). These were initially studied in pea seedlings in terms of epicotyl elongation and *CAB* gene regulation (Warpeha and Kaufman, 1989, 1990a, b; Warpeha *et al.*, 1989; Kaufman, 1993). The low fluence (LF) B response is evident both in etiolated seedlings and in seedlings grown in dim Rc, and has a fluence threshold of $<10^{-1}$ μmol m^{-2}. In contrast, the high fluence (HF) B response is evident only in seedlings grown in dim Rc and has a B-light threshold of 10^2–10^3 μmol m^{-2}. Growth in Rc not only saturates the phytochrome system (so that specific B-light action can be observed) but also causes the seedlings to attain a developmental stage necessary for expression of HF B response (Warpeha *et al.*, 1989; Warpeha and Kaufman, 1990b). In contrast to peas, etiolated *Arabidopsis* seedlings exhibit both the LF (Gao and

Kaufman, 1994; Folta and Kaufman, 1999) and HF B response (Anderson *et al.*, 1999) for *CAB* gene expression. The LF B response is maintained in *nph1*, *cry1*, and *cry2* mutant backgrounds (Gao and Kaufman, 1994; Folta and Kaufman, 1999), suggesting that yet another B photoreceptor may exist. In contrast, *nph1* mutants do not exhibit the HF B response for *CAB* gene expression (K. Folta and L. S. Kaufman, personal communication). So here we have a B response unrelated to phototropism that involves phototropin. Moreover, the response is at the level of gene expression, so now we can say that both types of photoreceptor kinase are capable of mediating alterations in gene expression.

VII. PHYTOCHROME

A. PHYTOCHROME IS A PHOSPHOPROTEIN

It has been known for some time that phytochrome is a phosphoprotein, since it could be ^{32}P labelled *in vivo* (Quail *et al.*, 1978), and contained an average of one phosphate per monomer (Hunt and Pratt, 1980). More recently, improved methodologies have been used to elucidate the *in vivo* phosphorylation status of phytochrome. Fast atom bombardment mass spectrometry was used to show that phyA is phosphorylated on Ser7 (Lapko *et al.*, 1997). This is almost certainly the site of phosphorylation detected earlier by McMichael and Lagarias (1990). Electrospray ionization mass spectroscopy was used to sequence oat phyA and examine the phosphorylation status of phyA purified either in the Pr or Pfr form. Phosphorylation at Ser7 was similar in Pr and Pfr. However, a new *in vivo* phosphorylation site was found at Ser598. Unlike the Ser7 site, this new site exhibits a Pr/Pfr difference; phosphorylation is $>1\%$ in Pr but nearly 10% in Pfr (Lapko *et al.*, 1999). Oat Pr is phosphorylated by PKA primarily on Ser17, whereas Ser598 is preferred in Pfr (McMichael and Lagarias, 1990; Lapko *et al.*, 1996). Phosphorylation at Ser598 with PKA, which lies in the 'hinge' region between the chromophore domain and the C-terminus, causes subtle conformational changes that inhibit protease accessibility.

B. PHYTOCHROME IS A PROTEIN KINASE

Borthwick and Hendricks (1960) postulated that phytochrome might exert its effects by virtue of being an enzyme regulating an important metabolic pathway. Despite considerable research on phytochrome, little substantive evidence accumulated to support the notion over the next 25 years. In the interim, photochrome was shown to be a phosphoprotein (see above), and so it must obviously be a substrate for a protein kinase in the plant. Then in the late 1980s, papers from Lagarias's laboratory used protein kinases as probes for light-induced conformational changes in phytochrome (Wong *et al.*, 1986;

McMichael and Lagarias, 1990). While interesting, what was much more provocative in these reports was the solid evidence that protein kinase activity was present in highly purified preparations of phytochrome (Wong *et al.*, 1986, 1989; McMichael and Lagarias, 1990). Among the properties of the 'phytochrome-associated' protein kinase were that it phosphorylated phytochrome and histone H1, with phosphoserine being the predominant phosphoamino acid formed. The Pr form of phytochrome was the preferred substrate over the Pfr form. The activity was stimulated by polycations, such as polylysine and histone H1 iself, and strongly inhibited by pyrophosphate (Wong *et al.*, 1986, 1989). Moreover, purified oat phytochrome could be affinity labeled with ATP analogs, and the labeling was stimulated by polycations (Wong and Lagarias, 1989). It seemed quite likely that the kinase found in purified phytochrome preparations was not just 'phytochrome associated' but might represent an intrinsic activity of phytochrome.

The proposal that phytochrome might be a protein kinase triggered similar experiments in other laboratories. At that time, one argument against phytochrome being a protein kinase was the observation that phytochrome had no obvious sequence relatedness to any of the known eukaryotic kinase catalytic domains (Grimm *et al.*, 1989). Moreover, two different groups reported that protein kinase activity was indeed present in purified phytochrome preparations (Grimm *et al.*, 1989; Kim *et al.*, 1989). In one case, the kinase activity was lost upon additional purification steps (Kim *et al.*, 1989). In the other case, the kinase was separated from phytochrome; a 60-kDa phosphorylated species was suspected to be the kinase and partial amino acid sequences were obtained (Grimm *et al.*, 1989). Interestingly, the sequence was shown much later to be that of avenacosidase, an oat β-glucosidase that is involved in defense responses (Gus-Mayer *et al.*, 1994). More recently, Biermann *et al.* (1994) found that anti-phytochrome immunoprecipitates of maize coleoptile extracts contained protein kinase activity that phosphorylated phytochrome. Phosphorylation was quite specific for Pr and they could detect no phosphorylation of other substrates (Biermann *et al.*, 1994). Interestingly, the Pr specificity was observed when photoconversion was accomplished by irradiation of either seedlings or the immunoprecipitates.

The next serious foray into whether phytochrome was a protein kinase began with the observations by Schneider-Poetsch and colleagues that the C-terminal domain of phytochrome exhibited sequence relatedness to the sensors of bacterial two-component systems (Schneider-Poetsch *et al.*, 1991; Schneider-Poetsch, 1992). That phytochrome had a HKRD was indeed noteworthy. Then, Kehoe and Grossman (1996) identified *rcaE* from the cyanobacterium *Fremyella*. The *rcaE* gene product, involved in sensing light quality during chromatic adaptation, had a region in its N-terminus that was related to the chromophore-binding domain of the phytochromes and a histidine kinase domain (HKD) at its C-terminus (Kehoe and Grossman, 1996). This led to the realization that the *Synechocystis* genome contained a phytochrome-like open reading frame. Significantly, when the recombinant apoprotein was reconstituted *in vitro* with chromophore, a holoprotein formed with the spectral properties of phytochrome (Hughes *et al.*, 1997; Lamparter *et al.*, 1997; Yeh *et al.*, 1997). Quite near the *Synechocystis* phytochrome gene (*cph1* – for cyanobacterial phytochrome) was a gene likely to encode the response regulator (*rcp1*) of the transmitter–receiver pair (Yeh *et al.*, 1997). Using reconstituted, recombinant Cph1, Yeh *et al.* (1997) showed that Cph1 was indeed a histidine kinase with greater activity towards itself and Rcp1 in the Pr form than in the Pfr form (Yeh *et al.*, 1997). Thus, Cph1 and Rcp1 seem to define a cyanobacterial, light-sensing, two-component system. These data, and the sequence relatedness between the HKD of Cph1 and the C-termini of plant phytochromes (Yeh *et al.*, 1997), rejuvenated interest in the question of whether phytochrome was a kinase.

So is plant phytochrome a histidine kinase? Is the plant phytochrome system a two-component system, with phytochrome being the transmitter paired with an unidentified receiver? One way to address the first question is to create mutants that should lack catalytic activity and then determine whether the mutants exhibit a phenotype in transgenic plants. Boylan and Quail (1996) mutated three conserved amino acids in phyA that correspond to amino acids that are important for histidine kinase activity in prokaryotes, one of these being the autophosphorylated histidine. Their results indicated that none of the three mutagenized amino acids is essential for phytochrome A activity in transgenic *Arabidopsis* seedlings. However, as they point out, phytochrome might function as a kinase independently of these three amino acids (Boylan and Quail, 1996). The answer was still uncertain.

Again we come back to question of whether phytochrome is a kinase. A direct biochemical analysis similar to that with recombinant Cph1 took advantage of the ability to express, affinity-purify, and reconstitute eukaryotic phyA with chromophore (Murphy and Lagarias, 1997). Oat and *Mesotaenium caldariorum* (a green alga) apoproteins were expressed in *Saccharomyces cerevisiae* and *Pichia pastoris*, respectively (Yeh and Lararias, 1998). Both oat and algal phyA became phosphorylated *in vitro* on serines. Autophosphorylation was R/FR dependent, inhibited by pyrophosphate, and stimulated by histone H1. Histone H1 was *trans*-phosphorylated as well. The velocity of autophosphorylation of oat phyA was concentration independent, as a monomolecular reaction would be. As expected, sequence alignments of eukaryotic phytochromes with Cph1 indicated that the HKRD of eukaryotic phytochromes and the HKD of Cph1 were related. However, nearly the same level of identity was found between the phytochrome PRD (see Fig. 1) and the Cph1 HKD. Thus it was proposed that the eukaryotic phytochromes evolved via a duplication and divergence of an ancestral HKD to yield the PRD and HKRD seen in extant plant phytochromes. Because of the relationship between the PRD and HKRD of eukaryotic phytochromes and the HKD of Cph1, Yeh and Lararias (1998) asked whether Rcp1, the Cph1 response regulator, could be phosphorylated by phyA. The answer is yes, but in a light-independent fashion. Moreover, mutating the phospho-accepting aspartate on Rcp1 to an alanine had no effect on labeling, and labeling was shown to be on Ser/Thr residues. Taken together, their data show that phyA (in contrast to Cph1) is not a histidine kinase, but rather a serine/threonine kinase with a histidine kinase ancestry. Because of the clustering of deleterious mutations in the core region (Quail, 1997) and the lack of effect of the three point mutants in the HKRD in transgenics (Boylan and Quail, 1996), Yeh and Lararias (1998) speculate that it is the PRD and not the HKRD that supplies a eukaryotic phytochrome with its protein kinase activity. It will be exciting as future work unveils the nature of the ATP binding site and the biochemistry of phosphoryl transfer in phytochrome.

C. DOWNSTREAM OF PHYTOCHROME

What are phytochrome's substrate proteins and binding partners? This is a pivitol question because it is presumably through its substrates and binding partners that phytochrome will transduce light signals. Thus far, three substrates and one interaction partner have been identified. The cryptochromes CRY1 and CRY2 (Ahmad *et al.*, 1998) and PKS1 (Fankhauser *et al.*, 1999) are the substrates so far identified. PIF3, a basic helix-loop-helix (bHLH) protein, is the interaction partner (Ni *et al.*, 1998).

Ahmad *et al.* (1998) began by testing whether reconstituted phyA could phosphorylate CRY1. Indeed it does, and labeling was approximately the same after irradiation with R or B but higher than in darkness. Phosphorylation is inhibited by pyrophosphate, stimulated by polylysine, and occurs on a C-terminal fragment of both CRY1 and CRY2. The major phosphorylation site lies within a short, serine-rich region which is conserved between CRY1 and CRY2. The CRY1 N-terminal fragment interacts with the C-terminal half of *Arabidopsis* PHYA in the yeast two-hybrid system. *In vivo* labeling of CRY1 in dark-adapted plants is increased by R and is FR reversible. Taken together, these observations argue persuasively that the cryptochromes are substrates for phyA. Next, Ahmad *et al.* (1998) sought to extend these observations to an *in vivo* response. If cryptochrome and phytochrome associate, they reasoned that a mutant cryptochrome might interfere with phytochrome by forming nonproductive complexes – in essence a dominant-negative phenotype. They tested three mutant alleles of *CRY1* and found an effect on hypocotyl elongation in Rc and on flowering time under short-day conditions in the two mutants that did express CRY1 protein, but not in one devoid of protein. Since phyB has been shown to be involved in both these responses, the impairment of the responses in the *cry* mutants is consistent with the idea of a direct interaction between CRY1 and phyB (Ahmad *et al.*, 1998).

Ahmad *et al.* (1998) present an insightful model for how phytochrome and cryptochrome might interact in terms of light activation and phosphorylation, predicated on the notion that phosphorylation is essential for CRY activity. This is a significant area for future work because solar light contains both B and R light, and because many plant responses to light exhibit co-action of the photoreceptors (Mohr, 1994). Perhaps the simplest way to think of co-action is as a synergistic effect of activating both photoreceptors. Historically, one problem with photobiological approaches to co-action is the overlap in the action spectra of the cryptochromes and phytochromes in the B region. B will activate phytochrome and cryptochrome but R will activate only phytochrome. Using a battery of photoreceptor mutants, CRY1 action on hypocotyl elongation and anthocyanin accumulation was shown to require Pfr (Ahmad and Cashmore, 1997). In other words, these responses exhibit co-action. Now that Ahmad *et al.* (1998) have shown that the cryptochromes are substrates for phytochrome, we can begin to envisage a biochemical mechanism for coaction. Figure 4 shows how cryptochrome phosphorylation by phytochrome may help

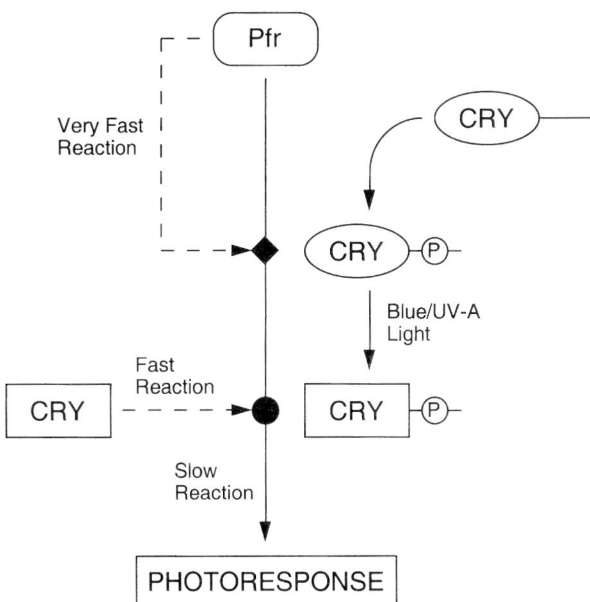

Fig. 4. Model for photoreceptor co-action in anthocyanin accumulation in milo seedlings. The central arrow represents the signaling chain leading from Pfr to pigment accumulation, adapted from Oelmüller and Mohr (1985). The dashed arrows on the left represent light-dependent reactions that govern responsiveness to Pfr. The proposed steps in cryptochrome activation are shown on the right. R light forms Pfr which very rapidly phosphorylates CRY (circled P). Activation of phosphorylated CRY by B/UV-A amplifies responsiveness to Pfr, allowing accumulation of anthocyanin. See text for additional details.

to explain one of the classic examples of photoreceptor co-action, that of anthocyanin accumulation in milo (*Sorghum vulgare*) seedlings (Oelmüller and Mohr, 1985). White light and B/UV-A trigger pigment to accumulate, but R by itself is totally ineffective in milo. Nevertheless, the effects of pulses of UV-A are fully reversible with FR, showing that phytochrome is required absolutely.

Based on these and other meticulous photobiological experiments, Oelmüller and Mohr (1985) proposed that Pfr plays two roles in anthocyanin accumulation in milo: (1) it is the effector that triggers the response, and (2) it establishes responsiveness to Pfr in a very fast reaction. Without B/UV-A the effects of Pfr remain cryptic, so activation of cryptochrome increases the responsiveness to Pfr in a fast reaction. Since there was a 3-h lag between the onset of illumination and the appearance of pigment, this was termed a slow reaction (Fig. 4). We can now predict that the very fast reaction by which R light establishes responsiveness to Pfr is the phosphorylation of CRY, which potentiates its activation. Upon illumination with B/UV-A, chromophore-dependent changes occur leading to full activation. This in turn will allow the 'slow reaction', which no doubt is induction of the requisite genes, to proceed. If phosphorylation of CRY is necessary but insufficient for activation, then a missense mutation at the phosphorylation site should lack activity, but apparently no such mutation has yet been identified. Future investigations of photoreceptor cross-talk will hopefully reveal how active CRY amplifies the signal emanating from Pfr, such as occurs during anthocyanin synthesis in milo.

PKS1 was identified in a yeast two-hybrid screen using the C-terminal 160 amino acids of *Arabidopsis* PHYA as bait (Frankhauser *et al.*, 1999). PKS1 is a novel protein of 439 amino acids that is basic and apparently cytoplasmic. *In vitro*, it binds to phyA and phyB in both the Pr and Pfr forms. A recombinant fusion protein containing the N-terminal half of PKS1 is indeed phosphorylated by recombinant oat phyA. Unlike phosphorylation of Rcp1 (Yeh and Lararias, 1998), labeling of PKS1 is Pr/Pfr dependent, with the Pfr having highest activity. The mobility of plant-derived PKS1 increases after alkaline phosphatase treatment, so it appears to be phosphorylated *in vivo*. As judged by hypocotyl elongation, transgenic *Arabidopsis* seedlings overexpressing PKS1 have reduced sensitivity to R but not B or FR, suggesting that PKS1 is a negative regulator of phyB signaling. One of the fascinating scenarios for PKS1 action mentioned by Frankhauser *et al.* (1999) is that PKS1 might be involved in sequestering phyB in the cytoplasm until triggered to enter the nucleus by light (Sakamoto and Nagatani, 1996; Kircher *et al.*, 1999a).

Mechanisms that influence nucleocytoplasmic partitioning of phytochromes are taking on greater significance because of the light-dependent nuclear import of both phyA and phyB (Kircher *et al.*, 1999a) and the identification of PIF3, a nuclear protein that interacts with phyB and phyA (Ni *et al.*, 1998). PIF3 contains a bHLH motif and so is a presumptive transcription factor. Genetically, PIF3 was shown to be involved with phyB signaling (Halliday *et al.*, 1999). Moreover, PIF3 interacts *in vitro* with phyB only when it is in the Pfr form (Ni *et al.*, 1999). Whether phosphorylation is involved in interaction of phyB with PIF3 is not yet known, nor is it known how this might affect the presumed DNA binding activity of PIF3. Regardless, there now exists a much more direct path between phytochrome and the nucleus than could have been imagined just a few years ago. Understanding this path is important because so many of the effects of phytochrome during photomorphogenesis are driven by changes in gene expression.

VIII. CONCLUSIONS

It is clear that protein phosphorylation plays a central role in transducing light signals. Light alters the phosphorylation status of several proteins, including the phytochromes and phototropin. Light triggers changes in gene expression, perhaps via the involvement of CKII in some cases. Light regulates the expression of genes encoding protein kinases; in most cases the signaling function of these kinases remains to be elucidated although clear hints exist on where to begin. From a protein phosphorylation perspective, perhaps the most profound recent advance is the identification of phytochrome and phototropin as protein kinases. A considerable amount remains to be learned about how the photoreceptor kinases translate the absorption of light into a change in their kinase activity at the molecular level. What are their substrates? What are the mechanisms of phosphoryl transfer? What proteins do they physically interact with? What are the implications of these associations for light sensing? What are the implications of these associations for integrating light, hormonal, and other environmental signals? What are the molecular mechanisms by which photoreceptors co-operate during the integration of spectrally complex cues? How can we use this knowledge to understand plants growing in a natural environment? Phytochrome is now thought to affect gene expression in a more direct fashion than could have been imagined just three years ago. What might be the target genes for nuclear phytochrome armed with an active kinase domain? The temptation is to continue listing the questions because there are so many of them that result from the recent steps forward. Imagine how many exciting questions there will be a few years hence, once we have answers to even a few of the ones just posed.

ACKNOWLEDGEMENTS

The author thanks the many colleagues who contributed to this review by providing reprints and preprints, and by discussing their research. Current work in the author's laboratory is funded by grant number IBN-9728907 from the National Science Foundation.

REFERENCES

Abler, M. L. and Green, P. J. (1996). Control of mRNA stability in higher plants. *Plant Molecular Biology* **32**, 63–78.

Ahmad, M. and Cashmore, A. R. (1993). *HY4* gene of *A. thaliana* encodes a protein with characteristics of a blue-light photoreceptor. *Nature* **366**, 162–166.

Ahmad, M. and Cashmore, A. R. (1996). Seeing blue: the discovery of cryptochrome. *Plant Molecular Biology* **30**, 851–861.

Ahmad, M. and Cashmore, A. R. (1997). The blue-light receptor cryptochrome 1 shows functional dependence on phytochrome A or phytochrome B in *Arabidopsis thaliana. Plant Journal* **11**, 421–427.

Ahmad, M., Jarillo, J., Smirnova, O. and Cashmore, A. (1998). The CRY1 blue light photoreceptor of *Arabidopsis* interacts with phytochrome A *in vitro. Molecular Cell* **1**, 939–948.

Anderson, M. B., Folta, K., Warpeha, K. M., Gibbons, J., Gao, J. and Kaufman, L. S. (1999). Blue light-directed destabilization of the pea *Lhcb1**4 transcript depends on sequences within the 5′ untranslated region. *Plant Cell* **11**, 1579–1589.

Batschauer, A. (1998). Photoreceptors of higher plants. *Planta* **206**, 479–492.

Biermann, B. J., Pao, L. I. and Feldman, L. J. (1994). Pr-specific phosphorylation *in vitro* by a protein kinase present in antiphytochrome maize immunoprecipitates. *Plant Physiology* **105**, 243–251.

Borthwick, H. A. and Hendricks, S. A. (1960). Photoperiodism in plants. *Science* **132**, 1223–1228.

Bowler, C. (1997). The transduction of light signals by phytochrome. *In* "Signal Transduction in Plants" (P. Aducci, ed.) pp. 137–152. Birkhäuser, Basel.

Boylan, M. T. and Quail, P. H. (1996). Are the phytochromes protein kinases? *Protoplasma* **195**, 12–17.

Briggs, W. R. and Huala, E. (1999). Blue-light photoreceptors in higher plants. *Annual Review of Cell and Developmental Biology* **15**, 33–62.

Cashmore, A. R., Jarillo, J. A., Wu, Y.-J. and Liu, D. (1999). Cryptochromes: blue light receptors for plants and animals. *Science* **284**, 760–765.

Chono, M., Nemoto, K., Yamane, H., Yamaguchi, I. and Murofushi, N. (1998). Characterization of a protein kinase gene responsive to auxin and gibberellin in cucumber hypocotyls. *Plant and Cell Physiology* **39**, 958–967.

Chory, J., Chatterjee, M., Cook, R. K., Elich, I. T., Fankhauser, C., Li, J., Nagpal, P., Neff, M., Pepper, A., Poole, D., Reed, J. and Vitart, V. (1996). From seed germination to flowering, light controls plant development via the pigment phytochrome. *Proceedings of the National Academy of Sciences of the United States of America* **93**, 12066–12071.

Christie, J. M., Reymond, P., Powell, G. K., Bernasconi, P., Raibekas, A. A., Liscum, E. and Briggs, W. R. (1998). *Arabidopsis* NPH1: a flavoprotein with the properties of a photoreceptor for phototropism. *Science* **282**, 1698–1701.

Christie, J. M., Salomon, M., Nozue, K., Wada, M. and Briggs, W. R. (1999). LOV (light, oxygen, or voltage) domains of the blue-light photoreceptor phototropin (nph1): binding sites for the chromophore flavin mononucleotide. *Proceedings of the National Academy of Sciences of the United States of America* **96**, 8779–8783.

Clack, T., Mathews, S. and Sharrock, R. A. (1994). The phytochrome apoprotein family in *Arabidopsis* is encoded by five genes: the sequences and expression of PHYD and PHYE. *Plant Molecular Biology* **25**, 413–427.

Datta, N., Chen, Y.-R. and Roux, S. J. (1985). Phytochrome and calcium stimulation of protein phosphorylation in isolated pea nuclei. *Biochemical and Biophysical Research Communications* **128**, 1403–1408.

Deeken, R. and Kaldenhoff, R. (1997). Light-repressible receptor protein kinase: a novel photo-regulated gene from *Arabidopsis thaliana*. *Planta* **202**, 479–486.

Doshi, A., Sopory, A. and Sopory, S. K. (1992). Regulation of protein phosphorylation by phytochrome in *Sorghum bicolor*. *Photochemistry and Photobiology* **55**, 465–468.

Ellard-Ivey, M., Hopkins, R. B., White, T. J. and Lomax, T. L. (1999). Cloning, expression and N-terminal myristoylation of CpCPK1, a calcium-dependent protein kinase from zucchini (*Curcurbita pepo* L.). *Plant Molecular Biology* **39**, 199–208.

Fallon, K. M. and Trewavas, A. J. (1994). Phosphorylation of a renatured protein from etiolated wheat leaf protoplasts is modulated by blue light and red light. *Plant Physiology* **105**, 252–258.

Fallon, K. M., Shacklock, P. S. and Trewavas, A. J. (1993). Detection *in vivo* of very rapid red light-induced calcium-sensitive protein phosphorylation in etiolated wheat (*Triticum aestivum*) leaf protoplasts. *Plant Physiology* **101**, 1039–1045.

Fankhauser, C. and Chory, J. (1997). Light control of plant development. *Annual Review of Cell and Developmental Biology* **13**, 203–229.

Fankhauser, C., Yeh, K.-C., Lagarias, J. C., Zhang, H., Elich, T. D. and Chory, J. (1999). PKS1, a substrate phosphorylated by phytochrome that modulates light signaling in *Arabidopsis*. *Science* **284**, 1539–1541.

Firn, R. D. (1994). Phototropism. *In* "Photomorphogenesis in Plants", 2nd edition (R. E. Kendrick and G. Kronenberg, eds) pp. 659–681. Kluwer Academic, Dordrecht.

Folta, K. M. and Kaufman, L. S. (1999). Regions of the pea *Lhcb1*4* promoter necessary for blue-light regulation in transgenic *Arabidopsis*. *Plant Physiology* **120**, 747–755.

Furuya, M. (1993). Phytochromes: their molecular species, gene families, and functions. *Annual Review of Plant Physiology and Plant Molecular Biology* **44**, 617–645.

Gallagher, S., Short, T. W., Ray, P. M., Pratt, L. H. and Briggs, W. R. (1988). Light-mediated changes in two proteins found associated with plasma membrane fractions from pea stem sections. *Proceedings of the National Academy of Sciences of the United States of America* **85**, 8003–8007.

Gao, J. and Kaufman, L. S. (1994). Blue-light regulation of the *Arabidopsis thaliana Cab1* gene. *Plant Physiology* **104**, 1251–1257.

Grimm, R., Gast, D. and Rüdiger, W. (1989). Characterization of a protein kinase activity associated with phytochrome from etiolated oat seedlings. *Planta* **178**, 199–206.

Gus-Mayer, S., Brunner, H., Schneider-Poetsch, H. A., Lottspeich, F., Eckerskorn, C., Grimm, R. and Rüdiger, W. (1994). The amino acid sequence previously attributed to a protein kinase or a TCP1-related molecular chaperone and co-purified with phytochrome is a beta-glucosidase. *FEBS Letters* **347**, 51–54.

Hager, A. and Brich, M. (1993a). Blue-light-induced phosphorylation of a plasma-membrane protein from phototropically sensitive tips of maize coleoptiles. *Planta* **189**, 567–576.

Hager, A. and Brich, M. (1993b). Redox dependence of the blue-light-induced phosphorylation of a 100-kDa protein on isolated plasma membranes from tips of coleoptiles. *Planta* **190**, 120–126.

Halliday, K. J., Hudson, M., Ni, M., Qin, M. and Quail, P. H. (1999). *poc1*: an *Arabidopsis* mutant perturbed in phytochrome signaling because of a T DNA insertion in the promoter of *PIF3*, a gene encoding a phytochrome-interaction bHLH protein. *Proceedings of the National Academy of Sciences of the United States of America* **96**, 5832–5837.

Hamada, T., Tanaka, N., Noguchi, T., Kimura, N. and Hasunuma, K. (1996). Phytochrome regulates phosphorylation of a protein with characteristics of a nucleoside diphosphate kinase in the crude membrane fraction from stem sections of etiolated pea seedlings. *Journal of Photochemistry and Photobiology B* **33**, 143–151.

Hanks, S. K. and Hunter, T. (1995). The eukaryotic protein kinase superfamily: kinase (catalytic) domain structure and function. *FASEB Journal* **9**, 576–596.

Hanks, S. K., Quinn, A. M. and Hunter, T. (1988). The protein kinase family: conserved features and deduced phylogeny of the catalytic domains. *Science* **241**, 42–52.

Harter, K., Frohnmeyer, H., Kircher, S., Kunkel, T., Mühlbauer, S. and Schäfer, E. (1994). Light induces rapid changes of the phosphorylation pattern in the cytosol of evacuolated parsley protoplasts. *Proceedings of the National Academy of Sciences of the United States of America USA* **91**, 5038–5042.

Holappa, L. D. and Walker-Simmons, M. K. (1997). The wheat protein kinase gene, TaPK3, of the PKABA1 subfamily is differentially regulated in greening wheat seedlings. *Plant Molecular Biology* **33**, 935–941.

Holtorf, H. and Apel, K. (1996). Transcripts of the two NADPH protochlorophyllide oxidoreductase genes *PorA* and *PorB* are differentially degraded in etiolated barley seedlings. *Plant Molecular Biology* **31**, 387–392.

Holtorf, H., Reinbothe, S., Reinbothe, C., Bereza, B. and Apel, K. (1995). Two routes of chlorophyllide synthesis that are differentially regulated by light in barley (*Hordeum vulgare* L.). *Proceedings of the National Academy of Sciences of the United States of America* **92**, 3254–3258.

Huala, E., Oeller, P. W., Liscum, E., Han, I.-S., Larsen, E. and Briggs, W. R. (1997). *Arabidopsis* NPH1: a protein kinase with a putative redox-sensing domain. *Science* **278**, 2120–2123.

Hughes, J., Lamparter, T., Mittmann, F., Hartmann, E., Gärtner, W., Wilde, A. and Börner, T. (1997). A prokaryotic phytochrome. *Nature* **386**, 663.

Hunt, R. E. and Pratt, L. H. (1980). Partial characterization of undegraded oat phytochrome. *Biochemistry* **19**, 390–394.

Iino, M., Briggs, W. R. and Schäfer, E. (1984). Phytochrome-mediated phototropism in maize seedling shoots. *Planta* **160**, 41–51.

Janoudi, A.-k., Gordon, W. R., Wagner, D., Quail, P. H. and Poff, K. L. (1997). Multiple phytochromes are involved in red-light-induced enhancement of first positive phototropism in *Arabidopsis thaliana*. *Plant Physiology* **113**, 975–979.

Jarillo, J. A., Ahmad, M. and Cashmore, A. R. (1998). NPL1 (Accession No. AF053941): a second member of the NPH serine/threonine kinase family of *Arabidopsis*. *Plant Physiology* **117**, 719.

Johnson, M., Baker, E., Colbert, J. and Green, P. (1998). Determinants of mRNA stability in plants. *In* "A Look Beyond Transcription: Mechanisms Determining mRNA Stability and Translation in Plants" (J. Bailey-Serres and D. Gallie, eds) pp. 40–53. American Society of Plant Physiologists, Rockville.

Kaufman, L. S. (1993). Transduction of blue-light signals. *Plant Physiology* **102**, 333–337.

Kehoe, D. M. and Grossman, A. R. (1996). Similarity of a chromatic adaptation sensor to phytochrome and ethylene receptors. *Science* **273**, 1409–1412.

Khanna, R. (1998). Light-regulated protein kinases from plants. PhD Dissertation, Purdue University.

Khanna, R. and Watson, J. C. (1997). cDNA sequence of PsPK3 (Accession No. U11553), a protein kinase from *Pisum sativum* L. *Plant Physiology* **114**, 1569.

Khanna, R., Lin, X. and Watson, J. (1999). Photoregulated expression of the *PsPK3* and *PsPK5* genes in pea seedlings. *Plant Molecular Biology* **39**, 231–242.

Khurana, J. P., Kochhar, A. and Tyagi, A. K. (1998). Photosensory perception and signal transduction in higher plants–molecular genetic analysis. *Critical Reviews in Plant Sciences* **17**, 465–539.

Kim, I.-S., Bai, U. and Song, P.-S. (1989). A purified 124-kDa oat phytochrome does not possess a protein kinase activity. *Photochemistry and Photobiology* **49**, 319–323.

Kircher, S., Kozma-Bognar, L., Kim, L., Adam, E., Harter, K., Schäfer, E. and Nagy, F. (1999a). Light quality-dependent nuclear import of the plant photoreceptors phytochrome A and B. *Plant Cell* **11**, 1445–1456.

Kircher, S., Wellmer, F., Nick, P., Rünger, A., Schäfer, E. and Harter, K. (1999b). Nuclear import of the parsley bZIP transcription factor CPRF2 is regulated by phytochrome photoreceptors. *Journal of Cell Biology* **144**, 201–211.

Knighton, D. R., Zheng, J., TenEyck, L. F., Ashford, V. A. and Xuong, N. H. (1991). Crystal structure of the catalytic subunit of cAMP-dependent protein kinase. *Science* **253**, 407–414.

Lagarias, J. C. and Lagarias, D. M. (1989). Self-assembly of synthetic phytochrome holoprotein *in vitro*. *Proceedings of the National Academy of Sciences of the United States of America* **86**, 5778–5780.

Lamparter, T., Mittmann, F., Gärtner, W., Hartmann, E., Börner, T. and Hughes, J. (1997). Characterization of recombinant phytochrome from the cyanobacterium *Synechocystis*. *Proceedings of the National Academy of Sciences of the United States of America* **94**, 11792–11797.

Lapko, V. N., Wells, T. A. and Song, P.-S. (1996). Protein kinase A-catalyzed phosphorylation and its effect on conformation in phytochrome A. *Biochemistry* **35**, 6585–6594.

Lapko, V. N., Jiang, X. Y., Smith, D. L. and Song, P.-S. (1997). Post-translational modification of oat phytochrome A: phosphorylation of a specific serine in a multiple serine cluster. *Biochemistry* **36**, 10595–10599.

Lapko, V. N., Jiang, X. Y., Smith, D. L. and Song, P.-S. (1999). Mass spectrometric characterization of oat phytochrome A: isoforms and post-translational modifications. *Protein Science* **8**, 1032–1044.

Lawton, M. A., Yamamoto, R. T., Hanks, S. K. and Lamb, C. J. (1989). Molecular cloning of plant transcripts encoding protein kinase homologs. *Proceedings of the National Academy of Sciences of the United States of America* **86**, 3140–3144.

Lee, Y., Lloyd, A. M. and Roux, S. J. (1999). Antisense expression of the CK2 α-subunit gene in *Arabidopsis*. Effect on light-regulated gene expression and growth. *Plant Physiology* **119**, 989–1000.

Lin, X. (1993). Molecular cloning and characterization of transcripts encoding protein kinase homologs from pea (*Pisum sativum* L.). PhD Dissertation, University of Maryland at College Park.

Lin, X. and Watson, J. C. (1992). cDNA sequence of PsPK5, a protein kinase homolog from *Pisum sativum* L. *Plant Physiology* **100**, 1072–1074.

Lin, X., Feng, X.-H. and Watson, J. C. (1991). Differential accumulation of transcripts encoding protein kinase homologs in greening pea seedlings. *Proceedings of the National Academy of Sciences of the United States of America* **88**, 6951–6955.

Liscum, E. and Briggs, W. R. (1995). Mutations in the *NPH1* locus of *Arabidopsis* disrupt the perception of phototropic stimuli. *Plant Cell* **7**, 473–485.

Ma, J., Khanna, R., Fukasawa-Akada, T., Poisso, J., Deitzer, G. and Watson, J. (1998). PK3At (accession n. AF082391): an *Arabidopsis* homolog of the PsPK3 protein kinase from *Pisum sativum*. *Plant Physiology* **118**, 712.

Mancinelli, A. L. (1994). The physiology of phytochrome action. *In* "Photomorphogenesis in Plants", 2nd edition (R. E. Kendrick and G. H. M. Kronenberg, eds) pp. 211–269. Kluwer Academic, Dordrecht.

McMichael, R. W., Jr and Lagarias, J. C. (1990). Phosphopeptide mapping of *Avena* phytochrome phosphorylated by protein kinases *in vitro*. *Biochemistry* **29**, 3872–3878.

Mohr, H. (1994). Coaction between pigment systems. *In* "Photomorphogenesis in Plants", 2nd edition (R. E. Kendrick and G. H. M. Kronenberg, eds) pp. 353–373. Kluwer Academic, Dordrecht.

Motchoulski, A. and Liscum, E. (1999). *Arabidopsis* NPH3: a NPH1 photoreceptor-interacting protein essential for phototropism. *Science* **286**, 961–964.

Murphy, J. T. and Lagarias, J. C. (1997). Purification and characterization of recombinant affinity peptide-tagged oat phytochrome A. *Photochemistry and Photobiology* **65**, 750–758.

Mustilli, A. C. and Bowler, C. (1997). Tuning in to the signals controlling photoregulated gene expression in plants. *EMBO Journal* **16**, 5801–5806.

Neuhaus, G., Bowler, C., Kern, R. and Chua, N.-H. (1993). Calcium/calmodulin-dependent and -independent phytochrome signal transduction pathways. *Cell* **73**, 937–952.

Ni, M., Tepperman, J. M. and Quail, P. H. (1998). PIF3, a phytochrome-interacting factor necessary for normal photoinduced signal transduction, is a basic helix-loop-helix protein. *Cell* **95**, 657–667.

Ni, M., Tepperman, J. M. and Quail, P. H. (1999). Binding of phytochrome B to its nuclear signalling partner PIF3 is reversibly induced by light. *Nature* **400**, 781–784.

Nozue, K., Kanegae, T., Imaizumi, T., Fukuda, S., Okamoto, H., Yeh, K.-C., Lagarias, J. C. and Wada, M. (1998). A phytochrome from the fern *Adiantum* with features of the putative photoreceptor NPH1. *Proceedings of the National Academy of Sciences of the United States of America* **95**, 15826–15830.

Oelmüller, R. and Mohr, H. (1985). Mode of coaction between blue/UV light and light absorbed by phytochrome in light-mediated anthocyanin formation in the milo (*Sorghum vulgare* Pers.) seedling. *Proceedings of the National Academy of Sciences of the United States of America* **82**, 6124–6128.

Otto, V. and Schäfer, E. (1988). Rapid phytochrome-controlled protein phosphorylation and dephosphorylation in *Avena sativa* L. *Plant Cell Physiology* **29**, 1115–1121.

Palmer, J. M., Short, T. W. and Briggs, W. R. (1993a). Correlation of blue light-induced phosphorylation to phototropism in *Zea mays* L. *Plant Physiology* **102**, 1219–1225.

Palmer, J. M., Short, T. W., Gallagher, S. and Briggs, W. R. (1993b). Blue light-induced phosphorylation of a plasma membrane-associated protein in *Zea mays* L. *Plant Physiology* **102**, 1211–1218.

Park, M. H. and Chae, Q. (1989). Intracellular protein phosphorylation in oat (*Avena sativa* L.) protoplasts by phytochrome action. *Biochemical and Biophysical Research Communications* **162**, 9–14.

Parker, K., Baskin, T. I. and Briggs, W. R. (1989). Evidence for a phytochrome-mediated phototropism in etiolated pea seedlings. *Plant Physiology* **89**, 493–497.

Pasentsis, K., Paulo, N., Algarra, P., Dittrich, P. and Thümmler, F. (1998). Characterization and expression of the phytochrome gene family in the moss *Ceratodon purpureus*. *Plant Journal* **13**, 51–61.

Poff, K. L., Janoudi, A.-K., Rosen, E. S., Orbovic, V., Konjevic, R., Rotin, M.-C. and Scott, T. K. (1994). The physiology of tropisms. *In* "*Arabidopsis*" (E. M.

Meyerwitz and C. R. Sommerville, eds) pp. 639–664. Cold Spring Harbor Press, Cold Spring Harbor.

Quail, P. (1994). Phytochrome genes and their expression. *In* "Photomorphogenesis in Plants", 2nd edition (R. Kendrick and G. Kronenberg, eds) pp. 71–104. Kluwer Academic, Dordrecht.

Quail, P. H. (1997). An emerging molecular map of the phytochromes. *Plant, Cell and Environment* **20**, 657–665.

Quail, P. H. (1998). The phytochrome family: dissection of functional roles and signalling pathways among family members. *Philosophical Transactions of the Royal Society of London B* **353**, 1399–1403.

Quail, P. H., Briggs, W. R. and Pratt, L. H. (1978). *In vivo* phosphorylation of phytochrome. *Carnegie Institution Yearbook* **77**, 342–344.

Quail, P. H., Boylan, M. T., Parks, B. M., Short, T. W., Xu, Y. and Wagner, D. (1995). Phytochromes: photosensory perception and signal transduction. *Science* **268**, 675–680.

Reed, J. W., Nagatani A., Elich, T. D., Fagan, M. and Chory, J. (1994). Phytochrome A and phytochrome B have overlapping but distinct functions in *Arabidopsis* development. *Plant Physiology* **104**, 1139–1149.

Reymond, P., Short, T. W. and Briggs, W. R. (1992a). Blue light activates a specific protein kinase in higher plants. *Plant Physiology* **100**, 655–661.

Reymond, P., Short, T. W., Briggs, W. R. and Poff, K. L. (1992b). Light-induced phosphorylation of a membrane protein plays an early role in signal transduction for phototropism in *Arabidopsis thaliana*. *Proceedings of the National Academy of Sciences of the United States of America* **89**, 4718–4721.

Romero, L. C., Biswal, B. and Song, P.-S. (1991). Protein phosphorylation in isolated nuclei from etiolated *Avena* seedlings. Effects of red/far-red light and cholera toxin. *FEBS Letters* **282**, 347–350.

Roux, S. (1994). Signal transduction in phytochrome responses. *In* "Photomorphogenesis in Plants", 2nd edition (R. E. Kendrick and G. H. M. Kronenberg, eds) pp. 187–210. Kluwer Academic, Dordrecht.

Rüdiger, W. and Briggs, W. R. (1995). Involvement of thiol groups in blue-light-induced phosphorylation of a plasma membrane-associated protein from coleoptile tips of *Zea mays* L. *Zeitschrift für Naturforschung* **50c**, 231–234.

Sakamoto, K. and Nagatani, A. (1996). Nuclear localization activity of phytochrome B. *Plant Journal* **10**, 859–868.

Salomon, M., Zacherl, M. and Rüdiger, W. (1996). Changes in blue-light-dependent protein phosphorylation during the early development of etiolated oat seedlings. *Planta* **199**, 336–342.

Salomon, M., Zacherl, M. and Rüdiger, W. (1997a). Asymmetric, blue light-dependent phosphorylation of a 116-kilodalton plasma membrane protein can be correlated with the first- and second-positive phototropic curvature of oat seedlings. *Plant Physiology* **115**, 485–491.

Salomon, M., Zacherl, M. and Rüdiger, W. (1997b). Exposure of oat seedlings to blue light results in amplified phosphorylation of the putative photoreceptor for phototropism and in higher sensitivity of the plants to phototropic curvature. *Plant Physiology* **115**, 493–500.

Sano, H. and Youssefian, S. (1994). Light and nutritional regulation of transcripts encoding a wheat protein kinase homolog is mediated by cytokinins. *Proceedings of the National Academy of Sciences of the United States of America* **91**, 2582–2586.

Schneider-Poetsch, H. A. W. (1992). Signal transduction by phytochrome: phytochromes have a module related to the transmitter modules of bacterial sensor proteins. *Photochemistry and Photobiology* **56**, 839–846.

Schneider-Poetsch, H. A. W., Braun, B., Marx, S. and Schaumburg, A. (1991). Phytochromes and bacterial sensor proteins are related by structural and functional homologies. *FEBS Letters* **281**, 245–249.

Schwechheimer, C., Zourelidou, M. and Bevan, M. W. (1998). Plant transcription factor studies. *Annual Review of Plant Physiology and Plant Molecular Biology* **49**, 127–150.

Seeley, K. A., Byrne, D. H. and Colbert, J. T. (1992). Red light-independent instability of oat phytochrome mRNA *in vivo*. *Plant Cell* **4**, 29–38.

Senger, H. and Schmidt, W. (1994). Diversity of photoreceptors. *In* "Photomorphogenesis in Plants", 2nd edition (R. Kendrick and G. Kronenberg, eds) pp. 301–326. Kluwer Academic, Dordrecht.

Shaffer, R., Ramsey, N., Samach, A., Cordon, S., Putterill, J., Carré, I. A. and Coupland, G. (1998). The late elongated hypocotyl mutation of *Arabidopsis* disrupts circadian rhythms and the photoperiodic control of flowering. *Cell* **93**, 1219–1229.

Sharma, V., Jain, P. K., Maheshwari, S. C. and Khurana, J. P. (1997a). Rapid blue-light-induced phosphorylation of plasma-membrane-associated proteins in wheat. *Phytochemistry* **44**, 775–780.

Sharma, V., Jain, P. K., Malik, M. K., Maheshwari, S. C. and Khurana, J. P. (1997b). Light- and calcium-modulated phosphorylation of proteins from wheat seedlings. *Phytochemistry* **44**, 781–786.

Shinomura, T., Nagatani, A., Hanzawa, H., Kubota, M., Watanabe, M. and Furuya, M. (1996). Action spectra for phytochrome A- and B-specific photoinduction of seed germination in *Arabidopsis thaliana*. *Proceedings of the National Academy of Sciences of the United States of America* **93**, 8129–8133.

Short, T. and Briggs, W. (1994). The transduction of blue light signals in higher plants. *Annual Review of Plant Physiology and Plant Molecular Biology* **45**, 143–171.

Short, T. W. and Briggs, W. R. (1990). Characterization of a rapid, blue-light mediated change in detectable phosphorylation of a plasma membrane protein from etiolated pea (*Pisum sativum* L.) seedlings. *Plant Physiology* **92**, 179–185.

Short, T. W., Porst, M. and Briggs, W. R. (1992). A photoreceptor system regulating *in vivo* and *in vitro* phosphorylation of a pea plasma membrane protein. *Photochemistry and Photobiology* **55**, 773–781.

Short, T. W., Reymond, P. and Briggs, W. R. (1993). A pea plasma membrane protein exhibiting blue light-induced phosphorylation retains photosensitivity following Triton solubilization. *Plant Physiology* **101**, 647–655.

Short, T. W., Porst, M., Palmer, J., Fernbach, E. and Briggs, W. R. (1994). Blue light induces phosphorylation at seryl residues on a pea (*Pisum sativum* L.) plasma membrane protein. *Plant Physiology* **104**, 1317–1324.

Singh, B. R. and Song, P.-S. (1990). Phytochrome and protein phosphorylation. *Photochemistry and Photobiology* **52**, 249–254.

Smith, H. (1995). Physiological and ecological function within the phytochrome family. *Annual Review of Plant Physiology and Plant Molecular Biology* **46,** 289–315.

Sommer, D., Wells, T. A. and Song, P.-S. (1996). A possible tyrosine phosphorylation of phytochrome. *FEBS Letters* **393**, 161–166.

Stowe-Evans, E. L., Harper, R. M., Motchoulski, A. V. and Liscum, E. (1998). NPH4, a conditional modulator of auxin-dependent differential growth response in *Arabidopsis*. *Plant Physiology* **118**, 1265–1275.

Sugano, S., Andronis, C., Green, R. M., Wang, Z.-H. and Tobin, E. M. (1998). Protein kinase CK2 interacts with and phosphorylates the *Arabidopsis* circadian clock associated 1 proteins. *Proceedings of the National Academy of Sciences of the United States of America* **95**, 12362–12366.

Tanaka, N., Ogura, I., Noguchi, T., Hirano, H., Yabe, N. and Hasunuma, K. (1998). Phytochrome-mediated light signals are transduced to nucleoside diphosphate kinase in *Pisum sativum* L. cv. Alaska. *Journal of Photochemistry and Photobiology B* **45**, 113–121.

Terzaghi, W. and Cashmore, A. (1995). Light-regulated transcription. *Annual Review of Plant Physiology and Plant Molecular Biology* **46**, 445–474.

Thompson, J. D., Higgins, D. G. and Gibson, T. J. (1994). CLUSTAL W: improving the sensitivity of progressive multiple sequence alignment through sequence weighting, position-specific gap penalties and weight matrix choice. *Nucleic Acids Research* **22**, 4673–4680.

Thompson, W. F. and White, M. J. (1991). Physiological and molecular studies of light-regulated nuclear genes in higher plants. *Annual Review of Plant Physiology and Plant Molecular Biology* **42**, 423–466.

Thümmler, F., Dufner, M., Kreisl, P. and Dittrich, P. (1992). Molecular cloning of a novel phytochrome gene of the moss *Ceratodon purpureus* which encodes a putative light-regulated protein kinase. *Plant Molecular Biology* **20**, 1013–1017.

Thümmler, F., Algarra, P. and Fobo, G. M. (1995a). Sequence similarities of phytochrome to protein kinases: implication for the structure, function and evolution of the phytochrome gene family. *FEBS Letters* **357**, 149–155.

Thümmler, F., Herbst, R., Algarra, P. and Ullrich, A. (1995b). Analysis of the protein kinase activity of moss phytochrome expressed in fibroblast cell culture. *Planta* **197**, 592–596.

Tong, C.-G., Kendrick, R. E. and Roux, S. J. (1996). Red light-induced appearance of phosphotyrosine-like epitopes on nuclear proteins from pea (*Pisum sativum* L.) plumules. *Photochemistry and Photobiology* **64**, 863–866.

von Arnim, A. and Deng, X.-W. (1996). Light control of seedling development. *Annual Review of Plant Physiology and Plant Molecular Biology* **47**, 215–243.

Wang, Z.-Y. and Tobin, E. M. (1998). Constitutive expression of the *CIRCADIAN CLOCK ASSOCIATED 1 (CCA1)* gene distrupts circadian rhythms and suppresses its own expression. *Cell* **93**, 1207–1217.

Wang, Z.-Y., Kenigsbuch, D., Sun, L., Harel, E., Ong, M. S. and Tobin, E. M. (1997). A Myb-related transcription factor is involved in the phytochrome regulation of an *Arabidopsis Lhcb* gene. *Plant Cell* **9**, 491–507.

Warpeha, K. M. F. and Kaufman, L. S. (1989). Blue-light regulation of epicotyl elongation in *Pisum sativum*. *Plant Physiology* **89**, 544–548.

Warpeha, K. M. F. and Kaufman, L. S. (1990a). Two distinct blue-light responses regulate epicotyl elongation in pea. *Plant Physiology* **92**, 495–499.

Warpeha, K. M. F. and Kaufman, L. S. (1990b). Two distinct blue-light responses regulate the levels of transcripts of specific nuclear-coded genes in pea. *Planta* **182**, 553–558.

Warpeha, K. M. F., Marrs, K. A. and Kaufman, L. S. (1989). Blue-light regulation of specific transcript levels in *Pisum sativum*. *Plant Physiology* **91**, 1031–1035.

Watson, J. C. (1989). Photoregulation of gene expression in plants. *In* "Plant Biotechnology" (S.-D. Kung and C. J. Arntzen, eds) pp. 161–205. Butterworths, Boston.

Wellmer, F., Kircher, S., Rünger, A., Frohnmeyer, H., Schäfer, E. and Harter, K. (1999). Phosphorylation of the parsley bZIP transcription factor CPRF2 is regulated by light. *Journal of Biological Chemistry* **274**, 29476–29482.

Whitelam, G. C., Patel, S. and Devlin, P. F. (1998). Phytochromes and photomorphogenesis in *Arabidopsis*. *Philosophical Transactions of the Royal Society of London B* **353**, 1445–1453.

Whitmarsh, A. J. and Davis, R. J. (1998). Structural organization of MAP-kinase signaling modules by scaffold proteins in yeast and mammals. *Trends in Biochemical Sciences* **23**, 481–485.

Wong, Y.-S. and Lagarias, J. C. (1989). Affinity labeling of *Avena* phytochrome with ATP analogs. *Proceedings of the National Academy of Sciences of the United States of America* **86**, 3469–3473.

Wong, Y.-S., Cheng, H.-C., Walsh, D. A. and Largarias, J. C. (1986). Phosphorylation of *Avena* phytochrome *in vitro* as a probe of light-induced conformational changes. *Journal of Biological Chemistry* **261**, 12089–12097.

Wong, Y.-S., McMichael, R. W., Jr and Lagarias, J. C. (1989). Properties of a polycation-stimulated protein kinase associated with purified *Avena* phytochrome. *Plant Physiology* **91**, 709–718.

Yeh, K.-C. and Lararias, J. C. (1998). Eukaryotic phytochromes: light-regulated serine/threonine protein kinases with histidine kinase ancestry. *Proceedings of the National Academy of Sciences of the United States of America* **95**, 13976–13981.

Yeh, K.-C., Wu, S.-H., Murphy, J. T. and Lararias, J. C. (1997). A cyanobacterial phytochrome two-component light sensory system. *Science* **277**, 1505–1508.

Calcium-Dependent Protein Kinases and their Relatives

ESTELLE M. HRABAK

Department of Plant Biology, University of New Hampshire, Durham, NH 03824, USA

I. INTRODUCTION

Calcium's role as a second messenger in signaling processes in plant cells is well established. Fluxes in cytoplasmic calcium have been correlated with the plant's response to a great number of stimuli, including cold (Monroy and Dhindsa, 1995; Plieth *et al.*, 1999), light (Shacklock *et al.*, 1992), plant hormones (Felle, 1988; Gehring *et al.*, 1990; Gilroy and Jones, 1992; Bush, 1996), touch or motion (Knight *et al.*, 1991, 1992), drought or salt stress (Knight *et al.*, 1997), anaerobic stress (Sedbrook *et al.*, 1996), fungal elicitors (Knight *et al.*, 1991), and hypo-osmotic shock (Takahashi *et al.*, 1997). This is not a complete list, but indicates the wide range of stimuli, both biotic and abiotic, that can lead to changes in cytoplasmic calcium concentration. Consistent with calcium's role as a signaling molecule, many cellular processes are known to require calcium. These include guard cell movement (Gilroy *et al.*, 1990), light-induced protoplast swelling (Shacklock *et al.*, 1992), pollen

Advances in Botanical Research Vol. 32
incorporating Advances in Plant Pathology
ISBN 0-12-005932-0

tube growth (Malho *et al.*, 1994; Pierson *et al.*, 1994), root hair tip growth (Schiefelbein *et al.*, 1992), and regulation of actin tension and cytoplasmic streaming (Williamson and Ashley, 1982; Grabski *et al.*, 1998). Calcium is also known to induce changes in gene expression (Braam, 1992; Neuhaus *et al.*, 1993).

In order for calcium to function as a second messenger in transducing signals, the concentration of calcium in the cytoplasm under resting conditions must be maintained at low levels (about 10–100 nM). In the resting state, calcium is sequestered in intracellular stores, such as the vacuole, endoplasmic reticulum (ER), chloroplast, or mitochondria. In plants, a large amount of calcium is also found in the cell wall. In these storage locales, calcium can reach millimolar concentrations (Trewavas, 1999). Calcium enters the cytoplasm through a variety of gated channels in organelle and plasma membranes and can increase rapidly in concentration in response to either extrinsic or intrinsic stimuli (reviewed in Webb *et al.*, 1996). Calcium ions are subsequently removed from the cytoplasm through the action of membrane transporters. Calcium fluxes through the cytoplasm can develop and then dissipate very rapidly, allowing for specific responses to be initiated, followed by resetting of the system in preparation for the next stimulus (Trewavas, 1999).

Interestingly, calcium is not readily diffusible in the cytoplasm (Allbritton *et al.*, 1992; Trewavas, 1999). Instead, calcium influx often creates localized domains of high calcium concentration near membranes, while the calcium concentration in the remainder of the cytoplasm changes much less rapidly (Etter *et al.*, 1996). The increased calcium concentration near the membrane can stimulate nearby calcium-induced channels to open and, in this manner, a calcium 'wave' can be propagated along the membrane surface (Trewavas, 1999). Calcium wave development and progression may occur on the surface of any membrane that contains calcium channels. Different stimuli may produce calcium waves which vary with respect to initiation time post-stimulus, peak shape, duration, and presence or absence of oscillations (Malho *et al.*, 1998; Trewavas, 1999). In animal cells, it has been demonstrated that distinctive calcium flux kinetics are generated by different stimuli and that these kinetics encode specific information that determines the cell's downstream response. The mechanism(s) for decoding the information in calcium waves is still not fully understood (McAinsh and Hetherington, 1998). However, it is likely that analogous processes also occur in plant cell calcium responses. Thus, calcium-binding proteins that are positioned on or very near membranes are likely candidates to serve as the immediate receptors for calcium signals.

In plants, the large calcium-dependent protein kinase (CDPK) family of enzymes may function as downstream targets of increased calcium levels. CDPKs are directly activated by calcium ions (Section IIIA). This is in contrast to the calcium/calmodulin-dependent protein kinase (CaMKII) and protein kinase C (PKC) families, which are the predominant calcium-stimulated protein kinases in animal or fungal cells. Both CaMKII and PKC require other factors in addition to calcium, such as calmodulin or lipids, for full activation. No functional CaMKII or PKC homologs have been identified in plants to date, while CDPKs have only been identified in plants and protists (Section IIIB). No protein kinases with a structure resembling the CDPKs have yet been identified in animal or fungal cells, even after complete sequencing of the genome of the yeast *Saccharomyces cerevisiae*. Thus, CDPKs apparently represent a novel group of kinases with a limited representation in eukaryotes.

After calcium activation of any of the aforementioned kinases, they are able to alter the phosphorylation states of other proteins. These subsequent phosphorylation events may be one relay in a multi-step signal transduction cascade or direct regulation of a target substrate in a one-step process. The common end results of these phosphorylation events are changes in the cell's gene expression or enzyme activity. Phosphatases, which may be calcium dependent also, function in concert with kinases to maintain balance within the system.

While calcium's role as a second messenger in plants has been known for many years, evidence is now beginning to appear that CDPKs are involved in transducing calcium signals. A potential role for CDPKs in drought stress-induced signal transduction has been described. Drought stress causes increased synthesis of abscisic acid (ABA) and ABA is known to increase the cytoplasmic calcium concentration (Gehring *et al.*, 1990; Assmann and Shimazaki, 1999). Two *Arabidopsis* CDPK genes are reported to be induced by drought or high-salt stress (Urao *et al.*, 1994b). An activated form of one of these *Arabidopsis* CDPKs induced expression from an ABA-responsive promoter in a heterologous maize protoplast system and a phosphatase reversed the induction (Sheen, 1996). These data indicate that activation of a drought stress-responsive promoter can be affected by a CDPK, although it is not known whether this effect is direct or indirect. A second role for CDPKs has been indicated in the classic response of barley aleurone cells to gibberellin (GA). GA treatment causes secretion of hydrolases, principally α-amylase, that break down stored starch to more mobile sugars. GA also increases cytosolic calcium levels. A CDPK is implicated in this process at some point between calcium influx and hydrolase secretion because injection of a peptide that acts as a competitive CDPK inhibitor blocks secretion of α-amylase (Ritchie and Gilroy, 1998). Finally, reorientation of pollen tube growth is known to involve changes in the intracellular calcium tip gradient. Recent work by Moutinho *et al.* (1998) using a fluorescent dye to image CDPK activity in pollen tubes demonstrated that a localized change in calcium concentration was followed by a localized increase in activity of a calcium-dependent protein kinase on or near the plasma membrane at the tip of the pollen tube. Elevated kinase activity was highly correlated with a change in the direction of pollen tube growth (Moutinho *et al.*, 1998).

This chapter covers calcium-dependent protein kinases, as well as their close relatives the CRKs or CDPK-related kinases. In Section II, an overview is given of the many publications on the purification and characterization of calcium-dependent protein kinase enzyme activity from plant extracts and the distinguishing characteristics of CDPKs are discussed. Sections III and IV contain data on the structural features of cloned CDPK and CRK genes, respectively, and their phylogenetic relationships. Sections V and VI review the current data on how CDPK gene expression is regulated and what proteins are known to be phosphorylated by CDPKs. Inevitably, some publications have not been included because of space limitations.

II. CALCIUM-DEPENDENT PROTEIN KINASE ACTIVITY IN PLANTS

All eukaryotic organisms contain protein kinases, a subset of which is regulated by calcium ions. In animals and fungi, the predominant calcium-stimulated kinases belong to two groups: the conventional protein kinase C types, which require both calcium and phospholipids for activity, and the calcium/calmodulin-dependent kinase II types, which require calcium and calmodulin (Hanks and Hunter, 1995). In contrast, the predominant calcium-stimulated kinase activity in plants is greatly stimulated by calcium alone and does not have an absolute requirement for either phospholipid or calmodulin for activation. The first reports of calcium-dependent protein kinase activity in plant extracts were published more than 15 years ago (Hetherington and Trewavas, 1982, 1984). Since then, calcium-dependent protein kinase activity has been detected and partially purified from many plant species, including monocots, dicots, and green algae.

Biochemical purification and characterization of CDPKs from plants is an active area of research and the general characteristics of CDPKs have become apparent from these studies. It is instructive to compare the results obtained from these studies with the characterization of cloned CDPKs (Sections III, V and VI). Table I highlights representative results from biochemical purification and analysis of plant CDPKs. The defining characteristics of CDPKs are the following.

1. Enzyme activity requires magnesium and a free calcium concentration of greater than 0.1 μM.
2. Enzyme activity is not affected by addition of calmodulin (CaM) and is usually not affected by phospholipids, indicating that these are not PKC or CaMKII type enzymes.
3. Calmodulin inhibitors (such as calmidazolium and W-7) affect enzyme activity, indicating that CDPKs contain a region with similarity to calmodulin.
4. Autophosphorylation is commonly observed and occurs most often on serine residues and less frequently on threonine residues.
5. The enzymes display broad pH optima.

TABLE I
Characteristics of some calcium-dependent protein kinases (CDPKs) purified from plant or algal extracts

Species	Notable features	References
Arachis hypogea (peanut)	In soluble fraction (membrane not studied) Does not require CaM; binds calcium directly	DasGupta (1994)
Avena sativa (oat)	Tightly plasma membrane-associated Requires Ca and lipid for full activity	Schaller *et al.* (1992)
Beta vulgaris (red beet)	Bound to plasma membrane (soluble not studied) Requires 4 Ca^{2+} ions for activation	Baizabal-Aguirre and de la Vara (1997)
Chara corralina (a green alga)	Binding to actin and associated organelles demonstrated using soybean CDPK antibody	McCurdy and Harmon (1992a)
Cucurbita pepo (zucchini)	Detected in crude extracts and plasma membrane Autophosphorylation requires Ca	Verhey *et al.* (1993)
Dunaliella tertiolecta (a green alga)	Exhibits calcium-dependent binding to microsomes Antibodies recognize soluble and microsomal proteins	Yuasa and Muto (1992); Yuasa *et al.* (1995)
Glycine max (soybean)	In soluble fraction (membrane not studied) Does not require CaM or lipid; binds Ca directly	Harmon *et al.* (1987); Putnam-Evans *et al.* (1990)
Hordeum vulgare (barley)	Membrane and soluble kinase activities detected	Klimczak and Hind (1990)
Mangifera indica (mango)	In soluble fraction Does not require CaM	Frylinck and Dubery (1998)
Mougeotia sp. (a green alga)	Does not require CaM or lipid Detected CDPK activity in other plants and algae	Roberts (1989)
Nicotiana tabacum (tobacco)	Plasma membrane-associated Inducible by sucrose	Iwata *et al.* (1998)
Oryza sativa (rice)	Requires Ca, but not CaM or lipid Membrane-associated	Abo-El-Saad and Wu (1995)
Pisum sativum (pea)	CDPK activity purified from nuclei Requires Ca, but not CaM or lipid	Li *et al.* (1991)
Solanum tuberosum (potato)	Both soluble and membrane-associated activities Autophosphorylation detected	MacIntosh *et al.* (1996)
Zea mays (corn)	Both soluble and membrane-associated activities	Battey (1990)

6. Histone IIIS and/or casein are usually good *in vitro* substrates.
7. CDPK activity can be membrane associated, cytoskeleton associated, or soluble.

The identity of purified enzymes as CDPKs is more definitive if immunological cross-reaction with a CDPK antibody is observed. Both monoclonal antibodies against the kinase domain of a soybean CDPK (Putnam-Evans *et al.*, 1990) and polyclonal antibodies against a number of CDPKs are available for these types of study. A positive Western blot provides greater certainty that the purified protein is a CDPK, although cross-reaction is not always observed owing to the diversity of CDPK sequences and the resulting variations in antigenic epitopes.

Direct binding of ^{45}Ca to CDPKs has been demonstrated in some cases (Table I). In addition, CDPKs often exhibit a calcium-dependent mobility shift during electrophoresis, which is thought to be indicative of a conformational change in the protein's structure, similar to that observed when calmodulin binds calcium (Zhang and Yuan, 1998). Depending on the species, CDPKs have been purified from microsomal or soluble fractions, or sometimes both. A subcellular location on or near cell membranes could be advantageous for a calcium-stimulated kinase, since evidence indicates that the highest calcium concentrations occur very near the membrane. Thus, membrane-associated CDPKs would be well positioned to detect and respond rapidly to variations in cytoplasmic calcium levels. Soluble CDPKs might phosphorylate cytoplasmic substrates or become associated with the membrane through interaction with other proteins.

In *in vitro* studies, several workers have observed that activity of certain CDPKs can be stimulated by phospholipids, such as phosphatidylinositol or lysophosphatidylcholine (Schaller *et al.*, 1992; Harper *et al.*, 1993; Binder *et al.*, 1994; Farmer and Choi, 1999). However, this property does not seem to be shared by all CDPKs (Table I). The mechanism of CDPK stimulation by phospholipids is not completely understood but might be due to direct activation of the enzyme by phospholipid binding. Modulation of kinase activity by phospholipid binding could be a mechanism for differential regulation of certain CDPKs based on their subcellular location.

Data about CDPKs purified from plant extracts can be confounded by two potential complications. First, it is possible that, in some cases, a mixture of CDPKs may have been co-purified. As CDPK genes from various species are cloned and analyzed, it is becoming apparent that many organisms have multiple CDPK isoforms (Section IIIB). These CDPK family members often have similar molecular weights and may have similar biochemical characteristics (Lee *et al.*, 1998). In such cases, it may be difficult to identify purification conditions under which multiple CDPK isoforms are separable from each other and the resulting data would represent the average of all CDPKs in the sample. Second, proteolysis is a common problem during isolation and purification of CDPKs (Suen and Choi, 1991; Schaller *et al.*, 1992; Binder *et al.*, 1994; Yuasa *et al.*, 1995). Many of the studies cited in Table I noted that two or more related calcium-stimulated kinases of differing molecular weights were detected in the purified enzyme preparation. While these may represent different gene products, it is also possible that they resulted from proteolysis, especially when the relative amounts of each protein differed from one experiment to the next. While these considerations do not invalidate the need for thorough biochemical characterization of CDPKs, they emphasize the challenge that faces those who wish to study the role of these enzymes in plants.

III. CDPK GENES AND GENE FAMILIES

The cloning of the first two CDPK genes, the full-length *SK5* complementary DNA (cDNA) clone from soybean (*Glycine max*) and the partial *DcPK431* cDNA clone from carrot (*Daucus carota*), were reported in 1991 (Harper *et al.*, 1991; Suen and Choi, 1991). Since that time, at least 75 complete or partial CDPK sequences from 24 plant and protist species have been reported. In the next two subsections, a discussion will follow on how the CDPK structure derived from sequence analysis of these genes has provided an explanation for their observed biochemical properties and how different CDPKs are related to each other.

A. STRUCTURE OF CDPKs

All calcium-dependent protein kinases have a four-part structure consisting of a kinase domain, an autoregulatory domain, a calmodulin-like domain, and a variable domain (Fig. 1). Each of these domains will be discussed in detail in this section. Overall, the predicted amino acid sequences of CDPKs from

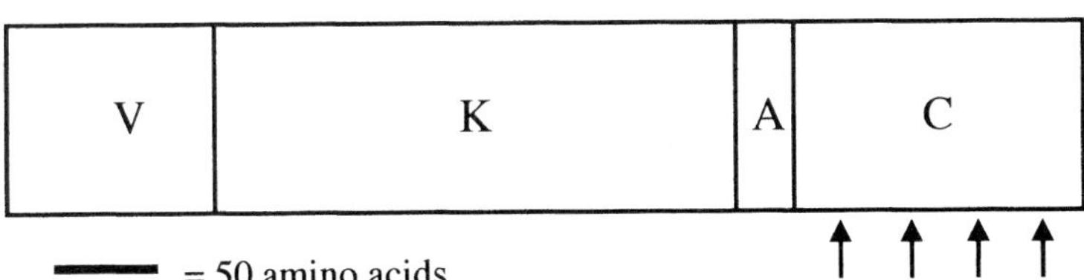

Fig. 1. Domain structure of calcium-dependent protein kinases. V = variable domain; K = kinase catalytic domain; A = autoregulatory domain; C = calmodulin-like domain. The sizes of the K, A, and C domains are conserved among most CDPKs. Variable domains range from 20 to nearly 200 amino acids, depending on the isoform. The figure depicts a CDPK isoform with a variable domain of 100 amino acids. Arrows indicate the approximate positions of the calcium-binding EF-hands.

different species are fairly well conserved in the kinase, autoregulatory, and calmodulin-like domains. The predicted structure of CDPKs provides an explanation for the calcium dependence and calmodulin independence of these enzymes described in Section II and Table I. A covalently attached calcium-binding domain with similarity to calmodulin is the distinguishing characteristic of CDPKs and has led to the proposal that CDPKs arose as a fusion between a calcium/calmodulin-dependent protein kinase gene and a calmodulin gene. The presence of a calmodulin-like domain also explains the observed sensitivity to calmodulin antagonists (Section II). A second related family of enzymes in plants, the CDPK-related kinases or CRKs, has a divergent calmodulin-like domain and is discussed in Section IV.

An alignment of the amino acid sequences of 75 CDPKs can be viewed online by going to the URL 'http://genetics.unh.edu/Faculty/Hrabak/' and clicking on the link 'CDPK alignment'. Table II summarizes selected information about these clones. CDPK names are based on the unified nomenclature for CDPKs proposed in 1996 (Hrabak *et al.*, 1996) which includes 66 CDPKs from the plant kingdom and nine from protists. This table does not contain information on a large number of CDPK expressed sequence tags (ESTs) from tomato (*Lycopersicon esculentum*), hybrid aspen (*Populus*), cotton (*Gossypium hirsutum*), corn (*Zea mays*), soybean (*Glycine max*), sorghum (*Sorghum*), pine (*Pinus taeda*), rice (*Oryza sativa*) and *Arabidopsis thaliana* in the GenBank EST database. Since the number of CDPK sequences is increasing rapidly, this table will soon be out of date, but is useful as a starting point for the current discussion.

In the following analysis, the published or annotated protein sequence has at times been altered to reflect what is more likely to be the correct and complete sequence. Corrections usually fell into one of the two following categories. First, the genomic sequences generated by the *Arabidopsis* Genome Initiative sequencing project are often not entirely accurate because computer programs are the primary source for intron prediction. The high degree of similarity of the catalytic, autoregulatory and calmodulin-like domains that is apparent when CDPK sequences are aligned was used to identify potential sequence discrepancies. Second, corrections were made to the variable domains of some published sequences to account for the existence of an extended open reading frame at the amino-terminus. In all cases, corrections resulted in CDPKs with sequences that more closely resembled other previously identified CDPKs. The molecular weights and variable domain sizes in Table II are based on these corrected sequences and thus may differ from published reports. Most CDPKs have a molecular weight between 54 and 72 kDa. This is close to, and often somewhat larger than, the masses reported for biochemically purified CDPKs (Section II) and may reflect the fact that CDPKs are often found to be susceptible to degradation during isolation (Suen and Choi, 1991; Schaller *et al.*, 1992; Binder *et al.*, 1994; Yuasa *et al.*, 1995).

TABLE II

Characteristics of calcium-dependent protein kinase from plants and protists

Species	CDPK name (published name, if different)	Accession number		Mol. wt (kDa)[a]	EF-hands[b]	Variable domain[c]	Myristoylation[d]	Reference (if published)
		cDNA	Genomic					
Arabidopsis thaliana (mouse-eared cress)	AtCPK1	L14771	AB008271	68.2	4	149	Y	Harper *et al.* (1993)
	AtCPK2	U31833	AC011560	72.2	4	185	Y	Hrabak *et al.* (1996)
	AtCPK3 (CDPK6)	U20623	U20625 and AL035394	59.3	4	77	N	Hong *et al.* (1996); Hrabak *et al.* (1996)
	AtCPK4	U31752	U55280 and AC007633 and AC006567	56.4	4	24	N	Hrabak *et al.* (1996)
	AtCPK5	U31834	AL022604	62.1	4	96	Y	Hrabak *et al.* (1996)
	AtCPK6 (ATCDPK3)	U31835 and D28582	AC002329	61.1	4	84	Y	Urao *et al.* (1994a); Hrabak *et al.* (1996)
	AtCPK7	U31836	–	60.3	3	58	Y	Hrabak *et al.* (1996)
	AtCPK8 (CDPK19)	U20624	U20627	59.9	3	56	Y	Hong *et al.* (1996); Hrabak *et al.* (1996)
	AtCPK9	U31751	AB024036	60.4	4	90	Y	Hrabak *et al.* (1996)
	AtCPK10 (ATCDPK1)	D21805	AC011809	61.4	3	62	Y	Urao *et al.* (1994b)
	AtCPK11 (ATCDPK2)	D21806	AC007887 and Y15964	55.9	4	26	N	Urao *et al.* (1994b)
	AtCPK12 (CDPK9)	U20388	U20626 and AB025633	55.4	4	21	N	Hong *et al.* (1996); Hrabak *et al.* (1996)
	AtCPK13	–	U54615 and AF049236	59.4	2	53	N	–
	AtCPK14	–	U90439 and AC002339	60.1	3	53	Y	–

TABLE II *Continued*

Species	CDPK name (published name, if different)	Accession number		Mol. wt (kDa)[a]	EF-hands[b]	Variable domain[c]	Myristoylation[d]	Reference (if published)
		cDNA	Genomic					
	AtCPK15	–	AL021890 and AL022140	62.6	4	101	Y	–
	AtCPK16	–	AC003952	64.8	4	107	Y	–
	AtCPK17	–	AB007727	58.5	4	72	Y	–
	AtCPK18	–	AL022373	60.2	4	70	Y	–
	AtCPK19	–	AC004392	62.9	4	97	N	–
	AtCPK20	X86963	AC005770	64.7	4	133	Y	–
	AtCPK21	–	AF118223	59.9	4	79	Y	–
	AtCPK22	–	AF118223	55.9	4	35	Y	–
	AtCPK23	–	AF118223	58.7	4	68	Y	–
	AtCPK24	–	AC007169 and AC007071	66.2	4	65	Y	–
	AtCPK25	–	AC007017	58.9	1	131	Y	–
	AtCPK26	–	AL035538	54.3	4	23	N	–
	AtCPK27	–	AF118223 and AF149414	54.9	4	27	Y	–
	AtCPK28	–	AB011474	59.0	4	61	Y	–
	AtCPK29	–	AC007396	60.5	4	84	N	–
Arachis hypogea (peanut)	AhCPK1	Y18055	–	Partial clone	4	–	–	–
Avena sativa (oat)	AsCPK6	–	–	Partial clone	–	–	–	Huttly and Phillips (1995)
	AsCPK7	–	–	Partial clone	–	–	–	Huttly and Phillips (1995)
	AsCPK8	–	–	Partial clone	–	–	–	Huttly and Phillips (1995)

TABLE II *Continued*

Species	CDPK name (published name, if different)	Accession number		Mol. wt (kDa)[a]	EF-hands[b]	Variable domain[c]	Myristoylation[d]	Reference (if published)
		cDNA	Genomic					
Chlamydomonas moewusii (green algae)	CmCPK1 (CCK1)	Z49233	–	66.2	4	151	N	Siderius *et al.* (1997)
Cucurbita pepo (zucchini)	CpCPK1	U90262	–	63.8	4	109	Y	Ellard-Ivey *et al.* (1999)
Daucus carota (carrot)	DcCPK1 (DcPK431)	X56599	–	60.1	4	80	Y	Suen and Choi (1991); Farmer and Choi (1999)
Eimeria maxima (protist)	EmCPK1	Z71756	–	Partial clone	4	–	–	Dunn *et al.* (1996)
Eimeria tenella (protist)	EtCPK1	Z71757	–	Partial clone	4	–	–	Dunn *et al.* (1996)
Fragaria x ananassa (strawberry)	FaCPK1	AF035944	–	63.7	4	111	Y	–
Glycine max (soybean)	GmCPK1 (CDPKα)	M64987	–	57.2	4	45	N	Harper *et al.* (1991)
	GmCPK2 (CDPKβ)	U69173	–	55.2	4	23	N	Lee *et al.* (1997)
	GmCPK3 (CDPKγ)	U69174	–	60.5	4	83	N	Lee *et al.* (1997)
Gossypium hirsutum (cotton)	GhCPK1	A1731660	–	Partial clone	–	72	Y	–
Ipomoea batatas (sweet potato)	IbCPK1	D87707	–	57.6	4	65	Y	–
Marchantia polymorpha (liverwort)	MpCPK1	–	AB017515	60.4	4	80	Y	Nishiyama *et al.* (1999)
Medicago sativa (alfalfa)	MsCPK1	–	–	Partial clone	–	–	–	Monroy and Dhindsa (1995)
	MsCPK2	–	–	Partial clone	–	–	–	Monroy and Dhindsa (1995)
Mesembryanthemum crystallinum (ice plant)	McCPK1	AF090835	–	60.0	4	85	Y	–
Nicotiana tabacum (tobacco)	NtCPK1 (NtCDPK1)	AF072908	–	61.6	4	92	Y	Yoon *et al.* (1999)

TABLE II *Continued*

Species	CDPK name (published name, if different)	Accession number		Mol. wt (kDa)[a]	EF-hands[b]	Variable domain[c]	Myristoylation[d]	Reference (if published)
		cDNA	Genomic					
Oryza sativa (rice)	OsCPK1 (spk)	D13436	–	60.6	2	72	N	Kawasaki *et al.* (1993)
	OsCPK2	X81394	–	59.5	4	84	Y	Breviario *et al.* (1995)
	OsCPK11	X81393	–	61.2	4	78	N	Breviario *et al.* (1995)
	OsCPK12	AF048691	–	61.3	4	78	N	–
Paramecium tetraurelia (protist)	PtCPK1 (PCaPK-α)	AF009562	AF009560	55.7	4	6	N	Kim *et al.* (1998)
	PtCPK2 (PCaPK-β)	–	AF009561	57.2	4	55	N	Kim *et al.* (1998)
Picea mariana (black spruce)	PmCPK1	AF051211	–	Partial clone	4	–	–	Perry and Bousquet (1998)
Plasmodium falciparum (protist)	PfCPK1 (PfCPK)	–	X67288	60.9	4	55	Y	Zhao *et al.* (1993)
	PfCPK2	X99763	–	59.1	4	71	Y	Farber *et al.* (1997)
	PfCPK3	–	Z98547	71.3	4	116	N	–
Solanum tuberosum (potato)	StCPK1	AF030879	–	63.7	–	110	Y	Lakatos *et al.* (1998)
	StCPK2	AF115406	–	Partial clone	–	–	–	–
Tortula ruralis (moss)	TrCPK1	U82087	–	63.7	4	111	Y	McKnight *et al.* (1997)
Toxoplasma gondii (protist)	TgCPK1	AF043629	–	56.6	4	50	Y	–
	TgCPK2 (TPK6)	AF118101	–	Partial clone	4	10	–	–
Vigna radiata (mung bean)	VrCPK1 (VrCCPK-1)	U08140	–	54.7	4	23	N	Botella *et al.* (1996)
Zea mays (corn)	ZmCPK1		L27484	60.0	4	91	Y	Estruch *et al.* (1994)
	ZmCPK2 (CRPK2)	U28376	–	58.1	4	64	N	Takezawa *et al.* (1996)
	ZmCPK3 (CRPK1)	L15390	–	Partial clone	4	–	–	Takezawa *et al.* (1996)

TABLE II *Continued*

Species	CDPK name (published name, if different)	Accession number		Mol. wt (kDa)[a]	EF-hands[b]	Variable domain[c]	Myristoylation[d]	Reference (if published)
		cDNA	Genomic					
	ZmCPK4 (ZmCDPK1)	D84408	–	Partial clone	4	–	–	Berberich and Kusano (1997)
	ZmCPK5	AJ007366	–	69.5	4	152	Y	–
	ZmCPK7 (ZmCDPK7)	D87042	–	61.1	4	90	Y	Saijo *et al.* (1997)
	ZmCPK8 (ZmZK8)	D87044	–	Partial clone	–	–	–	–
	ZmCPK9 (ZmCDPK9)	D85039 and D87043	–	59.4	4	82	Y	Saijo *et al.* (1997)
	ZmCPK19	D87045	–	Partial	–	–	–	–
	ZmCPK24	D87046	–	Partial	–	–	–	–

[a]Molecular weights were calculated as described in Wilkins *et al.* (1998).
[b]Number of EF-hands estimated by SMART (Shultz *et al.*, 1998).
[c]Number of amino acids in variable domain.
[d]Myristoylation motif predicted by PROSITE (Hofmann *et al.*, 1999) and experimental data (E. Hrabak, unpublished data).

The kinase domain of CDPKs contains all 12 of the classic subdomains of serine/threonine protein kinases as defined by Hanks and Hunter (1995). Once cloned and expressed, most CDPKs display the ability to autophosphorylate and/or to phosphorylate one or more kinase substrates, like histone IIIS or casein, *in vitro* in a calcium-dependent fashion. This indicates that these proteins do indeed function as calcium-dependent protein kinases.

The calmodulin-like domain of most CDPKs contains four calcium-binding EF-hands, although isoforms with one, two, and three EF-hands have also been identified (Table II). An EF hand is a helix-loop-helix structure in which six of the 12 residues in the loop domain act together to bind a single calcium ion (Zhang and Yuan, 1998). The number of EF-hands in each CDPK was predicted by computer using the SMART algorithm (Schultz *et al.*, 1998). This number usually, but not always, agrees with the published number of EF-hands. Discrepancies are due to the use of different algorithms to predict EF-hand motifs. For example, PROSITE (Hofmann *et al.*, 1999) often predicts fewer EF-hands than SMART. In any case, the numbers of EF-hands are predictions only and their ability to bind calcium requires experimental confirmation. In a comparison between the calmodulin-like domains of *Arabidopsis* CDPKs and native *Arabidopsis* calmodulins, the calcium-binding loop domains displayed less amino acid sequence divergence than the interloop regions (E. Hrabak, data not shown). In contrast, native calmodulins have highly conserved amino acid sequences (Zhang and Yuan, 1998). The sequence divergence of the calmodulin-like domain from native calmodulins may reflect changes which optimize the calmodulin-like domain to function in the context of the kinase molecule. Direct binding of ^{45}Ca, as well as a calcium-induced electrophoretic mobility shift, has been observed for many CDPKs (Yuasa *et al.*, 1995; Saijo *et al.*, 1997; Farmer and Choi, 1999; Yoon *et al.*, 1999). The role of the calcium-binding EF hands in CDPK activity has been investigated by two different approaches. Carboxy-terminal deletions that sequentially removed each of the four EF-hands from an *Arabidopsis* CDPK indicated that loss of even one EF-hand domain greatly decreased the ability of the enzyme to respond to calcium (Hong *et al.*, 1996). This approach has the drawback that the enzyme's conformation may be affected by the deletions. In a set of elegant experiments with a CDPK from *Plasmodium falciparum*, Zhao *et al.* (1994) performed site-directed mutagenesis of a highly conserved glutamate residue in each EF-hand. Their results indicated that all four EF-hands did not function equally and that mutations of the two EF-hands nearest to the catalytic domain had large, deleterious effects on calcium-binding and enzyme activity while mutations of the last two EF-hands had only minor effects. One recent study has provided evidence that different CDPKs, all with four predicted EF-hands, were activated at different calcium concentrations (Lee *et al.*, 1998). Three soybean CDPK isoforms (Table II) reached half-maximal activity at 0.06, 0.4, or 1 μM Ca^{2+} with the same substrate. Clearly, much more work needs to be done to understand the role of the calmodulin-like domain in regulation of CDPK function.

Regulation of CDPK activity occurs via interactions between the autoregulatory and calmodulin-like subdomains. The 30–35 amino acid autoregulatory domain (Fig. 1) contains a pseudosubstrate site (Harmon *et al.*, 1994; Harper *et al.*, 1994), analogous to the well-studied regulatory domain of calcium/calmodulin-dependent protein kinase II (Payne *et al.*, 1988). In fact, phylogenetic analyses place the plant CDPKs in the CaMK group of calcium/calmodulin-regulated enzymes, which includes CaMKII, myosin light chain kinase (MLCK) and the Snf1-like kinases (Hanks and Hunter, 1995). In the inactive state, the pseudosubstrate sequence in the autoregulatory domain is bound in the catalytic site of the enzyme and entry of other substrates is prevented. In the case of CDPKs, activation from this inhibited state is dependent on calcium binding. When intracellular calcium levels increase, calcium binds to EF-hands in the calmodulin-like domain. Calcium binding probably induces a conformational change in the enzyme, which removes the pseudosubstrate from the catalytic site. This conformational change may be analogous to the structural alteration observed upon interaction of free calmodulin with calcium (Zhang and Yuan, 1998). The following evidence from studies on CDPKs from *Arabidopsis thaliana* and *Glycine max* supports the role of the autoregulatory and calmodulin-like domains in regulating the activity of CDPKs. First, deletion of only the calmodulin-like domain resulted in an inactive enzyme that could not be activated by calcium, while deletion of both the autoregulatory and calmodulin-like domains created a constitutively active, calcium-insensitive enzyme (Harper *et al.*, 1994). These results indicate that both the calmodulin-like and autoregulatory domains are important for normal enzyme activation. Second, an inactive calmodulin-like domain deletion mutant was activated either by antibodies that recognize the autoregulatory domain or by addition of a calcium–calmodulin complex (Harper *et al.*, 1994; Huang *et al.*, 1996). Presumably, the antibody or the calcium/calmodulin acts by interfering with pseudosubstrate binding to the active site of the enzyme. Third, mutations within the autoregulatory domain that are predicted to disrupt the interaction between the pseudosubstrate and the catalytic domain resulted in increased activity of the enzyme in the absence of calcium (Harper *et al.*, 1994; Huang *et al.*, 1996). Fourth, peptide inhibitors of the related kinases CaMKII and MLCK, as well as CDPK autoregulatory domain peptides, inhibited CDPK activity (Harmon *et al.*, 1994; Harper *et al.*, 1994). This inhibition probably represents nonreversible binding of the free peptides to the catalytic site. As indicated by these experimental data, there is an intricate interplay between the autoregulatory and catalytic domains that controls calcium activation of CDPKs.

Many of the observed regulatory and catalytic functions of CDPKs can be explained by interactions between the kinase, autoregulatory, and calmodulin-like domains. The function of the variable domain, located at the amino-terminus of the enzyme, is not well understood. In contrast to the relatively high degree of sequence similarity observed for the other three domains of CDPKs, the variable domain is the least conserved region of the enzyme. There is little amino acid sequence conservation between the variable domains of different isoforms and the total number of amino acids in the variable domain ranges from about 20 to nearly 200 (Table II). One hypothesis to explain this observation is that the variable domain may help to 'customize' a CDPK for its specific role in the cell. Since CDPKs have been reported to be present in both soluble and membrane fractions, in nuclei, and associated with the cytoskeleton (Table I), some of the signals required to target the proteins to those locations may be located in the divergent variable domain. Alternatively, the variable domain could contain motifs for protein–protein interactions or post-translational modifications.

The variable domains of about half of the CDPKs listed in Table II contain potential acylation sites for covalent attachment of fatty acids. Since no transmembrane domains have been predicted or reported in CDPKs, attachment of fatty acid groups could provide sufficient hydrophobicity to achieve the observed membrane association of some CDPKs (Table I). Potential sites for attachment of myristate (C 14:0) and palmitate (C 16:0) residues are usually identified based on comparison to acylation motifs in animal and fungal systems (Duronio *et al.*, 1990; Johnson *et al.*, 1994; Lodge *et al.*, 1994). Myristate is most often attached to a glycine residue at position 2 in a reaction catalyzed by N-myristoyltransferase (Towler *et al.*, 1988; Johnson *et al.*, 1994). Myristoylation has been implicated in membrane binding and protein–protein interactions (Schultz *et al.*, 1988; Johnson *et al.*, 1994). The presence of potential myristoylation sites is usually identified using programs such as PROSITE (Hofmann *et al.*, 1999) and/or tested experimentally. The amino acid composition of the myristoylation consensus sequence recognized by fungal and animal N-myristoyl transferases has been well studied, but virtually nothing is known about the myristoylation reaction or its amino acid sequence requirements in plants. At least two CDPKs have been shown to be myristoylated in an *in vitro* wheat germ extract assay system (Ellard-Ivey *et al.*, 1999; S. Lu and E. Hrabak, unpublished data). Interestingly, neither of these CDPKs contains a myristoylation consensus sequence recognized by PROSITE, indicating that plant N-myristoyltransferases may have a different substrate specificity than their animal counterparts. Therefore, in predicting myristoylation of CDPKs, both the myristoylation consensus information from PROSITE (Hofmann *et al.*, 1999) and from Towler *et al.* (1988), in conjunction with the limited experimental data on myristoylated CDPKs in plants, have been used. Consistent with the observed amino terminal acylation of CDPKs, limited proteolysis data (Schaller *et al.*, 1992) indicate that a site required for membrane attachment is located near one terminus of the polypeptide. The role of myristoylation in CDPK function is currently under investigation in several laboratories.

Many CDPKs also contain a cysteine residue either immediately following the glycine at position 2 or within the first eight amino acids. Cysteine residues in this context are often palmitoylated in animal cells (Milligan *et al.*, 1995). However, the enzymes that carry out this reaction have not been identified and palmitoylation of proteins has not yet been described in plants. This motif remains to be investigated for its potential importance in membrane binding of CDPKs.

Bipartite nuclear localization signals (Robbins *et al.*, 1991) were identified by PROSITE in the variable domains of a few CDPKs (data not shown). While little information is currently available on the subcellular location of CDPKs, at least one CDPK has been purified from nuclei (Li *et al.*, 1991). The presence of nuclear localization signals on some CDPKs may provide the mechanism for this targeting.

B. CDPK FAMILY MEMBERS

Calcium-dependent protein kinase genes have been identified from a wide range of species within the plant kingdom, including angiosperms, gymnosperms, bryophytes, and algae. In plants, calcium-dependent protein kinases appear to be the largest family of protein kinases with a well-defined and conserved domain structure (Hardie, 1999). Although the receptor kinase family has more members, the within-family diversity is greater for the receptor kinases than for the CDPKs (Satterlee and Sussman, 1998). In addition to the apparently ubiquitous presence of CDPKs in the plant kingdom, CDPKs have been cloned from several species of protists: *Eimeria, Paramecium, Toxoplasma*, and *Plasmodium*. However, no CDPKs have as yet been identified in animals, fungi, or bacteria. This distribution indicates that CDPKs could have evolved either independently in both plants and protists or after the lineage that gave rise to the plants and protists split from the lineage that produced animals and fungi.

Arabidopsis thaliana currently holds the record for the most CDPKs identified in a single organism, probably because the genome sequence of this plant is nearing completion while other plant genome projects are not as far along. The 29 *Arabidopsis* CDPKs are distributed among all five chromosomes, with one known cluster of four contiguous CDPKs (containing AtCPK21, AtCPK22, AtCPK23, and AtCPK27) on chromosome 4 (data not shown). Based on the current status of the *Arabidopsis* genome sequencing project (about 75% complete as of July 1999), several more CDPKs probably remain to be identified. Sixteen of the *Arabidopsis* CDPKs are represented by cDNA clones (Table II; E. Hrabak, unpublished data), confirming that these genes are transcribed. Of the remaining 13 genomic sequences, AtCPK21 and AtCPK29 have good matches to ESTs (E. Hrabak, data not shown). Thus, even if some of the CDPK sequences predicted by the genome sequencing project are

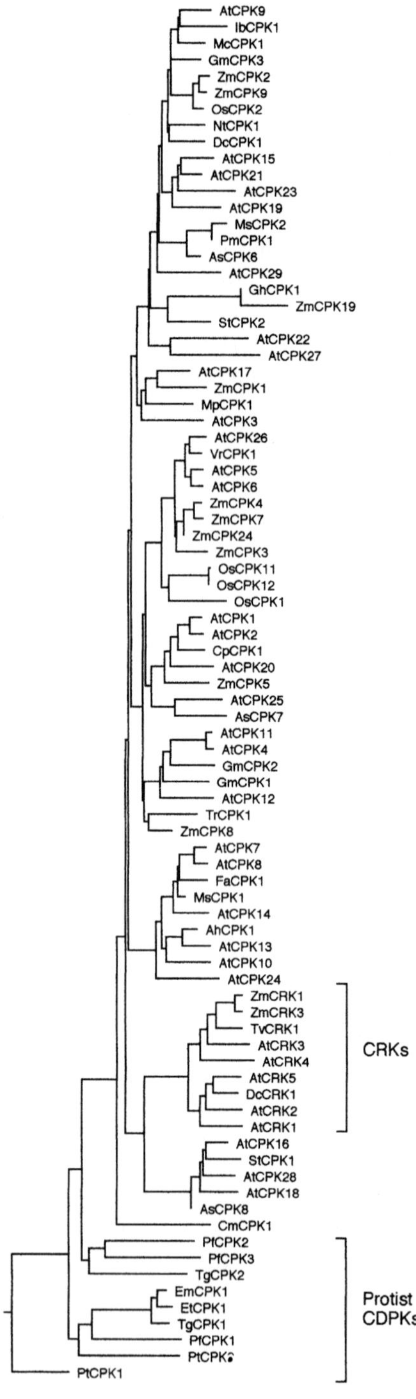
AtCPK9
IbCPK1
McCPK1
GmCPK3
ZmCPK2
ZmCPK9
OsCPK2
NtCPK1
DcCPK1
AtCPK15
AtCPK21
AtCPK23
AtCPK19
MsCPK2
PmCPK1
AsCPK6
AtCPK29
GhCPK1
ZmCPK19
StCPK2
AtCPK22
AtCPK27
AtCPK17
ZmCPK1
MpCPK1
AtCPK3
AtCPK26
VrCPK1
AtCPK5
AtCPK6
ZmCPK4
ZmCPK7
ZmCPK24
ZmCPK3
OsCPK11
OsCPK12
OsCPK1
AtCPK1
AtCPK2
CpCPK1
AtCPK20
ZmCPK5
AtCPK25
AsCPK7
AtCPK11
AtCPK4
GmCPK2
GmCPK1
AtCPK12
TrCPK1
ZmCPK8
AtCPK7
AtCPK8
FaCPK1
MsCPK1
AtCPK14
AhCPK1
AtCPK13
AtCPK10
AtCPK24
ZmCRK1
ZmCRK3
TvCRK1
AtCRK3
AtCRK4
AtCRK5
DcCRK1
AtCRK2
AtCRK1
CRKs
AtCPK16
StCPK1
AtCPK28
AtCPK18
AsCPK8
CmCPK1
PfCPK2
PfCPK3
TgCPK2
EmCPK1
EtCPK1
TgCPK1
PfCPK1
PtCPK2
PtCPK1
Protist CDPKs

nonfunctional, there are at least 18 expressed CDPKs in *Arabidopsis*. The reason for the large number of CDPK enzymes is not yet known but there are several possibilities. First, it seems likely that these enzymes are specialized in their expression patterns. For example, some isoforms may be present only in certain cell types or in particular subcellular locations or at specific times during plant development. Four of the *Arabidopsis* CDPKs have been studied in detail in the author's laboratory and each has a distinct spatial and temporal expression pattern (E. Hrabak, unpublished data). In addition, expression of many CDPK genes is altered in response to specific stimuli (Section V). These lines of evidence indicate that CDPK expression may be highly regulated. Second, these multiple enzyme isoforms may have evolved to perform similar or partially redundant functions in order to substitute for each other, should the need arise. If this is true, it indicates that CDPK activity may be critical for normal plant growth and development. Finally, each CDPK might interact with a unique set of substrates or respond to distinct types of calcium waves. Such specialization would necessitate a large number of enzymes with similar kinase activity but the ability to recognize different substrates or signals.

Multiple CDPK genes have also been identified in soybean, alfalfa, rice, potato, and corn, as well as most of the protists (Table II). As these plant and protist genomes are characterized further, it is expected that multiple CDPK genes will be the rule and not the exception. Indeed, many ESTs that represent CDPK genes can be found in database searches and likely represent additional members of the CDPK family in rice, corn, soybean, and other species.

The evolutionary relatedness of the CDPKs from plants and protists, as well as the CDPK-related kinases (Section IV), is shown in Fig. 2. This phylogenetic tree was constructed using only the conserved kinase domains of these proteins to exclude any potential bias from the other subdomains. Several features of this tree are noteworthy. First, all of the protist CDPKs cluster together, indicating that these sequences are more related to each other than to either the plant CDPKs or CRKs. Second, the plant CRKs, which contain no identifiable EF-hands in their carboxy-terminal domain (Section IV), are more closely related to each other than to any of the CDPKs. This is interesting because CRKs are probably functional kinases (kinase activity has been demonstrated for one of them) but are not regulated by calcium. Third, most of the plant CDPKs are clustered into smaller groups that are not dependent on species relationships. For instance, no strict clustering along taxonomic lines is observed between monocot, dicot, and algal CDPKs. It is not known whether the observed groupings have any functional significance, but some tantalizing hints can be found. For example, the subgroup of five CDPKs containing

Fig. 2. Phylogenetic tree containing all known CDPK and CRK sequences as of July 1999. The predicted amino acid sequences of the kinase domains were aligned using the Clustal (Thompson *et al.*, 1994) program.

AtCPK1 through to ZmCPK5 contains CDPKs with some of the largest amino-terminal domains, and all members of this subgroup also have myristoylation consensus sequences and four predicted EF-hands. Almost all members of the subgroup containing AtCPK7 through to AtCPK10 contain putative bipartite nuclear localization sequences (data not shown), although their actual subcellular locations are not known. Many of the isoforms in this group also contain fewer than four EF-hands. Expression of two closely grouped CDPKs, ZmCPK9 (from *Zea mays*) and OsCPK2 (from *Oryza sativa*), is known to be affected by light (Breviario *et al.*, 1995; Saijo *et al.*, 1997). The predictive significance of this observation cannot be addressed until the expression of the closely related CDPKs is known. One group of plant CDPKs (AtCPK16, StCPK1, AtCPK28, AtCPK18, and AsCPK8) is separated from the majority of the plant CDPKs by the CRK group. Interestingly, the four CDPKs in this group that have genomic clones also have an intron placement pattern that more closely resembles the placement of CRK introns than CDPK introns, and their sequences contain several other small insertions or deletions that are not found in the other CDPK sequences (E. Hrabak, unpublished data). As more data become available on the functions and substrates of CDPKs, these phylogenetic groupings may become a valuable predictive tool.

IV. CRK GENES AND GENE FAMILIES

CRKs have a structure similar to CDPKs (Fig. 1) but characterized by a divergent autoregulatory domain and a calmodulin-like domain containing degenerate consensus EF-hand sequences. Algorithms that identify protein domains (Schultz *et al.*, 1998) detect no EF-hand motifs in CRK amino acid sequences and CRKs do not bind calcium. Because of the lack of EF-hands, it has at times proven problematic to distinguish CRKs from calcium/calmodulin-dependent kinases. Both contain a kinase catalytic domain followed by a carboxy-terminal extension of about 200 amino acids. In CaMKII, this carboxy-terminus contains a kinase autoinhibitory domain, a calmodulin-binding domain, and a subunit association domain (Soderling, 1996). Binding of calcium/calmodulin to the calmodulin-binding domain causes CaMKII to autophosphorylate and become functionally active. The function of the carboxy-terminus of CRKs has not been investigated but a potential calmodulin-binding site (a basic amphipathic α helix) is found in most CRKs. At least one CRK binds calmodulin (Lu *et al.*, 1996), although the location of the binding site and the effect of calmodulin binding on CRK activity are not known. However, since at least one CDPK autoregulatory domain can also bind calmodulin (Huang *et al.*, 1996), CRKs cannot be definitively classified as CaMKIIs based on this characteristic. CRKs also have an N-terminal domain (usually about 150 amino acids in length) that precedes the kinase domain. This type of N-terminal structure is more typical of CDPKs and is not observed in CaMKIIs.

To date, the following CRKs have been identified: DcCRK1 from *Daucus carota* (Lindzen and Choi, 1995), ZmCRK1 and ZmCRK3 from *Zea mays* (Furumoto *et al.*, 1996; Lu *et al.*, 1996), TvCRK1 (a partial sequence) from *Tradescantia virginiana*, and AtCRK1, AtCRK2, AtCRK3, AtCRK4 (a partial sequence), and AtCRK5 from *Arabidopsis thaliana* (Table III). No CRKs have yet been reported from gymnosperms, algae, or protists. All of the CRKs except for AtCRK3 are represented by cDNA clones (Table III), indicating that these genes are transcribed. Since the *Arabidopsis* genome sequence is not yet complete, it is likely that additional CRK genes will be identified. However, it does not appear that the CRK gene family in *Arabidopsis* will be as large as the CDPK gene family.

An alignment of the known CRK amino acid sequences can be viewed by going to the URL 'http://genetics.unh.edu/Faculty/Hrabak/' and clicking on the link 'CRK alignment'. The CRK alignment reveals a number of interesting features. First, all of the CRKs with a full-length coding sequence contain myristoylation consensus sequences at their amino-termini, as predicted by PROSITE (Hofmann *et al.*, 1999). DcCRK1 has been shown to be myristoylated *in vitro* and is primarily associated with the membrane fraction of suspension-cultured carrot cells (J. Choi, personal communication). Second, the N-terminal variable domain of CRKs from all species contains a block of highly conserved amino acids (P/A)F(P/H)PPSPA(K/R)HI(K/R)A. This is noteworthy because the CDPKs do not have any analogous regions of similarity in the variable domain that are conserved across isoforms or across species. The function of this conserved CRK region is unknown. Third, all of the conserved subdomains typical of protein kinases (Hanks and Hunter, 1995) are found in CRK sequences, indicating that CRKs are likely to be bona fide protein kinases. One of the CRKs, ZmCRK3 expressed in *Escherichia coli*, has been demonstrated to be a functional protein kinase *in vitro* (Furumoto *et al.*, 1996). Since this enzyme assay was performed on a bacterial extract with no added calmodulin, it appears that CRKs can function in the absence of calmodulin. Fourth, although CRKs have a carboxy-terminal region similar in length to the calmodulin-like domain of CDPKs, no EF-hands are identified by protein domain prediction programs such as PROSITE (Hofmann *et al.*, 1999) or SMART (Schultz *et al.*, 1998). The lack of calcium-binding sites is consistent with the calcium-independent CRK kinase activity observed for bacterially expressed ZmCPK3, which phosphorylates casein and other kinase substrates in the presence of EGTA (Furumoto *et al.*, 1996). In addition, DcCRK1 does not bind calcium or show a gel mobility shift in response to calcium or EGTA (J. Choi, personal communication).

Little is known about the function of CRKs in plants. The maize CRKs are more highly expressed in roots than leaves by Northern analysis (Furumoto *et al.*, 1996). Other workers found ZmCRK1 expression in maize root tips, but

TABLE III

Characteristics of CRKs (CDPK-related kinases) cloned from plants

Species	CRK name	Accession number		Mol. wt (kDa)[a]	Variable domain[b]	Myristoylation[c]	Reference (if published)
		cDNA	Genomic				
Arabidopsis thaliana (mouse-eared cress)	AtCRK1	AF153351	AC004261	64.3	122	Y	–
	AtCRK2	AF153352	–	Partial clone	–	–	–
	AtCRK3	–	AC005819	66.6	142	Y	–
	AtCRK4	AF130252	AB016884	Partial clone	–	–	–
	AtCRK5	Y09418	–	67.0	147	Y	–
Daucus carota (carrot)	DcCRK1	X83869	–	67.2	147	Y	Lindzen and Choi (1995)
Tradescantia virginiana	TvCRK1	AF009337	–	Partial clone	–	–	–
Zea mays (corn)	ZmCRK1	D84507	–	68.8	173	Y	Furumoto *et al*. (1996); Lu *et al*. (1996)
		S82324					
	ZmCRK3	D48508 and D38452	–	67.4	155	Y	Furumoto *et al*. (1996)

[a]Molecular weights were calculated as described in Wilkins *et al*. (1998).
[b]Number of amino acids in variable domain.
[c]Myristoylation motif predicted by PROSITE (Hofmann *et al*., 1999) and experimental data (E. Hrabak, unpublished data).

did not investigate other tissues (Lu *et al.*, 1996). The carrot CRK, DcCRK1, is highly expressed in somatic embryos and to a lesser extent in mature plant tissues such as flowers, leaves, and roots (Lindzen and Choi, 1995).

A phylogenetic tree of the catalytic domains of CDPKs and CRKs (Fig. 2) indicates that CRKs form a distinct subgroup. More work is needed to determine the role of CRKs in plant cells, including cellular and subcellular localization studies, identification of CRK substrates, and further investigation into their ability to bind or be regulated by calmodulin and/or calcium.

V. REGULATION OF EXPRESSION OF CDPK GENES

A large number of CDPK genes is present in *Arabidopsis* and the same is likely to be true for other plants. This plethora of calcium-regulated kinases presents a challenge for researchers seeking to understand the role of these enzymes in plants. Many important cellular processes are calcium dependent (Section I) and phosphorylation is commonly used as a means of information transfer within plant cells. It is becoming clear that CDPKs are involved in these processes but their precise role is less well understood. CDPKs may be involved in signal transduction pathways that co-ordinate cytoplasmic calcium fluxes with changes in gene expression and/or they may be involved in direct regulation of enzyme activities by co-ordinated phosphorylation/dephosphorylation reactions. At the present time, we have only a rudimentary knowledge of when and how CDPK gene expression is regulated and, in the future, new information will certainly help to clarify the role of specific CDPKs. For instance, some CDPKs may have temporally or spatially restricted expression patterns or, as discussed in Section III, be targeted to different cellular compartments or to particular membranes, and thus may have limited or increased access to their substrates. In addition, CDPKs may function as calcium sensors that become active when different intracellular calcium concentrations are attained. Calcium concentration gradients are known to form in the cytoplasm in response to stimuli (Section I). In this section, current knowledge of expression patterns and the stimuli that induce CDPK gene expression are discussed. In some cases, the identity and the specificity of the probe used for Northern blot detection were not clearly specified by the authors. Since we now know that most plants probably express multiple CDPK proteins, it will be important in future experiments to use isoform-specific probes (for example, from the 5′ or 3′ untranslated region) or provide evidence that a probe made within the coding sequence does not recognize other CDPKs.

A. ORGAN OR TISSUE SPECIFIC

Northern blots are the most commonly used technique for detecting differential expression of genes in various plant organs or tissues. In many cases, expression was observed in most of the tissues examined. *ZmCPK4*, *ZmCPK7* and *ZmCPK9* messenger RNAs (mRNAs) were detected in both roots and leaves of maize seedlings using isoform-specific probes (Berberich and Kusano, 1997; Saijo *et al.*, 1997). OsCPK2- and OsCPK11-specific transcripts from 3-day-old rice (*Oryza sativa*) seedlings were detected in both roots and coleoptiles, although *OsCPK2* messages were more abundant in coleoptiles than in roots (Breviario *et al.*, 1995). In young zucchini (*Cucurbita pepo*) seedlings, the CpCPK1 transcript was detected in roots, hypocotyl, hook, and to a lesser extent, in cotyledons (Ellard-Ivey *et al.*, 1999). Potato *StCPK1* mRNA was detected in roots, leaves, flowers, and tubers, but not in berries. When flowers were examined in more detail, StCPK1 transcript was found in sepals, petals, and stamens, but not in carpels or immature flower buds (Lakatos *et al.*, 1998). Transcripts for three *Arabidopsis thaliana* CDPK genes were all detected in stems, roots, leaves, and flowers, although expression in flowers was generally higher than in these other tissues (Hong *et al.*, 1996). In these last three examples, the entire CDPK gene was used as a probe and its specificity was not determined. Thus it is perhaps not surprising that the StCPK1 probe also hybridized to mRNA from other solanaceous plants, as well as *Arabidopsis thaliana* (Lakatos *et al.*, 1998).

It is often very difficult to distinguish cell type-specific expression patterns using Northern blots. For instance, a CDPK that is reported in leaves, stems, roots, and flowers by Northern blotting could, in fact, be expressed only in the phloem cells that are found in each of these locations, although this would not be apparent from the results. In a few cases, Northern blots have been used to demonstrate that some CDPKs have very limited sites of expression. The OsCPK1 transcript was detected only in developing rice seeds and not in leaves, stems, or roots (Kawasaki *et al.*, 1993). ZmCPK1 was expressed specifically in maize pollen, beginning at the early trinucleate stage of development and continuing throughout pollen maturation and germination (Estruch *et al.*, 1994).

Limited cellular expression data are available for CDPK genes. Hong *et al.* (1996) used antibodies raised against two *Arabidopsis thaliana* CDPK isoforms in immunohistochemical experiments. The presence of cross-reacting proteins was detected in chloroplast-containing cells throughout the plant, as well as many other locales including phloem, stigma, endosperm and some embryonic stages. The caveat in these studies is that these antibodies were not isoform specific and thus this expression pattern represents the composite of an unknown number of CDPK proteins.

There is one report on alternative splice variants of a CDPK message in the bryophyte *Marchantia* that produces two proteins with only four amino acid differences between them (Nishiyama *et al.*, 1999). The authors propose that these two forms of the CDPK protein may be differentially expressed in male and female sex organs.

Future studies to determine the cellular and subcellular locations of CDPKs should greatly enhance our understanding of these enzymes. Such experiments should include localization of transcripts by *in situ* hybridization or *in situ* polymerase chain reaction (PCR) on tissue sections, as well as protein detection by immunological techniques using isoform-specific antibodies.

B. HORMONE INDUCIBLE

Induction of some CDPK genes by plant growth factors has been reported. Expression of NtCPK1, as detected by Northern blotting of RNA from detached leaves of *Nicotiana tabacum*, was induced by a 12-h treatment with 100 μM concentrations of gibberellin, abscisic acid, or cytokinin, but not by treatment with auxins (Yoon *et al.*, 1999). In contrast, a 12-h treatment with 500 μM indole acetic acid (IAA) induced expression of VrCPK1 from *Vigna radiata* (Botella *et al.*, 1996). The response to other hormones was not reported. Other workers have investigated the activity of CDPKs in response to hormone treatment by monitoring phosphorylation of a general kinase substrate such as histone. The ability of rice seed membrane calcium-stimulated kinase activity to phosphorylate histone increased within 10 min after exposure of imbibing seeds to 50 μM GA_3 (Abo-El-Saad and Wu, 1995). Finally, a barley ABA-responsive promoter linked to a green fluorescent protein (GFP) reporter gene was used to study the involvement of CDPKs in ABA or other related stress responses. Two of the four *Arabidopsis* CDPKs tested directly activated this ABA-responsive promoter in the absence of stress signals (Sheen, 1996) and a mutation in the kinase domain prevented this activation. These initial studies should serve as a stimulus for investigations into the hormone responsiveness of other CDPKs. These data should help to connect CDPKs to the signal transduction pathways for response to plant growth factors.

C. DROUGHT OR SALT STRESS INDUCIBLE

There is much interest in identifying genes that might be used to improve the tolerance of plants to drought or saline growth conditions, since this would not only improve performance under temporary stress conditions, but also extend the acreage of usable land for agriculture. Since the hormone ABA is synthesized in response to drought stress, there is the possibility that CDPKs that respond to drought or salt stress are also regulated by ABA. Several examples of salt-stress or drought-stress-inducible CDPK genes have been reported. Transcripts for two *Arabidopsis* CPK genes, *AtCPK10* and *AtCPK11*, were rapidly induced within 10 min after either dehydration or exposure to 250 mM NaCl (Urao *et al.*, 1994b). In mung bean (*Vicia faba*), strong induction of *VrCPK1* mRNA was observed in shoots within 2 h after treatment with 50 mM NaCl (Botella *et al.*, 1996). Finally, induction of

NtCPK1 gene expression was first detected 1 h after 100 mM NaCl treatment in *Nicotiana tabacum* (Yoon *et al.*, 1999).

D. INDUCTION BY OTHER STIMULI

Levels of several CDPK mRNAs are reported to be downregulated by light, including CpCPK1 from *Cucurbita pepo*, OsCPK2 from *Oryza sativa*, and ZmCPK7 and ZmCPK9 from *Zea mays*. Five-day-old dark-grown zucchini seedlings showed a decrease in *CpCPK1* mRNA in roots, hypocotyls, and hooks in comparison to light-grown seedlings (Ellard-Ivey *et al.*, 1999). OsCPK2 transcript levels were markedly reduced when seedlings that had been grown in the dark for 2.5 days were given 12 h of light before isolation of RNA from coleoptiles. Another CDPK from rice, OsCPK11, was not affected by light treatment (Breviario *et al.*, 1995). Two maize CDPK genes, *ZmCPK7* and *ZmCPK9*, were more highly expressed in 10-day-old etiolated seedlings than in light-grown seedlings (Saijo *et al.*, 1997). These results indicate a possible role for some CDPKs in germination or in response to fluctuations in light intensity.

The mRNAs for *Vigna radiata, VrCPK1*, and *Nicotiana tabacum*, *NtCPK1*, are induced by calcium chloride treatment. *Vigna radiata* stems that were immersed in 10 mM $CaCl_2$ for 6 h before RNA was isolated from leaves contained significantly more VrCPK1 transcript than water- or $MgCl_2$-treated controls (Botella *et al.*, 1996). In the case of tobacco, petioles of detached leaves were placed in a range of $CaCl_2$ solutions (from 10–250 mM) for 12 h before RNA extraction (Yoon *et al.*, 1999). In this experiment, the highest transcript levels were observed in the 100 mM $CaCl_2$ treatment. These results indicate that some CDPKs may be regulated both transcriptionally and translationally by calcium.

Both calcium and protein phosphorylation are required for acquisition of cold-induced freezing tolerance in plants (Monroy *et al.*, 1993). Exposure to cold temperatures and the resulting acclimation response has been correlated with an increase in expression of certain *Arabidopsis* CDPKs when studied with a modified differential display technique (Tahtiharju *et al.*, 1997). None of the partial CDPK sequences that were identified corresponds precisely to any of the 29 *Arabidopsis* CDPKs listed in Table II. All five of these CDPKs might represent new genes but this is unlikely based on the amount of the *Arabidopsis* genome remaining to be sequenced. Another possibility is that, since the differential display technique includes the use of PCR, sequence errors may have been introduced during the DNA amplification steps. In other experiments using alfalfa, expression of a potential CDPK, for which only a sequence fragment is available, was increased in response to cold treatment at 4°C (Monroy and Dhindsa, 1995). In addition, cold (5°C) or cycloheximide treatments were found to increase transcript levels of *Zea mays* ZmCPK4 in both roots and leaves within 3–6 h (Berberich and Kusano, 1997). It is not known whether the effect of cycloheximide was due to increased transcription, to changes in stability of transcription factors, or to direct stabilization of the ZmCPK4 transcript itself.

The final three examples represent phenomena that have been reported for only one CDPK at this time. First, a CDPK from *Vigna radiata*, VrCPK1, has been reported to be inducible by mechanical strain or bending (Botella *et al.*, 1996). This induction occurred within 30 min of application of a 30-s bending treatment. Interestingly, some calmodulin and calmodulin-related genes in *Arabidopsis* are also induced by mechanical stresses and calcium treatment, in a similar manner to VrCPK1 (Braam and Davis, 1990; Braam, 1992). Auxin is known to be involved in the plant's response to mechanical strain and VrCPK1 was also induced by auxin, albeit at a slower rate (Botella *et al.*, 1996). The protein synthesis inhibitor cycloheximide, which stimulates auxin-induced genes, also induced a large increase in *VrCPK1* mRNA (Botella *et al.*, 1996). Second, anoxia or anaerobiosis decreased expression of OsCPK2 from *Oryza sativa*, but did not affect transcript levels of OsCPK11 (Breviario *et al.*, 1995). Third, methyl jasmonate, fungal elicitors, and chitosan induced NtCPK1 expression in *Nicotiana tabacum*, indicating a potential role in defense against plant pathogens (Yoon *et al.*, 1999).

New technologies are emerging that allow comparison of global changes in expression of many genes simultaneously, such as DNA microarray hybridization with RNAs isolated from plants after different treatments (Jordan, 1998). Microarrays and similar techniques will enable researchers not only to evaluate changes in CDPK gene expression, but also to integrate information about other genes which are regulated by the same stimuli.

VI. CDPK SUBSTRATES

One important area of research is to correlate CDPKs with their substrates and signal transduction pathways. Expression data represent one approach for reaching this goal. The identification of downstream substrates is another critical area of research. In this section, data are presented about the variety of proteins in plants that are phosphorylated by CDPKs. In some cases, phosphorylation is a known regulatory mechanism of the target protein, which provides further support for the protein being a bona fide CDPK substrate. Some degree of caution must be used in the interpretation of these results because most of these studies were performed *in vitro* and often with heterologous CDPKs. CDPKs can phosphorylate proteins that are not true substrates, including proteins that are not synthesized in plants, such as casein. This is a useful property for studying these kinases since knowledge of real substrates is not required in order to perform biochemical analyses. This cautionary note should not be interpreted to mean that the substrates discussed below are not actually phosphorylated by a CDPK *in planta*, but merely to point out that it remains to be demonstrated that the CDPK and the proposed substrate are found in plant cells at the same time and location.

A number of soluble enzymes has been demonstrated to be regulated by reversible phosphorylation (reviewed in MacKintosh, 1998) and several of these enzymes are substrates for CDPKs. Leaf nitrate reductase (NR) activity is rapidly downregulated both in the dark and in response to osmotic stress. Inactivation is a two-step process involving phosphorylation of a specific serine residue, followed by binding of a 14-3-3 protein (Bachmann *et al.*, 1996; Huber *et al.*, 1996a; Moorhead *et al.*, 1996). 14-3-3 proteins recognize and bind to phosphorylated serine residues in many proteins and thereby modulate their activity (Muslin *et al.*, 1996; Yaffe *et al.*, 1997). Since CDPKs are serine/threonine protein kinases, it is not surprising to find that 14-3-3 proteins bind subsequent to CDPK phosphorylation events. Interestingly, 14-3-3 proteins also appear to bind to some CDPKs directly (VanderHoeven *et al.*, 1996; Camoni *et al.*, 1998; Moorhead *et al.*, 1999), presumably at an autophosphorylated serine, although this has not been proven yet. At least one CDPK has been purified based on its ability to phosphorylate NR and its identity as a CDPK has been confirmed by peptide sequencing (Douglas *et al.*, 1998).

The soluble enzyme sucrose-phosphate synthase (SPS) is involved in sucrose synthesis in plants. SPS catalyzes sucrose synthesis in source leaves and responds rapidly to changes in light intensity. In addition, SPS is upregulated in response to osmotic stress since a higher intracellular concentration of osmolytes facilitates uptake of water from the environment. SPS activity is tightly controlled by a variety of mechanisms, including phosphorylation (Huber and Huber, 1996). Phosphorylation of SPS has been detected on two separate serine residues. Phosphorylation at serine-158 (Ser158) was catalyzed by a stress-stimulated calcium-independent kinase and was responsible for modulation of SPS activity in response to changing light levels (McMichael *et al.*, 1993). However, Ser424 was phosphorylated under osmotic stress conditions, presumably by a CDPK that has been partially purified, and Ser424 phosphorylation increased SPS activity (Toroser and Huber, 1997). Like NR, SPS is bound by 14-3-3 proteins after phosphorylation (Moorhead *et al.*, 1999).

Sucrose synthase (SuSy) catalyzes a readily reversible reaction which either synthesizes or breaks down sucrose *in vitro*. The primary function of SuSy *in vivo* is probably the cleavage reaction which forms UDP-glucose, the precursor for synthesis of starch and cell wall components. SuSy exists in both soluble and plasma membrane-bound forms (Amor *et al.*, 1995; Carlson and Chourey, 1996; Sturm *et al.*, 1999). As with SPS, this important enzyme is regulated by several mechanisms, including phosphorylation. A partially purified CDPK from both soybean nodules (Zhang and Chollet, 1997) and maize leaves (Huber *et al.*, 1996b) has been demonstrated to phosphorylate SuSy. The phosphorylated enzyme became more active in catalyzing the sucrose cleavage reaction (Huber *et al.*, 1996b).

In addition to the aforementioned soluble proteins, a number of integral membrane proteins has been identified as potential CDPK substrates. Several studies have found that the plasma membrane H^+-adenosine triphosphate (ATP)ase is phosphorylated in a calcium-dependent manner (Schaller and Sussman, 1988; Lino *et al.*, 1998). The plasma membrane H^+-ATPase uses the energy of ATP to extrude protons from the cell and create the electrochemical gradient used for uptake of nutrients, establishment of turgor, and phloem loading. Using reconstituted proteoliposomes, Lino *et al.* (1998) demonstrated that the H^+-ATPase from beetroot (*Beta vulgaris*) underwent phosphorylation by a CDPK within the liposomes, which caused a decrease in ATPase activity. At least one phosphorylation site is located on the cytoplasmic carboxy-terminus of the ATPase (Sekler *et al.*, 1994). The carboxy-terminus is also referred to as the autoinhibitory domain because removal of this region by trypsin digestion or modification by point mutations resulted in higher enzyme activity (Palmgren *et al.*, 1991; Baunsgaard *et al.*, 1996; Morsomme *et al.*, 1998). A current model for regulation of the plasma membrane H^+-ATPase by phosphorylation proposes that a 14-3-3 protein binds to the phosphorylated proton pump resulting in increased proton pumping activity. Dephosphorylation disrupts the interaction with the 14-3-3 protein and deactivates the pump (Baunsgaard *et al.*, 1998). Phosphorylation of the H^+-ATPase during the response of tomato (*Lycopersion esculentum*) cells to a race-specific elicitor from *Cladosporium fulvum* may also be mediated in part by a CDPK (Xing *et al.*, 1996).

Phosphorylation is a common mechanism for regulation of channel gating (Levitan, 1994). Nodulin-26 is a voltage-sensitive ion channel found in soybean symbiosome membranes and is a member of the six transmembrane domain MIP (major intrinsic protein) family of transport channels (Weaver *et al.*, 1994). Nodulin 26 has been demonstrated to transport malate, although it is not known whether this is its *in vivo* function. Transport is inhibited by treatment of symbiosome membranes with alkaline phosphatase and stimulated by incubation with ATP, indicating that transport may be regulated by phosphorylation (Ouyang *et al.*, 1991). Calcium-dependent phosphorylation of nodulin-26 by a symbiosome membrane-localized CDPK has been demonstrated to occur on a serine residue located near the carboxy-terminus of the channel protein (Weaver and Roberts, 1992). This modification appears to alter the voltage-gating characteristics of nodulin-26, making the channel more voltage sensitive (Lee *et al.*, 1995).

Increases in cytoplasmic calcium concentration, as well as protein phosphorylation and dephosphorylation events, are proposed to be involved in guard cell movements that regulate stomatal aperture (MacRobbie, 1998; Assmann and Shimazaki, 1999). Chloride and potassium enter the guard cell vacuole during stomatal opening and exit during stomatal closure. Channels that regulate movement of these ions are regulated by calcium (MacRobbie, 1997). A purified *Arabidopsis* CDPK, AtCPK1, affected the activity of a channel in the guard cell vacuolar membrane of *Vicia faba* in a calcium-dependent manner. Patch-clamp experiments indicated that an anion channel, probably transporting chloride and malate ions, was activated at negative membrane potentials when vacuolar membranes were perfused with this kinase (Pei *et al.*, 1996). However, phosphorylation of the channel was not demonstrated, so it is not certain whether this is a direct or an indirect effect. Other workers have investigated KAT1, an inward rectifying potassium channel expressed in the guard cell plasma membrane. *Vicia faba* guard cells contain a calcium-dependent protein kinase (57 kD), found in both soluble and membrane fractions that phosphorylated *in vitro*-translated *Arabidopsis* KAT1 protein only when KAT1 was membrane localized (Li *et al.*, 1998). The effect of phosphorylation on KAT1 activity is not known.

Several other potential CDPK substrates have also been reported in the literature. These include: a slow-vacuolar channel in the barley aleurone tonoplast (Bethke and Jones, 1997), an activator of phosphoinositol 4-kinase from carrot cells (Yang and Boss, 1994), phosphoenolpyruvate carboxylase from *Zea mays* (Ogawa *et al.*, 1992), seed storage proteins (napins) from *Brassica*, *Ricinus*, *Raphanus*, and *Momordica* (Polya *et al.*, 1993; Neumann *et al.*, 1996a, b), myosin light chain from *Chara* (McCurdy and Harmon, 1992b), GT-1, a DNA-binding protein which interacts with the ribulose-1,5-bisphosphate carboxylase gene promoter in *Pisum sativum* (Marechal *et al.*, 1999), serine acetyltransferase from *Glycini max* (Yoo and Harmon, 1997) and tonoplast intrinsic protein from *Phaseolus vulgaris* (Johnson and Chrispeels, 1992).

VII. CONCLUSIONS

To date, calcium-dependent protein kinases have been identified in 25 species throughout the plant and protist kingdoms, but are apparently not present in bacteria, fungi, or animals. In plants, CDPKs are the major calcium-stimulated kinases and are therefore likely to be important in a variety of calcium-activated signaling processes. We have just begun to understand how CDPKs are specialized with respect to temporal and spatial expression patterns and how they respond to discrete stimuli or stresses. Much still remains to be learned about the roles of individual CDPKs, such as their specific substrates, modes of regulation, and subcellular locations. This task will be especially challenging in those organisms with very large CDPK gene families. In concert with protein phosphatases, CDPKs are undoubtedly part of the complex, integrated web of signaling networks that regulate the cell's metabolism and responses to stimuli. Complementary phosphorylation and dephosphorylation activities, involving kinases and phosphatases with different calcium activation thresholds or different subcellular locations, could provide the requisite specificity for the cell to respond appropriately to changing calcium concentrations. Understanding these signaling networks will continue to be a focus of future research.

ACKNOWLEDGEMENTS

The author wishes to thank G. Eric Schaller and Sheen Lu for helpful comments. Preparation of this manuscript was supported in part by a grant from the United States Department of Agriculture.

REFERENCES

Abo-El-Saad, M. and Wu, R. (1995). A rice membrane calcium-dependent protein kinase is induced by gibberellin. *Plant Physiology* **108**, 787–793.

Allbritton, N. L., Meyer, T. and Stryer, L. (1992). Range of messenger action of calcium ion and inositol 1,4,5-triphosphate. *Science* **258**, 1812–1815.

Amor, Y., Haigler, C. H., Johnson, S., Wainscott, M. and Delmer, D. P. (1995). A membrane-associated form of sucrose synthase and its potential role in synthesis of cellulose and callose in plants. *Proceedings of the National Academy of Sciences of the United States of America* **92**, 9353–9357.

Assmann, S. M. and Shimazaki, K.-I. (1999). The multisensory guard cell. Stomatal responses to blue light and abscisic acid. *Plant Physiology* **119**, 809–815.

Bachmann, M., Shiraishi, N., Campbell, W. H., Yoo, B.-C., Harmon, A. C. and Huber, S. C. (1996). Identification of Ser-543 as the major regulatory phosphorylation site in spinach leaf nitrate reductase. *Plant Cell* **8**, 505–517.

Baizabal-Aguirre, V. M. and de la Vara, L. E. G. (1997). Purification and characterization of a calcium-regulated protein kinase from beet root (*Beta vulgaris*) plasma membranes. *Physiologia Plantarum* **99**, 135–143.

Battey, N. H. (1990). Calcium-activated protein kinase from soluble and membrane fractions of maize coleoptiles. *Biochemical and Biophysical Research Communications* **170**, 17–22.

Baunsgaard, L., Venema, K., Axelsen, K. B., Villalba, J. M., Welling, A., Wollenweber, B. and Palmgren, M. G. (1996). Modified plant plasma membrane H^+-ATPase with improved transport coupling efficiency identified by mutant selection in yeast. *Plant Journal* **10**, 451–458.

Baunsgaard, L., Fuglsang, A. T., Jahn, T., Krothout, H. A. A. J., deBoer, A. H. and Palmgren, M. G. (1998). The 14-3-3 proteins associate with the plant plasma membrane H^+-ATPase to generate a fusicoccin binding complex and a fusicoccin responsive system. *Plant Journal* **13**, 661–671.

Berberich, T. and Kusano, T. (1997). Cycloheximide induces a subset of low temperature-inducible genes in maize. *Molecular and General Genetics* **254**, 275–283.

Bethke, P. C. and Jones, R. L. (1997). Reversible protein phosphorylation regulates the activity of the slow-vacuolar ion channel. *Plant Journal* **11**, 1227–1236.

Binder, B. M., Harper, J. F. and Sussman, M. R. (1994). Characterization of an *Arabidopsis* calmodulin-like domain protein kinase purified from *Escherichia coli* using an affinity sandwich technique. *Biochemistry* **33**, 2033–2041.

Botella, J. R., Arteca, J. M., Somodevilla, M. and Arteca, R. N. (1996). Calcium-dependent protein kinase gene expression in response to physical and chemical stimuli in mungbean (*Vigna radiata*). *Plant Molecular Biology* **30**, 1129–1137.

Braam, J. (1992). Regulated expression of the calmodulin-related TCH genes in cultured *Arabidopsis* cells: induction by calcium and heat-shock. *Proceedings of the National Academy of Sciences of the United States of America* **89**, 3213–3216.

Braam, J. and Davis, R. W. (1990). Rain-, wind-, and touch-induced expression of calmodulin and calmodulin-related genes in *Arabidopsis*. *Cell* **60**, 357–364.

Breviario, D., Morello, L. and Giani, S. (1995). Molecular cloning of two novel rice cDNA sequences encoding putative calcium-dependent protein kinases. *Plant Molecular Biology* **27**, 953–967.

Bush, D. S. (1996). Effects of gibberellic acid and environmental factors on cystosolic calcium in wheat aleurone cells. *Planta* **199**, 89–99.

Camoni, L., Harper, J. F. and Palmgren, M. G. (1998). 14-3-3 proteins activate a plant calcium-dependent protein kinase. *FEBS Letters* **430**, 381–384.

Carlson, S. J. and Chourey, P. S. (1996). Evidence for plasma membrane-associated forms of sucrose synthase in maize. *Molecular and General Genetics* **252**, 303–310.

DasGupta, M. (1994). Characterization of a calcium-dependent protein kinase from *Arachis hypogea* (groundnut) seeds. *Plant Physiology* **104**, 961–969.

Douglas, P., Moorhead, G., Hong, Y., Morice, N. and MacKintosh, C. (1998). Purification of a nitrate reductase kinase from *Spinacia oleracea* leaves, and its identification as a calmodulin-domain protein kinase. *Planta* **206**, 435–442.

Dunn, P. P., Bumstead, J. M. and Tomley, F. M. (1996). Sequence, expression and localization of calmodulin-domain protein kinases in *Eimeria tenella* and *Eimeria maxima*. *Parasitology* **113**, 439–448.

Duronio, R. J., Jackson-Machelski, E., Heuckeroth, R. O., Olins, P. O., Devine, C. S., Yonemoto, W., Slice, L. W., Taylor, S. S. and Gordon, J. I. (1990). Protein N-myristoylation in *Escherichia coli*. Reconstitution of a eukaryotic protein modification in bacteria. *Proceedings of the National Academy of Sciences of the United States of America* **87**, 1506–1510.

Ellard-Ivey, M., Hopkins, R. B., White, T. and Lomax, T. L. (1999). Cloning, expression and N-terminal myristoylation of *CpCPK1*, a calcium-dependent protein kinase from zucchini (*Cucurbita pepo* L.). *Plant Molecular Biology* **39**, 199–208.

Estruch, J. J., Kadwell, S., Merlin, E. and Crossland, L. (1994). Cloning and characterization of a maize pollen-specific calcium-dependent calmodulin-dependent protein kinase. *Proceedings of the National Academy of Sciences of the United States of America* **91**, 8837–8841.

Etter, E. F., Minta, A., Poenie, M. and Fay, F. S. (1996). Near-membrane [Ca^{2+}] transients resolved using the Ca^{2+} indicator FFP18. *Proceedings of the National Academy of Sciences of the United States of America* **93**, 5368–5373.

Farber, P. M., Graeser, R., Franklin, R. M. and Kappes, B. (1997). Molecular cloning and characterization of a second calcium-dependent protein kinase of *Plasmodium falciparum*. *Molecular and Biochemical Parasitology* **87**, 211–216.

Farmer, P. K. and Choi, J. H. (1999). Calcium and phospholipid activation of a recombinant calcium-dependent protein kinase (DcCPK1) from carrot (*Daucus carota* L.). *Biochimica et Biophysica Acta* **1434**, 6–17.

Felle, H. (1988). Auxin causes oscillations of cytosolic free calcium and pH in *Zea mays* coleoptiles. *Planta* **174**, 495–499.

Frylinck, L. and Dubery, I. A. (1998). Protein kinase activities in ripening mango, *Mangifera indica* L., fruit tissue III. Purification and characterisation of a calcium-regulated protein kinase. *Biochimica et Biophysica Acta* **1387**, 342–354.

Furumoto, T., Ogawa, N., Hata, S. and Izui, K. (1996). Plant calcium-dependent protein kinase-related kinases (CRKs) do not require calcium for their activities. *FEBS Letters* **396**, 147–151.

Gehring, C. A., Irving, H. R. and Parish, R. W. (1990). Effects of auxin and abscisic acid on cytosolic calcium and pH in plant cells. *Proceedings of the National Academy of Sciences of the United States of America* **87**, 9645–9649.

Gilroy, S. and Jones, R. L. (1992). Gibberellic acid and abscisic acid coordinately regulate cytoplasmic calcium and secretory activity in barley aleurone protoplasts. *Proceedings of the National Academy of Sciences of the United States of America* **89**, 3591–3595.

Gilroy, S., Read, N. D. and Trewavas, A. J. (1990). Elevation of cytoplasmic calcium by caged calcium or caged inositol triphosphate initiates stomatal closure. *Nature* **346**, 769–771.

Grabski, S., Arnoys, E., Busch, B. and Schindler, M. (1998). Regulation of actin tension in plant cells by kinases and phosphatases. *Plant Physiology* **116**, 279–290.

Hanks, S. K. and Hunter, T. (1995). The eukaryotic protein kinase superfamily: kinase (catalytic) domain structure and classification. *FASEB Journal* **9**, 546–596.

Hardie, D. G. (1999). Plant protein serine/threonine kinases: classification and functions. *Annual Review of Plant Physiology and Plant Molecular Biology* **50**, 97–131.

Harmon, A. C., Putnam-Evans, C. and Cormier, M. J. (1987). A calcium-dependent but calmodulin-independent protein kinase from soybean. *Plant Physiology* **83**, 830–837.

Harmon, A. C., Yoo, B.-C. and McCaffery, C. (1994). Pseudosubstrate inhibition of CDPK, a protein kinase with a calmodulin-like domain. *Biochemistry* **33**, 7278–7287.

Harper, J. F., Sussman, M. R., Schaller, G. E., Putnam-Evans, C., Charbonneau, H. and Harmon, A. C. (1991). A calcium-dependent protein kinase with a regulatory domain similar to calmodulin. *Science* **252**, 951–954.

Harper, J. F., Binder, B. M. and Sussman, M. R. (1993). Calcium and lipid regulation of an *Arabidopsis* protein kinase expressed in *Escherichia coli*. *Biochemistry* **32**, 3282–3290.

Harper, J. F., Huang, J.-F. and Lloyd, S. J. (1994). Genetic identification of an autoinhibitor in CDPK, a protein kinase with a calmodulin-like domain. *Biochemistry* **33**, 7267–7277.

Hetherington, A. and Trewavas, A. (1982). Calcium-dependent protein kinase in pea shoot membranes. *FEBS Letters* **145**, 67–71.

Hetherington, A. M. and Trewavas, A. (1984). Activation of a pea membrane protein kinase by calcium ions. *Planta* **161**, 409–417.

Hofmann, K., Bucher, P., Falquet, L. and Bairoch, A. (1999). The PROSITE database, its status in 1999. *Nucleic Acids Research* **27**, 215–219.

Hong, Y., Takano, M., Liu, C.-M., Gasch, A., Chye, M.-L. and Chua, N.-H. (1996). Expression of three members of the calcium-dependent protein kinase gene family in *Arabidopsis thaliana*. *Plant Molecular Biology* **30**, 1259–1275.

Hrabak, E. M., Dickmann, L. J., Satterlee, J. S. and Sussman, M. R. (1996). Characterization of eight new members of the calmodulin-like domain protein kinase gene family of *Arabidopsis thaliana*. *Plant Molecular Biology* **31**, 405–412.

Huang, J.-F., Teyton, L. and Harper, J. F. (1996). Activation of a Ca^{2+}-dependent protein kinase involves intramolecular binding of a calmodulin-like regulatory domain. *Biochemistry* **35**, 13222–13230.

Huber, S. C. and Huber, J. L. (1996). Role and regulation of sucrose-phosphate synthase in higher plants. *Annual Review of Plant Physiology and Plant Molecular Biology* **47**, 431–444.

Huber, S. C., Bachmann, M. and Huber, J. L. (1996a). Post-translational regulation of nitrate reductase activity: a role for Ca^{2+} and 14-3-3 proteins. *Trends in Plant Science* **1**, 432–438.

Huber, S. C., Huber, J. L., Liao, P.-C., Gage, D. A., McMichael, R. W., Jr, Chourey, P. S., Hannah, L. C. and Koch, K. (1996b). Phosphorylation of serine-15 of maize leaf sucrose synthase. Occurence *in vivo* and possible regulatory significance. *Plant Physiology* **112**, 793–802.

Huttly, A. K. and Phillips, A. L. (1995). Gibberellin-regulated expression in oat aleurone cells of two kinases that show homology to MAP kinase and a ribosomal protein kinase. *Plant Molecular Biology* **27**, 1043–1052.

Iwata, Y., Kuriyama, M., Nakakita, M., Kojima, H., Ohto, M.-A. and Nakamura, K. (1998). Characterization of a calcium-dependent protein kinase of tobacco leaves that is associated with the plasma membrane and is inducible by sucrose. *Plant and Cell Physiology* **39**, 1176–1183.

Johnson, D. R., Bhatnagar, R. S., Knoll, L. J. and Gordon, J. I. (1994). Genetic and biochemical studies of protein N-myristoylation. *Annual Review of Biochemistry* **63**, 869–914.

Johnson, K. D. and Chrispeels, M. J. (1992). Tonoplast-bound protein kinase phosphorylates tonoplast intrinsic protein. *Plant Physiology* **100**, 1787–1795.

Jordan, B. R. (1998). Large-scale expression measurement by hybridization methods: from high-density membranes to 'DNA chips'. *Journal of Biochemistry* **124**, 251–258.

Kawasaki, T., Hayashida, N., Baba, T., Shinozaki, K. and Shimada, H. (1993). The gene encoding a calcium-dependent protein kinase located near the *sbe1* gene encoding starch branching enzyme I is specifically expressed in developing rice seeds. *Gene* **129**, 183–189.

Kim, K., Messinger, L. A. and Nelson, D. L. (1998). Ca^{2+}-dependent protein kinases of *Paramecium*: cloning provides evidence of a multigene family. *European Journal of Biochemistry* **251**, 605–612.

Klimczak, L. J. and Hind, G. (1990). Biochemical similarities between soluble and membrane-bound calcium-dependent protein kinases of barley. *Plant Physiology* **92**, 919–923.

Knight, H., Trewavas, A. J. and Knight, M. R. (1997). Calcium signalling in *Arabidopsis thaliana* responding to drought and salinity. *Plant Journal* **12**, 1067–1078.

Knight, M. R., Campbell, A. K., Smith, S. M. and Trewavas, A. J. (1991). Transgenic plant aequorin reports the effects of touch and cold-shock and elicitors on cytoplasmic calcium. *Nature* **352**, 524–526.

Knight, M. R., Smith, S. M. and Trewavas, A. J. (1992). Wind-induced plant motion immediately increases cytosolic calcium. *Proceedings of the National Academy at Sciences of the United States of America* **89**, 4967–4971.

Lakatos, L., Hutvagner, G. and Banfalvi, Z. (1998). Potato protein kinase StCPK1: a putative evolutionary link between CDPKs and CRKs. *Biochimica et Biophysica Acta* **1442**, 101–108.

Lee, J. W., Zhang, Y., Weaver, C. D., Shomer, N. H., Louis, C. F. and Roberts, D. M. (1995). Phosphorylation of nodulin 26 on serine 262 affects its voltage-sensitive channel activity in planar lipid bilayers. *Journal of Biological Chemistry* **270**, 27051–27057.

Lee, J.-Y., Roberts, D. M. and Harmon, A. C. (1997). Isolation of two new CDPK isoforms (Accession Nos U69173 and U69174) from soybean (*Glycine max* L.). *Plant Physiology* **115**, 314.

Lee, J.-Y., Yoo, B.-C. and Harmon, A. C. (1998). Kinetic and calcium-binding properties of three calcium-dependent protein kinase isoenzymes from soybean. *Biochemistry* **37**, 6801–6809.

Levitan, I. B. (1994). Modulation of ion channels by protein phosphorylation and dephosphorylation. *Annual Review of Physiology* **56**, 193–212.

Li, H., Dauwalder, M. and Roux, S. J. (1991). Partial purification and characterization of a Ca^{2+}-dependent protein kinase from pea nuclei. *Plant Physiology* **96**, 720–727.

Li, J., Lee, Y.-R. J. and Assmann, S. M. (1998). Guard cells possess a calcium-dependent protein kinase that phosphorylates the KAT1 potassium channel. *Plant Physiology* **116**, 785–795.

Lindzen, E. and Choi, J. H. (1995). A carrot cDNA encoding an atypical protein kinase homologous to plant calcium-dependent protein kinases. *Plant Molecular Biology* **28**, 785–797.

Lino, B., Baizabal-Aguirre, V. M. and de la Vara, L. E. G. (1998). The plasma-membrane H^+-ATPase from beet root is inhibited by a calcium-dependent phosphorylation. *Planta* **204**, 352–359.

Lodge, J. K., Johnson, R. L., Weinberg, R. A. and Gordon, J. I. (1994). Comparison of myristoyl-CoA:protein N-myristoyl transferases from three pathogenic fungi: *Cryptococcus neoformans, Histoplasma capsulatum*, and *Candida albicans*. *Journal of Biological Chemistry* **269**, 2996–3009.

Lu, Y.-T., Hidaka, H. and Feldman, L. J. (1996). Characterization of a calcium/calmodulin-dependent protein kinase homolog from maize roots showing light-regulated gravitropism. *Planta* **199**, 18–24.

MacIntosh, G. C., Ulloa, R. M., Raices, M. and Tellez-Inon, M. T. (1996). Changes in calcium-dependent protein kinase activity during *in vitro* tuberization in potato. *Plant Physiology* **112**, 1541–1550.

MacKintosh, C. (1998). Regulation of cytosolic enzymes in primary metabolism by reversible protein phosphorylation. *Current Opinion in Plant Biology* **1**, 224–229.

MacRobbie, E. A. (1998). Signal transduction and ion channels in guard cells. *Philosophical Transactions of the Royal Society of London B* **353**, 1475–1488.

MacRobbie, E. A. C. (1997). Signalling in guard cells and regulation of ion channel activity. *Journal of Experimental Botany* **48**, 515–528.

Malho, R., Read, N. D., Pais, M. S. and Trewavas, A. J. (1994). Role of cytosolic free calcium in the reorientation of pollen tube growth. *Plant Journal* **5**, 331–341.

Malho, R., Moutinho, A., van der Luit, A. and Trewavas, A. J. (1998). Spatial characteristics of calcium signalling: the calcium wave as a basic unit in plant cell calcium signalling. *Philosophical Transactions of the Royal Society of London B* **353**, 1463–1473.

Marechal, E., Hiratsuka, K., Delgado, J., Nairn, A., Qin, J., Chait, B. T. and Chua, N.-H. (1999). Modulation of GT-1 DNA-binding activity by calcium-dependent phosphorylation. *Plant Molecular Biology* **40**, 373–386.

McAinsh, M. R. and Hetherington, A. M. (1998). Encoding specificity in Ca^{2+} signalling systems. *Trends in Plant Science* **3**, 32–36.

McCurdy, D. W. and Harmon, A. C. (1992a). Calcium-dependent protein kinase in the green alga *Chara*. *Planta* **188**, 54–61.

McCurdy, D. W. and Harmon, A. C. (1992b). Phosphorylation of a putative myosin light chain in *Chara* by calcium-dependent protein kinase. *Protoplasma* **171**, 85–88.

McKnight, S. L., Oliver, M. J. and Putnam-Evans, C. L. (1997). Nucleotide sequence of a full-length cDNA encoding a calmodulin-like domain protein kinase (CDPK) from the moss *Tortula ruralis*. *Plant Physiology* **115**, 313.

McMichael, R. W., Kochansky, J., Klein, R. R. and Huber, S. C. (1993). Identification of the major regulatory phosphorylation site in sucrose-phosphate synthase. *Archives of Biochemistry and Biophysics* **307**, 248–252.

Milligan, G., Parenti, M. and Magee, A. I. (1995). The dynamic role of palmitoylation in signal transduction. *Trends in Biological Science* **20**, 181–186.

Monroy, A. F. and Dhindsa, R. S. (1995). Low-temperature signal transduction: induction of cold acclimation-specific genes of alfalfa by calcium at 25°C. *Plant Cell* **7**, 321–331.

Monroy, A. F., Sarhan, F. and Dhindsa, R. S. (1993). Cold-induced changes in freezing tolerance, protein phosphorylation and gene expression. Evidence for a role of calcium. *Plant Physiology* **102**, 1227–1235.

Moorhead, G., Douglas, P., Morrice, N., Scarabel, M., Aitken, A. and MacKintosh, C. (1996). Phosphorylated nitrate reductase from spinach leaves is inhibited by 14-3-3 proteins and activated by fusicoccin. *Current Biology* **6**, 1104–1113.

Moorhead, G., Douglas, P., Cotelle, V., Harthill, J., Morrice, N., Meek, S., Deiting, U., Stitt, M., Scarabel, M., Aitken, A. and MacKintosh, C. (1999). Phosphorylation-dependent interactions between enzymes of plant metabolism and 14-3-3 proteins. *Plant Journal* **18**, 1–12.

Morsomme, P., Dambly, S., Maudoux, O. and Boutry, M. (1998). Single point mutations distributed in 10 soluble and membrane regions of the *Nicotiana plumbaginifolia* plasma membrane PMA2 H^+-ATPase activate the enzyme and modify the structure of the C-terminal region. *Journal of Biological Chemistry* **273**, 34837–34842.

Moutinho, A., Trewavas, A. J. and Malho, R. (1998). Relocation of a Ca^{2+}-dependent protein kinase activity during pollen tube reorientation. *Plant Cell* **10**, 1499–1509.

Muslin, A. J., Tanner, J. W., Allen, P. M. and Shaw, A. S. (1996). Interaction of 14-3-3 with signaling proteins is mediated by the recognition of phosphoserine. *Cell* **84**, 889–897.

Neuhaus, G., Bowler, C., Kern, R. and Chua, N.-H. (1993). Calcium/calmodulin-dependent and -independent phytochrome signal transduction pathways. *Cell* **73**, 937–952.

Neumann, G. M., Condron, R. and Polya, G. M. (1996a). Purification and sequencing of napin-like protein small and large chains from *Momordica charantia* and *Ricinus communis* seeds and determination of sites phosphorylated by plant Ca^{2+}-dependent protein kinase. *Biochimica et Biophysica Acta* **1298**, 223–240.

Neumann, G. M., Condron, R., Thomas, I. and Polya, G. M. (1996b). Purification and sequencing of multiple forms of *Brassica napus* seed napin small chains that are calmodulin antagonists and substrates for plant calcium-dependent protein kinase. *Biochimica et Biophysica Acta* **1295**, 23–33.

Nishiyama, R., Mizuno, H., Okada, S., Yamaguchi, T., Takenaka, M., Fukuzawa, H. and Ohyama, K. (1999). Two mRNA species encoding calcium-dependent protein kinases are differentially expressed in sexual organs of *Marchantia polymorpha* through alternative splicing. *Plant and Cell Physiology* **40**, 205–212.

Ogawa, N., Okumura, S. and Izui, K. (1992). A Ca^{2+}-dependent protein kinase phosphorylates phosphoenolpyruvate carboxylase in maize. *FEBS Letters* **302**, 86–88.

Ouyang, L.-J., Whelan, J., Weaver, C. D., Roberts, D. M. and Day, D. A. (1991). Protein phosphorylation stimulates the rate of malate uptake across the peribacteroid membrane of soybean nodules. *FEBS Letters* **293**, 188–190.

Palmgren, M. G., Sommarin, M., Serrano, R. and Larsson, C. (1991). Identification of an autoinhibitory domain in the C-terminal region of the plant plasma membrane H^+-ATPase. *Journal of Biological Chemistry* **266**, 20470–20475.

Payne, M. E., Fong, Y. L., Ono, T., Colbran, R. J., Kemp, B. E., Soderling, T. R. and Means, A. R. (1988). Calcium/calmodulin-dependent protein kinase II. Characterization of distinct calmodulin binding and inhibitory domains. *Journal of Biological Chemistry* **263**, 7190–7195.

Pei, Z. M., Ward, J. M., Harper, J. F. and Schroeder, J. I. (1996). A novel chloride channel in *Vicia faba* guard cell vacuoles activated by the serine/threonine kinase, CDPK. *EMBO Journal* **15**, 6564–6574.

Perry, D. J. and Bousquet, J. (1998). Sequence-tagged-site (STS) markers of arbitrary genes: development, characterization and analysis of linkage in black spruce. *Genetics* **149**, 1089–1098.

Pierson, E. S., Miller, D. D., Callaham, D. A., Shipley, A. M., Rivers, B. A., Cresti, M. and Hepler, P. K. (1994). Pollen tube growth is coupled to the extracellular calcium ion flux and the intracellular calcium gradient: effect of BAPTA-type buffers and hypertonic media. *Plant Cell* **6**, 1815–1828.

Plieth, C., Hansen, U.-P., Knight, H. and Knight, M. R. (1999). Temperature sensing by plants: the primary characteristics of signal perception and calcium response. *Plant Journal* **18**, 491–497.

Polya, G. M., Chandra, S. and Condron, R. (1993). Purification and sequencing of radish seed calmodulin antagonists phosphorylated by calcium-dependent protein kinase. *Plant Physiology* **101**, 545–551.

Putnam-Evans, C. L., Harmon, A. C. and Cormier, M. J. (1990). Purification and characterization of a novel calcium-dependent protein kinase from soybean. *Biochemistry* **29**, 2488–2495.

Ritchie, S. and Gilroy, S. (1998). Calcium-dependent protein phosphorylation may mediate the gibberellic acid response in barley aleurone. *Plant Physiology* **116**, 765–776.

Robbins, J., Dilworth, S. M., Laskey, R. A. and Dingwall, C. (1991). Two interdependent basic domains in nucleoplasmin nuclear targeting sequence: identification of a class of bipartite nuclear targeting sequence. *Cell* **64**, 615–623.

Roberts, D. M. (1989). Detection of a calcium-activated protein kinase in *Mougeotia* by using synthetic peptide substrates. *Plant Physiology* **91**, 1613–1619.

Saijo, Y., Hata, S., Sheen, J. and Izui, K. (1997). cDNA cloning and prokaryotic expression of maize calcium-dependent protein kinases. *Biochimica et Biophysica Acta* **1350**, 109–114.

Satterlee, J. S. and Sussman, M. R. (1998). Unusual membrane-associated protein kinases in higher plants. *Journal of Membrane Biology* **164**, 205–213.

Schaller, G. E. and Sussman, M. R. (1988). Phosphorylation of the plasma-membrane H^+-ATPase of oat roots by a calcium-stimulated protein kinase. *Planta* **173**, 509–518.

Schaller, G. E., Harmon, A. C. and Sussman, M. R. (1992). Characterization of a calcium- and lipid-dependent protein kinase associated with the plasma membrane of oat. *Biochemistry* **31**, 1721–1727.

Schiefelbein, J. W., Shipley, A. and Rowse, P. (1992). Calcium influx at the tip of growing root-hair cells of *Arabidopsis thaliana*. *Planta* **187**, 455–459.

Schultz, A. M., Henderson, L. E. and Oroszlan, S. (1988). Fatty acylation of proteins. *Annual Review of Cell Biology* **4**, 611–647.

Schultz, J., Milpetz, F., Bork, P. and Ponting, C. P. (1998). SMART, a simple modular architecture research tool: identification of signalling domains. *Proceedings of the National Academy of Sciences of the United States of America* **95**, 5857–5864.

Sedbrook, J. C., Kronenbusch, P. J., Borisy, G. G., Trewavas, A. J. and Masson, P. (1996). Transgenic aequorin reveals organ specific cytosolic Ca^{2+} responses to anoxia in *Arabidopsis thaliana* seedlings. *Plant Physiology* **111**, 243–257.

Sekler, I., Weiss, M. and Pick, U. (1994). Activation of the *Dunaliella acidophila* plasma membrane H^{+}-ATPase by trypsin cleavage of a fragment that contains a phosphorylation site. *Plant Physiology* **105**, 1125–1132.

Shacklock, P. S., Read, N. D. and Trewavas, A. J. (1992). Cytosolic-free calcium mediates red light induced photomorphogenesis. *Nature* **358**, 153–155.

Sheen, J. (1996). Ca^{2+}-dependent protein kinases and stress signal transduction in plants. *Science* **274**, 1900–1902.

Siderius, M., Henskens, H., Porto-leBlanche, A., Himbergen, J. van, Musgrave, A. and Haring, M. (1997). Characterisation and cloning of a calmodulin-like domain protein kinase from *Chlamydomonas moewusii* (Gerloff). *Planta* **202**, 76–84.

Soderling, T. R. (1996). Structure and regulation of calcium/calmodulin-dependent protein kinases II and IV. *Biochimica et Biophysica Acta* **1297**, 131–138.

Sturm, A., Lienhard, S., Schatt, S. and Hardegger, M. (1999). Tissue-specific expression of two genes for sucrose synthase in carrot (*Daucus carota* L.). *Plant Molecular Biology* **39**, 349–360.

Suen, K.-L. and Choi, J. H. (1991). Isolation and sequence analysis of a cDNA clone for a carrot calcium-dependent protein kinase: homology to calcium/calmodulin-dependent protein kinases and to calmodulin. *Plant Molecular Biology* **17**, 581–590.

Tahtiharju, S., Sangwan, V., Monroy, A. F., Dhindsa, R. S. and Borg, M. (1997). The induction of *kin* genes in cold-acclimating *Arabidopsis thaliana*. Evidence of a role for calcium. *Planta* **203**, 442–447.

Takahashi, K., Isobe, M. and Muto, S. (1997). An increase in cytosolic calcium ion concentration precedes hypoosmotic shock-induced activation of protein kinases in tobacco suspension culture cells. *FEBS Letters* **401**, 202–206.

Takezawa, D., Patil, S., Bhatia, A. and Poovaiah, B. W. (1996). Calcium-dependent protein kinase genes in corn roots. *Journal of Plant Physiology* **149**, 329–335.

Thompson, J. D., Higgins, D. G. and Gibson, T. J. (1994). CLUSTAL W: improving the sensitivity of progressive multiple sequence alignment through sequence weighting, position-specific gap penalties and weight matrix choice. *Nucleic Acids Research* **22**, 4673–4680.

Toroser, D. and Huber, S. C. (1997). Protein phosphorylation as a mechanism for osmotic-stress activation of sucrose-phosphate synthase in spinach leaves. *Plant Physiology* **114**, 947–955.

Towler, D. A., Gordon, J. I., Adams, S. P. and Glaser, L. (1988). The biology and enzymology of eukaryotic protein acylation. *Annual Review of Biochemistry* **57**, 69–99.

Trewavas, A. (1999). Le calcium, c'est la vie: calcium makes waves. *Plant Physiology* **120**, 1–6.

Urao, T., Katagiri, T., Mizoguchi, T., Yamaguchi-Shinozaki, K., Hayashida, N. and Shinozaki, K. (1994a). An *Arabidopsis thaliana* cDNA encoding Ca^{2+}-dependent protein kinase. *Plant Physiology* **105**, 1461–1462.

Urao, T., Katagiri, T., Mizoguchi, T., Yamaguchi-Shinozaki, K., Hayashida, N. and Shinozaki, K. (1994b). Two genes that encode Ca^{2+}-dependent protein kinases are induced by drought and high-salt stresses in *Arabidopsis thaliana*. *Molecular and General Genetics* **244**, 331–340.

VanderHoeven, P. C. J., Siderius, M., Krothout, H. A. A. J., Drabkin, A. V. and DeBoer, A. H. (1996). A calcium and free fatty acid-modulated protein kinase as putative effector of the fusicoccin 14-3-3 receptor. *Plant Physiology* **111**, 857–865.

Verhey, S. D., Gaiser, J. C. and Lomax, T. L. (1993). Protein kinases in zucchini. Characterization of calcium-requiring plasma membrane kinases. *Plant Physiology* **103**, 413–419.

Weaver, C. D. and Roberts, D. M. (1992). Determination of the site of phosphorylation of nodulin 26 by the calcium-dependent protein kinase from soybean nodules. *Biochemistry* **31**, 8954–8959.

Weaver, C. D., Shomers, N. H., Louis, C. F. and Roberts, D. M. (1994). Nodulin 26, a nodule-specific symbiosome membrane protein from soybean, is an ion channel. *Journal of Biological Chemistry* **269**, 17858–17862.

Webb, A. A. R., McAinsh, M. R., Taylor, J. E. and Hetherington, A. M. (1996). Calcium ions as intracellular second messengers in higher plants. *Advances in Botanical Research* **22**, 45–96.

Wilkins, M. R., Gasteiger, E., Bairoch, A., Sanchez, J.-C., Williams, K. L., Appel, R. D. and Hochstrasser, D. F. (1998). Protein identification and analysis tools in the ExPASy server. *In* "2-D Proteome Analysis Protocols" (A. J. Link, ed.) pp. 531–552. Humana Press, New Jersey.

Williamson, R. E. and Ashley, C. C. (1982). Free Ca^{2+} and cytoplasmic streaming in alga *Chara*. *Nature* **296**, 647–651.

Xing, T., Higgins, V. J. and Blumwald, E. (1996). Regulation of plant defense response to fungal pathogens: two types of protein kinases in the reversible phosphorylation of the host plasma membrane H^+-ATPase. *Plant Cell* **8**, 555–564.

Yaffe, M. B., Rittinger, K., Volinia, S., Caron, P. R., Aitken, A., Leffers, H., Gamblin, S. J., Smerdon, S. J. and Cantley, L. C. (1997). The structural basis for 14-3-3:phosphopeptide binding specificity. *Cell* **91**, 961–971.

Yang, W. and Boss, W. F. (1994). Regulation of phosphatidylinositol 4-kinase by the protein activator PIK-A49. Activation requires phosphorylation of PIK-A49. *Journal of Biological Chemistry* **269**, 3852–3857.

Yoo, B.-C. and Harmon, A. C. (1997). Regulation of recombinant soybean serine acetyltransferase by CDPK. *Plant Physiology* **114** (Supplement), 267.

Yoon, G. M., Cho, H. S., Ha, H. J., Liu, J. R. and Lee, H.-S. P. (1999). Characterization of *NtCDPK1*, a calcium-dependent protein kinase gene in *Nicotiana tabacum*, and the activity of its encoded protein. *Plant Molecular Biology* **39**, 991–1001.

Yuasa, T. and Muto, S. (1992). Ca^{2+}-dependent protein kinase from the halotolerant green alga *Dunaliella tertiolecta*: partial purification and Ca^{2+}-dependent association of the enzyme to the microsomes. *Archives of Biochemistry and Biophysics* **296**, 175–182.

Yuasa, T., Takahashi, K. and Muto, S. (1995). Purification and characterization of a Ca^{2+}-dependent protein kinase from the halotolerant green alga *Dunaliella tertiolecta*. *Plant and Cell Physiology* **36**, 699–708.

Zhang, M. and Yuan, T. (1998). Molecular mechanisms of calmodulin's functional versatility. *Biochemical Cell Biology* **76**, 313–323.

Zhang, X.-Q. and Chollet, R. (1997). Seryl-phosphorylation of soybean nodule sucrose synthase (nodulin-100) by a Ca^{2+}-dependent protein kinase. *FEBS Letters* **410**, 126–130.

Zhao, Y., Kappes, B. and Franklin, R. M. (1993). Gene structure and expression of an unusual protein kinase from *Plasmodium falciparum* homologous at its carboxyl terminus with the EF hand calcium-binding proteins. *Journal of Biological Chemistry* **268**, 4347–4354.

Zhao, Y., Pokutta, S., Maurer, P., Lindt, M., Franklin, R. M. and Kappes, B. (1994). Calcium-binding properties of a calcium-dependent protein kinase from *Plasmodium falciparum* and the significance of individual calcium-binding sites for kinase activation. *Biochemistry* **33**, 3714–3721.

Receptor-Like Kinases in Plant Development

KEIKO U. TORII[1] and STEVEN E. CLARK[2]*

[1]*Department of Botany, University of Washington, Seattle, WA 98195, USA*
[2]*Department of Biology, University of Michigan, Ann Arbor, MI 48109, USA*

*Author to whom correspondence should be addressed: Fax: 734 647 0884; e-mail: clarks@umich.edu

Advances in Botanical Research Vol. 32
incorporating Advances in Plant Pathology
ISBN 0-12-005932-0

I. INTRODUCTION

In animals, growth factor receptors with an intrinsic protein tyrosine kinase activity, known as receptor tyrosine kinases (RTKs), play key roles in cellular processes to co-ordinate the development of multicellular organs (for review see Ullrich and Schlessinger, 1990). Because plant cells are encapsulated by cell walls, the idea of plant cell–cell communication via the recognition of polypeptide ligands and transmembrane receptors at the cell surface was once viewed with skepticism. The molecular cloning of the maize *ZmPK1* gene provided a breakthrough, showing that higher plants possess a gene product structurally analogous to receptor protein kinases (Walker and Zhang, 1990). Since the discovery of *ZmPK1*, numerous genes encoding receptor-like proteins with cytoplasmic kinase domains, referred to as RLKs (receptor-like kinases), have been identified. Recent advances in molecular genetics, especially in the model plant *Arabidopsis*, have provided evidence that some RLKs regulate the development of higher plants (Becraft *et al*., 1996; Lee *et al*., 1996; Torii *et al*., 1996; Clark *et al*., 1997; Li and Chory, 1997). In this chapter, structural features, biological and biochemical functions, and modes of action of the plant RLKs will be described, with particular emphasis on RLKs that play a crucial role in development. In addition, the latest progress in uncovering the components the RLK-mediated signal transduction pathways, including both potential ligands and downstream targets, will be highlighted. Finally, the importance of such signaling mechanisms in plant cell–cell interaction will be discussed.

II. PLANT RLKs: CLASSIFICATIONS BASED ON STRUCTURAL FEATURES

A common feature of plant RLKs is that each has a cytoplasmic protein kinase catalytic domain, a single membrane-spanning region, an N-terminal signal sequence, and an extracellular domain that varies in structure. Plant RLKs are classified into several groups based on the structural features of the predicted extracellular domain (Fig. 1 and Table I).

1. *S-domain class*. S-RLKs possess an extracellular S-domain homologous to the self-incompatibility-locus glycoproteins (SLG) of *Brassica oleracea* (Nasrallah *et al*., 1988). The S-domain consists of 12 conserved cysteine residues (10 of which are absolutely conserved) in a consensus $CX_5CX_5CX_7CXCX_nCX_7CX_nCX_3CX_3CXCX_nC$. In addition, the S-domain possesses the PTDT-box, which has a conserved WQSFDXPTDTΦL sequence (X = nonconserved amino acid; Φ = aliphatic amino acid) (Walker, 1994). In *Brassica*, the *S-RLK* gene is physically linked to the *S* locus, so that SLG and S-RLK are proposed to function together in the self-incompatibility recognition between pollen and stigma (Stein *et al*.,

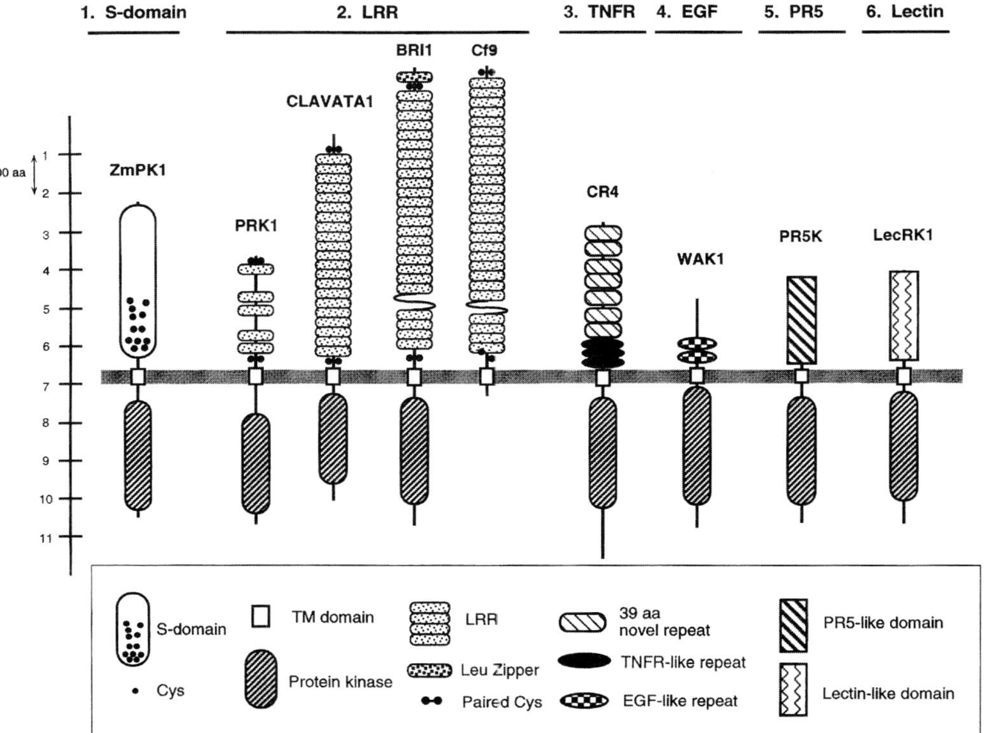

Fig. 1. Structural topology of plant RLKs and their classification. Each RLK protein is drawn approximately to scale. The N-terminal signal peptides are not included. Domains represented are listed below the diagrams of the RLKs. Cys, cysteine; TM, transmembrane; LRR, leucine-rich repeat; Leu Zipper, leucine zipper; aa, amino acids; TNFR, tumor necrosis factor receptor; EGF, epidermal growth factor; PR5, pathogenesis-related protein 5.

TABLE I
The plant RLKs

Gene products	Plant species	Biological function (if not known, expression patterns)	Reference
S-domain class			
SRK6	*Brassica oleracea*	Self-incompatibility recognition	Stein *et al.* (1993)
SRK910	*Brassica napus*	Self-incompatibility recognition	Goring and Rothstein (1992)
SFR2	*Brassica oleracea*	(Pathogenesis/wound induced)	Pastuglia *et al.* (1997)
ARK1	*Arabidopsis thaliana*	(Leaf; cell expansion?)	Tobias *et al.* (1992)
ARK2		(Cotyledon, leaf, sepal)	Dwyer *et al.* (1994)
ARK3		(Root–hypocotyl boundary, base of the lateral roots, axillary buds, and flower pedicels)	Dwyer *et al.* (1994)
RLK1		(Rossettes)	Walker (1993)
RLK4		(Root–hypocotyl boundary, base of the lateral roots, base of the petiole)	Walker (1993)
ZmPK1	*Zea mays*	(Seedling roots and shoots, silks)	Walker and Zhang (1990)
KIK1		(Husks, etiolated shoots)	Braun *et al.* (1997)
OsPK10	*Oryza sativa*	(Ubiquitous; upregulated by light)	Zhao *et al.* (1994)
LRR class			
BRI1	*Arabidopsis thaliana*	BR perception	Li and Chory (1997)
CLAVATA1		Meristem development	Clark *et al.* (1997)
ERECTA		Organ elongation	Torii *et al.* (1996)
PRK1	*Petunia inflata*	Pollen development	Mu *et al.* (1994); Lee *et al.* (1996)
SERK	*Daucus carota*	Correlation with embryogenic potential	Schmidt *et al.* (1997)
Xa21	*Oryza sative*	Resistance to *Xanthomonas oryzae*	Song *et al.* (1995)

TABLE I *Continued*

Gene products	Plant species	Biological function (if not known, expression patterns)	Reference
LePRK1, 2	*Liyospersicon esculentum*	Pollen–pistil recognition?	Muschietti *et al.* (1998)
RKF1	*Arabidopsis thaliana*	(Anther-specific)	Takahashi *et al.* (1998)
RPK1		(Osmotic-stress induced)	Hong *et al.* (1997)
LRRPK		(Light-repressed)	Deeken and Kaldenhoff (1997)
TMK1		(Ubiquitous?)	Chang *et al.* (1992)
RLK5		(Rosettes > roots; shoot apex)	Walker (1993)
LTK1, 2, 3	*Zea mays*	(LTK2, 3: endosperm-specific)	Li and Wurtzel (1998)
OsTMK1	*Oryza sativa*	(Gibberellin-induced; stem elongation?)	van der Knaap *et al.* (1996)
TNFR class			
CR4	*Zea mays*	Epidermal cell specification	Becraft *et al.* (1996)
EGF class			
WAK1, 2, 3, 4	*Arabidopsis thaliana*	Disease/stress response? (leaf)	He *et al.* (1996)
PR5 class			
PR5K	*Arabidopsis thaliana*	Disease/stress response?	Wang *et al.* (1996)
Lectin class			
LecRK1	*Arabidopsis thaliana*	(Stem, leaf, silique, flower; increased expression at beginning of stationary phase in cell culture)	Harvé *et al.* (1996)
Other class			
CrRLK1	*Catharanthus roseus*	(Cultured cells?)	Schulze-Muth *et al.* (1996)
RKF2	*Arabidopsis thaliana*	(Ubiquitous)	Takahashi *et al.* (1998)
RKF3	*Arabidopsis thaliana*	(Ubiquitous)	Takahashi *et al.* (1998)

Modified and updated from Braun and Walker (1996).

1991). However, isolation of several *S-RLK* genes from self-compatible plant species and their expression in vegetative tissues indicate that S-RLKs may play a developmental role in addition to self-incompatibility recognition. Such S-RLKs include: ZmPK1 and KIK1 from maize, and ARK1, ARK2, ARK3, RLK1, and RLK4 from *Arabidopsis* (Walker and Zhang, 1990; Tobias *et al.*, 1992; Walker, 1993; Dwyer *et al.*, 1994; Braun *et al.*, 1997; Table I). In addition, SFR2, a S-RLK of *Brassica oleracea*, is implicated in plant defense response signaling pathways (Pastuglia *et al.*, 1997; Table I).

2. *LRR class*. To date, LRR-RLKs comprise the largest class of plant RLKs. LRRs (leucine-rich repeats) are tandem repeats of approximately 24 amino acids with conserved leucines. LRRs have been found in a variety of proteins with diverse functions, from yeast, flies, human, and plants, and are implicated in protein–protein interactions (Kobe and Deisenhofer, 1994). This class was first represented by the isolation of genes coding for TMK1 and RLK5 (Chang *et al.*, 1992; Walker 1993). Several LRR-RLKs have been shown to play critical roles in development. Those include ERECTA which regulates organ shape, CLAVATA1 which controls cell differentiation at the shoot meristem, BRI1 which is involved in brassinosteroid (BR) signaling, and pollen receptor-like kinase1 (PRK1) which is essential for pollen development (Mu *et al.*, 1994; Torii *et al.*, 1996; Clark *et al.*, 1997; Li and Chory, 1997; Table I). In addition, many other LRR-RLKs have been found to exhibit developmental-specific expression patterns (Schmidt *et al.*, 1997; Li and Wurtzel, 1998; Muschietti *et al.*, 1998; Takahashi *et al.*, 1998; Table I). The rice gene *Xa21* confers resistance to *Xanthomonas oryzae pv oryzae* (Song *et al.*, 1995). Therefore, LRR-RLKs also play a role in disease resistance. Interestingly, the tomato *Cf9*, *Cf4*, and *Cf2* disease-resistance gene products, which confer a race-specific resistance to *Cladosporium fulvum* carrying *Avr9*, *Avr4*, and *Avr2*, respectively, contain extracellular LRR domains but only very short cytoplasmic domains (Jones and Jones, 1997; Thomas *et al.*, 1997; Fig. 1).

3. *TNFR class*. The maize *CR4* (*CRINKLY4*) gene product is the sole member of this class (Becraft *et al.*, 1996; Fig. 1). The TNFR (tumor necrosis factor receptor)-like repeat motif has a conserved arrangement of six cysteines. A role for CR4 in specification of epidermal cell fate will be discussed in detail in Section III.D.

4. *EGF class*. The cell wall-associated receptor kinases (WAKs) represent the epidermal growth factor (EGF) class (Fig. 1). The EGF-like repeat motif is characterized by a conserved arrangement of six cysteines (Rebay *et al.*, 1991). The EGF-like repeats are found in variety of animal extracytoplasmic receptor domains and are known to play a role in protein–protein interactions (Rebay *et al.*, 1991; for review see Engel, 1989). In *Arabidopsis*, four WAKs (WAK1 to WAK4) have been identified, and all of them have

extracellular EGF-like repeats (He *et al.*, 1996; Ellard-Ivey *et al.*, 1997). Some specific WAKs contain an additional extracellular stretch with similarity to tenascins, collagens, or extensines (Ellard-Ivey *et al.*, 1997). Reverse-genetic experiments suggest that WAKs may be involved in pathogenic responses (He *et al.*, 1998; see Section VII).

5. *PR class*. The *Arabidopsis* PR5K (PR5-like receptor kinase) is the only known example of this class. The extracellular domain of PR5K exhibits sequence similarity to PR5 (pathogenesis related protein 5), whose expression is induced upon pathogen attack (Wang *et al.*, 1996). The structural similarity between the PR5K receptor domain and PR5 suggests a role for PR5K in pathogenesis response.
6. *Lectin class*. The *Arabidopsis LecRK1* gene product possesses an extracellular domain homologous to carbohydrate-binding proteins of the legume family (Harvé *et al.*, 1996). Although the biological function of LecRK1 is not yet known, its structural feature suggests that LecRK1 may be involved in the perception of oligosaccharide-mediated signal transduction. DNA gel blot analysis suggests that *Arabidopsis* may have several genes coding for lectin-RLKs (Harvé *et al.*, 1996).
7. *Others*. Other RLKs possess extracellular domains sharing no homology to known motifs, such as CrRLK1 of *Catharanthus roseus* and RKF3 of *Arabidopsis* (Schulze-Muth *et al.*, 1996; Takahashi *et al.*, 1998). The predicted extracellular domain of *Arabidopsis* RKF2 possesses two novel repeats with a consensus $YX_3QCX_3LX_4CRXCX_4RX_2L$, whose function is unknown (Takahashi *et al.*, 1998).

As classified above, a huge structural diversity in the extracellular receptor domains perhaps reflects their functional diversity. However, in this chapter, we will focus on those RLKs that play a defined role or are implicated in development.

III. RECEPTOR-LIKE KINASES REGULATING PLANT DEVELOPMENT

A. ERECTA AND THE REGULATION OF ORGAN SHAPE

Given a lack of free cell migration, plant cells must co-ordinate with each other during organogenesis in order to form organs with determinate shape. Thus cell–cell communication must play a crucial role during organ formation. The *Arabidopsis* ecotype Landsberg *erecta* (L*er*) carries a mutation in the *ERECTA* locus that confers upright stature, compact inflorescence with flowers clustering at the top, short pedicels, and short, blunt siliques (Fig. 2). To understand a role for the *ERECTA* gene in regulating organ shape, additional *er* mutant alleles were isolated from other ecotypes (Torii *et al.*, 1996). Phenotypic characterization of new alleles suggested that the *ERECTA* gene

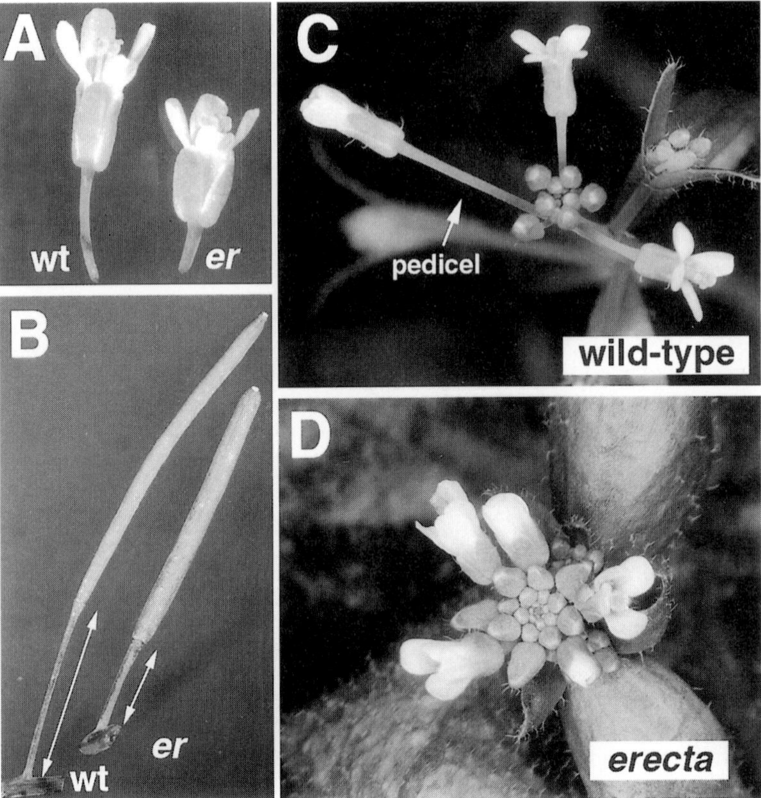

Fig. 2. The *Arabidopsis erecta* mutants. (A) Mature flowers of wild-type (left) and *erecta* mutants (*er*, right). Compared to the wild-type flower, the *erecta* flower is short. (B) Fully expanded siliques of wild-type (wt, left) and *erecta* mutants (*er*, right). Note that the *erecta* silique is short, blunt, and robust. In addition, the *erecta* pedicel (double arrow) is shorter and thicker than the wild-type pedicel. (C–D) Top view of the inflorescence of the 3-week-old wild-type (ecotype Columbia) plants (C) and *erecta* (*erecta-104*) plants (D). Note that the buds and flowers cluster at the top in *er* mutants, mainly due to elongation defects of stem and pedicels. Images C and D are shown at the same magnifications.

regulates the elongation of organs originating from the shoot meristem (Torii *et al.*, 1996). A precise analysis of stems and pedicels of *er* mutant plants indicated that the short stem/pedicel phenotype is mainly due to a decrease in cell number (Komeda *et al.*, 1998; K. U. Torii and S. E. Clark, unpublished). Thus, the *ERECTA* gene is likely to be involved in a process that regulates the number of cells constituting mature organs.

The *ERECTA* gene was isolated utilizing the transfer DNA (T-DNA) tagged allele *er-104*. The predicted 976 amino acid ERECTA protein shares common motifs of the LRR-RLK, namely an N-terminal signal sequence, 20 tandem LRRs (19 complete plus one incomplete), a membrane-spanning region, and a cytoplasmic protein kinase domain (Torii *et al.*, 1996). Expression patterns of the *ERECTA* messenger RNA (mRNA) and *ERECTA* promoter activity suggest that *ERECTA* acts during the process of organ formation (Torii *et al.*, 1996; Yokoyama *et al.*, 1998). Both the *ERECTA* mRNA and the *ERECTA promoter: GUS* activity were detected in the vegetative meristem, with higher levels in the entire inflorescence meristem and developing flower primordia, and were significantly reduced in the mature flower (Yokoyama *et al.*, 1998). The molecular nature of the *ERECTA* gene product as a LRR-RLK and its expression patterns suggest that ERECTA is a component of the signaling mechanism specifying organ shape at the supercellular level.

B. CLAVATA1 AND THE MAINTENANCE OF THE SHOOT APICAL MERISTEM

The shoot apical meristem (SAM) is a population of undifferentiated cells that gives rise to the above-ground portion of the plants. The cells at the center of the meristem are maintained as undifferentiated 'stem' cells, while those on the flanks of the meristem acquire a determinate fate and are incorporated into developing organ primordia (Clark 1997; Lyndon, 1998). The fate of the cells is determined by their position, not by their lineage, therefore the positional information provided by cell–cell interactions appears to be the major determinant of the meristem cell fate (Clark, 1997). In dicotyledonous plants, the SAM consists of three clonally distinct layers, namely the L1, L2, and L3, corresponding to the epidermal, subepidermal, and internal cells, respectively (Lyndon, 1998). Studies on genetic mosaics have provided evidence that cells in different cell layers co-ordinate with each other during development. For instance, the analysis of periclinal chimeras of tomato wild-type and *fasciated* mutant plants demonstrated that the genotype of the L3 layer determines floral meristem size and hence carpel number (Szymkowiak and Sussex, 1992).

The *Arabidopsis clavata* (*clv*) mutant plants accumulate a mass of undifferentiated cells in all three layers in the central region of the SAM. Consequently, the *clv* plants develop enlarged SAMs, fasciated stems, and increased floral organ whorls and organ numbers (Clark *et al.*, 1993; Fig. 3). The phenotype is very conspicuous in siliques, which become 'club-like' in shape owing to additional carpels and organs developing inside the gynoecium (Fig. 3A). Therefore, the wild-type *CLV* genes are thought to act to restrict the proliferation of the undifferentiated cells in the center, or alternatively, to promote the transition of these cells from undifferentiated to determine state on the flanks. There are three known *CLV* loci: *CLV1*, *CLV2*, and *CLV3*. Loss-of-function mutations in any of the *CLV* loci confer a similar phenotype, suggesting that these gene products function in the same pathway (Clark *et al.*, 1993, 1995; Kayes and Clark, 1998). In particular, genetic analysis indicated

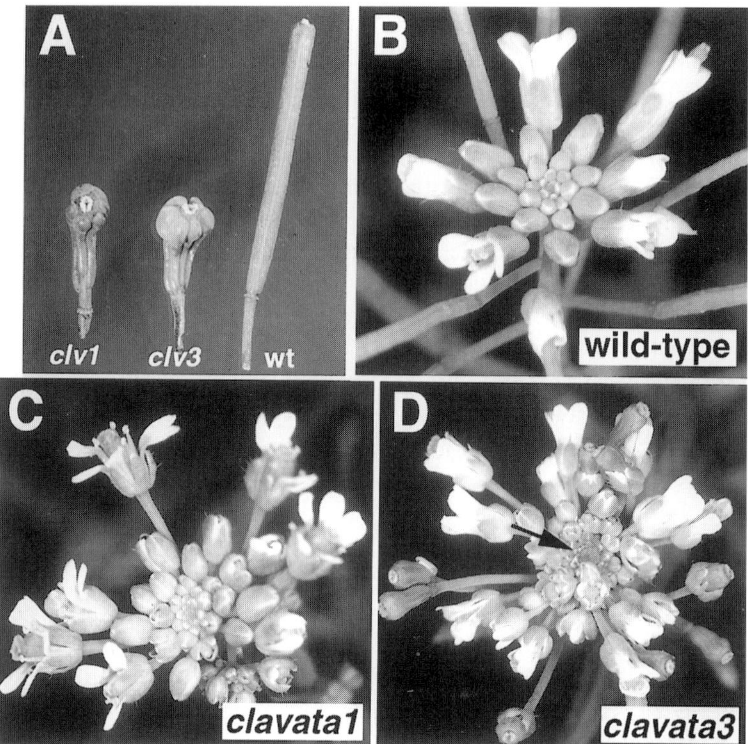

Fig. 3. The *Arabidopsis clavata* mutants. (A) Fully expanded siliques of *clv1* (*clv1–4*), *clv3* (*clv3–2*) and wild type (ecotype L*er*; note that L*er* has the *erecta* mutation). The *clv* siliques have extra numbers of carpels and become characteristic 'club-like' in shape, due to an organ formation and the accumulation of undifferentiated cells inside of gynoecia. (B–D) Top view of the inflorescence of the 4-week-old wild-type (L*er*) (B), *clv1* (*clv1–4*) (C), and *clv3* (*clv3–2*) (D) plants. Note that the *clv* flowers have an extra number of organs and the *clv* SAMs are enlarged (arrow in panel D). Images B–D are shown at the same magnifications.

that the *CLV1* and *CLV3* gene products function in a close association (Clark *et al.*, 1995; for detail see Section VIII.B.2). In contrast, *CLV2* appears to have additional roles other than regulating meristem size, as the *clv2* mutations confer additional defects, such as a high frequency of valveless gynoecia formation, elongated pedicels, and short-day suppression of some phenotypes (Kayes and Clark, 1998).

The *CLV1* gene was cloned by a map-based approach and was shown to encode a 980 amino acid LRR-RLK (Clark *et al.*, 1997). The predicted CLV1 protein possesses an N-terminal signal sequence, 21 tandem LRRs in the extracellular domain, a membrane-spanning region, and a cytoplasmic serine kinase domain. The expression pattern of *CLV1* is restricted predominantly to the L3 cells in a central region of the SAM (Clark *et al.*, 1997). Thus *CLV1* appears to act in a noncell autonomous manner, as all three layers of the SAM, including superficial L1 layer, accumulate undifferentiated cells in the *clv1* mutant plants (Clark *et al.*, 1993). That *CLV1* codes for a transmembrane RLK suggests the existence of a soluble ligand(s), which freely moves in the apoplastic space and co-ordinates cell differentiation between the cell layers. The genetic, molecular, and biochemical evidence supports the hypothesis that the *CLV3* gene product acts as a ligand (see Section VIII.B.2).

C. BRI1 AND BRASSINOSTEROID SIGNALING

Brassinosteroids (BRs) are growth-promoting steroids found in plants. BRs were first purified from the pollen of *Brassica napus* over 20 years ago (Grove *et al.*, 1979). To date, over 40 BR compounds have been identified from as many as 36 plant species, including monocots, dicots, gymnosperms, and ferns (Clouse, 1996; Fujioka and Sakurai, 1997; Altmann, 1998). Genetic evidence for the role of BRs in development was only recently uncovered. Several *Arabidopsis* mutants defective in BR-biosynthesis pathways have been isolated. These mutants include *deetiolated2* (*det2*), *constitutive photomorphogenic and dwarfism* (*cpd*), *dwarf1/diminuto1/cabbage1* (*dwf1/dim1/cbb1*), *dwf4*, and *dwf5–dwf9* (Feldmann *et al.*, 1989; Chory *et al.*, 1991; Kauschmann *et al.*, 1996; Li *et al.*, 1996; Szekeres *et al.*, 1996; Azpiroz *et al.*, 1998; reviewed in Altmann, 1998; Fig. 4). All of these mutants display a characteristic dwarf phenotype with curled, round, dark-green leaves (see Fig. 4), very short stems, reduced fertility, and delayed senescence. The genes for the *DET2*, *CPD*, *DWF1*, *DWF4*, and *DWF7* loci have been identified. They were all shown to encode potential BR biosynthetic enzymes (Li *et al.*, 1996; Szekeres *et al.*, 1996; Choe *et al.*, 1998; Klahre *et al.*, 1998; Choe *et al.*, 1999). For instance, *DET2* encodes a steroid-5α-reductase, which catalyzes an early step of brassinolide biosynthesis (Li *et al.*, 1996). *CPD* and *DWF4* encode cytochrome P450 proteins which function as steroid hydroxylases (Szekeres *et al.*, 1996; Choe *et al.*, 1998). All of the BR-deficient mutants can be rescued by feeding brassinolides or appropriate intermediates.

To identify components of the BR-signaling pathway, several research groups performed genetic screens to isolate *Arabidopsis* mutants with phenotypes identical to the BR-biosynthetic mutants, but that did not show any response to exogenously applied BRs. Despite the large-scale screens, all 20 mutants isolated by three groups fell into alleles of one complementation group, *brassinosteroid insensitive1/cabbage2/brassinosteroid insensitive1* (*bri1/cbb2/bin1*) (Clouse, 1996; Kauschmann *et al.*, 1996; Li and Chory, 1997; Fig. 4C). Thus *BRI1* is most likely to be the only nonreductant component of the BR-signaling pathway whose mutation confers characteristic dwarf phenotype (Fig. 4C). In addition to the morphological resemblance to the BR-biosynthetic

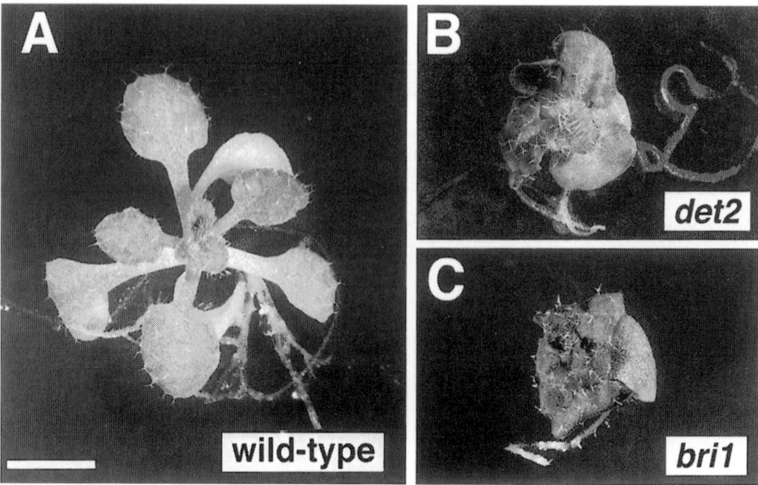

Fig. 4. The *Arabidopsis* mutants defective in BR biosynthesis and BR perception. (A–C) The 2-week-old wild-type (ecotype Columbia) plant (A), *det2* mutant plant (B), and *bri1* (*bri1–101*) mutant plant (C) grown on normal MS media. Note that both the *det2* and *bri1* plants display very similar phenotypes, including curled, round leaf with short petioles and dark-green color. Images are shown at the same magnifications. Bar, 2.5 mm. Plant materials are courtesy of Dr Jianming Li (University of Michigan).

mutants, *bri1* mutants are also defective in BR-induced gene expression (Kauschmann *et al.*, 1996).

The cloned *BRI1* gene was shown to encode an LRR-RLK (Li and Chory, 1997) Fig. 1) The predicted BRI1 protein possesses an N-terminal signal sequence, a single membrane-spanning domain, a cytoplasmic protein kinase domain, and extracytoplasmic receptor domain with 25 LRRs (Li and Chory, 1997). The LRR domain is interrupted by a 70 amino-acid 'loop-out' island. Consequently, the extracellular domain may form two tandem LRR domains, one with 21 LRRs and the other with four LRRs (Fig. 1). BRI1 also possesses a putative leucine zipper motif at the N-terminal region, which may function as a dimerization motif (Li and Chory, 1997). *BRI1* is expressed ubiquitously, and its expression level is not significantly altered between light- and dark-grown seedlings (Li and Chory, 1997).

That *BRI1* encodes an LRR-RLK is very surprising, given that known mammalian steroid receptors are ligand-dependent DNA-binding transcription factors (Beato, 1989) and that both the plant and animal steroid biosynthetic pathways are functionally conserved and interchangeable (Li *et al.*, 1997). Despite the *Arabidopsis* genome sequencing projects and intensive 'gene hunts' by several researchers, plant counterparts for mammalian intracellular steroid receptors, such as the glucocorticoid receptor, have not been found.

Interestingly, there are several lines of evidence indicating the existence of cell-surface steroid-binding sites in mammalian cells, and signal transduction through such membrane steroid-binding sites involves protein phosphorylation and calcium flux (Blackmore *et al.*, 1991; Tesarik *et al.*, 1993; Mendoza *et al.*, 1995). To date, none of the mammalian cell-surface steroid receptors has been identified. It is fascinating to hypothesize that they may be analogous to RLKs and that further analysis of BRI1-mediated BR signal transduction pathways may contribute to a better understanding of mammalian steroid signaling.

D. CR4, RLK WITH A TNFR-LIKE RECEPTOR DOMAIN, IS REQUIRED FOR SPECIFICATION OF THE EPIDERMIS

The epidermis is the outermost cell layer that covers plant organs. The epidermal cell layer is essential for the separation of mature organs. In the leaf, the epidermal cells form thick cell walls with a cuticle layer and epicuticular wax, which play a role in water retention and the prevention of wetting and pathogen invasion (Fahn, 1990). Studies on periclinal chimeras have shown that epidermal cell fate is in part determined by positional cues: leaf epidermal cells that are mispositioned to the lower mesophyll layers will adopt mesophyll cell fate (Poethig, 1987). Thus, cell–cell communication is important in specifying epidermal, as well as mesophyll, cell fates.

The maize *CR4* (*CRINKLY4*) gene is required for proper development of the epidermis (Becraft *et al.*, 1996). *cr4* mutant plants have crinkly leaves with a rough surface (Fig. 5A). Morphological analysis revealed that the defects are largely due to a loss of proper epidermal cell patterning, although a strong *cr4* allele displays morphological anomalies in mesophyll layers as well (Becraft *et al.*, 1996; P. Becarft, personal communication; Fig. 5). The epidermis of *cr4* contains nearly normal epidermal cell types, but cells become enlarged and disorganized (Becraft *et al.*, 1996; Fig. 5C). Strikingly, cell adhesion to adjacent organs occurs occasionally, leading to unfurled and fused leaves. In such cases, the epidermal layer becomes several cells thick and forms interconnecting bridges of tissues. Thus, the important functions of the epidermis to restrict cell division patterns to anticlinal planes and to prevent surface dedifferentiation are both severely compromised by the *cr4* mutation (Becraft *et al.*, 1996). In the *cr4* endosperm, the differentiation of aleurone cells is also strongly compromised. In addition, flower organs are affected, but the roots appear normal (Becraft *et al.*, 1996).

The *CR4* gene was cloned utilizing *Mu1* transposon-tagged alleles (Becraft *et al.*, 1996). *CR4* encodes a 901 amino acid RLK with a single transmembrane domain, an N-terminal signal peptide, a cytoplasmic protein kinase catalytic domain, and an extracellular domain representing a novel class. The recombinant CR4 kinase domain possesses protein kinase activity with serine/threonine specificity (Braun *et al.*, 1997; P. Becraft, personal communication). The extracellular domain contains seven novel repeats of a 39-amino

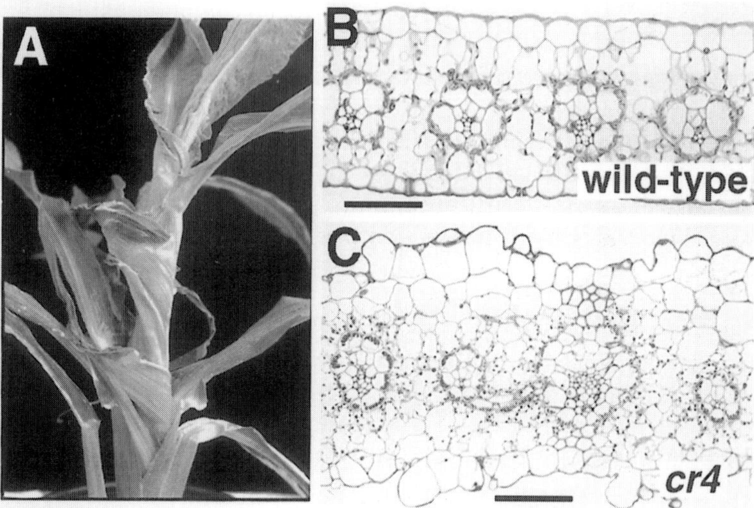

Fig. 5. The maize *cr4* mutants. (A) Appearance of the *cr4* leaves. Note that leaves are crinkly and rough. (B) Transverse section of a wild-type leaf. (C) Transverse section of a *cr4* leaf. A collapse of the epidermal layer is visible. The *cr4* leaf retains a normal Krunts structures and mesophyll tissue, although the cells are somewhat disorganized. Bars, 225 μm. Photographs are courtesy of Dr Philip Becraft (Iowa State University).

acid motif and a domain with similarity to mammalian TNFR (Becraft *et al.*, 1996; Fig. 1).

TNFRs play a crucial role in mammalian inflammatory response and cell death. The TNR family possesses an extracellular, canonical motif of cysteine-rich repeats, each of which contains about six cysteines in 40 amino acids (Smith *et al.*, 1994). A crystallographic study revealed that 80% of the interaction between TNFR and its ligand occurs within the second repeat (Banner *et al.*, 1993). Strikingly, the highest sequence similarity between CR4 and TNFR resides within this second repeat, with all conserved cysteines at the correct position (Becraft *et al.*, 1996). This leads to the fascinating possibility that CR4 binds TNF-like peptide hormones yet to be found in plants.

E. PRK1 AND POLLEN DEVELOPMENT

In an attempt to isolate gene products mediating recognition processes in pollen development, complementary DNA (cDNA) clones predominantly expressed in *Petunia inflata* germinating pollen tubes were isolated. One of the clones, *PRK1* (*pollen receptor-like kinase1*), encodes a 720 amino acid RLK with an extracellular domain containing five LRRs (Mu *et al.*, 1994). The LRRs of PRK1 have a very distinct, nontandem arrangement: the first and second LRRs, as well as the third and fourth LRRs, are separated by long stretches of 55 and 48 amino acids, respectively (Fig. 1). The second and third, as well as the fourth and fifth LRRs are spaced by four amino acids (Fig. 1). Therefore, the tertiary structure of the PRK LRR domain may be quite different from that of tandemly repeated LRRs found in CLAVATA1, ERECTA, Xa21, BRI1, and many others. The kinase domain of PRK1 displays high sequence similarity to most other plant RLKs. However, signature sequences for serine/threonine specificity are not conserved in PRK1 (for detail, see Section VI.A). Consistent with this observation, a recombinant PRK1 kinase domain autophosphorylated both serine and tyrosine residues. Thus PRK1 appears be a dual-specificity kinase (Mu *et al.*, 1994). A native PRK1 protein from *P. inflata* pollen was found in microsomal fractions, consistent with a membrane localization for PRK (Mu *et al.*, 1994).

The expression of *PRK1* is specific to pollen and pollen tubes (Mu *et al.*, 1994). *PRK1* expression was first detected at the binucleate stage and reached its highest level at pollen maturation (Mu *et al.*, 1994). In order to gain insights into the *in vivo* function of PRK1 during pollen development, transgenic *P. inflata* plants that express a fragment of *PRK1* in the antisense orientation under the control of a pollen-specific promoter were generated. Three such lines segregated normal and aborted pollen at a 1:1 ratio (Lee *et al.*, 1996). All developing pollen looked normal at the uninucleate stage, but half of the pollen did not undergo mitosis and eventually collapsed (Lee *et al.*, 1996). Because pollen is haploid, the result suggests that the downregulation of *PRK1* in pollen by the antisense transgene is lethal. Indeed, the transgenic plants failed to transmit the transgene when used as a male. Therefore, PRK1 plays an essential role in postmitotic development of microspores.

Interestingly, the transgene was transmitted in an extremely low frequency even when transgenic plants were used as a female for crosses, suggesting that the antisense *PRK1* transgene is also detrimental for female gametogenesis (Lee *et al.*, 1997). Subsequent, careful analysis of the transgenic plants revealed that embryo sac development is also affected by the downregulation of *PRK1* (Lee *et al.*, 1997). The *PRK1* expression was also detected in wild-type ovules. Thus, PRK1 may represent a signaling component that is required for normal gametogenesis.

IV. RECEPTOR-LIKE KINASES POSSIBLY INVOLVED IN DEVELOPMENT – IMPLICATIONS FROM EXPRESSION ANALYSIS

A. SERK AND THE EMBRYOGENIC COMPETENCE OF SOMATIC CELLS

Somatic embryogenesis is defined as a process where somatic or asexual cells develop into plantlets through stages morphologically analogous to those of zygotic embryogenesis (for review see Zimmerman, 1993). In carrot, somatic embryogenesis is achieved through two sequential, critical steps: (1) the generation of embryogenic callus cells by incubating seedling hypocotyl explants in auxin-containing medium; and (2) the formation of somatic embryos by eliminating the exogenous application of auxin. Hence, an acquisition of 'embryogenic' potential can be separated from somatic embryogenesis per se. Taking advantage of this tissue culture system, cDNAs preferentially expressed in embryogenic cell cultures were isolated (Schmidt *et al.*, 1997).

One clone, named *Somatic Embryogenesis Receptor Kinase* (*SERK*), codes for a 553 amino acid LRR-RLK (Schmidt *et al.*, 1997). The extracellular domain of SERK contains five tandem LRR followed by a proline-rich region, which includes the SPPPP signature O-glycosylation motif found in hydroxyproline-rich glycoproteins. A recombinant SERK protein autophosphorylated *in vitro*.

Correlation of SERK expression and embryogenic competence was carefully determined by cell-tracking experiments (Schmidt *et al.*, 1997). A videotape recording of more than 20 000 immobilized cells revealed that 'enlarged' cells are the only cell type which can develop into somatic embryos. The expression of the *SERK* gene and the *SERK-promoter::Luciferase-reporter* transgene was strictly restricted to such 'enlarged' cell types. This observation established that the *SERK* expression marks embryonic competence and thus predicts somatic embryo formation. Once the somatic embryo reached mid- to late globular stage, the *SERK* expression disappeard. In zygotic embryogenesis, *SERK* expression was first detected after pollination and persisted up to the globular stage. Thus the *SERK* gene represents a signaling mechanism correlated with the embryonic fates, and perhaps such a mechanism is common to both somatic and zygotic embryogenesis.

B. LePRK1 AND LePRK2, LRR-RLKs IMPLICATED IN POLLEN–PISTIL INTERACTIONS

To isolate RLKs that may play a role in the pollen–pistil recognition process, degenerate oligonucleotide reverse transcription–polymerase chain reaction (RT-PCR) was performed using polyA+ RNA from mature tomato pollen. Two genes encoding RLKs, namely *LePRK1* and *LePRK2* for *L. esculentum Pollen Receptor Kinase*, were isolated (Muschietti *et al.*, 1998). Predicted LePRK1 and LePRK2 proteins possess an N-terminal signal peptide and a membrane-spanning region. The extracellular domains of LePRK1 and LePRK2 have poorly conserved, six and five tandem LRRs, respectively. The recombinant LePRK1 and LePRK2 kinase domains autophosphorylated *in vitro* (Muschietti *et al.*, 1998). Overall, the extracellular domains of LePRK1 and LePRK2 show 50% amino acid identity, and no cross-hybridization was observed when DNA gel blot analysis was performed under high stringency (Muschietti *et al.*, 1998). Thus, they are not multi-copy genes.

Both LePRK1 and LePRK2 are specifically expressed in mature and germinating pollen; no clear expression was observed in immature pollen. This is consistent with a hypothesis that both LePRK1 and LePRK2 function in a pollen–pistil recognition process rather than during pollen development, as proposed for *Petunia* PRK1. In fact, both the LePRK1 and LePRK2 proteins are membrane associated and localized in germinating pollen tubes (Muschietti *et al.*, 1998). Further, LePRK2 was found to be phosphorylated in pollen membrane fractions, and LePRK2 was partially dephosphorylated by incubating the pollen membrane fraction with tomato-style extracts (Muschietti *et al.*, 1998). Thus LePRK2 may be a receptor involved in pollen–pistil interaction.

C. RKF1, A FLOWER-SPECIFIC RLK

A degenerate oligo-PCR strategy was taken to isolate any RLKs expressed in an *Arabidopsis* flower cDNA library. Three clones, *RKF1–3*, were isolated, but subsequent expression analysis revealed that only *RKF1* exhibits a flower-specific expression (Takahashi *et al.*, 1998). RKF1 possesses 13 tandem LRRs in the extracellular domain. A bacterially produced recombinant RKF1 protein has serine/threonine-specific autophosphorylation (Takahashi *et al.*, 1998). Detailed expression study of *RKF1* by *in situ* hybridization and histochemical GUS analysis of transgenic plants expressing *RKF1-promoter::GUS* construct revealed that *RKF1* promotor activity is highest in anthers (Takahashi *et al.*, 1998). Thus, RKF1 may play a role in microsporogenesis and pollen maturation, as proposed for PRK1 (Lee *et al.*, 1996), or in pollen–pistil interaction, as proposed for LePRK1 and LePRK2 (Muschietti *et al.*, 1998).

V. RLKs WITH RESPONSE TO ENVIRONMENTAL CUES

As sessile organisms, higher plants adopted sophisticated sensory mechanisms to respond to environmental cues, such as light, temperature, drought, and pathogen attack. Detecting changes in the external environments and modulating the endogenous developmental programs are essential for the survival of plants. Recent expression studies suggest that some RLKs are likely to be involved in responding to environmental factors.

A. RPK1 AND OSMOTIC STRESS

The *Arabidopsis RPK1* gene was isolated by a degenerate oligonucleotide-PCR strategy and by subsequent screening of *Arabidopsis* leaf cDNA library. *RPK1* encodes a 540 amino acid RLK with extracellular domain consisting of five poorly conserved LRRs (Hong *et al.*, 1997). *RPK1* is expressed ubiquitously in all organs examined, namely flower, stem, leaf, and root. However, a rapid induction of *RPK1* expression within 1 h was observed when *Arabidopsis* leaves were treated with abscisic acid (ABA) (Hong *et al.*, 1997). Similar kinetics of

RPK1 induction were observed when plants were subjected to dehydration, salt, and cold treatments (Hong *et al.*, 1997). Thus, *RPK1* appears to be upregulated by osmotic stress, because all the treatments affect the osmotic response. Interestingly, dehydration induction of *RPK1* was not impaired by ABA-deficient (*aba1*) or ABA-insensitive (*abi1,2,3*) mutations (Hong *et al.*, 1997). Therefore, although *RPK1* expression is upregulated by exogenous ABA treatment, the induction of *RPK1* by dehydration seems ABA independent. Intermediate signaling components (e.g. MAPKKK, MAPK), potential downstream transcription factors, and target cis-acting DNA responsive elements for the osmotic stress response have been defined (for review see Shinozaki *et al.*, 1998). It would be very interesting to know whether RPK1 has any link to the previously defined osmotic-stress signaling pathway.

B. LRRPK AND LIGHT REGULATION

To isolate light-regulated genes, a subtractive cDNA screen was performed using blue-irradiated and red-light irradiated *Arabidopsis* cDNA libraries. One gene isolated in this screen, designated *Light-Repressible Receptor Protein Kinase* (*LRRPK*), encodes a putative LRR-RLK and showed high expression in darkness (Deeken and Kaldenhoff, 1997). *LRRPK* expression was negatively regulated by light; pulse irradiation of blue, red, far-red, and ultraviolet (UV) light to *Arabidopsis* seedlings drastically reduced the steady-state level of the *LRRPK* transcripts (Deeken and Kaldenhoff, 1997). In mature plants, *LRRPK* was most highly expressed in roots. Because the 5′ upstream region of *LRRPK* has four repeats of the *PHYA* repressor element, which is found in the promoter region of *PHYA* and is thought to act as a cis-element for light-repression, the *LRRPK* expression may be regulated in a similar manner with phytochrome A (Deeken and Kaldenhoff, 1997). Interestingly, a recombinant phytochrome A protein was recently demonstrated to possess a serine/threonine protein kinase catalytic activity (Yeh and Lagarias, 1998). Thus, phosphorylation by a serine/threonine protein kinase may play an important role in light signal transduction.

VI. STRUCTURAL/FUNCTIONAL PROPERTIES OF PLANT RLKs

A. PROTEIN KINASE DOMAIN

Based on the amino-acid sequence alignment of the protein kinase catalytic domains by Hanks and Quinn (see: www.sdsc.edu/kinases; Hanks and Quinn, 1991), plant RLKs are most related to each other and thus comprise a distinct family of protein kinases. Predicted amino-acid sequences of RLKs indicate that they possess serine/threonine substrate specificity. Two notable signature sequences for serine/threonine kinases are the consensus DLKPEN in subdomain VIb and G(T/S)XX(Y/F)IAPE in subdomain VIII, as opposed to DLAARN and FPIKWMAPE in tyrosine kinase, respectively (Hanks and Quinn 1991; Walker 1994). Plant RLKs have nearly conserved DϕKXSN in subdomain VIb and GTXGYϕAPE in subdomain VIII (Fig. 6; ϕ = aliphatic amino acids; X = nonconserved amino acids); both of which are more similar to serine/threonine indicative sequences. To date, all but one of the plant RLKs have been shown to autophosphorylate on serine and/or threonine residues *in vitro* when expressed as a recombinant protein in *Escherichia coli* (Chang *et al.*,

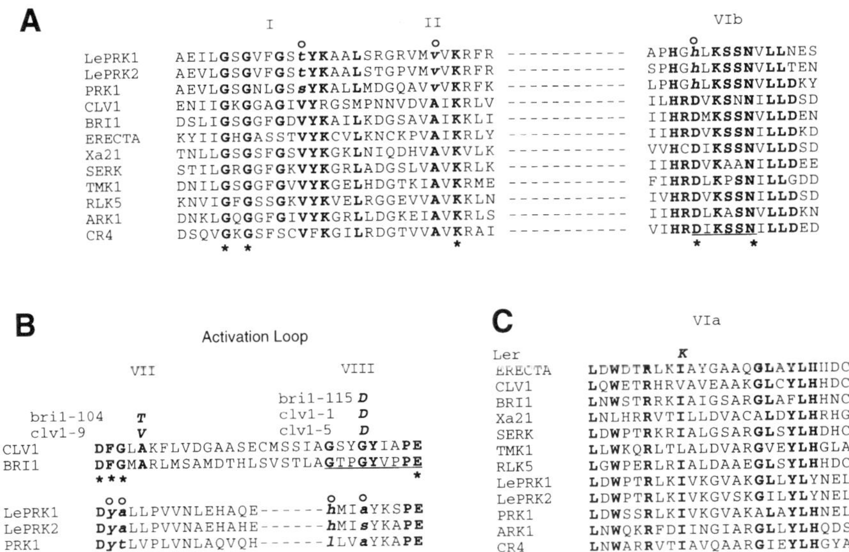

Fig. 6. Alignments of kinase domains from representative plant RLKs. (A) The alignments at subdomains I, II, and VIb. Note that the RLKs expressed in pollen have characteristic substitutions (lower case in italic and bold) in well-conserved amino acid residues (bold), including an invariant Asp in subdomain VIb (asterisk). Sites of the substitutions are highlighted with open circles. A signature sequence for serine/threonine kinases is underlined. Dash (–) indicates a gap in the alignments. (B) Top: mapping the lesions in *clv1* and *bri1* alleles highlights the importance of the activation loop. The activation loops of CLV1 and BRI1 are aligned with the amino-acid substitutions (bold, italics) in each *clv1* and *bri1* mutant allele. Highly conserved amino acids among protein kinases are in bold, and invariant amino acids are highlighted with asterisks. Roman numerals indicate kinase subdomains. One of the signature sequences for serine/threonine specificity is underlined. Bottom: RLKs expressed in pollen have substitutions in an invariant DFG motif (asterisks) and conserved glycine residues within the activation loop (lower case in italic; highlighted with open circles). Note that the latter substitutions destroy the signature sequence for serine/threonine specificity. (C) The alignments at subdomain VIa. Highly conserved amino acids among plant RLKs are in bold. An I-to-K substitution in L*er* is indicated in italic.

1992; Goring and Rothstein, 1992; Stein and Nasrallah, 1993; Horn and Walker, 1994; Zhao *et al.*, 1994; Harvé *et al.*, 1996; Schulze-Muth *et al.*, 1996; Tobias and Nasrallah, 1996; Braun *et al.*, 1997; Williams *et al.*, 1997; Stone *et al.*, 1998; Takahashi *et al.*, 1998; Wang *et al.*, 1998; P. Becraft, personal communication; J. Li, personal communication).

The one exception, *Petunia inflata* PRK1, has been shown to have a 'dual specificity', autophosphorylating both serine and tyrosine residues *in vitro* (Mu *et al.*, 1994). Interestingly, many conserved amino acid residues among RLKs are altered in the PRK1 kinase domain, emphasizing a distinct characteristic of PRK1 as a protein kinase. Most importantly, an invariant aspartic acid within the 'catalytic loop (HRDLKXSN)' in subdomain VIb is substituted by histidine (Fig. 6A). This substitution is significant because the invariant aspartic acid is the kinase catalytic base, which accepts the proton from the substrate at the site of phosphorylation (Hanks and Quinn, 1991; Taylor and Radzio-Andzelm, 1994; see also: www.sdsc.edu/kinases/pkr/3D/xray/2cpk/). In addition, an invariant DFG motif of a 'Mg^{2+} loop' in subdomain VII is changed to DYT (Fig. 6B). Further, conserved glycines in subdomain VIII are substituted to leucine and alanine in PRK1 (Fig. 6B). These substitutions destroy one of the serine/threonine signature sequences described above (Fig. 6B). In a 'glycine-rich loop' in subdomain I, a highly conserved valine which plays a role in anchoring adenosine triphosphate (ATP) is substituted to serine in PRK1 (Fig. 6A). Finally, a highly conserved alanine in subdomain II is replaced by valine (Fig. 6A). Interestingly, two additional RLKs, LePRK1 and LePRK2, also possess amino acid changes in the identical positions; namely V-to-T in subdomain I, A-to-V in subdomain II, D-to-H in subdomain VIb, DFG-to-DYA in subdomain VII, and G-to-H and G-to-A (LePRK1) or to S (LePRK2) in subdomain VII (Muschietti *et al.*, 1998; Fig. 6A, B). These amino acid alterations indicate that the pollen-specific RLKs may have a distinct biochemical property and therefore represent a distinct subclass of plant RLKs.

In many animal PTKs, ligand-induced activation of the kinase domain is mediated by receptor dimerization, which allows two facing kinase domains to transphosphorylate each other (Ullrich and Schlessinger, 1990). Biochemical experiments on two RLKs, RLK5 and CLV1, have shown that the nature of autophosphorylation *in vitro* appears to be transphosphorylation (Horn and Walker, 1994; Williams *et al.*, 1997). For instance, an inactive form of the CLV1 kinase could be phosphorylated in the presence of the active CLV1 kinase; conversely, autophosphorylation of the active CLV1 kinase was reduced when incubated with increasing amounts of the inactive form of CLV1 kinase (Williams *et al.*, 1997). The finding suggests that the plant RLKs may function as dimers, just like RTKs in animals.

RLKs with known biological function (i.e. whose genetic mutations confer phenotypic defects) provide an excellent opportunity to understand the *in vivo* functional significance of the kinase domain. The kinase domain appears to be essential for the function of BRI1, CLV1 and ERECTA. For example, a nonsense mutation in *bri1–107* and a frame-shift mutation in *er1–2*, that remove the C-terminal region of the kinase domain after subdomain VIII and VII, respectively, result in severe phenotypes (Li and Chory, 1997; F. Tax, personal communication). Analysis of molecular lesions in *bri1* and *clv1* mutant alleles has shown that most of the missense mutations in the kinase domain lie within the activation loop (Fig. 6B). The activation loop is a short segment which begins at the conserved DFG motif in subdomain VII and ends at the conserved APE motif in subdomain VIII (Johnson *et al.*, 1996). *bri1–104* and *clv1–9* have missense mutations in the identical position, replacing conserved A to T and V, respectively; *bri1–115*, *clv1-1*, and *clv1-5* have identical G-to-D substitutions within the activation loop (Fig. 6B) (Clark *et al.*, 1997; Li and Chory 1997; for review see Lease *et al.*, 1998). Phosphorylation of the activation loop is known to convert an inactivate RTK to an active form (Johnson *et al.*, 1996). It is therefore likely that such point mutations in *bri1* and *clv1* severely compromise the proper substrate recognition and perturb signal transduction.

Landsberg *erecta* has I-to-K substitution in a subdomain VIa. Although this subdomain does not have any invariant amino acids and does not participate in substrate/Mg^{2+} ATP interaction, isoleucine at the site of the mutation is highly conserved among plant RLKs (Fig. 6C). Because the subdomain VIa is predicted to form a hydrophobic α-helix, it perhaps plays a significant structural role, and the L*er* mutation may perturb the proper structure, leading to a loss of the biological activity.

There are only limited reports on phosphorylation analysis of native RLKs using plant microsomal membrane fractions. When purified *Arabidopsis* microsomal membranes were subjected to a protein kinase assay, several polypeptides showed autophosphorylation. Such phosphorylation was primarily on serine and threonine residues (Schaller and Bleecker, 1993). More specific experiments were performed with tomato LePRK2. LePRK2 was shown to be phosphorylated in pollen microsomal fraction and was partially dephosphorylated in the presence of tomato-style extracts, suggesting that phosphorylation/dephosphorylation events occur during pollen–pistil interaction (Muschietti *et al.*, 1998). *In vivo* biochemical analysis of other RLKs is necessary for a further understanding of the biochemical properties of the kinase domain.

B. EXTRACELLULAR RECEPTOR DOMAIN

The basic structural features of plant RLKs are summarized in Section II. In this section, we will mainly focus on the receptor domains whose functional significance has been revealed by genetic and biochemical experiments.

1. *LRR Domain*

The structural features of the plant LRRs, including those of RLKs, are comprehensively described in a review by Jones and Jones (1997). The LRRs of the LRR-RLKs have a consensus sequence that is distinct from the consensus for LRRs found in cytoplasmic receptors, such as tobacco N, *Arabidopsis* RPM1 and RPS2, and flax L6 (Jones and Jones, 1997; Fig. 7). However, the numbers of LRRs vary significantly among different RLKs. For instance, SERK, PRK1, LePRK1, and RPK1 each have five LRRs; TMK1 has 11 LRRs which are segmented into arrays of nine and two; RLK5, CLV1, ERECTA, and BRI1 have >20 consecutive LRRs (Fig. 7; Chang *et al.*, 1992; Walker, 1993; Mu *et al.*, 1994; Torii *et al.*, 1996; Clark *et al.*, 1997; Hong *et al.*, 1997; Li and Chory, 1997; Schmidt *et al.*, 1997; Muschietti *et al.*, 1998). This suggests that the LRR domain of each RLK may associate with a wide variety of ligands.

Based on the studies of the *Xa21* and *Cf* family of disease-resistance genes, the LRR domain appears to be a determinant of ligand-binding specificity (for detail see Sections VII and VIIIB1; Thomas *et al.*, 1997; Wang *et al.*, 1998). In addition, the LRR domains of CLV1 and ERECTA are essential for their biological functions. *clv1–4* and *clv1–8* alleles have G-to-D and D-to-N substitutions within the sixth and 10th LRR, respectively, causing a strong phenotype (Clark *et al.*, 1997). *er-103* has a M-to-I substitution in the 10th LRR that results in a weak phenotype (Torii *et al.*, 1996). Perhaps these

```
Cf9          L--L--L--LDLS-N-L-G-IP--     27 (23,4)
BRI1         Φ--Φ--L--L-LS-N-ΦSG-IP--     25 (21,4)
XA21         Φ--Φ--L--L-L--N-LSG-IP--     23
CLV1         Φ--Φ--L--L-Φ--N-ΦTG-IP--     23
RLK5         L--L--L--L-L--N-LSG-IP--     21
ERECTA       LG-L--L--L-L--N-L-G-IP--     20
TMK1         L--L--L--L-L--N-Φ-G-IP--     11 (9,2)
RKF1         Φ--L--L--Φ-L--N-L-GTIP--     13
LTK1, 2      Φ--Φ--L--L-L--N-Φ-G-----      8 (1,7)
LePRK1             L--Φ-Φ-NN-Φ-GPIP--      6
LePRK2             L--L-Φ-NN-L-GPIP--      5
PRK1         L--L--Φ--L-L-NN---G-IP        5  (1, 2, 2)
SERK         L--L--L--L-L-NN-ΦSGPIP--      5


Extracytoplasmic LRR consensus:

             L--L--L--L-L--N-L-G-IP--

Cytoplasmic LRR consensus:

             L--L--L--L-L--(N/C/T)-(-)L--IP--
```

Fig. 7. Alignments of the consensus sequences of LRRs found in representative LRR-RLKs. –, nonconserved amino acid; ϕ, aliphatic amino acids. The numbers on the right indicate the total numbers of repeats, including both complete and incomplete repeats. Numbers in the parenthesis indicate the existence of a gap, or a 'loop-out' island that breaks the consecutive repeats (see also Fig. 1). At the bottom are the consensus sequence for extracellular and cytoplasmic LRRs, adopted from Jones and Jones (1997).

missense mutations perturb the proper ligand–receptor interactions, and compromise activation of the receptors. Alternatively, such mutations may interfere with receptor dimer formation (see Section VI.B.2).

In contrast, the importance of the LRR domain for BRI1 function remains ambiguous, as no *bri1* mutations reside within the LRR domain (Li and Chory, 1997). It is possible that the lesions within BRI1 LRR may not alter association of BRI1 with BR. The importance of the 70 amino-acid 'loop-out' island of the BRI1 extracellular domain in ligand recognition will be discussed in Section VIII.B.1. Alternatively, the mutations in the BRI1 LRR domain may lead to an additional phenotype other than the characteristic dwarf (see Fig. 4).

2. *Putative Dimerization Modules*

First we will look at *paired cysteines* adjacent to the LRR. The LRRs of plant RLKs are often flanked by paired cysteines spaced by six amino acids (CX_6C). Figure 8 shows the alignments of the cysteine pairs from representative LRR-RLKs. In addition to the spacing, some degree of sequence conservation is observed for the first paired cysteines, with a consensus sequence CXWXGV(S/T)C (Fig. 8). The presence of cysteine pairs is altered in some RLKs. For instance, SERK does not have an amino-terminal pair, while XA21 lacks a second pair, although it does possess several cysteines immediately adjacent to the transmembrane domain (Song *et al.*, 1995; Schmidt *et al.*, 1997). LePRK1 and LePRK2 have a pair of amino-terminal cysteines spaced by 13 amino acids (Muschietti *et al.*, 1998; Fig. 8).

Paired cysteines probably participate to form disulfide bonds that may play a role in intermolecular assembly. The CLV1 protein has been shown to be linked by intermolecular disulfide bonds to another protein(s) (Trotochaud *et al.*, 1999). In the presence of a reducing agent (DTT), CLV1 from cauliflower extracts migrated in sodium dodecyl sulfate–polyacrylamide gel electrophoresis (SDS–PAGE) at ~105 kDa, which corresponds to the predicted monomer size. In the absence of DTT, the CLV1 band shifted to ~185 kDa, which may be a homo- or heterodimeric form. Interestingly, monomeric forms of the mutant

	1st		2nd
CLV1	CSFSGVSC	--LRR--	CLPHRVSC
BRI1	CTFDGVTC	--LRR--	CGYPLPRC
ERECTA	CVWRGVSC	--LRR--	CGSWLNSPC
TMK1	CKWTHIVC	--LRR--	CLSSPGEC
RLK5	CKWLGVSC	--LRR--	CVDLDGLC
PRK1	CFHHQLSC	--LRR--	CGKPLESAC
RPK1	CSWYGVSC	--LRR--	
Xa21	CTWVGVVC	--LRR--	
SERK		--LRR--	CGPVTGRPC
LePRK1	CDKKTDRPNWDNVIC	--LRR--	CDGPFSKC
LePRK2	CVKDNNKPKWNNLFC	--LRR--	CGPPLAKSC

Fig. 8. Alignments of the paired cysteines found in representative LRR-RLKs. 1st and 2nd indicates the first and second pairs of cysteines which are located immediately before and after the LRR domain, respectively (see also Fig. 1). –LRR– indicates an existence of the LRR domain between two pairs of cysteines.

CLV1 proteins were detected in *clv1–4* and *clv1–8* extracts (Trotochaud *et al.*, 1999). Because both *clv1–4* and *clv1–8* contain missense mutations in the LRR domain (Clark *et al.*, 1997; see Section VI.B.1), alterations in the LRR region may reduce dimerization efficiency. Generating transgenic *Arabidopsis* plants, such as those that express mutated forms of CLV1 with cysteines replaced by another amino acid, may provide insights into the functional significance of the cysteine pairs.

The second putative dimerization module is the *cysteine-rich S domain.* Recently, *Brassica* S_{29}SLG, a secreted glycoprotein with an S domain involved in self-incompatibility recognition, was found as monomers and homodimers in stigmatic protein extracts (Doughty *et al.*, 1998). The homodimerization of S_{29}SLG was reversible by reducing agents, indicating that dimers are linked via disulfide bridges. As the S domain possesses 12 conserved cysteines, it is likely that some of the cysteine residues may function as a dimerization module. Although Doughtly *et al.* (1998) also reported the existence of S_{29}SLG antigen in microsomal membrane fractions of stigmatic extracts, it is not certain whether it represents S-RLKs or whether the S-RLKs form intermolecular association with SLGs via disulfide bonds.

VII. *IN VIVO* MODE OF ACTION OF PLANT RLKs

In animals, co-expression of an RTK along with a mutant form of an RTK containing a truncated or inactive kinase domain is known to confer a 'dominant negative' effect (e.g. see Ueno *et al.*, 1991). This is due to the formation of heterodimers of wild-type and mutant RTKs, which prevents proper transphosphorylation and in turn fails to transduce signals. In contrast, mutant forms of RTK lacking an extracellular ligand-binding domain are known to confer 'constitutively active' effects (e.g. see Basler *et al.*, 1991). Such *in vivo* modes of action for plant RLKs have not been clearly demonstrated, although several reports provide some implications.

A reverse-genetic approach was taken to understand the *in vivo* function of the *Arabidopsis ARK1* gene, which codes for an S-RLK (Tobias and Nasrallah, 1996). While constitutive, high-level expression of ARK1 resulted in a severe growth retardation, overexpression of the ARK1 kinase-inactive form, which is referred to as a dominant-negative form, did not lead to any visible phenotypes, although a significant amount of transgene expression was detected (Tobias and Nasrallah, 1996). This result can be explained if ARK1 has a redundant role in development, or if the kinase-inactive forms of this receptor do not act in a dominant-negative manner.

Reverse-genetic experiments on the *Arabidopsis WAK1* gene provided evidence that the mutant forms of plant RLKs may act *in vivo* in a similar manner to that of animal RTKs. Specifically, transgenic *Arabidopsis* plants that expressed the kinase-deleted WAK1 or WAK1 in an antisense orientation showed susceptibility to a low concentration of a salicylic acid analog. However, transgenic plants that ectopically express the full-length or the receptor-deleted WAK1 showed an increased tolerance to salicylic acid (He *et al.*, 1998). Together, these observations revealed a role for WAK1 in survival during the pathogenesis response. In addition, the results indicated that mutant forms of WAK1, which lacked the kinase domain or extracellular receptor domain, function in a dominant-negative or constitutive-active manner, respectively (He *et al.*, 1998).

While a cytoplasmic kinase catalytic domain is absolutely required for the animal RTK function (Hunter and Lindberg, 1994), two recent studies of the *Arabidopsis CLV1* and rice *Xa21* genes imply distinct mode of actions for plant RLKs. As discussed in Section VI.A, mutations resulting in defective kinase domains are detrimental for RLK function. Among such mutations, *clv1–1* acts in a semidominant manner, implying that the mutant clv1–1 protein heterodimerizes with wild-type protein and interferes with transphosphorylation, as reported for the animal RTKs (Clark *et al.*, 1993; see Section VI.A and Fig. 6B). However, two *clv1* mutations, *clv1–6* and *clv1–7*, that truncate the entire protein kinase domain, were shown to confer a weak phenotype (Clark *et al.*, 1993, 1994). This is unlikely to be due to readthrough, as no CLV1 protein was detected in *clv1–6* extracts using anti-CLV1 antisera directed to the kinase domain (Trotochaud *et al.*, 1999). Therefore, the CLV1 kinase domain per se may be partially redundant and perhaps able to be functionally replaced by other cytoplasmic protein kinases. A similar mode of action was suggested for *Xa21D*, a member of the *Xa21* multigene family. *Xa21D* codes for a secreted LRR receptor-like protein lacking the transmembrane and kinase domains due to a retrotransposon insertion (Wang *et al.*, 1998). Nevertheless, the introduction of *Xa21D* into a susceptible rice strain conferred partial disease resistance to *X. oryzae* (Wang *et al.*, 1998). Based on these results, Wang and co-workers proposed a model whereby the secreted XA21D protein forms a heterodimer with the endogenous LRR-RLK lacking recognition specificity. The kinase domain of the endogenous LRR-RLK will be activated upon ligand (AvrXA21) binding to the XA21D.

A possible mechanism for *clv1–6* and *clv1–7* action lies in the function of the tomato Cf receptor family. The *Cf* gene products, which lack cytoplasmic kinase domains (Fig. 1), may associate with cytoplasmic protein kinases (Jones and Jones, 1997). In fact, the *Drosophila* Toll protein, which is a transmembrane receptor containing extracellular LRR domain but lacking a cytoplasmic kinase domain, is known to associate with Pelle, a cytoplasmic serine/threonine kinase, in a phosphorylation-dependent manner (Shelton and Wasserman, 1993; Shen and Manley, 1998). It is intriguing that the *Drosophila* Pelle kinase is structurally most related to and categorized within the plant RLK superfamily (see www.sdsc.edu/kinases/pkr/pk_catalytic/pk_hanks_class.html). Additional studies, together with the identification of signaling components as described in the next section, may provide a paradigm for the mode of action of plant RLKs.

VIII. RLK SIGNALING

In the last few years, great advances have been made in uncovering components of RLK-mediated signaling pathways. Studies on the *Arabidopsis* CLV1 protein and other RLKs are beginning to provide an emerging framework of signaling events, whereby the activated ligand/receptor complex transduces signals via intermediate components, leading to gene expression (Fig. 9).

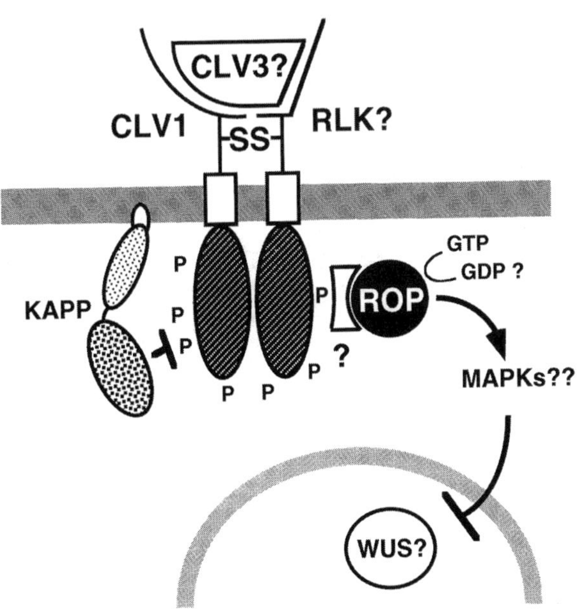

Fig. 9. Model for the CLV1–receptor complex signaling. The model presumes that CLV3 acts as a ligand and CLV1 forms a heterodimer with another RLK via disulfide bonds (S–S). Upon binding CLV3, the CLV1 receptor heterodimer is activated and autophosphorylates the kinase domains (hatched ellipse). This activation leads to an association with KAPP, a negative-regulator that attenuates CLV1 receptor activity. The association and activation of Rop may in turn transduce a signal perhaps via a MAPK cascade, and finally, regulate the activity of the transcription factors, most likely the WUS protein. As described in Section VIII.A.1, KAPP possesses three domains, an N-terminal membrane anchor (open ellipse), KI domain (small dotted), and protein phosphatase C domain (large dotted). Currently, there is no evidence for whether the Rop interacts with the RLKs directly, or via association with an 'adaptor' protein (?). Nor is it known whether the GTPase activity of Rop is required for the association with the CLV1 receptor complex. Modified from Trotochaud *et al.* (1999).

A. DOWNSTREAM COMPONENTS

1. KAPP, a Kinase-Associated Protein Phosphatase

To date, KAPP (kinase-associated protein phosphatase) is the only downstream component whose interaction with RLKs has been proven to be functionally relevant.

KAPP was first discovered as a protein that interacted with the protein kinase catalytic domain of RLK5, by *in vitro* interactive (far-western) screening (Stone *et al.*, 1994). The isolated partial-length cDNA encoded a domain referred to as the KI (kinase interaction) domain. The KI domain encompasses a region called FHA (fork-head associated) domain, a domain with unknown function found in a subset of fork-head family transcription factors and some other nuclear-localized proteins (Hofmann and Bucher, 1995). The association of the KI domain with RLK5 appears to be phosphorylation dependent, since the KI domain did not interact with the kinase inactive form of RLK5 (Stone *et al.*, 1994). The subsequently cloned full-length cDNA was shown to code for a novel protein, named KAPP, that possesses an N-terminal type I signal anchor, the KI domain, and a functional C-terminal type 2C serine/threonine protein phosphatase (PP2C) domain (Stone *et al.*, 1994). The identification of KAPP leads to an attractive scenario by which membrane-anchored KAPP associates with the cytoplasmic kinase domain of active RLK5 *in vivo* to attenuate the signal strength by dephosphorylation. *KAPP* is a single copy gene and is expressed in all tissues examined, including tissues where RLK5 is barely expressed (Stone *et al.*, 1994), suggesting that KAPP may also associate with other RLKs *in vivo*. Subsequent biochemical experiments demonstrated that KAPP associates *in vitro* with many, but not all, other RLKs (Braun *et al.*, 1997). For instance, KAPP did not interact with the kinase domain of CR4 or ZmPK1 (Braun *et al.*, 1997).

Interaction of KAPP and CLV1 was tested both *in vitro* and *in vivo*, to address the biological significance of KAPP-RLK interactions. KAPP exhibited interaction with CLV1, weaker interaction with the clv1–1 mutant protein and no interaction with the kinase inactive form of CLV1 (Williams *et al.*, 1997; Stone *et al.*, 1998). KAPP was shown to dephosphorylate CLV1 *in vitro* (Williams *et al.*, 1997). Immunoprecipitation experiments using anti-CLV1 antisera successfully co-precipitated the endogenous KAPP protein from cauliflower protein extracts, providing robust evidence that KAPP and CLV1 indeed interact *in vivo* (Stone *et al.*, 1998). The next important step was to address the effects of altered KAPP levels on the biological function of CLV1. This was achieved by two complementary approaches. Williams *et al.* (1997) generated KAPP overexpression lines in a wild-type background and showed that the overexpressed KAPP confers a weak *clv1*-like phenotype. Stone *et al.* (1998) generated KAPP sense suppression lines in a *clv1–1* mutant background, and showed that a decreased level of KAPP suppressed the *clv1* phenotype (Fig. 10). *clv1–1* is a partial loss-of-function allele and has reduced

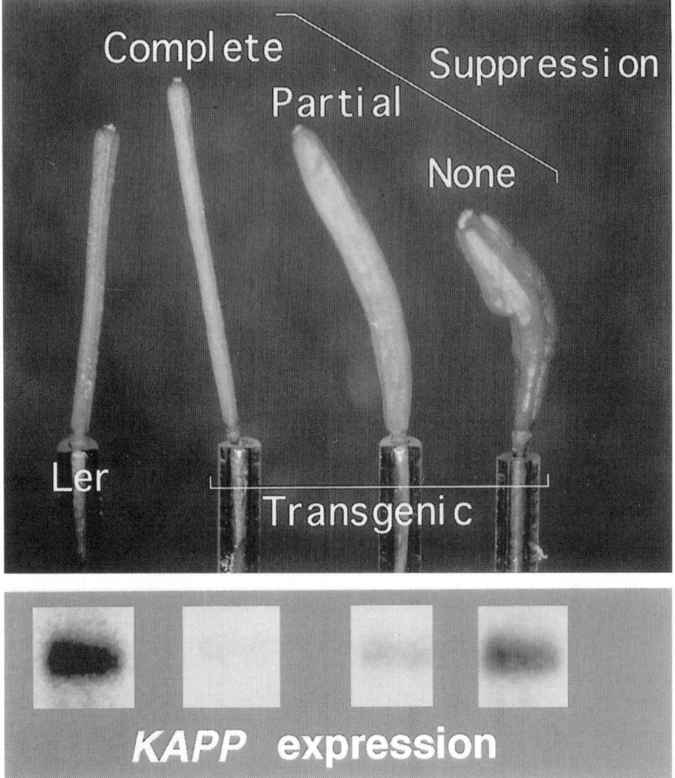

Fig. 10. KAPP modulates the CLV1 activity *in vivo*. Introduction of the *CAMV35S::KAPP* transgene into the *clv1–1* plants resulted in the suppression of the *clv*-phenotype due to a sense suppression of KAPP. Transgenic lines were categorized into three phenotypic classes: those with no phenotype (none; silique is indistinguishable from *clv1–1* silique), those with intermediate suppression (partial; silique has subtle *clv*-like phenotype), and those with complete suppression (complete; silique looks like that of wild-type). RNA gel blot analysis demonstrated that a degree of suppression of the *clv1* phenotype correlates with the degree of suppression of the KAPP expression. Reproduced and partly modified from Stone *et al.* (1998) by courtesy of the American Society of Plant Physiologists.

kinase activity (Clark *et al.*, 1997; Williams *et al.*, 1997; for a molecular lesion in *clv1–1*, see Fig. 6B). Both results clearly demonstrate that KAPP functions in the CLV1 signaling pathway as a negative regulator. Recent biochemical analysis revealed that KAPP associates exclusively with the putative active form of CLV1 (Trotochaud *et al.*, 1999), consistent with a proposed role for KAPP in modulating CLV1 signal strength upon activation of the CLV1 pathway.

It seems somewhat unlikely that the modulated KAPP activity would affect only the CLV1 signaling pathway, because KAPP is expected to interact with other RLKs *in vivo*. Perhaps, careful characterization of transgenic plants will address whether KAPP regulates the function of other RLKs, such as BRI1 and ERECTA. In fact, additional phenotypes associated with altered KAPP expression were reported (Williams *et al.*, 1997; Stone *et al.*, 1998). The biochemical experiments by Trotochaud *et al.* (1999) support this notion. The elution profile of KAPP (300–670 kDa) was shown to be much broader than that of the active CLV1 complex (~450 kDa), when *Arabidopsis* meristem protein extracts were subjected to a gel-filtration analysis (Trotochaud *et al.*, 1999). Importantly, the KAPP elution profile was not altered by *clv* mutations that eliminated the formation of active CLV1 complex (see below). Therefore, the majority of KAPP found in the high molecular range associates with other RLKs.

2. Rops: Rho-type GTPases, a Possible Downstream Target?

That KAPP is a negative regulator of CLV1 implies the existence of another downstream target protein(s) that relays signals leading to a regulation of gene expression. Recent biochemical studies put Rops, Rho GTPase-related proteins from plants, as potential candidates (Trotochaud *et al.*, 1999). The Rop subfamily of GTPases is proposed to play a critical role in the plant signal transduction pathways leading to polarized cell growth (Li *et al.*, 1998). *Arabidopsis* possesses 10 reported *Rop* genes, each of which exhibits different developmental expression patterns (Li *et al.*, 1998). By immunoprecipitation analysis, a Rop-related protein was found to be associated with the ~450-kDa active CLV1 complex but not with a ~185-kDa inactive complex (Trotochaud *et al.*, 1999). In animal cells, members of the Ras GTPase superfamily are known to relay RTK signals through activation of MAPK pathways. It is very fascinating to propose that Rops may take the place of Ras in plant RLK signal transduction. Indeed, increasing evidence emphasizes important roles for the MAPK cascades in plant development and environmental response (Hirt, 1997; Mizoguchi *et al.*, 1997). Detailed biochemical experiments, such as testing whether the activated Rop can in turn activate a MAPK cascade, should link a bridge between RLKs and downstream signaling cascades. Very recently, transgenic tobacco cells expressing Cf9 were reported to elicit a rapid activation of MAPKs in an Avr9-dependent manner, indicating that pathogen ligand/plant receptor recognition pathways are linked to the MAPK cascades (Romeis *et al.*, 1999). The intermediate signaling components of RLKs regulating development are probably very similar to those in disease resistance.

3. Thioredoxin Family

In order to identify potential targets of the S-RLK during *Brassica* self-incompatibility response, yeast two-hybrid analysis was performed using the kinase domain of S-locus receptor-like kinase (SRK)-910 as a bait (Bower *et al.*, 1996). Two genes, *THL-1* and *THL-2*, that encode members of the thioredoxin family, were identified from the screen. THL-1 and THL-2 exhibited a specific interaction to the active kinase domain of SRK-910: no interaction was observed with the kinase-inactive form of SRK-910 or kinase domains from other plant RLKs, namely RLK4 and RLK5 (Bower *et al.*, 1996). The THLs were phosphorylated *in vitro* by SRK-910. At this moment, the *in vivo* significance of the interaction of SRK-910 and THLs is not known. As the thioredoxins are known to modulate enzyme activity through their redox activity, THLs may regulate the kinase activity of SRK-910 *in vivo*. Alternatively, SRK-910 may modulate THL enzymatic activity, which in turn activates the further step of a signaling cascade.

4. *Transcription Factors and Regulation of Gene Expression: The Final Targets of RLK Signaling?*

Given that changing developmental fate, whether by cell division, cell proliferation, or cell elongation, requires gene expression, it is most likely that signals mediated by RLKs eventually regulate gene expressions through the activation of transcription factors. The *Arabidopsis WUSCHEL (WUS)* gene product is a strong candidate for a target of the CLV1-mediated signaling pathway. Loss-of-function *wus* mutant plants cannot replenish stem cells constituting the SAM. As a consequence, in *wus* plants the SAM terminates at a very early stage, and flower meristems produce a decreased number of floral organs (Laux *et al.*, 1996). Thus, the *WUS* gene is required for maintaining stem cells, a role antagonistic to the proposed role for *CLV1*. Double mutant analysis showed that *wus* is epistatic to *clv1*, leading to two contrasting possibilities: either WUS is upstream and activates CLV1 function, or CLV1 is upstream and suppresses WUS function (Laux *et al.*, 1996). The overlapping pattern of gene expression for *WUS* and *CLV1* and the identification of WUS as a homeodomain transcription factor suggest that the CLV3/CLV1 signaling pathway may lead to a suppression of WUS function (Mayer *et al.*, 1998).

Downstream transcription factors for other RLKs, including those with known biological functions (e.g. BRI1, ERECTA, CR4), are enigmatic. Perhaps, in-depth characterization of these mutants, as well as isolating additional mutants that exhibit genetic interactions with these RLKs, will address such a question. A gene (*BRU1*) whose expression is upregulated by BR has been identified (Zurek and Clouse, 1994). A transcription factor that binds to a presumptive BR-upregulated cis element, if it exists, could potentially be a target of the BRI1-mediated signaling pathway.

B. LIGANDS FOR RLKs

The structures of the receptor domains of some RLKs, namely the LRR motif, the TNFR-like domain, and the EGF-like domains, suggest that their respective ligands may be proteins, as similar extracellular domains in animals all bind proteinaceous ligands (Rebay *et al.*, 1991; Kobe and Deisenhofer, 1994; Smith *et al.*, 1994). The ligands for plant RLKs are presumed to be secreted proteins rather than membrane-bound, cell-surface proteins owing to a presence of cell walls that set apart plasma membranes of two adjacent cells (Becraft, 1998). Ligands for some RLKs may not be polypeptides: a lectin-like extracellular domain in LecRK1 implies a possible role for polysaccharides in some RLK-mediated signaling pathways. In addition, BRI1 may act as a steroid receptor. In this section, we will describe the latest progress in seeking ligand-like molecules in RLK pathways and provide further speculation.

1. Brassinosteroids, Ligands for BRI1?

As discussed in the Section IIIC, genetic evidence places BRI1 as a sole, nonredundant component of BR perception, supporting a hypothesis that BRI1 itself may be a BR receptor. The molecular nature of BRI1 as a LRR-RLK, however, may bring some arguments. This is because LRR domains have hitherto been demonstrated to be involved in protein–protein interactions (Kobe and Deisenhofer, 1994), and it seems somewhat unlikely that a large LRR domain with 25 repeats is the binding surface for such small molecules as BRs. Unlike CLV1 or ERECTA, the LRR domain of BRI1 is interrupted by a 70 amino acid 'loop-out' island (see Fig. 1). This 'loop-out' island is essential for the BRI1 function, because a G-to-E substitution within this island in the *bri1–113* mutation confers severe phenotype (Li and Chory, 1997). If this 'loop-out' island is a binding site for BRs, then the extracellular LRR domain as well as the N-terminal LZ motif perhaps interact with other 'nonredundant' components of the BR receptor complex.

An alternative hypothesis is that BR binds to a soluble 'BR-binding protein' which in turn binds to the BRI1 LRR domain (Li and Chory, 1997). In such a case, the 'loop-out' island may play more of a structural role than a direct role in ligand binding. That some other LRR receptor-like proteins, namely tomato Cf9 and Cf4, possess a 'loop-out' island in homologous positions within the extracellular domain supports this possibility (Jones *et al.*, 1994; Jones and Jones, 1997; Thomas *et al.*, 1997; see Fig. 1). Amino acid sequence comparison of Cf9 and Cf4 revealed that their 'loop-out' islands are 100% identical and thus unlikely to be a determinant for ligand recognition, as they presumably interact with different ligands (Thomas *et al.*, 1997; see Section VIII.B.3). However, the 'loop-out' islands of Cfs and BRI1 do not share sequence similarity and differ significantly in size. Future biochemical experiments, such as a ligand-binding assay using a heterologous system, should ultimately provide an answer to the question: Is BRI1 a BR receptor?

2. CLAVATA3, a Potential Ligand for CLAVATA1

Several lines of genetic evidence indicate that *CLV3* and *CLV1* function in the same pathway and that CLV3 and CLV1 proteins may act in close association.

First, *clv1* and *clv3* confer identical phenotypes (Fig. 3). Second, the mutations are mutually epistatic to each other: *clv1/clv3* double mutant plants do not exhibit phenotypes more severe than the strongest allele (Clark *et al.*, 1995). Lastly, *clv1* and *clv3* exhibit trans-heterozygous interactions: *clv1*/+ *clv3*/+ double-heterozygous plants display a *clv1* phenotype, although *clv1* and *clv3* per se are recessive mutations (Clark *et al.*, 1995). Because *CLV1* encodes an LRR-RLK, it is quite reasonable to speculate that CLV3 may function as a ligand.

CLV3 has been cloned utilizing both a T-DNA tagged and *En-1* tagged *clv3* alleles (Fletcher *et al.*, 1999). The *CLV3* gene encodes a small, 96 amino-acid novel protein with no apparent sequence motifs, outside an N-terminal signal sequence which is predicted to be cleaved to form a mature peptide of 78 amino acids (Fletcher *et al.*, 1999). Thus CLV3 appears to be a secreted protein, supporting a hypothesis that CLV3 acts as a ligand for CLV1 receptor-like kinase.

In fact, a recent biochemical study showed that the *CLV3* gene product is required for the activation of the CLV1 receptor complex (Trotochaud *et al.*, 1999). In wild-type meristems, CLV1 is present in two protein complexes; one ~185 kDa and the other ~450 kDa (Trotochaud *et al.*, 1999). Because the mature CLV1 protein is ~105 kDa, the 185-kDa complex most likely constitutes the 'core' receptor complex that includes the hetero- or homo-dimeric form of CLV1. The ~450-kDa complex represents an active signaling complex, which contains downstream factors (e.g. KAPP and Rop GTPase) and perhaps the ligand (Trotochaud *et al.*, 1999). In *clv3* mutants, only the ~185-kDa core receptor complex was detected. Thus the ~450-kDa active signaling complex does not form in the absence of the CLV3 protein (Trotochaud *et al.*, 1999). The data strongly suggest that CLV3 functions to promote the assembly of an active CLV1-receptor complex (Fig. 9).

In addition to the biochemical evidence, cell biological studies on genetic mosaics provide insight into *CLV3* action (Fletcher *et al.*, 1999). Periclinal chimeras were generated by spontaneous reversion of the *En-1* transposon insertion in *clv3–7*, and a restoration of *CLV3* activity in one cell layer was shown to be sufficient to rescue the *clv3* phenotype (Fletcher *et al.*, 1999). Thus *CLV3* acts in a noncell autonomous manner, consistent with its possible role as a diffusable ligand. *CLV3* mRNA expression is restricted to the very center of the meristem and predominantly in the superficial L1 and L2 layers, while the *CLV1* expression pattern is slightly broader and in the underlying L2 and L3 layers (Clark *et al.*, 1997; Fletcher *et al.*, 1999). The observed *CLV3* and *CLV1* expression patterns fit well with the view that the CLV3–CLV1, or the ligand–receptor, interaction mediates a cell–cell communication which is required for co-ordinating growth of different cell layers during development of the shoot apical meristem. Of course, direct biochemical experiments, such as examining the binding of CLV3 to microsomal membranes expressing CLV1 in a heterologous system, are necessary to prove that CLV3 acts as a ligand.

3. CR4 Ligand and the Role of Proteolysis in RLK Signaling

The highest sequence similarity between CR4 and TNFR resides within the TNFR ligand-binding domain, which reinforces the possibility that CR4 may indeed bind TNF-like ligands (Becraft *et al.*, 1996). If the ligand molecule for CR4 is encoded by a nonredundant gene and if it is not shared by other CR4-like RLKs with distinct functions, then a loss-of-function mutation of the ligand should confer nearly identical phenotypes as *cr4*. Genetic evidence puts the maize *DEK1* locus as a candidate gene coding for another component of the CR4 signaling pathway, possibly the ligand. Recessive *dek1* mutant plants display a *cr4*-like phenotype and interact with the *cr4* mutation in a dominant manner (P. Becraft, personal communication). These findings suggest that DEK1 and CR4 proteins function in a linear pathway and may interact physically.

CR4 seems to act cell-autonomously, consistent with the molecular identity of CR4 as a transmembrane receptor (P. Becraft, personal communication). It has been reported that *DEK1* acts noncell-autonomously, suggesting its possible molecular nature as a diffusable ligand (Neuffer, 1994). However, in a recent mosaic analysis, *DEK1* appeared to act cell-autonomously in leaves (P. Becraft, personal communication). At this point it is unclear what the role of DEK1 is in the CR4 pathway.

Recently, an *Arabidopsis* mutant with a similar phenotype to *cr4* was isolated (Tanaka *et al.*, 1998). The loss-of-function mutant, *ale1* (*abnormal leaf-shape1*), has defects predominantly in epidermal cell differentiation. Consequently, cotyledons and leaves of *ale1* wrinkle and adhere to each other. The *ALE1* gene was cloned by transposon tagging and shown to encode a member of the subtilisin family of serine proteases (H. Tanaka and Y. Machida, personal communication). Members of this family are known as 'subtilisin-like prohormone convertases (SPCs)'. Mammalian SPCs function as endoproteolytic processing enzymes of the major secretory pathway and generate biologically active mature peptide hormones, such as insulin (for review see Steiner, 1998; Wilson and White, 1998). It is fascinating to hypothesize that ALE1 cleaves the *Arabidopsis* CR4 ligand to generate a mature form. If so, loss-of-function mutations in *Arabidopsis* CR4 and its ligand will presumably give phenotypes similar to those of *ale1*. Indeed, an additional locus conferring *ale1*-like phenotype was isolated from *Arabidopsis* (H. Tanaka and Y. Machida, personal communication).

Proteolytic cleavage appears to be an essential step for the activation of ligands for *Drosophila* LRR receptors. In the *Drosophila* Toll signaling pathway, a serine proteinase encoded by the *Easter* gene cleaves the extracellular Spätzel protein to generate mature, active ligand. This ligand binds to the extracellular LRR domain of the transmembrane receptor Toll, and activates a signal cascade (for review see Anderson, 1998). It would be interesting to know whether the generation of the active form of CLV3 requires proteolytic cleavage, beyond the removal of the signal peptides.

4. What are the Ligands for Other RLKs?

If CLV3 acts as a ligand, does it suggest that the ligand molecules for other LRR-RLKs, such as ERECTA, PRK1, TMK1, and RLK5, are homologous to CLV3? Although a paradigm for ligands is far from being established, several studies on disease resistance in tomato and self-incompatibility (SI) proteins of *Brassica* and *Papaver* fuel speculation that ligand-like molecules for structurally similar receptors may share no resemblance in a primary structure but may adopt a similar tertiary structure. Many such ligand-like molecules appear to form a compact, folded, and stable structure with disulfide bonds contributing to structural integrity. Finally, several peptide growth factors have recently been identified. In this section, we will discuss molecules that may potentially act as ligands for RLKs.

Tomato disease-resistant proteins Cf9 and Cf4 confer a race-specific resistance to fungal pathogens *C. fulvum* that express the *Avr9* and *Avr4* genes, respectively (Jones and Jones, 1997). Cf9 and Cf4 are highly homologous extracytoplasmic LRR proteins with >90% amino acid identity (Jones *et al.*, 1994; Thomas *et al.*, 1997). In contrast, the fungal-produced Avr9 and Avr4, although both small secreted proteins, show no similarity in primary sequence and differ significantly in size. The mature Avr9 protein has 28 amino acids, while the mature Avr4 protein has 86 amino acids (Van den Ackerveken *et al.*, 1992; Jones and Jones, 1997; Joosten *et al.*, 1997). Interestingly, both Avr9 and Avr4 peptides contain a patch of cysteine residues, which may participate in the formation of a stable tertiary structure via disulfide bonds. In fact, ^{1}H-nuclear magnetic resonance (NMR) analysis revealed that the Avr9 peptide forms a compact structure containing three antiparallel β sheets connected by three disulfide bridges (Vervoort *et al.*, 1997).

This discussion is reinforced by recent studies on *Papaver rhoeas* SI proteins, which are small (~15 kDa) secreted proteins that play a role in self-incompatibility. Although *P. rhoeas* SI proteins are highly polymorphic among each other, they are predicted to adopt a similar secondary structure, again with conserved cysteines which have been shown to be critical for their biological activity (Walker *et al.*, 1996; Kakeda *et al.*, 1999). To date, receptor counterparts for the *P. rhoeas* SI proteins are not known.

Recently, the gene encoding a pollen coat protein (PCP-A1) that binds to SLG of *Brassica* was identified (Doughty *et al.*, 1998). PCP-A1 is a small, secreted hydrophilic protein with a mature peptide of 55 amino acids which shows some similarity to defensin proteins. PCP-A1 possesses eight conserved cysteines which are predicted to form four intramolecular disulfide bonds (Doughty *et al.*, 1998). Far-western experiments using stigmatic extracts and radiolabelled PCP-A1 showed that PCP-A1 is capable of binding to both the monomeric and dimeric forms of SLG, and probably binds to SRK as well (Doughty *et al.*, 1998). Although PCP-A1 is unlinked to the S locus and thus is probably not itself a determinant of SI recognition specificity, PCP-A1 could function as a co-activator that promotes homo- or hetero-dimerization of SLGs and/or SRKs.

In contrast to this, the recently revealed CLV3 mature protein has no cysteines, and thus is unlikely to form a structure related to any of the disulfide-linked ligand-like molecules discussed above (Fletcher *et al.*, 1999). As anticipated, a DNA database search did not reveal any sequences homologous to CLV3. A hunt for ligand molecules for other RLKs should therefore be achieved through a conventional genetic and biochemical approach.

To date, several peptide growth factors have been identified in plants, although their receptors have not been found. Such 'peptide hormones' include systemin, ENOD40, phytosulfokine, and CDT-1 (Pearce *et al.*, 1991; Matsubayashi and Sakagami, 1996; Charon *et al.*, 1997; Furini *et al.*, 1997; Matsubayashi *et al.* 1997; for review see Franssen, 1998). Among them, ENOD40 (10–13 amino acids) and phytosulfokine (pentapeptides) dramatically promote cell division and cell proliferation when overexpressed (in the case of ENOD40) or when purified or synthetic peptides were added to cell cultures (in the case of phytosulfokine) (Charon *et al.*, 1997; Matsubayashi *et al.*, 1997). It would be of great interest to understand their *in vivo* functions during normal plant development, as they may turn out to be ligands for RLKs. Obviously, further studies of both receptors and ligands should greatly facilitate our understanding of receptor-mediated signaling cascades regulating plant development.

IX. CONCLUSIONS AND PERSPECTIVES

Increasing evidence now clearly emphasizes a significant role for RLK-mediated signaling cascades in plant development. The molecular identity of *CLAVATA1*, *ERECTA*, and *CR4* as RLKs, for instance, indicates that the RLK-mediated signaling cascades represent a substantial portion of a 'long sought' signaling mechanisms for cell–cell communication regulating morphogenesis, whose existence has been implicated by genetic mosaic and clonal analysis. That *BRI1* encodes a LRR-RLK revealed a novel mechanism for steroid perception in plants. As recent advances in plant molecular genetics have promoted the rapid increase in our knowledge about the biological functions of RLKs, the versatile nature of the RLK superfamily has become manifest: some RLKs regulate morphogenesis, while others play roles in disease resistance, pollen–pistil recognition, or response to environmental cues, including light, drought, and wounding. Consistently, recent progress in express sequence tags and genome sequencing places RLKs as a very abundant class. For instance, *Arabidopsis* has at least 100, probably more, RLKs. The true number of the RLKs in *Arabidopsis* will ultimately be unveiled at the completion of the genome sequencing projects. Functional-genomic studies of RLKs, such as generating knock-out lines for each RLK, may provide a comprehensive view in understanding the breadth of biological activities for RLKs.

Biochemical analysis has shown that most of the plant RLKs phosphorylate serine/threonine residues, in contrast to the animal RTKs which phosphorylate tyrosine. *In vivo* mode of action of plant RLKs indicated that the plant RLKs may function as homodimers in a similar manner to the animal RTKs, but may also act in a novel manner. Studies of XA21D, Cfs, and CLV1 suggest that the plant RLKs may heterodimerize with extracellular receptors that lack a cytoplasmic kinase domain, with cytoplasmic kinases, or with the other RLKs that do not have recognition specificity. It is interesting that TGF-β receptor serine/threonine kinases in animals are known to act as hetero-oligomers, including a 'primary receptor' that binds a ligand and a 'transducer' that lacks recognition specificity (Massagué, 1996). Our continued efforts in understanding plant RLK complex formation will bring a defined view on the RLKs' mode of action.

New discoveries from the studies of CLV1 now shed light on one mechanism of RLK activation and downstream signaling: that the association of CLV3, a potential ligand, to the CLV1 receptor appears to activate the receptor complex that in turn binds KAPP, a negative regulator, and Rop GTPase which may relay signals to downstream targets (Fig. 9). Perhaps the signal may be incorporated into a MAPK cascade that eventually regulates transcription factors (Fig. 9). Because plants do not seem to have proteins containing SH2-SH3 domains, the interaction motifs of the RLKs with downstream targets are probably distinct from those of the animal RTKs. Further biochemical characterization of the CLV1–receptor complex will provide a clear answer as to how the downstream targets associate with the receptors. It is not certain at this moment whether the CLV1 signaling mechanism can be extrapolated to the other RLKs. Detailed analysis of the signaling mechanisms mediated by the other RLKs will promote synergistic effects to establish a paradigm for RLK signaling in plants. Understanding how the RLK-mediated signaling cascades cross-talk with other signaling pathways will eventually determine the dynamic arrays of plant cell signaling in regulating development.

NOTE ADDED IN PROOF

Recently, *Arabidopsis* RLK5 was shown to function in developmentally regulated, floral organ abscission processes. Therefore, RLK5 has been renamed HAESA, a Latin word meaning to adhere to (Jinn *et al.*, 2000).

ACKNOWLEDGEMENTS

The authors would like to thank Philip Becraft, Yasunori Machida, Jianming Li, Frans Tax, and Amy Trotochaud for sharing unpublished or prepublication data; Philip Becraft for providing photographic images for Fig. 5; and Jianming Li for providing plant materials for Fig. 4. Works in the authors' laboratory are supported by US-NSF and US-DOE to S.E.C.

REFERENCES

Altmann, T. (1998). Recent advances in brassinosteroid molecular genetics. *Current Opinion in Plant Biology* **1**, 378–383.

Anderson, K. V. (1998). Pinning down positional information: dorsal–ventral polarity in the *Drosophila* embryo. *Cell* **95**, 439–442.

Azpiroz, R., Wu, Y., LoCascio, J. C. and Feldmann, K. A. (1998). An *Arabidopsis* brassinosteroid-dependent mutant is blocked in cell elongation. *Plant Cell* **10**, 219–230.

Banner, D. W., D'Arcy, A., Janes, W., Gentz, R., Schoenfeld, H.-J., Broger, C., Loetscher, H. and Lesslauer, W. (1993). Crystal structure of the soluble human 55 kd TNF receptor–human TNF complex: implications for TNF receptor activation. *Cell* **73**, 431–445.

Basler, K., Christen, B. and Hafen, E. (1991). Ligand-independent activation of sevenless receptor tyrosine kinase changes the fate of cells in the developing *Drosophila* eye. *Cell* **64**, 1069–1081.

Beato, M. (1989). Gene regulation by steroid hormones. *Cell* **56**, 335–344.

Becraft, P. W. (1998). Receptor kinases in plant development. *Trends in Plant Science* **3**, 384–388.

Becraft, P. W., Stinard, P. S. and McCarthy, D. R. (1996). *CRINKLY4*, a TNFR-like receptor kinase involved in maize epidermal differentiation. *Science* **273**, 1406–1409.

Blackmore, P. F., Neulen, J., Lattanzoi, F. and Beebe, S. J. (1991). Cell surface-binding sites for progesterone mediate calcium uptake in human sperm. *Journal of Biological Chemistry* **266**, 18655–18659.

Bower, M. S., Matias, D. D., Fernandes-Carvalho, E., Mazzurco, M., Gu, T., Rothstein, S. J. and Goring, D. R. (1996). Two members of the thioredoxin-h family interact with the kinase domain of a Brassica *S* locus receptor kinase. *Plant Cell* **8**, 1641–1650.

Braun, D. M. and Walker, J. C. (1996). Plant transmembrane receptors: new pieces in the signaling puzzle. *Trends in Biochemical Sciences* **21**, 70–73.

Braun, D. M., Stone, J. M. and Walker, J. C. (1997). Interaction of the maize and *Arabidopsis* kinase interaction domains with a subset of receptor-like protein kinases: implications for transmembrane signaling in plants. *Plant Journal* **12**, 83–95.

Chang, C., Schaller, G. E., Patterson, S. E., Kwok, S. F., Meyerowitz, E. M. and Bleecker, A. B. (1992). The *TMK1* gene from *Arabidopsis* codes for a protein with structural and biochemical characteristics of a receptor protein kinase. *Plant Cell* **4**, 1263–1271.

Charon, C., Johansson, C., Kondrosi, E., Kondrosi, A. and Crespi, M. (1997). *enod40* induces dedifferentiation and division of root cortical cells in legumes. *Proceedings of the National Academy of Sciences of the United States of America* **94**, 8901–8906.

Choe, S., Dilkes, B. P., Fujioka, S., Takatsuto, S., Sakurai, A. and Feldmann, K. A. (1998). The *DWF4* gene of *Arabidopsis* encodes a cytochrome P450 that mediates multiple 22α-hydroxylation steps in brassinosteroid biosynthesis. *Plant Cell* **10**, 231–243.

Choe, S., Noguchi, T., Fujioka, S., Takatsuto, S., Tissier, C. P., Gregory, B. D., Ross, A. S., Tanaka, A., Yoshida, S., Tax, F. and Feldmann, K. A. (1999). The *Arabidopsis dwf7/ste1* mutant is defective in Δ^7 sterol C-5 desaturase step leading to brassinosteroid biosynthesis. *Plant Cell* **11**, 207–221.

Chory, J., Nagpal, P. and Peto, C. A. (1991). Phenotypic and genetic analysis of *det2*, a new mutant that affects light regulated seedling development in *Arabidopsis*. *Plant Cell* **3**, 445–459.

Clark, S. E. (1997). Organ formation at the vegetative shoot meristem. *Plant Cell* **9**, 1067–1076.

Clark, S. E., Running, M. P. and Meyerowitz, E. M. (1993). *CLAVATA1*, a regulator of meristem and flower development in *Arabidopsis*. *Development* **119**, 397–418.

Clark, S. E., Running, M. P. and Meyerowitz, E. M. (1995). *CLAVATA3* is a specific regulator of shoot and floral meristem development affecting the same processes as *CLAVATA1*. *Development* **121**, 2057–2067.

Clark, S. E., Williams, R. W. and Meyerowitz, E. M. (1997). The *CLAVATA1* gene encodes a putative receptor kinase that controls shoot and floral meristem size in *Arabidopsis*. *Cell* **89**, 575–585.

Clouse, S. D. (1996). Molecular genetic studies confirm the role of brassinosteroids in plant growth and development. *Plant Journal* **10**, 1–8.

Deeken, R. and Kaldenhoff, R. (1997). Light-repressible receptor protein kinase: a novel photo-regulated gene from *Arabidopsis thaliana*. *Planta* **202**, 479–486.

Doughty, J., Dixon, S., Hiscock, S. J., Willis, A. C., Parkin, I. A. P. and Dickinson, H. G. (1998). PCP-A1, a defensin-like *Brassica* pollen coat protein that binds the *S* locus glycoprotein, is the product of gametophytic gene expression. *Plant Cell* **10**, 1333–1347.

Dwyer, K. G., Kandasamy, M. K., Mahosky, D. I., Acciai, J., Kudish, B., Miller, J. E., Nasrallah, M. E. and Nasrallah, J. B. (1994). A superfamily of *S* locus-related sequences in *Arabidopsis*: diverse structure and expression patterns. *Plant Cell* **6**, 1829–1843.

Ellard-Ivey, M., Demura, T., Lomax, T. and Carpita, N. (1997). Connections: the hard wiring of the plant cell for perception, signaling, and response. *Plant Cell* **9**, 2105–2117.

Engel, J. (1989). EGF-like domains in extracellular matrix proteins: localized signals for growth and differentiation? *FEBS Letters* **251**, 1–7.

Fahn, A. (1990). Epidermis. *In* "Plant Anatomy", pp. 152–185. Pergamon Press, Oxford.

Feldmann, K. A., Marks, M. D., Christianson, M. L. and Quatrono, R. S. (1989). A *dwarf* mutant of *Arabidopsis* generated by T-DNA insertion. *Science* **243**, 1351–1354.

Fletcher, J. C., Brand, U., Running, M. P., Simon, R. and Meyerowitz, E. M. (1999). Signaling of cell fate decisions by *CLAVATA3* in *Arabidopsis* shoot meristems. *Science* **283**, 1911–1914.

Franssen, H. J. (1998). Plants embrace a stepchild: the discovery of peptide growth regulators. *Current Opinion in Plant Biology* **1**, 384–387.

Fujioka, S. and Sakurai, A. (1997). Biosynthesis and metabolism of brassinosteroids. *Physiologia Planterum* **100**, 710–715.

Furini, A., Koncz, C. and Salamini, C. F. (1997). High level transcription of a member of repeated gene family confers dehydration tolerance to callus tissue of *Craterostigma plantagineum*. *EMBO Journal* **16**, 3599–3608.

Goring, D. R. and Rothstein, S. J. (1992). The *S*-locus receptor kinase gene in a self-incompatible *Brassica napus* line encodes a functional serine/threonine kinase. *Plant Cell* **4**, 1273–1281.

Grove, M. D., Spencer, G. F., Rohwedder, W. K., Mandava, N., Worley, J. F. and Warhen, J. D. (1979). A unique plant growth promoting steroid from *Brassica napus* pollen. *Nature* **281**, 216–217.

Hanks, S. K. and Quinn, A. M. (1991). Protein kinase catalytic domain sequence database: identification of conserved features of primary structure and classification of family members. *Methods in Enzymology* **200**, 38–61.

Harvé, C., Dabos, P., Galaud, J.-P., Rougé, P. and Lescure, B. (1996). Characterization of an *Arabidopsis thaliana* gene that defines a new class of putative plant receptor kinases with an extracellular lectin-like domain. *Journal of Molecular Biology* **258**, 778–788.

He, Z.-H., Fujiki, M. and Kohorn, B. D. (1996). A cell wall-associated, receptor-like protein kinase. *Journal of Biological Chemistry* **271**, 19789–19793.

He, Z.-H., He, D. and Kohorn, B. D. (1998). Requirement for the induced expression of a cell wall associated receptor kinase for survival during the pathogen attack. *Plant Journal* **14**, 55–63.

Hirt, H. (1997). Multiple roles of MAP kinases in signal transduction in plants. *Trends in Plant Science* **2**, 11–15.

Hofmann, K. and Bucher, P. (1995). The FHA domain – a putative nuclear signaling domain found in protein kinases and transcription factors. *Trends in Biochemical Sciences* **20**, 347–349.

Hong, S. W., Jon, J. H., Kwak, J. M. and Nam, H. G. (1997). Identification of a receptor-like protein kinase gene rapidly induced by abscisic acid, dehydration, high salt, and cold treatments in *Arabidopsis thaliana*. *Plant Physiology* **113**, 1203–1212.

Horn, M. A. and Walker, J. C. (1994). Biochemical-properties of the autophosphorylation of RLK5, a receptor-like protein-kinase from *Arabidopsis thaliana*. *Biochimica et Biophysica Acta* **1208**, 65–74.

Hunter, T. and Lindberg, R. A. (1994). Receptor protein-tyrosine kinases. *In* "Protein Kinases" (J. R. Woodgett, ed.) pp. 178–211. IRL Press, New York.

Jinn, T. L., Stone, J. M. and Walker, J. C. (2000). HAESA, an *Arabidopsis* leucine-rich repeat receptor kinase, controls floral organ abscission. *Genes and Development* **14**, 108–117.

Johnson, L. N., Noble, M. E. M. and Owen, D. J. (1996). Active and inactive protein kinases: structural basis for regulation. *Cell* **85**, 149–158.

Jones, D. A. and Jones, J. D. G. (1997). The role of leucine-rich repeats in plant defences. *In* "Advances in Botanical Research Incorporating Advances in Plant Pathology", Vol. 24, pp. 89–167. Academic Press, London.

Jones, D. A., Thomas, C. M., Hammond-Kosack, K. E., Balintkurti, P. J. and Jones, J. D. G. (1994). Isolation of the tomato *Cf-9* gene for resistance to *Cladosporium fulvum* by transposon tagging. *Science* **266**, 789–793.

Joosten, M. H. A. J., Vogelsang, R., Cozijnsen, T. J., Verberne, M. C. and De Wit, P. J. G. M. (1997). The biotrophic fungus *Cladosporium fulvum* circumvents *Cf-4*-mediated resistance by producing unstable AVR4 elicitors. *Plant Cell* **9**, 367–379.

Kakeda, K., Jordan, N. D., Conner, A., Ride, J., Franklin-Tong, V. E. and Franklin, F. C. H. (1999). Identification of residues in a hydrophilic loop of the *Papaver rhoeas* S protein that play a crucial role in recognition of incompatible pollen. *Plant Cell* **10**, 1723–1731.

Kauschmann, A., Jessop, A., Koncz, C., Szekeres, M., Willmitzer, L. and Altmann, T. (1996). Genetic evidence for an essential role of brassinosteroids in plant development. *Plant Journal* **9**, 701–713.

Kayes, J. M. and Clark, S. E. (1998). *CLAVATA2*, a regulator of meristem and organ development in *Arabidopsis*. *Development* **125**, 3843–3851.

Klahre, U., Noguchi, T., Fujioka, S., Takatsuto, S., Yokota, T., Nomura, T., Yoshida, S. and Chua, N.-H. (1998). The *Arabidopsis DIMINUTO/DWARF1* gene encodes a protein involved in steroid synthesis. *Plant Cell* **10**, 1677–1680.

Kobe, B. and Deisenhofer, J. (1994). The leucine-rich repeat – a versatile binding motif. *Trends in Biochemical Sciences* **19**, 415–421.

Komeda, Y., Takahashi, T. and Hanzawa, Y. (1998). Development of inflorescences in *Arabidopsis thaliana*. *Journal of Plant Research* **111**, 283–288.

Laux, T., Mayer, K. F., Berger, J. and Jürgens, G. (1996). The *WUSCHEL* gene is required for shoot and floral meristem integrity in *Arabidopsis*. *Development* **122**, 87–96.

Lease, K., Ingham, E. and Walker, J. C. (1998). Challenges in understanding RLK function. *Current Opinion in Plant Biology* **1**, 388–392.

Lee, H.-S., Karunanandaa, B., McCubbin, A., Gilroy, S. and Kao, T.-h. (1996). PRK1, a receptor-like kinase of *Petunia inflata*, is essential for postmeiotic development of pollen. *Plant Journal* **9**, 613–624.

Lee, H.-S., Chung, Y. Y., Das, C., Karunanandaa, B., van Went, J. L., Mariani, C. and Kao, T.-h. (1997). Embryo sac development is affected in *Petunia inflata* plants transformed with an antisense gene encoding the extracellular domain of receptor kinase PRK1. *Sexual Plant Reproduction* **10**, 341–350.

Li, H., Wu, G., Ware, D., Davis, K. R. and Yang, Z. (1998). *Arabidopsis* Rho-related GTPases: differrential gene expression in pollen and polar localization in fission yeast. *Plant Physiology* **118**, 407–417.

Li, J. and Chory, J. (1997). A putative leucine-rich repeat receptor kinase involved in brassinosteroid signal transduction. *Cell* **90**, 929–938.

Li, J., Nagpal, P., Vitart, V., McMorris, T. C. and Chory, J. (1996). A role for brassinosteroids in light-dependent development of *Arabidopsis*. *Science* **272**, 398–401.

Li, J., Biswas, M. G., Chao, A., Russel, D. W. and Chory, J. (1997). Conservation of function between mammalian and plant steroid 5α-reductase. *Proceedings of the National Academy of Sciences of the United States of America* **94**, 3554–3559.

Li, Z. and Wurtzel, E. T. (1998). The *ltk* gene family encodes novel receptor-like kinases with temporal expression in developing maize endosperm. *Plant Molecular Biology* **37**, 749–761.

Lyndon, R. F. (1998). A source of cells: the meristem. *In* "The Shoot Apical Meristem – Its Growth and Development", pp. 16–42. Cambridge University Press, Edinburgh.

Massagué, J. (1996). TGFβ signaling: receptors, transducers and Mad proteins. *Cell* **85**, 947–950.

Matsubayashi, Y. and Sakagami, Y. (1996). Phytosulfokine, sulfted peptides that induce the proliferation of single mesophyll cells of *Asparagus officinalis* L. *Proceedings of the National Academy of Sciences of the United States of America* **93**, 7623–7627.

Matsubayashi, Y., Takagi, L. and Salagami, Y. (1997). Phytosulfokine-a, a sulfated pentapeptide, stimulates the proliferation of rice cells by means of specific high- and low-affinity binding sites. *Proceedings of the National Academy of Sciences of the United States of America* **94**, 13357–13382.

Mayer, K. F. X., Schoof, H., Haecker, A., Lenhard, M., Jürgens, G. and Laux, T. (1998). Role of *WUSCHEL* in regulating stem cell fate in the *Arabidopsis* shoot meristem. *Cell* **95**, 805–815.

Mendoza, C., Soler, A. and Tesarik, J. (1995). Nongenomic steroid action: independent targeting of a plasma membrane calcium channel and a tyrosine kinase. *Biochemical and Biophysical Research Communications* **210**, 518–523.

Mizoguchi, T., Ichimura, K. and Shinozaki, K. (1997). Environmental stress response in plants: the role of mitogen-activated kinases. *Trends in Biotechnology* **15**, 15–19.

Mu, J.-H., Lee, H.-S. and Kao, T.-h. (1994). Characterization of a pollen-expressed receptor-like kinase gene of *Petunia inflata* and the activity of its encoded kinase. *Plant Cell* **6**, 709–721.

Muschietti, J., Eyal, Y. and McCormick, S. (1998). Pollen tube localization implies a role in pollen–pistil interactions for the tomato receptor-like protein kinases LePRK1 and LePRK2. *Plant Cell* **10**, 319–330.

Nasrallah, J. B., Yu, S.-M. and Nasrallah, M. E. (1988). Self-incompatibility genes of *Brassica oleracea*: expression, isolation, and structure. *Proceedings of the National Academy of Sciences of the United States of America* **85**, 5551–5555.

Neuffer, M. G. (1994). Chimeras for genetic analysis. *In* "The Maize Handbook" (M. Freeling and V. Walbot, eds) pp. 258–262. Springer, New York.

Pastuglia, M., Roby, D., Dumas, C. and Cock, J. M. (1997). Rapid induction by wounding and bacterial infection of an *S* gene family receptor-like kinase gene in *Brassica oleracea*. *Plant Cell* **9**, 49–60.

Pearce, G., Strydom, D., Johnson, S. and Ryan, C. (1991). A polypeptide from tomato leaves induces wound-inducible proteinas inhibitor proteins. *Science* **253**, 895–898.

Poethig, R. S. (1987). Clonal analysis of cell lineage patterns in plant development. *American Journal of Botany* **74**, 581–594.

Romeis, T., Piedras, P., Zhang, S., Klessing, D. F., Hirt, H. and Jones, J. D. G. (1999). Rapid Avr9- and Cf-9-dependent activation of MAP kinases in tobacco cell culture and leaves: convergence of resistance gene, elicitor, wound, and salicylate responses. *Plant Cell* **11**, 273–287.

Schaller, G. E. and Bleecker, A. B. (1993). Receptor-like kinase-activity in membranes of *Arabidopsis thaliana*. *FEBS Letters* **333**, 306–310.

Schmidt, E. D. L., Guzzo, F., Toonen, M. A. J. and de Vries, S. C. (1997). A leucine-rich repeat containing receptor-like kinase marks somatic plant cells competent to form embryos. *Development* **124**, 2049–2062.

Schulze-Muth, P., Irmler, S., Schröder, G. and Schröder, J. (1996). Novel type of receptor-like protein kinases from a higher plant (*Catharanthus roseus*) cDNA, gene, intramolecular autophosphorylation and identification of a threonine important for auto- and substrate phosphorylation. *Journal of Biological Chemistry* **271**, 26684–26689.

Shelton, C. A. and Wasserman, S. A. (1993). *pelle* encodes a protein kinase required to establish developmental polarity in the *Drosophila* embryo. *Cell* **72**, 515–525.

Shen, B. and Manley, J. L. (1998). Phosphorylation modulates distinct interactions between the Toll receptor, Pelle kinase and Tube. *Development* **125**, 4719–4728.

Shinozaki, K., Yamaguchi-Shinozaki, K., Mizoguchi, T., Urao, T., Katagiri, T., Uno, Y., Iuchi, S., Seki, M., Ito, T., Hirayama, T. and Mikami, K. (1998). Molecular responses to water stress in *Arabidopsis thaliana*. *Journal of Plant Research* **111**, 345–351.

Smith, C. A., Farrah, T. and Goodwin, R. G. (1994). The TNF receptor superfamily of cellular and viral proteins: activation, costimulation, and death. *Cell* **76**, 959–962.

Song, W.-Y., Wang, G.-L., Chen, L.-L., Kim, H.-S., Pi, L.-Y., Holsten, T., Gardner, J., Wang, B., Zhai, W.-X., Zhu, L.-H., Fauquet, C. and Ronald, P. (1995). A receptor kinase-like protein encoded by the rice disease resistance gene, *Xa21*. *Science* **270**, 1804–1806.

Stein, J. and Nasrallah, J. (1993). A plant receptor-like gene, the *S*-locus receptor kinase of *Brassica oleracea* L, encodes a functional serine threonine kinase. *Plant Physiology* **101**, 1103–1106.

Stein, J. C., Howlett, B., Boyes, D. C., Nasrallah, M. E. and Nasrallah, J. B. (1991). Molecular-cloning of a putative receptor protein-kinase gene encoded at the self-incompatibility locus of *Brassica oleracea*. *Proceedings of the National Academy of Sciences of the United States of America* **88**, 8816–8820.

Steiner, D. F. (1998). The proprotein convertases. *Current Opinion in Chemical Biology* **2**, 31–39.

Stone, J. M., Collinge, M. A., Smith, R. D., Horn, M. A. and Walker, J. C. (1994). Interaction of a protein phosphatase with an *Arabidopsis* serine/threonine receptor kinase. *Science* **266**, 793–795.

Stone, J. M., Trotochaud, A. E., Walker, J. C. and Clark, S. E. (1998). Control of meristem development by CLAVATA1 receptor kinase and kinase-associated protein phosphatase interactions. *Plant Physiology* **117**, 1217–1225.

Szekeres, M., Nemeth, K., Koncz-Kalman, Z., Mathur, J., Kauschmann, A., Altmann, T., Redei, G. P., Nagy, F., Schell, J. and Koncz, C. (1996). Brassinosteroids rescue the deficiency of CYP90, a cytochrome P450, controlling cell elongation and de-etiolation in *Arabidopsis*. *Cell* **85**, 171–182.

Szymkowiak, E. J. and Sussex, I. M. (1992). The internal meristem layer (L3) determines floral meristem size and carpel number in tomato periclinal chimeras. *Plant Cell* **4**, 1089–1100.

Takahashi, T., Mu, J.-H., Gasch, A. and Chua, N.-H. (1998). Identification by PCR of receptor-like protein kinases from *Arabidopsis* flowers. *Plant Molecular Biology* **37**, 587–596.

Tanaka, H., Onouchi, H., Tsukaya, H., Machida, C. and Machida, Y. (1998). The *ABNORMAL LEAF-SHAPE* gene regulates epidermal differentiation in *Arabidopsis*. International Conference of *Arabidopsis* Research, University of Wisconsin-Madison, WI, USA.

Tesarik, J., Moos, J. and Mendoza, C. (1993). Stimulation of protein tyrosine phosphorylation by a progesterone receptor on the cell surface of human sperm. *Endocrinology* **133**, 328–335.

Thomas, C. M., Jones, D. A., Parniske, M., Harrison, K., Balint-Kurti, P. J., Hatzixanthis, K. and Jones, J. D. G. (1997). Characterization of the tomato *Cf-4* gene for resistance to *Cladosporium fulvum* identifies sequences that determine recognitional specificity in Cf-4 and Cf-9. *Plant Cell* **9**, 2209–2224.

Tobias, C. M. and Nasrallah, J. B. (1996). An S-locus-related gene in *Arabidopsis* encodes a functional kinase and produces two classes of transcripts. *Plant Journal* **10**, 523–531.

Tobias, C. M., Howlett, B. and Nasrallah, J. B. (1992). An *Arabidopsis thaliana* gene with sequence similarity to the S-locus receptor kinase of *Brassica oleracea* – sequence and expression. *Plant Physiology* **99**, 284–290.

Torii, K. U., Mitsukawa, N., Oosumi, T., Matsuura, Y., Yokoyama, R., Whittier, R. F. and Komeda, Y. (1996). The *Arabidopsis ERECTA* gene encodes a putative receptor protein kinase with extracellular leucine-rich repeats. *Plant Cell* **8**, 735–746.

Trotochaud, A. E., Hao, T., Wu, G., Yang, Z. and Clark, S. E. (1999). The *Arabidopsis* CLV1 receptor-like kinase requires CLV3 for its assembly into a signaling complex that includes KAPP and a Rho GTPase. *Plant Cell* **11**, 393–405.

Ueno, H., Colbert, H., Escobedo, J. A. and Williams, L. T. (1991). Inhibition of PDGF β receptor signal transduction by coexpression of a truncated receptor. *Science* **252**, 844–847.

Ullrich, A. and Schlessinger, J. (1990). Signal transduction by receptors with tyrosine kinase activity. *Cell* **61**, 203–212.

Van den Ackerveken, G. F. J. M., Van Kan, J. A. L. and De Wit, P. J. G. M. (1992). Molecular analysis of the avirulence gene *Avr9* of the fungal pathogen *Cladosporium fulvum* fully suports the gene-for-gene hypothesis. *Plant Journal* **2**, 359–366.

van der Knaap, E., Sauter, M., Wilford, R. and Kende, H. (1996). Identification of a gibberellin-induced receptor-like kinase in deepwater rice (Accession no. Y07748) (PGR96-100). *Plant Physiology* **112**, 1397.

Vervoort, J., vandenHooven, H. W., Berg, A., Vossen, P., Vogelsang, R., Joosten, M. H. A. and deWit, P. J. G. M. (1997). The race-specific elicitor AVR9 of the tomato pathogen *Cladosporium fulvum*: a cystine knot protein sequence-specific H-1 NMR assignments, secondary structure and global fold of the protein. *FEBS Letters* **404**, 153–158.

Walker, E. A., Ride, J. P., Kurup, S., Franklin-Tong, V. E., Lawrence, M. J. and Franklin, F. C. H. (1996). Molecular analysis of two functional homologues of the S_3 allele of the *Papaver rhoeas* incompatibility gene isolated from different populations. *Plant Molecular Biology* **30**, 983–994.

Walker, J. C. (1993). Receptor-like protein kinase genes of *Arabidopsis thaliana*. *Plant Journal* **3**, 451–456.

Walker, J. C. (1994). Structure and function of the receptor-like kinases of higher plants. *Plant Molecular Biology* **26**, 1599–1609.

Walker, J. C. and Zhang, R. (1990). Relationship of a putative receptor protein kinase from maize to the S-locus glycoproteins of *Brassica*. *Nature* **345**, 743–746.

Wang, G.-L., Ruan, D.-L., Song, W.-Y., Sideris, S., Chen, L., Pi, L.-Y., Zhang, S., Zhang, Z., Fauquet, C., Gaut, B. S., Whalen, M. C. and Ronald, P. C. (1998). *Xa21D* encodes a receptor-like molecule with a leucine-rich repeat domain that determines race-specific recognition and is subjected to adaptive evolution. *Plant Cell* **10**, 765–779.

Wang, X., Zafian, P., Choudhary, M. and Lawton, M. (1996). The PR5K receptor protein kinase from *Arabidopsis thaliana* is structurally related to a family of plant defense protein. *Proceedings of the National Academy of Sciences of the United States of America* **93**, 2598–2602.

Williams, R. W., Wilson, J. M. and Meyerowitz, E. M. (1997). A possible role for kinase-associated protein phosphatase in the *Arabidopsis* CLAVATA1 signaling pathway. *Proceedings of the National Academy of Sciences of the United States of America* **94**, 10467–10472.

Wilson, H. E. and White, A. (1998). Prohormones: their clinical relevance. *Trends in Endocrinology and Metabolism* **9**, 396–402.

Yeh, K. C. and Lagarias, J. C. (1998). Eukaryotic phytochromes: light-regulated serine/threonine protein kinases with histidine kinase ancestry. *Proceedings of the National Academy of Sciences of the United States of America* **95**, 13976–13981.

Yokoyama, R., Takahashi, T., Kato, A., Torii, K. U. and Komeda, Y. (1998). The *Arabidopsis ERECTA* gene is expressed in the shoot apical meristem and organ primordia. *Plant Journal* **15**, 301–310.

Zhao, Y., Feng, X.-H., Watson, J. C., Bottino, P. J. and Kung, S.-D. (1994). Molecular-cloning and biochemical-characterization of a receptor-like serine/threonine kinase from rice. *Plant Molecular Biology* **26**, 791–803.

Zimmerman, J. L. (1993). Somatic embryogenesis: a model for early development in higher plants. *Plant Cell* **5**, 1411–1423.

Zurek, D. M. and Clouse, S. D. (1994). Molecular cloning and characterization of a brassinosteroid-regulated gene from elongating soybean (*Glycine max* L.) epicotyls. *Plant Physiology* **104**, 161–170.

A Receptor Kinase and the Self-Incompatibility Response in *Brassica*

J. M. COCK

Reproduction et Dévelopment des Plantes, UMR 5667 CNRS-INRA-ENSL, Ecole Normale Supérieure de Lyon, 46 allée d'Italie, 69364 Lyon Cedex 07, France

I. INTRODUCTION

One of the most remarkable features of flowering plants is the wide range of reproductive strategies that they employ to promote outbreeding. In contrast to the situation in animals, dioecy, where a species is comprised of individuals

Advances in Botanical Research Vol. 32
incorporating Advances in Plant Pathology
ISBN 0-12-005932-0

of two different sexes, is relatively rare (approximately 5% of angiosperm species). The majority of flowering plant species is hermaphrodite and possesses flowers in which the male and female organs are in close proximity. Despite this proximity, however, outcrossing is favoured in a large number of these species as a result of the action of self-incompatibility (SI) systems that permit the pistil to reject self-pollen.

SI systems can be classified as either heteromorphic, where there are marked morphological differences between plants carrying different *S* alleles, or homomorphic, where there is no obvious morphological difference. The number of genotypic variants in heteromorphic SI systems is limited, probably because of constraints on the number of functionally interacting, morphological forms that can be generated. In contrast, homomorphic systems often exhibit large numbers of alleles and this has theoretical advantages with regard to the success of cross-pollinations and the economical use of pollen.

Homomorphic SI systems can be further classified into gametophytic or sporophytic systems depending on the genetic behaviour of the pollen. In gametophytic systems, the phenotype of a pollen grain is determined by its own haploid (gametophytic) genome and, therefore, each pollen grain carries determinants corresponding to one *S* allele. Currently, the best characterized gametophytic SI system is that of the Solanaceae where ribonucleases mediate self-pollen rejection.

In sporophytic SI systems, the pollen phenotype is determined by the male parent's diploid (sporophytic) genome so that pollen may carry SI determinants corresponding to two different alleles. Homomorphic sporophytic systems have been described only in seven angiosperm families and are, therefore, less widespread than gametophytic systems. This review will concentrate on one particular sporophytic SI system in the genus *Brassica*, and on the role of a putative receptor kinase, the *S* locus receptor kinase, in the recognition of self-pollen. Other aspects of the *Brassica* SI system and SI systems in general have been extensively reviewed (de Nettancourt, 1971; Nasrallah *et al.*, 1994b; McCubbin and Kao, 1996; Nasrallah, 1997a; McCormick, 1998).

II. SELF-INCOMPATIBILITY IN *BRASSICA*

A. A CELL–CELL INTERACTION AT THE STIGMA SURFACE

Following another dehiscence, pollen grains must migrate to the female part of the flower, the pistil, in order to complete the life cycle. In *Brassica*, this transfer is usually mediated by insect vectors, in particular the honeybee (Williams and Leung, 1983). Arrival of the pollen grain on the surface of the stigma marks the start of a series of events that, if successful, lead to a double fertilization involving the male gametes carried by the pollen grain and the two female gametes held in the embryo sac within the ovules (Fig. 1). The first of these events is the adhesion of the pollen grain to the surface of the stigmatic papillar cells. The stigma surface in *Brassica* is of the dry type (Dickinson, 1995). Initially the pollen is attracted electrostatically to the stigma surface but, on making contact, adhesion is reinforced by the formation of a meniscus between the pollen grain and the papillar cell. The meniscus consists, in part, of pollen coat material that flows out onto the stigma surface. In the meniscus, the pollen coat material changes its physicochemical properties (is 'converted'; Elleman and Dickinson, 1990).

After adhering to the stigma, a self-compatible pollen grain will hydrate, taking up water from the adjacent papillar cell, and germinate. The germinating pollen grain emits a pollen tube that penetrates the cell wall of the papillar cell at the site of initial contact between the papilla and the pollen grain (Elleman and Dickinson, 1990). The pollen tube does not normally penetrate the cell wall completely, but migrates down the length of the papillar cell between two distinct layers of the pecto-cellulose cell wall (Elleman *et al.*, 1988). On arriving at the base of the papillar cell, the pollen tube re-emerges from the papillar cell wall and grows in the intercellular spaces between

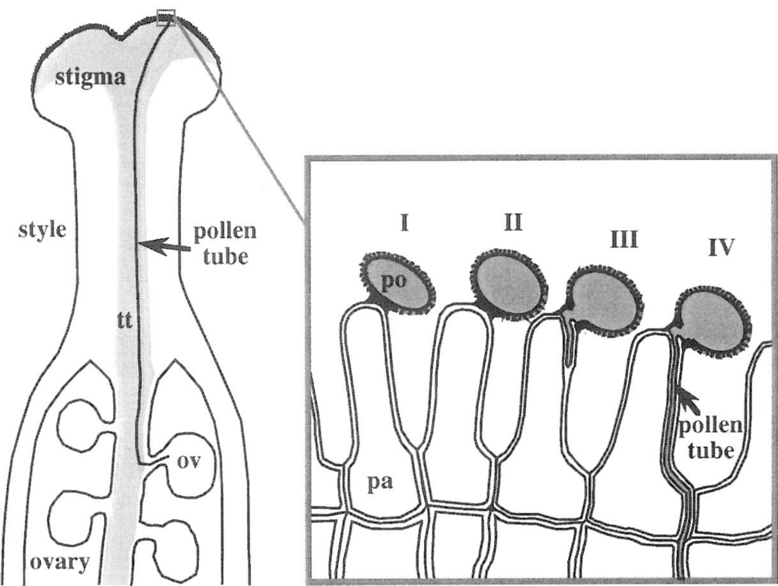

Fig. 1. Schematic representation of pollination in *Brassica*. A *Brassica* pistil is represented on the left. The detail to the right depicts the events that occur at the stigma surface during pollination. I, pollen adhesion; II, pollen hydration; III, pollen germination and penetration of the papillar cell wall; IV, growth of the pollen tube towards an ovule in the ovary; tt, transmitting tissue; ov, ovule; po, pollen grain; pa, papillar cell on the stigma surface.

parenchyma cells of the transmitting tissue. The pollen tube continues to grow in this manner carrying the gametes down the style to an ovule in the ovaries.

B. GENETIC CONTROL OF SELF-POLLEN REJECTION

SI in *Brassica* is controlled by a single, highly polymorphic genetic locus, the *S* locus (for self-incompatibility locus). Current estimates suggest that there are at least 80 different allelic forms of the *S* locus in *Brassica oleracea* (V. Ruffio, personal communication). *S*-locus alleles are referred to as haplotypes because of the complex nature of this locus (see Section III below).

Rejection of self-pollen in self-incompatible *Brassica* occurs at the stigma surface. The exact stage at which fertilization is aborted depends on the genotype of the plant and, in particular, on the *S* haplotype involved (Sarker *et al.*, 1988; Dickinson, 1995; Fig. 1). In *Brassica* plants carrying early-acting *S* haplotypes, self-pollen fails to hydrate completely (arrest during stage II in Fig. 1). In plants carrying later-acting haplotypes, however, the pollen grains may hydrate and even germinate but the majority of the pollen tubes will fail to penetrate the stigma surface (arrest during stage III in Fig. 1). Using a novel assay based on the flotation force of Archimedes, Luu *et al.* (1997) have shown that the initial step of pollen adhesion is not influenced by the self-incompatibility response.

Interaction between *S* haplotypes is a common feature of sporophytic SI systems and such interactions are observed in *Brassica*, on both the male and the female side. *Brassica S* haplotypes can be ordered in a non-linear dominance hierarchy (Thompson and Taylor, 1966). The dominance hierarchy in stigmas is different to that on the male side, indicating that dominance operates by different mechanisms in the two organs. Two major groups of haplotypes can be distinguished in the dominance hierarchy: class I haplotypes tend to be dominant and class II haplotypes to be recessive. There is also an approximate correlation between these two classes and the SI phenotype; class I haplotypes tend to be early acting and to confer a strong SI response whereas self-pollen is arrested later in the presence of class II haplotypes which confer a weaker SI response.

In addition to variation associated with genetic differences at the *S* locus and interaction between *S* alleles, the self-incompatibility response is influenced by a number of unlinked suppressor and modifier genes (Nasrallah, 1989; Nasrallah *et al.*, 1992; Eukere *et al.*, 1997). Mutations at suppressor gene loci result in an alteration in the expression level of *S*-locus genes, indicating that suppressor genes probably encode epistatic regulators of *S*-gene expression. Modifier genes, in contrast, modify the strength of the SI response without detectable effects on the expression of the known *S*-locus genes. Therefore, these genes either have more subtle effects on *S*-gene action or are involved in events upstream or downstream of the recognition step of the SI response. The importance of loci other than the *S* locus in the control of SI has been underlined by a recent study that has demonstrated a significant variability in the strength of the SI response between different *Brassica* lines homozygous for the same *S* allele (Ruffio-Châble, 1997).

In addition to the genetic effects of variation at the *S* locus and of suppressor and modifier genes, the strength of the SI response can be influenced by environmental conditions such as high temperature or high levels of humidity (Hinata *et al.*, 1994).

C. ADAPTING SI FOR USE AS A BREEDING TOOL

The previous section described a number of genetic and environmental factors that can influence the strength of the SI response. Variability in the strength of the SI response can cause problems under field conditions when SI is used as a breeding tool to produce hybrid seed. The high uniformity of crops grown from F_1 hybrid seed allows efficient and low-cost farming, provided that the level of contamination with self-seed is not too great. Variability in the strength of the SI response can, however, lead to significant levels of self-seed contamination and this has been one of the main reasons for the tendency, in recent years, to move away from the use of SI in *Brassica* breeding programmes and towards the use of male sterility. Nonetheless, there are also problems associated with the use of male sterility and self-incompatibility remains an important tool for breeders. Furthermore, the fact that the genotype of the plant has a large influence on the strength of the SI response suggests that there is potential for improving this system. In this respect it is important to consider SI in a evolutionary context. Although out-crossing will usually confer a selective advantage on a population, suitable sexual partners for an individual plant may not always be available in the immediate environment. One of the advantages of possessing hermaphroditic flowers in this case is that isolated plants can produce self-seed, provided that the system which promotes out-crossing is not 100% efficient. Therefore, it is likely that, in some instances, a weak SI system that promoted out-crossing but allowed a low level of auto-fertilization would have been selected. A better understanding of the mechanism of SI in *Brassica*, from both a physiological and a molecular point of view, should facilitate the adaptation of this system to meet breeders' requirements.

III. THE MOLECULAR MECHANISM OF SI IN *BRASSICA*

A. CHEMICAL INHIBITORS AND SOME CLUES ABOUT THE MECHANISM OF SI

Several groups have used pharmacological approaches to dissect the SI system in *Brassica*. Sarker *et al.* (1988) showed that self-pollen grains were rapidly hydrated on stigmas that had been treated with cycloheximide. This indicates that maintenance of the SI system requires active protein synthesis. Application of tunicamycin, an inhibitor of protein glycosylation, also resulted in a breakdown of the SI response suggesting the involvement of a glycoprotein (Sarker *et al.*, 1988).

Experiments using the protein phosphatase inhibitors okadaic acid and microcystin have given conflicting results. Scutt *et al.* (1993) reported a breakdown of the SI response after treatment with between 1 and 5 μM okadaic acid depending on the S haplotype. However, Rundle *et al.* (1993) observed no effect on SI in mature flowers but reported an inhibition of pollen tube growth following cross-pollination when flowers were treated with as low as 1 μM okadaic acid or 2 μM microcystin. A breakdown of SI was only observed in the latter study when immature stigmas (just before anthesis) were treated with okadaic acid or microcystin.

Callose is often deposited in the papillar cells during the SI response and there has been some debate as to whether this is essential for self-pollen rejection (Dickinson, 1995). Sulaman *et al.* (1997) have recently shown that callose is not required for the SI response by expressing a callose degrading enzyme, β-1,3-glucanase, in the stigma of transgenic *B. napus*.

B. GENES AT THE *S* LOCUS: *SLG* AND *SRK*

Investigations into the molecular mechanism of the SI response in *Brassica* have concentrated on the characterization of genes located at the *S* locus. The *S* locus has been shown to have a complex structure, extending over several kilobases and including several different genes, two of which, *SLG* (Nasrallah *et al.*, 1987) and *SRK* (Stein *et al.*, 1991), have been implicated in the SI response. *S*-locus glycoprotein (SLG) is found almost exclusively in stigmas, although low levels can also be detected in the transmitting tissue of the style and ovary (Kleman-Mariac *et al.*, 1995). *S*-locus receptor-like kinase (SRK) is also expressed almost exclusively in stigmas. The only other tissue in which *SLG* and *SRK* transcripts have been detected (at a very low abundance) is anthers (Sato *et al.*, 1991; Stein *et al.*, 1991; Delorme *et al.*, 1995), but protein products have not been detected in this organ and recent evidence indicates that the *SRK* transcripts in anthers correspond to the non-coding strand of the gene (Cock *et al.*, 1997). Both proteins are absent in immature, self-compatible stigmas and accumulate prior to anthesis, at the time when the stigma acquires the ability to reject self-pollen (Nasrallah *et al.*, 1985; Stein *et al.*, 1996). Comparison of *SLG* and *SRK* alleles from different *S* haplotypes has shown that both genes are highly polymorphic. Moreover, sequence similarities indicate that alleles of both genes can be divided into two classes which correlate with the class I and class II haplotypes defined on the basis of SI phenotype.

SLG encodes the *S*-locus glycoprotein, which is found in the cell wall of the stigma papillar cells (Kandasamy *et al.*, 1989), whereas *SRK* encodes the *S*-

locus receptor kinase, an integral plasma-membrane glycoprotein (Delorme *et al.*, 1995; Stein *et al.*, 1996; Cabrillac *et al.*, 1999). SRK is predicted to have a similar structure to animal receptor kinases. Based on its deduced amino acid sequence, the SRK protein consists of an extracellular domain that closely resembles SLG, a single membrane-spanning domain and a cytoplasmic kinase domain. The kinase domain of SRK has been expressed in *Escherichia coli* and has been shown to possess a serine/threonine protein kinase activity (Goring and Rothstein, 1992). The resemblance between SRK and animal receptor kinases suggests a model in which SRK recognizes a pollen-borne ligand and transmits a signal to the interior of the papillar cell, initiating the SI response.

C. STIGMAS EXPRESS A COMPLEX MIXTURE OF *SRK*- AND *SLG*-ENCODED PROTEINS

Both *SRK* and *SLG* are complex genes that can encode multiple transcripts. For example, in the class I S_3 haplotype, stigmas contain at least seven different *SRK* transcripts, including transcripts from both strands of the gene (Delorme *et al.*, 1995; Cock *et al.*, 1997). These transcripts have been shown to code for at least two different proteins, the integral SRK protein, and a soluble, truncated protein, eSRK (for extracellular SRK), that corresponds to the extracellular domain of SRK and which, therefore, resembles SLG (Giranton *et al.*, 1995; Fig. 2). The first intron of SRK_3 possesses a stop codon just after the 5′ GT splice site, a feature common to all *SRK* alleles sequenced so far. eSRK is encoded by alternative transcripts that terminate within the first intron (Giranton *et al.*, 1995).

In the class II S_{15} haplotype the situation is even more complicated (Fig. 2). Two different *SLG* genes, *SLGA* and *SLGB*, have recently been identified in this haplotype (Cabrillac *et al.*, 1999). In both *SLGA* and *SLGB*, the transcribed regions are divided into two exons by a single intron. The second exon of *SLGA* encodes a putative membrane-spanning domain so that transcripts which include, or not, the second exon encode either membrane-anchored (mSLGA for membrane-anchored SLGA) or soluble forms of SLGA, respectively. The coding region of *SLGB* does not include a predicted membrane-spanning domain and this gene is therefore predicted to encode only soluble SLGB proteins. Immunological analysis of stigma proteins has identified two membrane-associated proteins of 105 and 65 kDa that probably correspond to SRK and mSLGA, respectively. Four soluble proteins could be separated by IEF and matrix-assisted laser desorption ionization time-of-flight mass spectroscopy (MALDI-TOF-MS) was used to show that three of these proteins were encoded by *SLGB* and the fourth by *SLGA*. No eSRK protein was detected in stigmas of the S_{15} haplotype.

The range of proteins encoded by the S_{15} haplotype is particularly complex and the situation is somewhat simpler in other class II *S* haplotypes. DNA blot analysis indicates that the S_2 haplotype includes alleles of *SRK* and *SLGA* but

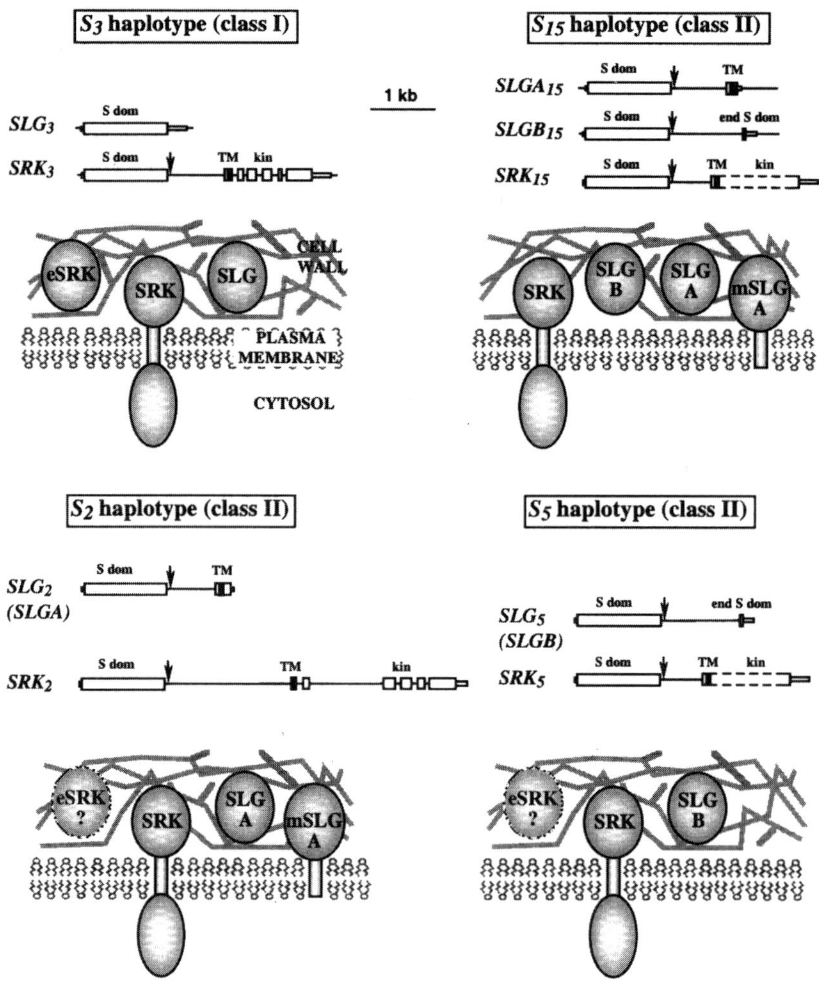

Fig. 2. Variation in the complement of *SLG* and *SRK* genes and their corresponding gene products in stigmas of plants carrying four different *S* haplotypes. One class I (strong, dominant) and three class II (weak, recessive) haplotypes are shown. Two different *SLG* genes, *SLGA* and *SLGB*, have been identified in class II haplotypes. The S_2 haplotype includes an allele of *SLGA* but not *SLGB*, whereas the S_5 haplotype includes an allele of *SLGB* but not *SLGA*. SRK and a soluble, truncated form of SRK (eSRK) are encoded by the *SRK* gene, *SLGA* encodes both secreted and membrane-anchored proteins (SLGA and mSLGA, respectively), and *SLGB* encodes only secreted SLG proteins (SLGB). In the gene diagrams, transcribed regions and coding regions are indicated by narrow boxes and thick boxes, respectively. Membrane-spanning domains (TM) are indicated by black boxes. Arrows indicate stop codons near the 5′ ends of introns that potentially allow the production of truncated proteins from alternatively spliced transcripts that retain all or part of the intron sequences. S dom, S domain; kin, kinase domain.

not of *SLGB*. Conversely, *SRK* and *SLGB* alleles were detected in the S_5 haplotype but no *SLGA* allele (Cabrillac *et al.*, 1999). These differences are reflected in the spectra of proteins encoded by these two haplotypes (Fig. 2).

The absence of *SLGA* from the S_5 haplotype raises a question about the functional role of the *SLGA*-encoded protein mSLG in the SI response. Tantikanjana *et al.* (1993) first reported the identification of a membrane-anchored form of SLG in the S_2 haplotype (mSLG$_2$). The fact that this protein was encoded by a class II haplotype but that no corresponding protein was encoded by the class I S_6 haplotype initially indicated that its presence or absence may explain the phenotype differences between haplotypes of the two classes. However, the absence of *SLGA*, and consequently of an mSLGA protein (Fig. 2), from plants carrying the class II S_5 haplotype suggests that this is not the case.

eSRK was detected in stigmas expressing the class I S_3 haplotype but not in stigmas carrying the class II S_{15} haplotype (Giranton *et al.*, 1995; Cabrillac *et al.*, 1999). It will be interesting to analyse other class I and II haplotypes to determine whether the presence of this protein correlates with the strong, dominant SI phenotype conferred by class I alleles.

D. *SLG*, *SRK* AND RECOGNITION OF SELF-POLLEN BY THE PISTIL

The *S* locus location and polymorphic nature of *SLG* and *SRK*, taken together with the expression patterns and subcellular localizations of their encoded proteins, suggest that these genes are involved in the recognition of self-pollen. However, despite a considerable effort in recent years, it has been difficult to obtain direct proof of their role in SI.

Both transgenic approaches and genetic analysis of lines carrying suppressor genes have been used in an effort to determine the function of *SLG* (Toriyama *et al.*, 1991; Nasrallah *et al.*, 1992; Nishio *et al.*, 1992; Shiba *et al.*, 1995; Conner *et al.*, 1997). A correlation has been reported between loss of *SLG* expression and the acquisition of a self-compatible phenotype using both approaches but the interpretation of these results is often complicated by the fact that *SRK* expression may also have been affected. Simultaneous effects on the two genes are highly likely because of their shared sequence similarity and co-suppression of *SLG* and *SRK* has been described on more than one occasion (Conner *et al.*, 1997; Stahl *et al.*, 1998). Currently, the most convincing evidence of a role for *SLG* in SI is that SLG protein was not detected in self-compatible plants carrying a mutant allele of the suppressor gene *SCF1* (Nasrallah *et al.*, 1992). *SRK* transcripts were present at normal levels in these plants.

In a study that argued against a role for SLG in the SI response, Gaude *et al.* (1995) showed that the level of expression of *SLG* is not correlated with the strength of the SI response. SLG was abundant in stigmas of a self-compatible line carrying the S_{15} haplotype but weakly expressed in self-incompatible S_2

plants. This study indicates that, if SLG does play a role in the SI response, it must be capable of functioning at a low abundance.

Further evidence against SLG being involved in self-pollen recognition has come from the analysis of class II *S* haplotypes. The absence of *SLGA* and *SLGB* from the S_5 and S_2 haplotypes, respectively (Fig. 2), indicates that either these genes are not required for the SI response, or they carry out redundant functions. Moreover, functional redundancy would seem to be unlikely because $SLGA_{15}$ and $SLGB_{15}$ are both more similar to *SLG* alleles from other haplotypes than they are to each other (Cabrillac *et al.*, 1999).

Several studies have investigated the role of SRK in the recognition of self-pollen. Non-functional *SRK* alleles have been identified in two self-compatible *Brassica* lines but in neither case was the self-compatible phenotype shown to be a direct result of the mutation and it is possible that other genes are affected in these lines (Goring *et al.*, 1993; Nasrallah *et al.*, 1994a). Efforts to express novel *SRK* alleles or dominant negative forms of SRK in transgenic plants have run into problems due to insufficient expression and co-suppression (Stein *et al.*, 1991; Conner *et al.*, 1997). The most convincing evidence that SRK is involved in the SI response has been obtained by Stahl *et al.* (1998), who observed haplotype-specific breakdown of SI in a *B. napus* plant transformed with a kinase-defective *SRK* gene. However, the breakdown of SI in these experiments was partial and the observations were made on a single transgenic line. It will be important to demonstrate a similar phenotype in additional independent transformants.

E. INVESTIGATION OF SRK FUNCTION AT THE MOLECULAR LEVEL: EVIDENCE FOR A RECEPTOR KINASE COMPLEX *IN VIVO*

SRK has recently been expressed in insect cells using a baculovirus vector (J.-L. Giranton, J. M. Cock, C. Dumas and T. Gaude, manuscript in preparation). Under these conditions SRK exhibited a constitutive kinase activity and this activity was used to demonstrate 'trans-auto-phosphorylation' between SRK molecules in a membrane environment. This is an important observation because it supports the analogy between SRK and animal receptors that has hitherto been based mainly on their structural resemblance. The role of oligomerization and trans-phosphorylation in animal receptor kinase function is well established (Heldin, 1995).

The relevance of the above observations to SRK function *in planta* has been investigated in cross-linking and velocity sedimentation experiments. The results of these experiments indicate that SRK is part of a complex *in vivo* (Fig. 3). The sizes of the cross-linked complexes suggest both interaction between SRK proteins (dimerization) and interaction of SRK with other proteins (J.-L. Giranton, J. M. Cock, C. Dumas and T. Gaude, manuscript in preparation). Only a small proportion of the SLG and eSRK proteins sediments with the SRK complex, the rest behaving as monomers. This suggests that only a fraction of the pool of these proteins is associated with the complex *in vivo* (although an alternative explanation could be that SLG and eSRK are weakly associated with the complex and dissociate during extraction).

The biochemical evidence for SRK dimerization is supported by genetic data. The dominant negative effect of a kinase-inactive form of SRK in transgenic *B. napus* suggests that the mutant form of SRK physically interacts with the corresponding wild-type SRK (Stahl *et al*., 1998). An analogous situation has been described for the *Arabidopsis CLAVATA1 (CLV1)* gene that is also predicted to encode a membrane-localized receptor kinase. The

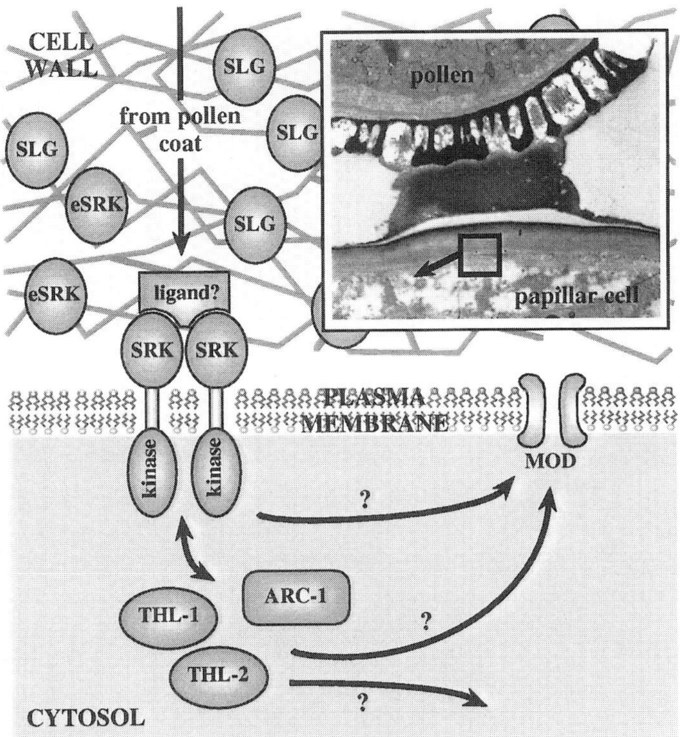

Fig. 3. A model for signalling during the SI response in *Brassica*. The schema represents a view of the events that occur at the surface of a papillar cell following the arrival of a self-pollen grain (see inset). SRK is thought to function as a receptor and to bind a putative ligand derived from the pollen coat of the self-pollen grain. The binding complex is shown as a dimer of SRK molecules but probably includes other proteins including a subset of the SLG and eSRK proteins that are secreted into the cell wall. Very little is known about how the signal is transduced downstream of SRK but it is likely that this involves protein phosphorylation. Three proteins, ARC1, THL-1 and THL-2, have been shown to interact with the kinase domain of SRK and may be involved in signal transduction. One possible downstream effector of the SI response is the aquaporin MOD (see text for details).

clv1–1 mutation is semidominant, suggesting that CLV1 also functions as a multimer (Clark *et al.*, 1995). SRK and CLAVATA1 are both members of the receptor-like kinase (RLK) superfamily in plants (see Section IV.E). Dominance relationships between *S* haplotypes indicate that interactions occur between *S*-locus determinants in both the pollen and the pistil. An interesting possibility is that dominance in the pistil might be mediated by interaction between SRK proteins encoded by different haplotypes within a receptor complex.

F. POLYMORPHISM AT THE *S* LOCUS AND HAPLOTYPE SPECIFICITY

Both *SLG* and *SRK* are highly polymorphic. S domains of alleles of the same gene can share as little as 57% similarity at the amino acid level. The S domain of *SRK* exhibits a significantly higher level of polymorphism than the kinase domain. Taken together with its extracellular location, this suggests that the S domain mediates pollen recognition and that *S*-allele specificity is determined by structural motifs within this domain.

When *SLG* and *SRK* were first identified, it was hoped that sequence analysis and comparison of alleles would allow identification of the residues that determine *S*-allele specificity, but the identification of such residues is complicated by the very high level of polymorphism. Polymorphism is not distributed uniformly, however, and the S domains of both *SLG* and *SRK* can be divided up into four conserved and variable regions based on these differences. Short stretches of extremely polymorphic sequences occur within the variable domains and these have been referred to as hypervariable domains (Nasrallah *et al.*, 1987; Kusaba *et al.*, 1997).

The highly polymorphic and generally hydrophilic nature of the hypervariable domains initially suggested that they could be involved in determining *S*-allele specificity. However, if we assume that *SLG* is involved in pollen recognition, the existence of pairs of *SLG* alleles with identical hypervariable domains argues against a role for these regions in determining specificity (Kusaba *et al.*, 1997; Cabrillac *et al.*, 1999). It has even been suggested that the hypervariable domains may represent regions of minor functional importance that are subjected to fewer selective constraints and are therefore free to vary to a greater degree (Nasrallah, 1997b). As mentioned above, however, there are doubts as to the role of *SLG* in the SI response and a more profitable approach to this problem may be to compare alleles of *SRK*. It would be particularly interesting to analyse independent isolates of alleles with the same *S*-haplotype specificity. An analysis of allelic variation between such isolates could allow phenotypically silent amino acid polymorphisms to be distinguished from those that affect specificity.

Sequences involved in ligand binding at recognition loci often exhibit a higher rate of non-synonymous (Ka) than synonymous (Ks) mutation and this has been taken to indicate that diversifying selection has been acting on these loci (e.g. Hughes and Nei, 1988). Based on this observation, analysis of the Ka/Ks ratio has been used to substantiate the existence of potential ligand-binding sites in other putative recognition molecules including the *S*-RNases that mediate gametophytic SI in the Solanaceae (Clark and Kao, 1991; Parniske *et al.*, 1997). This method has been less effective in *Brassica* where comparison of *SLG* and *SRK* alleles has failed to identify regions where Ka exceeds Ks, probably indicating that these genes are not subject to strong diversifying selection (Hinata *et al.*, 1995; Charlesworth and Awadalla, 1998).

G. THE MALE COMPONENT OF THE SELF-INCOMPATIBILITY RESPONSE

Self-incompatibility in *Brassica* is sporophytic. This means that the SI determinants carried by the pollen correspond not just to the *S* haplotype present in its own haploid (gametophytic) genome but to both of the *S* haplotypes present in the diploid (sporophytic) genome of the male parent plant. It is not known exactly how the pollen grain acquires the SI determinants but three models have been proposed. In the first model, the 'male' *S*-locus gene is expressed in diploid, progenitor cells of the pollen (the pollen mother cells for example) before meiosis and its gene product is inherited by the developing microspores (Pandey, 1959).

The second model proposes that the male determinants are encoded by the haploid genome of the developing pollen grains in a non-autonomous manner so that determinants are shared between microspores in the same locule (Doughty *et al.*, 1998).

The third model proposes that the 'male' *S*-locus gene is expressed in other diploid cells within the anther, most probably the cells of the tapetum that line the locule in which the microspores develop (Heslop-Harrison, 1975). The product of the 'male' *S*-locus gene, the SI determinant, would then be transferred onto the developing microspores. This last model is supported by the fact that, during pollen development, the cells of the tapetum undergo a sort of programmed cell death. Histological observation of anthers has shown that material from the degenerating tapetal cells is deposited on the surface of the developing pollen grains (reviewed in Piffanelli *et al.*, 1998). This material fills the spaces between the baculae of the outer (exine) layer and constitutes the pollen coat.

The pollen coat can be purified from pollen grains either by extracting with cyclohexane or by simply allowing the coat to flow onto a surface such as a glass slide. Stephenson *et al.* (1997) have recently developed a bioassay in which isolated pollen coat is applied to the stigma surface before pollination. A lower percentage of cross-pollen hydrated when they were placed on stigmas that had been pretreated with an application of 'self' pollen coat (i.e. pollen coat extracted from pollen of the same plant as the stigma) than when the cross-pollen was applied to untreated control stigmas. Conversely, when 'cross' pollen coat was applied to stigmas before application of self-pollen there was a significant increase in the percentage of pollen grains that hydrated. These experiments provide evidence that the male SI determinant is localized in the pollen coat.

The male determinant factor in pollen coat preparations has not yet been purified, but fractionation experiments indicate that it is a water-soluble molecule with a molecular mass of less than 10 kDa (Stephenson *et al.*, 1997). Interestingly, the fraction that retains activity in the bioassay also contains PCP-1A (previously PCP7), a highly basic, 7-kDa protein related to plant defensins that has been shown to bind to SLG *in vitro* (Doughty *et al.*, 1993). PCP-A1 is encoded by a gene that is unlinked to the *S* locus and it is, therefore, unlikely to represent the male component of the SI response. However, as *PCP-A1* is one member of a large gene family which may include as many as 40 genes in *Brassica* (Stanchev *et al.*, 1996) it is possible that the male component is encoded by an *S*-locus-linked member of this family.

An alternative approach to identifying the male component is to look for genes that are both linked to the *S* locus and expressed in anthers. Several groups have used long-range restriction mapping and systematic sequence analysis to investigate the structure of the *S* locus and to search for novel *S*-locus genes (Yu *et al.*, 1996; Boyes *et al.*, 1997; Suzuki *et al.*, 1997). Comparison of different *S* haplotypes has shown that the *S* locus is in a region of the genome that has undergone extensive rearrangements and sequence divergence (Boyes *et al.*, 1997). Comparison of *S* haplotypes from *B. oleracea* and *B. rapa* indicates that a significant proportion of the rearrangements at the *S* locus is ancient and preceded speciation in the *Brassica* genus. The polymorphic region surrounding the *S*-locus genes may extend up to 570 kbp (Boyes and Nasrallah, 1993) and several genes have been identified in this region. Although there are large-scale differences between different *S* haplotypes, many of the genes identified do not appear to be polymorphic and are therefore unlikely to encode the male component of the SI response. It is possible, however, that these genes, which include *298*, *299*, *BcRL1*, *BcRK1* and *BcSL1* from *B. rapa* (Boyes *et al.*, 1997; Suzuki *et al.*, 1997), and SLL_1 and SLL_2 from *B. napus* (for *S*-locus-linked genes; Yu *et al.*, 1996), are involved in some other aspect of the SI response. SLL_1 and SLL_2 are particularly interesting in this respect because they are both expressed specifically in the sexual organs of the plant (SLL_1 in anthers and SLL_2 in anthers and stigmas), and expression of SLL_1 is correlated with the presence of a functional SI response (Yu *et al.*, 1996).

Another gene that was identified by chromosome walking at the *S* locus is the *S* locus anther gene, *SLA* (Boyes and Nasrallah, 1995). When this gene was initially characterized, several pieces of evidence suggested that it might encode the male component of the SI response:

1. Its *S*-locus location: *SLA* is situated 500 bp downstream of *SLG* in the S_2 haplotype.

2. Its expression pattern: *SLA* encodes two antisense transcripts that accumulate specifically in anthers at different stages of development.
3. Its putative protein products: one of the *SLA* transcripts includes two short open reading frames that could encode short peptides.
4. Its apparently polymorphic nature: an *SLA* probe only hybridized to DNA from a small number of haplotypes.
5. The presence in a self-compatible *B. napus* lines of an *SLA* allele interrupted by a large insertion.

However, a more recent study carried out in our laboratory indicates that *SLA* exhibits a very low level of polymorphism and is only present in a limited number of *S* haplotypes (Pastuglia *et al.*, 1997b). Moreover, a number of cauliflower lines was identified that carried an allele of *SLA* which was interrupted by a retrotransposon (similar to the allele found in *B. napus*) but which exhibited a self-incompatible phenotype. *SLA* does not, therefore, appear to be required for the SI response and it is hence unlikely that it encodes the male component. This study underlines the importance of directly demonstrating the functional roles of genes at the *S* locus. However, this remains technically difficult, in general, because of problems associated with transformation and mutagenesis in *Brassica*.

H. PUTATIVE DOWNSTREAM TARGETS OF SIGNAL TRANSDUCTION VIA SRK AND THEIR ROLE IN POLLEN REJECTION

Based on genetic analysis, both the male and female components involved in self-pollen recognition are predicted to be encoded by the *S* locus. However, other components of the SI response, for example molecules that act downstream of the pollen recognition system and mediate the inhibition of self-pollen, may be encoded by genes outside the *S* locus. Two approaches have been used to identify such downstream components of the SI system: (1) identification of proteins that interact directly with SRK, and (2) characterization of so-called modifier loci that are not linked to the *S* locus but which modify the SI response.

Experiments using the yeast two-hybrid system and/or *in vitro* binding assays have shown that the kinase domain of SRK can interact with a number of different proteins (Fig. 3). These include two thioredoxin-h-like proteins (THL-1 and THL-2) and arm repeat containing protein 1 (ARC1) from *Brassica* (Bower *et al.*, 1996; Gu *et al.*, 1998) and an *Arabidopsis* protein, the kinase-associated protein phosphatase (KAPP; Braun *et al.*, 1997).

ARC1 is perhaps the most interesting of these proteins because (1) it is expressed specifically in stigmas, (2) it associates only with the phosphorylated form of the SRK kinase domain, and (3) it is specifically phosphorylated by the SRK kinase domain *in vitro*. ARC1 contains arm repeats similar to those found in the *Drosophila* armadillo protein, β-catenin and a number of other proteins.

In some of these proteins the arm repeat motif has been shown to be involved in protein–protein interactions (Rubinfeld *et al.*, 1993; Orsulic and Peifer, 1996) and it seems likely that the arm repeats in ARC1 mediate its interaction with the phosphorylated kinase domain of SRK.

In contrast to the stigma-specific expression pattern observed for *ARC1, THL-1* and *THL-2* messenger RNAs (mRNAs) were detected in all the plant organs tested. *THL-2* mRNA was, however, more abundant in floral organs than in other tissues (Bower *et al.*, 1996). In the yeast two-hybrid system THL-1 and THL-2 interacted with the kinase domain of SRK but not with an SRK kinase mutant or with the kinase domains of two other putative plant receptor kinases, receptor-like kinase 4 (RLK4) and RLK5. However, conflicting results were obtained in co-immunoprecipitation experiments where THL-1 bound to both active and kinase-inactive forms of the SRK kinase domain. THL-1 was weakly phosphorylated by the SRK kinase domain *in vitro*.

The physiological significance of the interaction between SRK and KAPP is not clear as an orthologue of KAPP has not yet been described in *Brassica*. However, considering the high degree of synteny between the *Arabidopsis* and *Brassica* genomes, it is probable that such a gene exists. In *Arabidopsis*, there is evidence that KAPP acts as a negative regulator of CLAVATA1, a putative receptor kinase involved in controlling the size of apical and floral meristems (Williams *et al.*, 1997). A study of protein–protein interaction using the far-western technique has shown that KAPP can interact *in vitro* with the kinase domains of several RLKs (including SRK) from a range of different species (Braun *et al.*, 1997). Based on these data, it has been suggested that KAPP may play a broad role as a negative regulator of RLKs (Williams *et al.*, 1997). A KAPP homologue may therefore carry out the same function in the SRK pathway.

The identification of proteins that interact with SRK represents an important step towards elucidating, at a molecular level, the pathway that leads from self-pollen recognition to self-pollen rejection. It will be important to determine the functional role of these proteins in the SI response.

As mentioned above, the characterization of modifier and suppressor genes provides an alternative approach to identifying molecules involved in the SI response. Ikeda *et al.* (1997) used differential display reverse transcription-polymerase chain reaction (DDRT-PCR) analysis to identify an aquaporin-like gene that is very closely linked, and probably identical to, the *Brassica MOD* gene (Hinata *et al.*, 1983). mRNA for the aquaporin was present in stigmas of self-incompatible *MOD*/*MOD* plants but was not detected in self-compatible plants homozygous for a mutant *mod* allele. As pointed out by Ikeda *et al.* (1997), implication of an aquaporin in the SI response would be consistent with the observation that pollen hydration is an important control step in the acceptance of pollen grains by the stigma (Dickinson, 1995). However, the role of the aquaporin gene in this process is not very clear because loss of expression is correlated with the acquisition of self-compatibility. Ikeda *et al.* (1997) propose two models to explain this observation. In the first, the aquaporin would serve to channel water away from the stigma surface during the SI response so that less water is available to hydrate self-pollen grains. In order to test this model, more needs to be known about osmotic gradients at the stigma surface, bearing in mind that the aquaporin is predicted to act as a water channel rather than as a pump. The second model suggests that the aquaporin acts as a channel for a small molecule(s) that inhibits pollen germination. In both models, the aquaporin is predicted to be a downstream target of SRK and to be activated during the SI response (Fig. 3).

IV. EVOLUTIONARY ASPECTS

A. EVOLUTION OF SI SYSTEMS IN THE ANGIOSPERMS

In the last 15 years, proteins involved in the pollen–pistil recognition step have been identified for a number of different homomorphic SI systems. One of the most striking results of these studies is that very often the proteins involved in self-pollen recognition in plants of one botanical family bear no resemblance to those of a plant from another family. For example, *SLG* and *SRK* in the Brassicaceae are unrelated to the *S*-RNases that mediate SI in the Solanaceae. Similarly, the *S*-locus genes identified in the grass *Phalaris* and in poppy encode additional classes of protein, a thioredoxin-like and a novel protein, respectively (reviewed by McCubbin and Kao, 1996).

Whitehouse (1950) suggested that the angiosperms acquired SI early in their evolution and that this was a determining factor in their rise to dominance over other plant groups. This model was an attempt to explain data from the fossil record that indicate a very rapid evolutionary divergence and rise to dominance of the angiosperms during the Cretaceous period. Whitehouse's model implies that the diversity of SI systems present today would have evolved from this ancient SI system as suggested by Pandey (1959). However, the occurrence of unrelated SI systems in different plant families argues against this model and suggests that SI has arisen independently more than once in the angiosperms, as was suggested by Bateman (1952).

The number of both synonymous and non-synonymous substitutions per site has been estimated to be higher in the *S*-RNase alleles of the Solanaceae than in the *Brassica SLG* and *SRK* genes, suggesting that SI may have evolved more recently in the Brassicaceae than in the Solanaceae (Hinata *et al.*, 1995). This is supported by the fact that gametophytic SI systems involving *S*-RNase have been found in a wide range of families including the Solanaceae, the Scrophulariaceae and the Rosacea, whereas receptor-kinase-based sporophytic systems have not yet been found outside the Brassicaceae (Matton *et al.*, 1994). This indicates a more ancient origin for the *S*-RNase system. Recently, molecular data have been used to re-access both the age of the SI system in *Brassica* and the time of origin of the angiosperms. The estimated age of 40–50 million years for the *Brassica* SI system (Uyenoyama, 1995) would place its origins considerably later than the emergence of the angiosperms, considering that the monocot/dicot divergence is now thought to have occurred approximately 200 million years ago (Martin *et al.*, 1989; Wolfe *et al.*, 1989). These observations provide further support for models in which SI has evolved several times during angiosperm evolution.

B. EVOLUTION OF SELF-COMPATIBILITY WITHIN THE BRASSICACEAE

Arabidopsis thaliana, like *B. oleracea*, is a member of the Brassicaceae but this species does not possess a self-incompatibility system and is, therefore, self-fertile. A recent study by Conner *et al.* (1998) has revealed a remarkably high level of synteny between the region surrounding the *S* locus in *Brassica* and a region near *ETR1* on chromosome 1 of *A. thaliana*. However, although many of the markers flanking the *S* locus correspond to sequences at the *Arabidopsis* locus, no genes with similarity to *SLG* and *SRK* have been found in this region. Conner *et al.* (1998) proposed that the orthologues of *SLG* and *SRK* were deleted from this region in the *Arabidopsis* lineage as part of the process leading to the evolution of autogamy in this species. They provide several arguments in favour of the hypothesis that *SLG* and *SRK* were deleted (rather than that they were acquired at the *S* locus in the *Brassica* lineage). These include the occurrence of self-incompatible species in *Arabis*, a genus that is very closely related to *Arabidopsis*, and the estimated age of the *Brassica* SI system which suggests that it originated before the divergence of the *Brassica* and *Arabidopsis* lineages.

C. BALANCING SELECTION AND TRANS-SPECIFIC EVOLUTION OF *S*-LOCUS GENES

SI is of interest from an evolutionary point of view because of the extreme level of polymorphism at the *S* locus. Both *SLG* and *SRK* are highly polymorphic and similar levels of polymorphism have been found for genes associated with other recognition systems, including the S-RNases involved in gametophytic SI and genes of the major histocompatibility complex (MHC). The highly polymorphic nature of these genes does not appear to be due to an intrinsic hypermutability but is best explained by models that invoke a mechanism involving balancing selection (or overdominant selection in the case of the MHC). In situations where allelic diversity is selected, new alleles have a strong selective advantage. This advantage is reduced as the allele becomes established in the population. As a result, it is possible to maintain a large number of alleles in the population over long periods, a process referred to as balancing selection. One of the predictions of this model is that individual alleles stay in the population for a long period. Comparison of both MHC (Figueroa *et al.*, 1988; Lawlor *et al.*, 1988) and *S*-locus (Ioeger *et al.*, 1990; Dwyer *et al.*, 1991; Sakamoto *et al.*, 1998) sequences has identified alleles that are more similar to alleles from a related species than to alleles from the same species. This indicates that these alleles existed before the divergence of the species, a phenomenon that has been referred to as trans-specific evolution (Arden and Klein, 1982).

D. INVESTIGATION OF THE ORIGIN OF SI IN *BRASSICA*: FUNCTIONAL ANALYSIS OF THE S-GENE FAMILY

A common feature of the genes that have been implicated in the control of SI in different taxonomic families is that they are all members of gene families (Dwyer *et al.*, 1989; Taylor *et al.*, 1993; Li *et al.*, 1997). These gene families are not limited to the taxonomic group that possesses each SI system. On the contrary, they invariably include members from a wide range of plants, indicating that they are of ancient origin. This has been taken to indicate that each SI system evolved independently by recruiting and adapting pre-existing genes which originally carried out other functions.

In *Brassica, SLG* and *SRK* are members of the S-gene family. Proteins encoded by members of the S-gene family are characterized by the presence of an S domain that corresponds to the extracellular domain of SRK (or to the entire mature SLG protein). To date, the S domain has been shown to occur in three different configurations: secreted glycoproteins consisting solely of an S domain (e.g. SLG) and integral membrane proteins consisting of an N-terminal S domain and a membrane-spanning domain followed or not by a cytosolic kinase domain (e.g. mSLG and SRK, respectively). Comparison of the S domains of different members of this gene family has allowed the identification of several conserved residues including 10 highly conserved cysteine residues (Walker, 1994). Members of the S-gene family have been identified in *Brassica, Arabidopsis*, carrot, corn and rice (Dwyer *et al.*, 1989, 1994; Walker and Zhang, 1990; van Engelen *et al.*, 1993). Moreover, there is evidence that this family may include a large number of genes, at least in *Arabidopsis* and *Brassica* (Dwyer *et al.*, 1989; Kumar and Trick, 1994; Buehler *et al.*, Genbank accession no. AC004255, 1998).

Although quite a number of S-family genes has been characterized at the sequence level, very little is known about the functions carried out by this family. Two members of the family, *SLR1* and *SLR2* (*SLR* for *S*-locus-related), are closely related to *SLG* at the sequence level. *SLR1* and *SLR2* are linked to each other but are not linked to the *S* locus (Lalonde *et al.*, 1989; Trick and Flavell, 1989; Boyes *et al.*, 1991). Both genes are expressed specifically in stigmas and are predicted to encode secreted glycoproteins similar to SLG. This has been confirmed for the SLR1 protein which has been shown to accumulate in the cell wall (Umbach *et al.*, 1990).

The structure of *SLR2* resembles class II *SLGB* alleles in that it is composed of two exons. The second exon does not include a transmembrane domain and the gene is predicted to encode only secreted proteins (Tantikanjana *et al.*, 1996). There is also a high degree of similarity between *SLR2* and *SLGB* at the sequence level and phylogenetic analyses indicate that these two genes diverged recently at approximately the same time as the speciation event between *B. oleracea* and *B. rapa* (Uyenoyama, 1995). *SLR1*, in contrast, is more distantly related to *SLG* and is probably not derived from an ancestor with an SI function.

Luu *et al.* (1999) have recently shown that pollen adheres less strongly to stigmas in which *SLR1* expression has been reduced by an antisense approach. Reduced adhesion was also observed when SLR1 was masked by treatment with an anti-SLR1 antibody. These results suggest that *SLG* and *SRK* may have evolved from genes that were originally involved in compatible pollen–pistil interaction, specifically in the adhesion step. This is supported by the observation that treatment with anti-SLG antibodies also has a significant effect on the strength of pollen adhesion, suggesting that SLG continues to play a role in pollen adhesion (Luu *et al.*, 1999).

Apart from *SLR1* and the *S*-locus genes, the only other members of the S-gene family for which we have information concerning their possible function are two small families of receptor-like kinases, the *SFR* (for S-family receptor) and *ARK* (for *Arabidopsis* receptor kinase) gene from *Brassica* and *Arabidopsis*, respectively. Phylogenetic comparison based on S-domain sequences indicates that the *SFR* and *ARK* genes are more closely related to *SRK* and *SLG* than is *SLR1* (Sakamoto *et al.*, 1998). Transcripts of one of the *Brassica* genes, *SFR2*, accumulate rapidly and to a high level following wounding or bacterial infection, suggesting that this gene may play a role in the plant's defence response (Pastuglia *et al.*, 1997a). This observation is interesting because of the many similarities at the genetic, physiological and molecular level between the self-incompatibility response and the response of the plant to invasion by a pathogen (Hodgkin *et al.*, 1988; Pastuglia *et al.*, 1997a). Reporter gene and RNA gel blot experiments have shown that the three *ARK* genes have very precise expression patterns in different *Arabidopsis* tissues (Tobias *et al.*, 1992; Dwyer *et al.*, 1994). This has been taken to indicate that these genes are involved in developmental processes within the plant. This conclusion is supported by the fact that developmental defects were observed in plants in which *ARK1* was expressed ectopically from a 35S promoter (Tobias and Nasrallah, 1996). More recently, however, both *ARK1* and *ARK3* transcripts have also been shown to accumulate following both wounding and bacterial infection, suggesting that, like *SFR2*, these genes may play a role in the plant's defence response (M. Pastuglia, D. Roby, C. Dumas, J. M. Cock, manuscript in preparation). One possible explanation for these apparently conflicting results is that some of the members of these two small families may have overlapping roles during both development and the defence response. This would be analogous to the situation in animals where wounding transiently up-regulates receptors for platelet-derived growth factor and epidermal growth factor, both of which also carry out specific functions during development (Antoniades *et al.*, 1991; Wenczak *et al.*, 1992).

In summary, functional analysis of genes such as *SLR1* and the *SFR* and *ARK* families has provided several clues as to the possible origin of the self-incompatibility genes in the Brassicaceae. The role of SLR1 in pollen adhesion suggests that *SLG* and *SRK* may have evolved from a gene involved in compatible pollen–pistil interactions. Analysis of *SFR* and *ARK* genes indicates that the S-gene family may have evolved by adaptation of a basic cell–cell recognition system to carry out a number of different functions. A more complete characterization of this family, with regard both to function and evolutionary relationships, should shed more light on the evolutionary origins of SI in the Brassicaceae.

E. *SRK* AND THE PLANT RECEPTOR-LIKE KINASE GENE SUPERFAMILY

In plants, the receptor-like kinase (RLK) gene superfamily includes a large number of genes that are predicted to encode putative transmembrane proteins with a cytosolic serine/threonine protein kinase domain (Walker, 1994). *SRK* is a member of the RLK superfamily as are other membrane-spanning kinases of the S-gene family. However, not all RLKs possess an extracellular S domain and at least eight different classes of RLK can be defined based on extracellular domain structure. These include proteins with leucine-rich repeats (LRRs), epidermal growth factor (EGF) repeats, tumour necrosis factor receptor (TNFR) homology, lectin-like and thaumatin-like domains and other novel sequences in their predicted extracellular domains (see references in Pastuglia *et al.*, 1997a; Torii and Clark, 2000, this volume). In addition, a small number of putative cytosolic kinases including Pto [see the chapter by Sessa and Martin (2000) in this volume] is closely related to the RLK superfamily at the sequence level and could, therefore, be considered to be RLKs that lack extracellular and membrane-spanning domains.

The members of the RLK superfamily are referred to as receptor-like kinases because none of their protein products has yet been shown to function as a receptor and to bind a ligand. A possible exception is Pto, which has been shown to bind a bacterial avirulence protein, avrPto (Sessa and Martin, 2000). However, as mentioned above, Pto is unusual in that it is a cytosolic kinase with neither a membrane-spanning nor an extracellular domain (although Pto may be associated with the membrane as it contains a potential myristoylation site).

Phylogenetic analysis of the kinase domains of several different members of the RLK superfamily indicates that they share a common evolutionary origin (Fig. 4). The RLK superfamily forms a group that is distinct from both serine/threonine and tyrosine receptor protein kinases from animals and other plant cytosolic kinases such as CDC2, MAPK and shaggy/GSK-3 homologues. The similarity between RLK kinase domains suggests that the different members of this superfamily function in a similar manner. If this is the case, then information gained from analysis of SRK, which is perhaps the best characterized

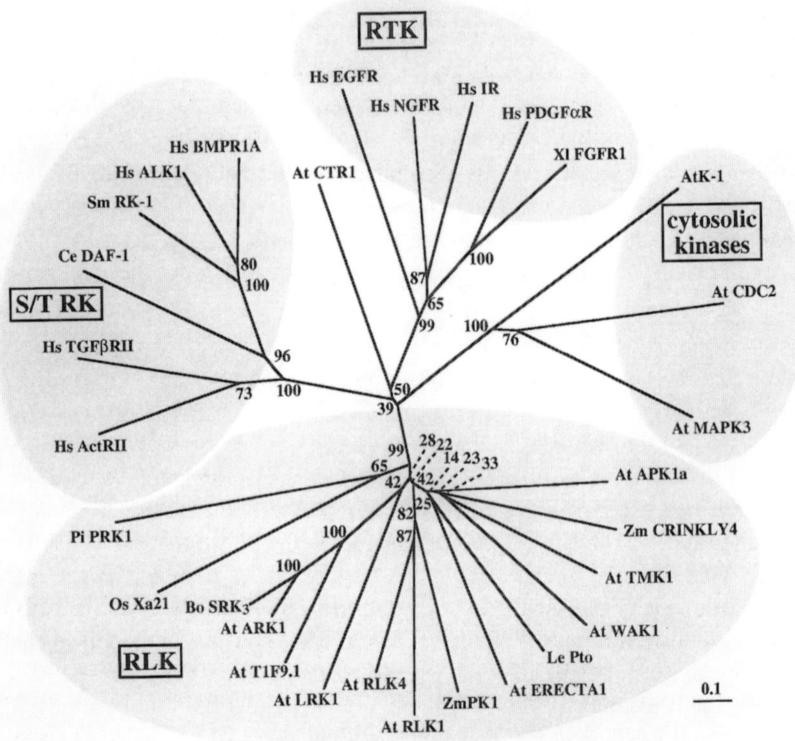

Fig. 4. Neighbour-joining tree based on the kinase domains of a range of receptor kinases and cytosolic kinases. The sequences include plant receptor-like kinases (RLK) with S (SRK_3, ARK1, T1F9.1, RLK1, RLK4, ZmPK1), LRR (PRK1, Xa21, ERECTA1, TMK1), EGF-repeat (WAK1), TNFR-like (CRINKLY4) or lectin-like (LRK1) domains (or no extracellular domain: Pto, APK1a), animal serine/threonine receptor (S/T RK) kinases (ActRII, TGFβRII, DAF-1, RK-1, ALK1, BMPR1A), animal tyrosine receptor (RTK) kinases (EGFR, NGFR, IR, PDGFαR, FGFR1) and plant cytosolic kinases (CDC2, MAPK3, AtK-1, CTR1, the latter two sequences being shaggy/GSK-3 and Raf homologues, respectively). Prefixes refer to the species of origin: Bo, *B. oleracea*; At, *Arabidopsis thaliana*; Pi, *Petunia inflata*; Le, *Lycopersicum esculentum*; Zm, *Zea mays*; Os, *Oryza sativa*; Hs, *Homo sapiens*; Xl, *Xenopus laevis*; Ce, *Caenorhabditis elegans*; Sm, *Schistosoma mansoni*. The numbers next to the branches are bootstrap values (%) with 2000 repeats.

member of this superfamily, should be useful in the analysis of cell–cell interactions mediated by other members of the superfamily.

The extreme differences between extracellular domains of the different classes of RLK are of interest from an evolutionary point of view. The most parsimonious explanation for these differences is that these genes have evolved from an ancestral kinase by multiple, independent recombination events involving genes possessing the different extracellular domains. This hypothesis suggests that the extracellular domains would have initially evolved independently of the kinase domain. This is consistent with the observation that both genes which encode secreted S-domain proteins (e.g. *SLR1*) and genes which encode RLKs (e.g. *SRK*) have been shown to carry out functional roles in *Brassica*. There is also evidence within the S-gene family of a reversal of the above process, in which *RLK* genes have been modified to encode secreted and/or membrane-anchored forms of their extracellular domains (Tantikanjana *et al.*, 1993; Cock *et al.*, 1995).

V. CONCLUSION

SI in flowering plants is controlled by systems of great complexity and subtlety that are exquisitely adapted to promote out-crossing and assure reproductive success. Historically, the function of these systems has intrigued many biologists, including Charles Darwin, who understood the importance of out-crossing to plant vigour and who carried out several pioneering studies on SI. Interest has been heightened in recent years as information about the molecular nature of different SI systems has become available. Identification of SRK was a major step forward in the understanding of SI in *Brassica*. More recent advances such as biochemical characterization of SRK and the identification of potential downstream targets of the SI response are starting to provide a clearer picture of the mechanisms of both self-pollen recognition and rejection.

A significant number of questions remains to be answered concerning SI in *Brassica*. The most obvious question concerns the nature of the male component, the putative ligand for the SRK receptor. This is an important question both with regard to the mechanism of SI and, in a wider context, because it would provide a means of testing the hypothesis that the members of the RLK superfamily function as receptors. A major advantage of the SI system, in this respect, is the fact that the male component of the SI response is predicted to be encoded by a gene located at the *S* locus. It seems probable that a combination of biochemical and map-based approaches will lead to the identification of this gene in the near future.

NOTE ADDED IN PROOF

While this manuscript was going to press, several major advances were made in the characterization of the SI system in *Brassica*. These included the identification of the male component of the SI response which was shown to be small, cysteine-rich protein related to PCP-A1 (*SCR* for *S* locus cysteine-rich gene: Schopfer *et al.*, 1999; Takayama *et al.*, 2000). On the female side, *ARC1* was shown to be required for the SI response (Stone *et al.*, 1999) and a gain-of-function approach demonstrated that *SRK* alone is sufficient to confer *S*-haplotype specificity in the pistil (Takasaki *et al.*, 2000).

ACKNOWLEDGEMENTS

The author would like to thank Philippe Heizmann (Université Lyon I) for his comments on the manuscript, Marc Robinson (ENS, Lyon) and Christophe Blanchet (IBCP, Lyon) for help with the phylogenetic analysis in Fig. 4 and Thierry Gaude (RDP, Lyon) for the electron micrograph in Fig. 3. Work carried out the Lyon laboratory was supported by the Institut National de la Recherche Agronomique, the Centre National de la Recherche Scientifique and the Ecole Normale Supérieure de Lyon.

REFERENCES

Antoniades, H. N., Galanopoulos, T., Neville-Golden, J., Kiritsy, C. P. and Lynch, S. E. (1991). Injury induces *in vivo* expression of the platelet-derived growth factor (PDGF) and PDGF receptor mRNAs in skin epithelial cells and PDGF mRNA in connective tissue fibroblasts. *Proceedings of the National Academy of Sciences of the United States of America* **88**, 565–569.

Arden, B. and Klein, J. (1982). Biochemical comparison of major histocompatibility complex molecules from different subspecies of *Mus musculus*: evidence for trans-specific evolution of alleles. *Proceedings of the National Academy of Sciences of the United States of America* **79**, 2342–2346.

Bateman, A. J. (1952). Self-incompatibility systems in angiosperms. *Heredity* **6**, 285–310.

Bower, M. S., Matias, D. D., Fernandes-Carvalho, E., Mazzurco, M., Gu, T., Rothstein, S. J. and Goring, D. R. (1996). Two members of the thioredoxin-h family interact with the kinase domain of a *Brassica S*-locus receptor kinase. *Plant Cell* **8**, 1641–1650.

Boyes, D. C. and Nasrallah, J. B. (1993). Physical linkage of the *SLG* and *SRK* genes at the self-incompatibility locus of *Brassica oleracea*. *Molecular and General Genetics* **236**, 369–373.

Boyes, D. C. and Nasrallah, J. B. (1995). An anther-specific gene encoded by an *S* locus haplotype of *Brassica* produces complementary and differentially regulated transcripts. *Plant Cell* **7**, 1283–1294.

Boyes, D. C., Chen, C. H., Tantikanjana, T., Esch, J. J. and Nasrallah, J. B. (1991). Isolation of a second *S*-locus-related cDNA from *Brassica oleracea*: genetic relationships between the *S* locus and two related loci. *Genetics* **127**, 221–228.

Boyes, D. C., Nasrallah, M. E., Vrebalov, J. and Nasrallah, J. B. (1997). The self-incompatibility (*S*) haplotypes of *Brassica* contain highly divergent and rearranged sequences of ancient origin. *Plant Cell* **9**, 237–247.

Braun, D. M., Stone, J. M. and Walker, J. C. (1997). Interaction of the maize and *Arabidopsis* kinase interaction domains with a subset of receptor-like protein kinases: implications for transmembrane signaling in plants. *Plant Journal* **12**, 83–95.

Cabrillac, D., Delorme, V., Garin, J., Ruffio-Châble, V., Giranton, J. L., Dumas, C., Gaude, T. and Cock, J. M. (1999). The S_{15} self-incompatibility haplotype in *Brassica oleracea* includes three S gene family members expressed in stigmas. *Plant Cell*, **11**, 971–986.

Charlesworth, D. and Awadalla, P. (1998). Flowering plant self-incompatibility: the molecular genetics of *Brassica* S-loci. *Heredity*, **81**, 1–9.

Clark, A. G. and Kao, T.-h. (1991). Excess nonsynonymous substitution at shared polymorphic sites among self-incompatibility alleles of Solanaceae. *Proceedings of the National Academy of Sciences of the United States of America* **88**, 9823–9827.

Clark, S. E., Running, M. P. and Meyerowitz, E. M. (1995). *CLAVATA3* is a specific regulator of shoot and floral meristem development affecting the same processes as *CLAVATA1*. *Development* **121**, 2057–2067.

Cock, J. M., Stanchev, B., Delorme, V., Croy, R. R. D. and Dumas, C. (1995). *SLR3*: a modified receptor kinase gene that has been adapted to encode a putative secreted glycoprotein similar to the *S* locus glycoprotein. *Molecular and General Genetics* **248**, 151–161.

Cock, J. M., Swarup, R. and Dumas, C. (1997). Natural antisense transcripts of the *S* locus receptor kinase gene and related sequences in *Brassica oleracea*. *Molecular and General Genetics* **255**, 514–524.

Conner, J. A., Tantikanjana, T., Stein, J. C., Kandasamy, M. K., Nasrallah, J. B. and Nasrallah, M. E. (1997). Transgene-induced silencing of *S*-locus genes and related genes in *Brassica*. *Plant Journal* **11**, 809–823.

Conner, J. A., Conner, P., Nasrallah, M. E., Nasrallah, J. B. (1998). Comparative mapping of the *Brassica S* locus region and its homeolog in *Arabidopsis*: implications for the evolution of mating systems in the Brassicaceae. *Plant Cell* **10**, 801–812.

Delorme, V., Giranton, J. L., Hatzfeld, Y., Friry, A., Heizmann, P., Ariza, M. J., Dumas, C., Gaude, T. and Cock, J. M. (1995). Characterisation of the *S* locus genes, *SLG* and *SRK*, of the *Brassica* S_3 haplotype: identification of a membrane-localised protein encoded by the *S* locus receptor kinase gene. *Plant Journal* **7**, 429–440.

de Nettancourt D. (1977). "Incompatibility in Angiosperms". Springer, Berlin.

Dickinson, H. (1995). Dry stigmas, water and self-incompatibility in *Brassica*. *Sexual Plant Reproduction* **8**, 1–10.

Doughty, J., Hedderson, F., McCubbin, A. and Dickinson, H. (1993). Interaction between a coating-borne peptide of the *Brassica* pollen grain and stigmatic-*S* (self-incompatibility)-locus-specific glycoproteins. *Proceedings of the National Academy of Sciences of the United States of America* **90**, 467–471.

Doughty, J., Dixon, S., Hiscock, S. J., Willis, A. C., Parkin, I. A. P. and Dickinson, H. G. (1998). PCP-1A, a defensin-like *Brassica* pollen coat protein that binds the *S* locus glycoprotein, is the product of gametophytic gene expression. *Plant Cell* **10**, 1333–1347.

Dwyer, K. G., Chao, A., Cheng, B., Chen, C.-H. and Nasrallah, J. B. (1989). The *Brassica* self-incompatibility multigene family. *Genome* **31**, 969–972.

Dwyer, K. G., Balent, M. A., Nasrallah, M. E. and Nasrallah, J. B. (1991). DNA sequences of self-incompatibility genes from *Brassica campestris* and *B. oleracea*: polymorphism predating speciation. *Plant Molecular Biology* **16**, 481–486.

Dwyer, K. G., Kandasamy, M. K., Mahosky, D. I., Acciai, J., Kudish, B. I., Miller, J. E., Nasrallah, M. E. and Nasrallah, J. B. (1994). A superfamily of *S*-locus-related sequences in *Arabidopsis*: diverse structures and expression patterns. *Plant Cell* **6**, 1829–1843.

Elleman, C. J. and Dickinson, H. G. (1990). The role of the exine coating in pollen-stigma interactions in *Brassica oleracea*. *New Phytologist* **114**, 511–518.

Elleman, C. J., Willson, C. E., Sarker, R. H. and Dickinson, H. G. (1988). Interaction between the pollen tube and stigmatic cell wall following pollination in *Brassica oleracea*. *New Phytologist* **109**, 114–117.

Eukere, U., Bowman, C., Marshall, D. and Lydiate, D. (1997). The genetic control of self-incompatibility in oilseed rape (*Brassica napus*). ISHS Symposium on Crucifers, Tenth Crucifer Genetics Workshop, Rennes, France.

Figueroa, R., Günther, E. and Klein, J. (1988). MHC polymorphism predating speciation. *Nature* **335**, 265–267.

Gaude, T. Rougier, M., Heizmann, P., Ockendon, D. J. and Dumas, C. (1995). Expression of the *SLG* gene is not correlated with the self-incompatibility phenotype in the class II *S* haplotypes of *Brassica oleracea*. *Plant Molecular Biology* **27**, 1003–1014.

Giranton, J.-L., Ariza, M. J., Dumas, C., Cock, J. M. and Gaude, T. (1995). The *S* locus receptor kinase gene encodes a soluble glycoprotein corresponding to the SRK extracellular domain in *Brassica oleracea*. *Plant Journal* **8**, 101–108.

Goring, D. R. and Rothstein, S. J. (1992). The *S*-locus receptor kinase gene in a self-incompatible *Brassica napus* line encodes a functional serine threonine kinase. *Plant Cell* **4**, 1273–1281.

Goring, D. R., Glavin, T. L., Schafer, U. and Rothstein, S. J. (1993). An *S* receptor kinase gene in self-compatible *Brassica napus* has a 1-bp deletion. *Plant Cell* **5**, 531–539.

Gu, T., Mazzurco, M., Sulaman, W., Matias, D. D. and Goring, D. R. (1998). Binding of an arm repeat protein to the kinase domain of the *S*-locus receptor kinase. *Proceedings of the National Academy of Sciences of the United States of America* **95**, 382–387.

Heldin, C.-H. (1995). Dimerization of cell surface receptors in signal transduction. *Cell* **80**, 213–223.

Heslop-Harrison, J. (1975). Incompatibility and the pollen stigma reaction. *Annual Review of Plant Physiology* **26**, 403–425.

Hinata, K., Okasaki, K. and Nishio, T. (1983). "Proceedings of the Sixth International Rapeseed Conference", Vol. 1, p. 354. Groupe Consultatif International de Recherche sur le Colza, Paris.

Hinata, K., Isogai, A. and Isuzugawa, K. (1994). Manipulation of sporophytic self-incompatibility in plant breeding. *In* "Genetic Control of Self-Incompatibility and Reproductive Development in Flowering Plants" (E. G. Williams, ed.) pp. 102–115. Kluwer Academic, Dordrecht.

Hinata, K., Watanabe, M., Yamakawa, S., Satta, Y. and Isogai, A. (1995). Evolutionary aspects of the S-related genes of the *Brassica* self-incompatibility system: synonymous and nonsynonymous base substitutions. *Genetics* **140**, 1099–1104.

Hodgkin, T., Lyon, G. D. and Dickinson, H. G. (1988). Recognition in flowering plants: a comparison of the *Brassica* self-incompatibility system and pathogen interactions. *New Phytologist* **110**, 557–569.

Hughes, A. L. and Nei, M. (1988). Pattern of nucleotide substitution at major histocompatibility complex class-I loci reveals overdominant selection. *Nature* **335**, 167–170.

Ikeda, S., Nasrallah, J. B., Dixit, R., Preiss, S. and Nasrallah, M. E. (1997). An aquaporin-like gene required for the *Brassica* self-incompatibility response. *Science* **276**, 1564–1566.

Ioeger, T. R., Clark, A. G. and Kao, T.-h. (1990). Polymorphism at the self-incompatibility locus in Solanaceae predates speciation. *Proceedings of the National Academy of Sciences of the United States of America* **87**, 9732–9735.

Kandasamy, M. K., Paolillo, D. J., Faraday, C. D., Nasrallah, J. B. and Nasrallah, M. E. (1989). The S-locus specific glycoproteins of *Brassica* accumulate in the cell wall of developing stigma papillae. *Developmental Biology* **134**, 462–472.

Kleman-Mariac, C., Rougier, M., Cock, J. M., Gaude, T. and Dumas, C. (1995). S-locus glycoproteins are expressed along the path of pollen tubes in *Brassica* pistils. *Planta* **196**, 614–621.

Kumar, V. and Trick, M. (1994). Expression of the S-locus receptor multigene family in *Brassica oleracea. Plant Journal* **6**, 807–813.

Kusaba, M., Nishio, T., Satta, Y., Hinata, K. and Ockendon, D. (1997). Striking sequence similarity in inter- and intra-specific comparisons of class I *SLG* alleles from *Brassica oleracea* and *Brassica campestris*: implications for the evolution and recognition mechanism. *Proceedings of the National Academy of Sciences of the United States of America* **94**, 7673–7678.

Lalonde, B. A., Nasrallah, M. E., Dwyer, K. D., Chen, C.-H., Barlow, B. and Nasrallah, J. B. (1989). A highly conserved *Brassica* gene with homology to the S-locus-specific glycoprotein structural gene. *Plant Cell* **1**, 249–258.

Lawlor, D. A., Ward, F. E., Ennis, P. D., Jackson, A. P. and Parham, P. (1988). HLA-A and B polymorphisms predate the divergence of humans and chimpanzees. *Nature* **335**, 268–271.

Li, X., Paech, N., Nield, J., Hayman, D. and Langridge, P. (1997). Self-incompatibility in the grasses: evolutionary relationship of the *S* gene from *Phalaris coerulescens* to homologous sequences in other grasses. *Plant Molecular Biology* **34**, 223–232.

Luu, D.-T., Heizmann, P. and Dumas, C. (1997). Pollen-stigma adhesion in kale is not dependent on the self-(in)compatibility genotype. *Plant Physiology* **115**, 1221–1230.

Luu, D.-T., Marty-Mazars, D., Trick, M., Dumas, C. and Heizmann, P. (1999). Pollen-stigma adhesion in *Brassica* involves SLG and SLR1 glycoproteins. *Plant Cell* **11**, 251–262.

Martin, W., Gierl, A. and Saedler, H. (1989). Molecular evidence for pre-Cretaceous angiosperm origins. *Nature* **339**, 46–48.

Matton, D. P., Nass, N., Clarke, A. E. and Newbigin, E. (1994). Self-incompatibility: how plants avoid illegitimate offspring. *Proceedings of the National Academy of Sciences of the United States of America* **91**, 1992–1997.

McCormick, S. (1998). Self-incompatibility and other pollen-pistil interactions. *Current Opinion in Plant Biology* **1**, 18–25.

McCubbin, A. G. and Kao, T.-h. (1996). Molecular mechanisms of self-incompatibility. *Current Opinion in Biotechnology* **7**, 150–154.

Nasrallah, J. B. (1997a). Signal perception and response in the interactions of self-incompatibility in *Brassica. Essays in Biochemistry* **32**, 143–160.

Nasrallah, J. B. (1997b). Evolution of the *Brassica* self-incompatibility locus: a look into *S*-locus gene polymorphisms. *Proceedings of the National Academy of Sciences of the United States of America* **94**, 9516–9519.

Nasrallah, J. B., Doney, R. C. and Nasrallah, M. E. (1985). Biosynthesis of glycoproteins involved in the pollen-stigma interaction of incompatibility in developing flowers of *Brassica oleracea* L. *Planta* **165**, 100–107.

Nasrallah, J. B., Kao, T.-h., Chen, C. H., Goldberg, M. L. and Nasrallah, M. E. (1987). Amino-acid sequence of glycoproteins encoded by three alleles of the *S*-locus of *Brassica oleracea. Nature* **326**, 617–619.

Nasrallah, J. B., Rundle, S. J. and Nasrallah, M. E. (1994a). Genetic evidence for the requirement of the *Brassica S*-locus receptor kinase gene in the self-incompatibility response. *Plant Journal* **5**, 373–384.

Nasrallah, J. B., Stein, J. C., Kandasamy, M. K. and Nasrallah, M. E. (1994b). Signaling the arrest of pollen tube development in self-incompatible plants. *Science* **266**, 1505–1508.

Nasrallah, M. E. (1989). The genetics of self-incompatibility in *Brassica* and the effects of suppressor genes. *In* "Plant Reproduction: From Floral Induction to Pollination. Proceedings of the 12th Annual Riverside Symposium in Plant Physiology, Vol. 1. Current Topics in Plant Physiology, Vol. 1" (E. Lord and G. Bernier, eds) pp. 146–155.

Nasrallah, M. E., Kandasamy, M. K. and Nasrallah, J. B. (1992). A genetically defined *trans*-acting locus regulates *S*-locus function in *Brassica*. *Plant Journal* **2**, 497–506.

Nishio, T., Toriyama, K., Sato, T., Kandasamy, M. K., Paolillo, D. J., Nasrallah, J. B. and Nasrallah, M. E. (1992). Expression of S-locus glycoprotein genes from *Brassica oleracea* and *B. campestris* in transgenic plants of self-compatible *B. napus* cv Westar. *Sexual Plant Reproduction* **5**, 101–109.

Orsulic, S. and Peifer, M. (1996). An *in vivo* structure–function study of *armadillo*, the β-catenin homologue, reveals both separate and overlapping regions of the protein required for cell adhesion and for *wingless* signaling. *Journal of Cell Biology* **134**, 1283–1300.

Pandey, K. K. (1959). Evolution of gametophytic and sporophytic systems of self-incompatibility in angiosperms. *Evolution* **14**, 98–115.

Parniske, M., Hammond-Kosack, K. E., Golstein, C., Thomas, C. M., Jones, D. A., Harrison, K., Wulff, B. B. H. and Jones, J. D. G. (1997). Novel disease resistance specificities result from sequence exchange between tandemly repeated genes at the *Cf-4/9* locus of tomato. *Cell* **91**, 821–832.

Pastuglia, M., Roby, D., Dumas, C. and Cock, J. M. (1997a). Rapid induction by wounding and bacterial infection of an *S* gene family receptor-like kinase gene in *Brassica oleracea*. *Plant Cell* **9**, 49–60.

Pastuglia, M., Ruffio-Châble, V., Delorme, V., Gaude, T., Dumas, C. and Cock, J. M. (1997b). A functional *S* locus anther gene is not required for the self-incompatibility response in *Brassica oleracea*. *Plant Cell* **9**, 2065–2076.

Piffanelli, P., Ross, J. H. E. and Murphy, D. J. (1998). Biogenesis and function of the lipidic structures of pollen grains. *Sexual Plant Reproduction* **11**, 65–80.

Rubinfeld, B., Souza, B., Albert, I., Muller, O., Chamberlain, S. H., Masiarz, F. R., Munnmitsu, S. and Polakis, P. (1993). Association of the APC gene product with β-catenin. *Science* **262**, 1741–1734.

Ruffio-Châble, V. (1998). Complexity of self-incompatibility phenotype in *Brassica*: its measure and some thoughts about its genetic control. *In* "Proceedings of the International Symposium on Brassicas" (T. Grégoire and A. A. Monteiro, eds) pp. 281–288. Acta Horticulturae, ISHS, Leiden.

Rundle, S. J., Nasrallah, M. E. and Nasrallah, J. B. (1993). Effects of inhibitors of protein serine/threonine phosphatases on pollination in *Brassica*. *Plant Physiology* **103**, 1165–1171.

Sakamoto, K., Kusaba, M. and Nishio, T. (1998). Polymorphism of the *S*-locus glycoprotein gene (*SLG*) and the *S*-locus related gene (*SLR1*) in *Raphanus sativus* L. and self-incompatible ornamental plants of the Brassicaceae. *Molecular and General Genetics* **258**, 397–403.

Sarker, R. H., Elleman, C. J. and Dickinson, H. G. (1988). Control of pollen hydration in *Brassica* requires continued protein synthesis, and glycosylation is necessary for intraspecific incompatibility. *Proceedings of the National Academy of Sciences of the United States of America* **85**, 4340–4344.

Sato, T., Thorness, M. K., Kandasamy, M. K., Nishio, T., Hirai, M., Nasrallah, J. B. and Nasrallah, M. E. (1991). Activity of an *S*-locus gene promoter in pistils and anthers of transgenic *Brassica*. *Plant Cell* **3**, 867–876.

Schopfer, C. R., Nasrallah, M. E. and Nasrallah, J. B. (1999). The male determinant of self-incompatibility in *Brassica*. *Science* **286**, 1697–1700.

Scutt, C. P., Fordham-Skelton, A. P. and Croy, R. R. D. (1993). Okadaic acid causes breakdown of self-incompatibility in *Brassica oleracea*: evidence for the involvement of protein phosphatases in the incompatible response. *Sexual Plant Reproduction* **6**, 282–285.

Sessa, G. and Martin, G. B. (2000). Protein kinases in the plant defense response. *In* "Advances in Botanical Research" (J. A. Callow, ed.) pp. 381–406. Academic Press, London.

Shiba, H., Hinata, K., Suzuki, A. and Isogai, A. (1995). Breakdown of self-incompatibility in *Brassica* by the antisense RNA of the *SLG* gene. *Proceedings of the Japanese Academy* **71**, 81–83.

Stahl, R. J., Arnoldo, M., Glavin, T. L., Goring, D. R. and Rothstein, R. J. (1998). The self-incompatibility phenotype in *Brassica* is altered by the transformation of a mutant *S* locus receptor kinase. *Plant Cell* **10**, 209–218.

Stanchev, B. S., Doughty, J., Scutt, C. P., Dickinson, H. and Croy, R. R. D. (1996). Cloning of *PCP1*, a member of a family of pollen coat protein (*PCP*) genes from *Brassica oleracea* encoding novel cysteine-rich proteins involved in pollen–stigma interactions. *Plant Journal* **10**, 303–313.

Stein, J. C., Howlett, B., Boyes, D. C., Nasrallah, M. E. and Nasrallah, J. B. (1991). Molecular cloning of a putative receptor protein kinase gene encoded at the self-incompatibility locus of *Brassica oleracea*. *Proceedings of the National Academy of Science of the United States of America* **88**, 8816–8820.

Stein, J. C., Dixit, R., Nasrallah, M. E., and Nasrallah, J. B. (1996). SRK, the stigma-specific *S* locus receptor kinase of *Brassica*, is targeted to the plasma membrane in transgenic tobacco. *Plant Cell* **8**, 429–445.

Stephenson, A. G., Doughty, J., Dixon, S., Elleman, C., Hiscock, S. and Dickinson, H. G. (1997). The male determinant of self-incompatibility in *Brassica oleracea* is located in the pollen coating. *Plant Journal* **12**, 1351–1359.

Stone, S. L., Arnold, M. and Goring, D. R. (1999). A breakdown of *Brassica* self-incompatibility in *ARC1* antisense transgenic plants. *Science* **286**, 1729–1731.

Sulaman, W., Arnoldo, M., Yu, K., Tulsieram, L., Rothstein, S. J. and Goring, D. R. (1997). Loss of callose in the stigma papillae does not affect the *Brassica* self-incompatibility phenotype. *Planta* **203**, 327–331.

Suzuki, G., Watanabe, M., Kai, N., Matsuda, N., Toriyama, K., Takayama, S., Isogai, A. and Hinata, K. (1997). Three members of the *S* multigene family are linked to the *S* locus of *Brassica*. *Molecular and General Genetics* **256**, 257–264.

Takasaki, T., Hatakeyama, K., Suzuki, G., Watanabe, M., Isogai, A. and Hinata, K. (2000). The *S* receptor kinase determines self-incompatibility in *Brassica* stigma. *Nature* **403**, 913–916.

Takayama, S., Shiba, H., Iwano, M., Shimosato, H., Che, F. S., Kai, N., Watanabe, M., Suzuki, G., Hinata, K. and Isogai A. (2000). The pollen determinant of self-incompatibility in *Brassica campestris*. *Proceedings of the National Academy of Sciences of the United States of America* **97**, 1920–1925.

Tantikanjana, T., Nasrallah, M. E., Stein, J. C., Chen, C.-H. and Nasrallah, J. B. (1993). An alternative transcript of the *S*-locus glycoprotein gene in a class-II pollen-recessive self-incompatibility haplotype of *Brassica oleracea* encodes a membrane-anchored protein. *Plant Cell* **5**, 657–666.

Tantikanjana, T., Nasrallah, M. E. and Nasrallah, J. B. (1996). The *Brassica S* gene family: molecular characterization of the *SLR2* gene. *Sexual Plant Reproduction* **9**, 107–116.

Taylor, C. B., Bariola P. A., delCardayre S. B., Raines R. T. and Green P. J. (1993). RNS2: a senescence-associated RNase of *Arabidopsis* that diverged from the S-

RNase before speciation. *Proceedings of the National Academy of Sciences of the United States of America* **90**, 5118–5122.

Thompson, K. F. and Taylor, J. P. (1966). Non-linear relationships between *S* alleles. *Heredity* **21**, 345–362.

Tobias, C. M. and Nasrallah, J. B. (1996). An *S*-locus related gene in *Arabidopsis* encodes a functional kinase and produces two classes of transcripts. *Plant Journal* **10**, 523–531.

Tobias, C. M., Howlett, B. and Nasrallah, J. B. (1992). An *Arabidopsis thaliana* gene with sequence similarity to the *S*-locus receptor kinase of *Brassica oleracea*-sequence and expression. *Plant Physiology* **99**, 284–290.

Torii, K. U. and Clark, S. E. (2000). Receptor-like kinases in plant development. *In* "Advances in Botanical Research" (J. A. Callow, ed.) pp. 225–267. Academic Press, London.

Toriyama, K., Stein, J. C., Nasrallah, M. E. and Nasrallah, J. B. (1991). Transformation of *Brassica oleracea* with an *S*-locus gene from *B. campestris* changes the self-incompatibility phenotype. *Theoretical and Applied Genetics* **81**, 769–776.

Trick, M. and Flavell, R. B. (1989). A homologous *S* genotype of *Brassica oleracea* expresses two *S*-like genes. *Molecular and General Genetics* **218**, 112–117.

Umbach, A. L., Lalonde, B. A., Kandasamy, M. K., Nasrallah, J. B. and Nasrallah, M. N. (1990). Immunodetection of protein glycoforms encoded by two independent genes of the self-incompatibility multigene family of *Brassica*. *Plant Physiology* **93**, 739–747.

Uyenoyama, M. K. (1995). A generalized least-squares estimate for the origin of sporophytic self-incompatibility. *Genetics* **139**, 739–747.

van Engelen, F. A., Hartog, M. V., Thomas, T. L., Taylor, B., Sturm, A., van Kammen, A. and de Vries, S. C. (1993). The carrot secreted glycoprotein gene *EP1* is expressed in the epidermis and has sequence homology to the *Brassica* S-locus glycoproteins. *Plant Journal* **4**, 855–862.

Walker, J. C. (1994). Structure and function of the receptor-like protein kinases of higher plants. *Plant Molecular Biology* **26**, 1599–1609.

Walker, J. C. and Zhang, R. (1990). Relationship of a putative receptor protein kinase from maize to the S-locus glycoproteins of *Brassica*. *Nature* **345**, 743–746.

Wenczak, B. A., Lynch, J. B. and Nanney, L. B. (1992). Epidermal growth factor receptor distribution in burn wounds. *Journal of Clinical Investigation* **90**, 2392–2401.

Whitehouse, H. L. K. (1950). Multiple-allelomorph incompatibility of pollen and style in the evolution of the angiosperms. *Annals of Botany* **54**, 199–216.

Williams, P. H. and Leung, H. (1983). Hybrid crucifers. *In* "Handbook of Experimental Pollination Biology" (C. E. Jones and R. J. Little, eds) pp. 514–519. Scientific and Academic Editions, New York.

Williams, R. W., Wilson, J. M. and Meyerowitz, E. M. (1997). A possible role for kinase-associated protein phosphatase in the *Arabidopsis* CLAVATA1 signaling pathway. *Proceedings of the National Academy of Sciences of the United States of America* **94**, 10467–10472.

Wolfe, K. H., Gouy, M., Yang, Y.-W., Sharp, P. M. and Li W.-H. (1989). Date of the monocot–dicot divergence estimated from chloroplast DNA sequence data *Proceedings of the National Academy of Sciences of the United States of America* **86**, 6201–6205.

Yu K., Schafer U., Glavin T. L., Goring D. R., and Rothstein S. J. (1996). Molecular characterization of the *S* locus in two self-incompatible *Brassica napus* lines. *Plant Cell* **8**, 2369–2380.

Plant Mitogen-Activated Protein Kinase Signalling Pathways in the Limelight

S. JOUANNIC, A.-S. LEPRINCE, A. HAMAL, A. PICAUD, M. KREIS and Y. HENRY

Institut de Biotechnologie des Plantes, Laboratoire de Biologie du Développement des Plantes, Bâtiment 630, UMR/CNRS 8618, Université de Paris-Sud, 91405 Orsay, France

'In nature's infinite book of secrecy, a little I can read'

Shakespeare

I. INTRODUCTION

Plants are exposed to a variety of abiotic (i.e. temperature, water, light, rain, wind, gas, mineral salts, ions) and biotic (i.e. herbivorous, pathogens, hormones) stimuli. Their fitness depends upon their ability to sense changes in the surrounding environment, and their proper development requires the integration of signalling pathways that recognize and respond to spatial and temporal information. Changes occurring either in external or in internal stimuli lead to cell responses by modulating the expression of the gene

Advances in Botanical Research Vol. 32
incorporating Advances in Plant Pathology
ISBN 0-12-005932-0

repertoire programme involved in plant growth and development. This means that development is partly factual, i.e. linked to environmental effects.

Cellular regulation by extracellular factors is one of the most intensely studied areas in cell biology today, and an interesting question is how and when the conserved signalling modules are involved in the modulation of development processes. Plants have evolved different mechanisms for the modulation of cellular processes in response to various signals. These regulatory mechanisms involve cascades of biochemical events. A stimulus perceived by the cell generates a signal which is channelled through a signal transduction pathway, permitting an intracellular biochemical change to occur. Such a stimulus–response coupling process requires the recognition of the stimulus by a receptor and the transduction of a message that will trigger the appropriate response. Recent progress using various organisms has revealed pathways that link extracellular signals with the signalling machinery driving both cell cycle progression, morphogenesis and stress responses. In eukaryotes the enzymatic esterification of amino acid residues by a phosphate group, i.e. protein phosphorylation, by protein kinase plays a fundamental role in these processes. Classically a protein kinase possesses a catalytic domain of about 250–300 residues, surrounded by an amino- and a carboxy-terminal extension which represents 20–80% of the kinase length. Within their catalytic domain, protein kinases possess 12 conserved stretches of amino acids, referred to as subdomains, that show invariant or nearly invariant residues (Hanks *et al.*, 1988; Hanks and Quinn, 1991). These subdomains play different roles, which are the binding of the adenosine triphosphate (ATP) phosphatidyl groups, the binding of the substrate and the transfer of the γ-phosphate to the serine (Ser)/threonine (Thr) and/or tyrosine (Tyr) residues of the substrate (Hanks and Hunter, 1995).

An elaborate network of cascades, involving sequential protein phosphorylation and dephosphorylation events, mediated by protein kinases and phosphatases, respectively, and regulatory factors, has been found to operate in many organisms. This mechanism of signal transfer has been shown to occur through many pathways, including the MAP kinase module. Mitogen-activated protein (MAP) kinase cascade components constitute families of protein kinases which participate in cascades ubiquitously observed from yeast to vertebrates and plants. The MAP kinase modules are both highly conserved and extremely versatile. These components of the signalling machinery have been shown to play cardinal roles in the mediation and the response to a variety of stimuli. In the model systems, paralogous MAP kinase module elements are co-expressed within single cells or tissues. Most MAP kinase-mediated signal transduction pathways channel various extracellular or intracellular stimuli and activate signal transmission from membrane receptors to intracellular targets, thus generating specific cellular responses (Pelech and Sanghera, 1992). All these mechanisms are controlled by a network of positively and negatively acting elements.

In vertebrates, MAP kinases activated by mitogens, cytokines, and stresses [ultraviolet (UV) light, ionizing radiations, starvation, osmotic stress, heat shock, oxidative stress] are involved in various cellular responses such as integrin-mediated cell attachment, cell differentiation and cell cycle regulation (Marshall, 1994; Kyriakis and Avruch, 1996; Moriguchi *et al.*, 1996; Lewis *et al.*, 1998). Biochemical, genetic and molecular approaches have revealed the existence and physiological function of conserved signalling proteins in plants that have similarities with more thoroughly characterized molecules from model systems in other organisms.

The aim of this review is to detail how MAP kinase module elements are integrated in mitogen and stress responses, with particular emphasis on model species and plants. In this chapter, we describe the module, detail a few model MAP kinase pathways in well-developed non-plant systems and attempt to review the information currently available on plant MAP kinase signalling pathways, including recent data. This allows us to detail conserved features within the MAP kinase module. Additional data should be obtained from the Protein Kinase Resource (Smith *et al.*, 1997) (http://www.sdsc.edu/kinases/pk_home.html).

II. THE MITOGEN-ACTIVATED PROTEIN KINASE MODULE

A. THE MAP KINASE MODULE

The original meaning of MAPK was microtubule-associated protein-2 kinase (Ray and Sturgill, 1987), but mitogen-activated protein kinase is currently used (Rossomando *et al.*, 1989). Similarly, the term 'MAP kinase module' was first used to refer to the MAPK–MAP2K pair of proteins (Errede and Levin, 1993), and later its meaning was extended to the three protein kinases involved, namely MAPK, MAP2K and MAP3K (see Table I for abbreviations).

Classically, a response to a change in the environment of a cell proceeds through the sequential activation of the cytoplasmic kinases MAP3K, MAP2K and MAPK. In several pathways, MAP3K kinases (MAP4Ks) are also involved. Figure 1 schematically shows a MAP kinase cascade. Such a cascade plays a central role in mediating processes implicated in a switch-like response (Huang and Ferrel, 1996). This suggests that the module is able to filter out noise and to respond over a narrow range of stimuli (Huang and Ferrel, 1996). The signal is transmitted to the MAP kinase kinase kinase (MAP3K or MEKK) which phosphorylates and activates the MAP kinase kinase (MAP2K or MEK), which activates a MAP kinase (MAPK) by phosphorylation. In recent years, it has been demonstrated that three types of receptors act upstream of MAP kinase cascades the heterotrimeric G protein-coupled receptors (Lopez-Ilasaca *et al.*, 1997), the receptor tyrosine kinases (RTK) and the membrane-anchored histidine kinase. In mammals, flies, worms and yeast,

TABLE I
List of abbreviations and synonyms

AGC	assembly of protein kinases consisting of the cyclic nucleotide-dependent family (PKA, PKG), the PKC family and the ribosomal S6 kinase family
ERK	extracellular signal-regulated kinase (MAPK)
ERMK	elicitor-responsive MAPK
GCK	germinal centre kinase (MAP4K)
GDP	guanosine diphosphate
GEF	guanosine exchange factor
GPCR	G-protein-coupled receptor
grb2	growth factor receptor-bound protein 2 (adaptor protein)
GTP	guanosine triphosphate
JAK	Janus family of tyrosine kinase, first known as just another kinase
JNK	amino-terminal c-Jun kinase (MAPK)
JNKK	JNK kinase (MAP2K)
LRR	leucine-rich repeats
MAPK	mitogen-activated protein kinase
MAP2K	MAPKK = MKK = MAPK kinase = MAP kinase kinase
MAP3K	MAPKKK = MAP2K kinase = MAP kinase kinase kinase
MAP4K	MAPKKKK = MAP3K kinase = MAP kinase kinase kinase kinase
MEK	MAP/ERK kinase, a MAP2K
MEKK	MEK kinase (MAP3K)
MKP	MAPK phosphatase
MLK	mixed lineage kinase (MAP3K)
MOS	moloney murine sarcoma virus (or msv) protein kinase
NCK	non-catalytic region of tyrosine kinase (adaptor protein)
NEK	never in mitosis-A related kinase (=NRK)
NIMA	never in mitosis-A
PAK	p21Ras-activated protein kinase (MAP4K)
PH	pleckstrin homology domain (association to membrane)
PKA, C, G	protein kinase A, C or G
PP	protein phosphatase
PTK	protein tyrosine kinase
RAF	cellular homologue of oncogene products from murine sarcoma virus, or cellular homologue of v-raf, the transforming gene from an avian retrovirus (MAP3K)
RAS	small GTP-binding proteins of the ras proto-oncogene family
RST	regulator of sterile twelve
RTK	receptor tyrosine kinase

SAPK	stress-activated protein kinase (MAPK)
SEK	SAPK and ERK activating kinase (MAP2K)
SH2	src-homology 2 (polyproline linker) domain
SH3	src-homology 3 (polyproline linker) domain
SIPK	salicylic-acid-induced protein kinase (MAPK)
SOS	son of sevenless (a GEF)
src	cellular homologue of Rous avian sarcoma virus oncoprotein (non-receptor tyrosine kinase)
STAT	signal transducer and activator of transcription
T-DNA	transfer-DNA from *Agrobacterium*
WIPK	wounding-induced protein kinase (MAPK)

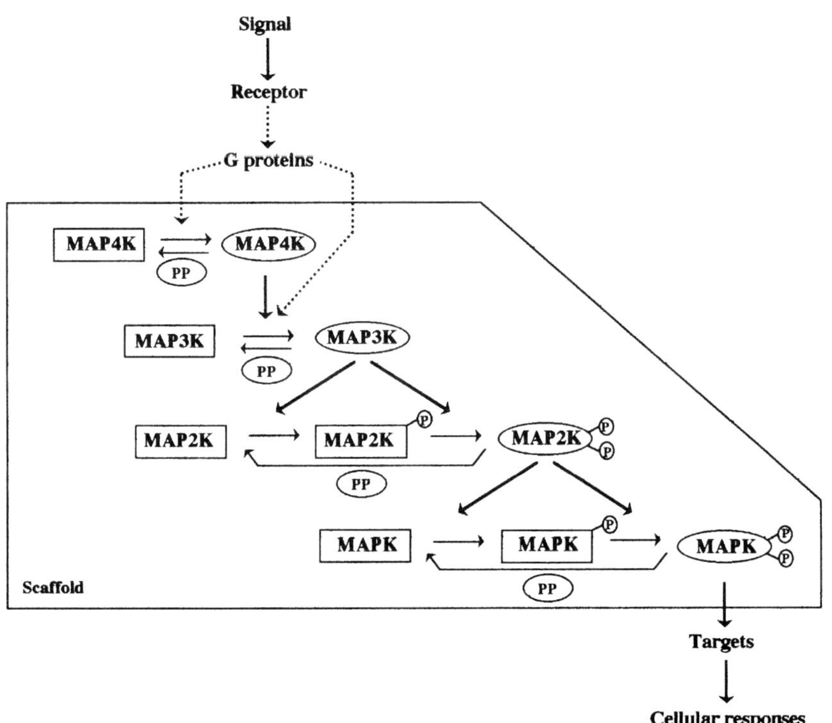

Fig. 1. General scheme showing the mitogen-activated protein kinase cascade. Inactive forms are square boxed, active forms are oval boxed; PP indicates a phosphatase and the phosphates are circled.

small GTP-binding proteins (G-proteins), also known as RAS homologues, regulate evolutionary conserved MAPK pathways (Mösch *et al.*, 1996). This activation is followed by the recruitment to the plasma membrane of a protein complex involving the MAP kinase module. In yeast and animals, Ste20-like MAP4Kinases also function upstream of MAP kinase pathways, which might be regulated by direct recruitment to GTPases (Sells and Chernoff, 1997).

Following their activation, MAP kinases phosphorylate target substrates exclusively on serine or threonine residues (Ferrel, 1996) with a proline residue in the P + 1 position within the substrate recognition consensus motif Ψ-x-[TS]-P, with Ψ indicating P or an aliphatic residue (Clark-Lewis *et al.*, 1991; Marshall, 1994). MAPK substrates include transmembrane and cytoplasmic proteins, enzymes, cytoskeletal substrates, and also signalling components that may be involved in feedback regulation (MAP2Ks, MAP3Ks and upstream regulators), and other protein kinases such as MAPK-activated protein kinases (Rouse *et al.*, 1994; Kültz, 1998). MAPK targets also include a number of nuclear proteins such as c-Myc (Seth *et al.*, 1991) and c-Jun (Alvarez *et al.*, 1991; Pulverer *et al.*, 1991; Derijard *et al.*, 1994; Musti *et al.*, 1997) transcription factors which are positively or negatively regulated.

B. MAP KINASE PATHWAYS FROM MODEL ORGANISMS

Focused studies on model organisms have been important for advancing the topic of MAP kinase modules. Genetic and molecular analyses of developmental processes involving MAP kinase modules have revealed complex patterns of signalling. In addition to the metazoan systems (mammals, *Drosophila melanogaster*, *C. elegans* and *D. discoideum*), *Saccharomyces cerevisiae* with at least five cascades has provided a significant contribution to the dissection of MAP kinase cascades (Errede and Levin, 1993; Herskowitz, 1995; Gustin *et al.*, 1998). In many cases the distinctions between MAP kinase family members are subtle, but functionally they represent distinct signalling pathways. Several of these pathways need not respond directly to extracellular signals, but may be activated or inactivated secondarily in response to the activation by other signalling pathways.

1. The Mating Pheromone Cascade

The regulation of the mating pheromone pathway in the budding yeast *S. cerevisiae* has been studied genetically through the isolation of sterile (Ste) mutants that are deficient in mating (Wittenberg and Reed, 1996). Yeast cells respond to mating pheromones by activating a signal transduction pathway which is a model MAP kinase pathway (see Fig. 2). It consists of sequential events leading to conjugation between haploid cells of opposite mating types a and α, and it is concluded by the transcriptional induction of mating-specific genes, cell cycle arrest and morphological changes. Each haploid cell type

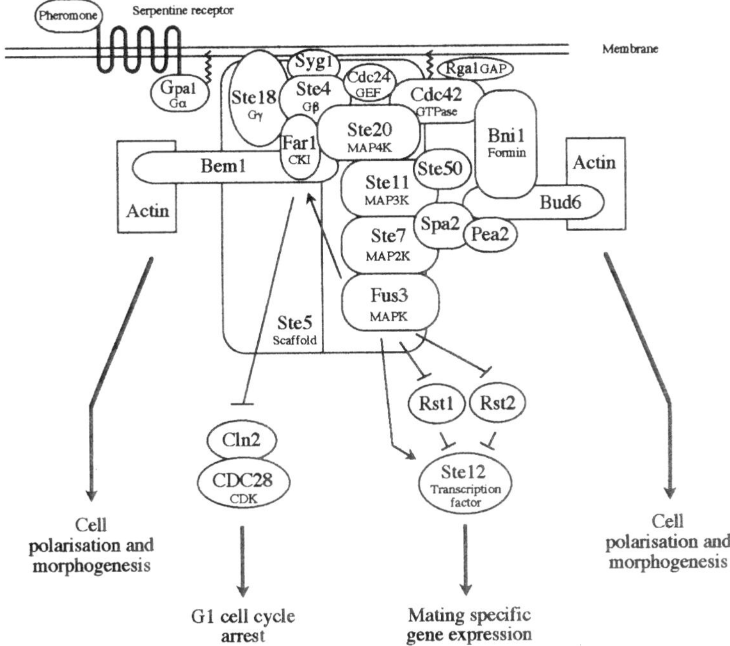

Fig. 2. The yeast mating-response MAP kinase cascade: an example to illustrate the complexity of a MAP kinase signalling pathway. For abbreviations see Table I.

secretes a unique signalling molecule, namely the a-factor and the α-factor. To initiate mating, pheromone from the opposite a and α cell type binds respectively to the serpentine receptors Ste2p and Ste3p. Ligand binding triggers modification of the receptor proteins, including both ubiquitination and conformational changes. Binding activates the pheromone signal transduction pathway that is common to both cell types by activating the Gα subunit (GDP-to-GTP exchange) from a heterotrimeric guanine nucleotide binding protein (Gpa1) which associates with the receptor. Free β–γ dimer subunits of the heterotrimeric G protein (respectively, Ste4 and Ste18) released in response to pheromone binding may activate GTP exchange on the small monomeric G protein of the Rho family, Cdc42, by interaction with the guanyl nucleotide exchange factor (GEF) Cdc24 and its GTPase activating protein (Rga 1).

G-proteins couple the receptor to downstream elements by stimulating a phosphorylation cascade. The activated small G-protein, Cdc42, probably recruits and activates the MAP4K Ste20 to the site of bud emergence (Leberer *et al.*, 1992). Leeuw *et al.* (1998) identified a binding site (i.e SSLAPLVKLA) for the G-protein β-subunit (Ste4) in the non-catalytic carboxy-terminal region of Ste20. A role for G$\beta\gamma$ membrane localization may be both to recruit the Ste5 scaffold protein to the membrane (Pryciak and Huntress, 1998) and to bring

Ste20 close to Ste5, an initial step in the mating cascade. Experimental results indicate that Bem1 [a src-homology-3 (SH3) domain-containing protein], Ste4 and the protein kinases, namely Ste20, Ste 11 and Ste7, are involved in sequential transmission of the signal. Furthermore, they are organized into a multi-protein complex (see Fig. 2) detected *in vivo* (Leeuw *et al.*, 1995; Schultz *et al.*, 1995; Whiteway *et al.*, 1995), by the dimeric scaffold protein Ste5 (Choi *et al.*, 1994; Yablonski *et al.*, 1996). This complex ensures the control of the signalling pathway. Pryciak and Huntress (1998) showed that pheromones stimulate the translocation of part of the Ste5 population to the cell surface, through $G\beta\gamma$ function. Haploid yeast cells are able to sense the origin of the pheromone source, and to direct protusions towards this source, through a polarized cell growth mediated by cytoskeletal elements (actin, myosin). The involvement of Bem1 which binds Cdc24 and interacts with Ste5 and actin (Leeuw *et al.*, 1995) demonstrates that cytoskeletal reorganization is required for proper signal transduction. A complex containing Bem1, Ste5, Far1, Cdc24, Ste20, Cdc42, Ste4 and actin could be crucial for activating the downstream signalling pathway (Nern and Arkowitz, 1998).

Ste20 phosphorylates the upstream element of the MAP kinase cascade, MAP3K Ste11, which in turn phosphorylates and activates a MEK homologue, the MAP2K Ste7. There are two types of Ste7 regulatory phosphorylation. One, catalysed by Ste11, is required for catalytic competence, and as a consequence, a hyperphosphorylation within an amino-terminal domain by downstream MAPK (in a feedback reaction) is associated with attenuation of Ste7 activity (Errede and Ge, 1996), probably by altering their association. Downstream, Ste7 phosphorylates and activates alternatively both the MAPKs Fus3 and Kss1. Fus3 regulates mating whereas Kss1 is involved in the filamentation pathway (Madhani *et al.*, 1997). Fus3 has a dual task through its two substrates: the Ste12 transcriptional activator which activates mating-specific gene expression, and Far1, a cyclin-dependent kinase inhibitor required for cell cycle arrest in the G1 phase by binding the cyclin-dependent protein kinase Cdc28. Moreover, in response to pheromone, a fraction of Far1p moves from the nucleus to the cytoplasm, and functions as an adaptor that binds $G\beta\gamma$ and interacts with Bem1, Cdc24 and Cdc42 (Butty *et al.*, 1998). Ste12 binds inside pheromone-inducible promoters on pheromone response elements (PRE) domains and leads to gene activation (Madhani and Fink, 1998). Recently, two newly identified proteins, Rst1 and Rst2, were found to physically associate with Fus3 and Ste12 in unstimulated cells, and thus repress the mating (Tedford *et al.*, 1997). Upon pheromone stimulation, Fus3 phosphorylates Rst1, Rst2 and Ste12, and thus liberates Ste12 to interact with the transcription machinery of particular genes. This signal transduction pathway induces: (1) cell cycle arrest in late G1 phase at the 'start' position, (2) transcriptional activation of mating-specific genes involved in conjugation between the a and α cell types, and (3) acquisition of polarized growth towards a mating partner.

2. The HOG (High Osmolarity) Pathway in Yeast

Exposure of the budding yeast *S. cerevisiae* to an extracellular osmotic shock rapidly activates a MAP kinase cascade (Brewster *et al.*, 1993). The HOG pathway (see Fig. 3) appears to be controlled by a two-component histidine kinase which is homologous to prokaryotic two-component transmembrane sensor systems (Ota and Varshawsky, 1993; Maeda *et al.*, 1994). In bacteria, these signal transduction systems consist of two proteins, a sensor and a response regulator. The yeast two-component system involves as a first component the Sln1p sensor composed of transmembrane segments (ligand-binding domain), a histidine kinase transmitter domain and a receiver domain. The second component is the histidine phosphorelay Ypd, and the third component is the response regulator Ssk1, containing a receiver domain. In the absence of an extracellular signal, Sln1 autophosphorylates at histidine 576, and the phosphatidyl group is transferred to an aspartate (1144), then to a histidine residue (64) on Ypd1 and finally to an aspartate (554) residue of Ssk1 (Maeda *et al.*, 1994). This constitutes an example of a classical multi-step

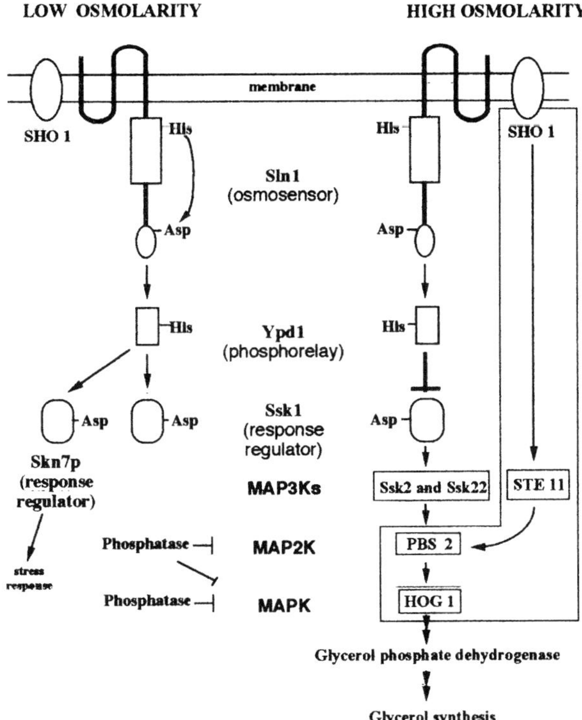

Fig. 3. The yeast high osmolarity MAP kinase response pathway. This figure has been adapted from Posas and Saito (1998).

phosphorelay. The response-regulator protein Ssk1 is then inactivated, thus the downstream MAP kinase cascade is not activated. Ypd1 may also act as phosphodonor for Skn7, which is suspected to participate in modulating HOG-pathway activity (Ketela *et al.*, 1998). Under low osmolarity, the second osmosensor, Sho1, is not activated. Sho1 has four predicted transmembrane segments and a carboxy-terminal cytoplasmic region containing a SH3 domain.

Exposure of the yeast cells to high salt concentrations disrupts the phosphorylation cascade. Sln1 histidine kinase activity is suppressed, and accumulating unphosphorylated Ssk1 binds an amino-terminal segment and activates Ssk2 and/or Ssk22 MAP3Ks (Maeda *et al.*, 1994). Ssk2/Ssk22 increases its catalytic activity by autophosphorylation on threonine in the catalytic domain and activates MAP2K Pbs2 (a MEK homologue), which in turn phosphorylates and activates the MAPK homologue Hog1 (Posas and Saito, 1998). The phosphorylation of Hog1 by Pbs2 is required for its nuclear translocation (Ferrigno *et al.*, 1998) which occurs very rapidly but is transient. As a consequence, this induces the transcription of genes participating in glycerol synthesis required for maintaining the osmotic balance. Schüller *et al.* (1994) have identified stress-response elements as specific targets of the HOG pathway. Under high osmolarity, Pbs2 can also be activated by the transmembrane osmosensor Sho1 through MAP3K Ste11 (Boguslawski, 1992). This pathway requires a mechanism of protein–protein interaction between the Sho1 SH3 domain and the Pbs2 SH3-binding site (Maeda *et al.*, 1995). Recent results from Posas and Saito (1997) are consistent with the formation of a multi-protein complex that includes the Sho1 osmosensor, the MAP3K Ste11, the MAP2K Pbs2 and the MAPK Hog1, in which Pbs2 may serve as a scaffolding protein. Both Sln1 and Sho1 appear to function as osmosensors, allowing an optimal adaptation to a fluctuating environment.

3. *Mammalian Mitogen-Response and Stress-Response Pathways*

Mammalian cells respond to changes in their environment by inducing MAP kinase cascades (for a review see Lewis *et al.*, 1998), which regulate different biological responses. Several mammalian MAP kinase cascade element genes have recently been characterized, suggesting the existence of more MAP kinase pathways in metazoans than was previously suspected. The archetype is the ERK (extracellular signal-regulated kinase) pathway, highly conserved through metazoan evolution, which responds primarily to mitogenic stimuli via Ras small G-proteins. ERK signalling stimulated by growth and differentiation factors (Boulton *et al.*, 1991) is involved in the control of mitosis, meiosis and differentiation. The signal for ERK activation is relayed from the cell surface to the nucleus through a well-characterized signalling pathway. The binding of growth factors to their specific receptor induces interaction with a monomeric G-protein Ras. The GTP binding to Ras activates the Raf (or Mos) MAP3K by recruitment to the membrane. Raf activates a MEK (MAP2K) which targets an Erk1 or Erk2 MAPK (Woodgett *et al.*, 1996). Activation of MAPK takes place in the cytoplasm and the activated enzyme moves to the nucleus to phosphorylate transcription factors. Following phosphorylation, Erk2 dimerizes, which is part of the mechanism required for nuclear translocation (Canagarajah *et al.*, 1997; Khokhlatchev *et al.*, 1998). Extended activation and nuclear retention of ERKs is required for transcription regulation. Recent results indicate that ERKs are also involved in cytoskeletal reorganization during cell division and cell differentiation, and in the regulation of various cytoplasmic proteins involved in signal transduction pathways (Lewis *et al.*, 1998).

Studies of the ERK pathway have given rise to the idea that this pathway is not a simple linear pathway but is likely to be one strand in a complex web of signalling pathways involved in cell-to-cell communication. ERK cascades are activated in response to various signals that stimulate cell surface receptors, including at least receptor tyrosine kinases (RTKs) and G-protein-coupled receptors (GPCRs).

Upon tyrosine phosphorylation of the RTK receptor, Shc (src homologous and collagen homologous) functions as a linker to recruit Grb2–Sos1 complexes to the activated receptor (VanderKuur *et al.*, 1995). A Grb2 SH2 domain mediates Shc binding while its flanking SH3 domains recruit Sos1 (He *et al.*, 1995), which promotes Ras-GTP formation (Campbell *et al.*, 1998), thus initiating the MAP kinase cascade through activation of Raf. Many components of the RTK–Ras–MAPK pathway are organized into supra-molecular assemblies (see Fig. 4).

The Janus tyrosine kinases (JAKs) and the signal transducers and activators of transcription (STATs) are required for cytokine signalling through the JAK/STAT pathway (for a review see Moutoussamy *et al.*, 1998), conserved between mammals and *Drosophila* (Binari and Perrimon, 1994; Hou *et al.*, 1996; Ihle, 1996; Yan *et al.*, 1996; Horvath and Darnel, 1997). JAKs are mediators of ERK activation by cytokines (Ihle and Kerr, 1995; Winston and Hunter, 1995; Bonni *et al.*, 1997). Tyrosine residues of activated receptors are phosphorylated by the active JAK, providing binding sites for the coupling of Ras-regulating proteins.

Signals such as growth factors, hormones and other agonists that stimulate GPCR or RTK can cause diacylglycerol (DAG) production. Many mammalian signalling pathways activate protein kinase C (PKC) by production of its second messenger. PKC has been implicated both in yeast Bck1 phosphorylation (Herskowitz, 1995) and in mammalian cell proliferation though MAP kinase pathways. Stimulation of MKK/ERK through Gqα-coupled receptors is independent of G$\beta\gamma$ or Ras, and activates Raf/MKK/ERK through a PKC-dependent mechanism (Lewis *et al.*, 1998). The mitogenic pathway mediated by GPCR may also involve Giα-coupled receptors which activate ERK through the Gi$\beta\gamma$ subunit complex and PTK, via Ras activation (Wan *et al.*, 1996).

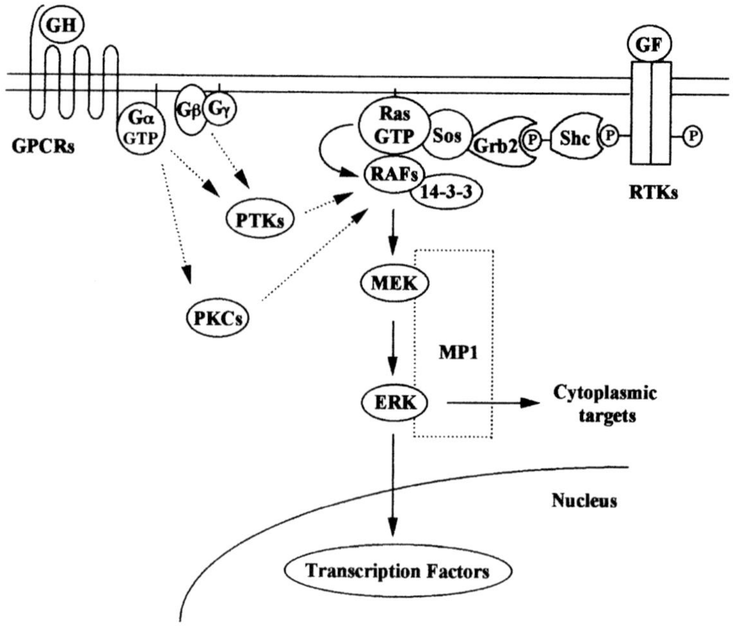

Fig. 4. Schematic representation of signalling through the mammalian RAF/MEK/ERK pathway.

Despite intensive research, the mechanisms that couple GPCRs and the ERK cascade are still poorly characterized (see Fig. 4).

Mammalian MAPKs are involved in signalling associated not only with mitogens (ERKs), but also with environmental stress responses (Robinson and Cobb, 1997). SAPKs (or JNKs) and p38 signal transduction pathways mediate responses to various cellular stresses, including inflammatory cytokines, and to apoptosis (Lewis *et al.*, 1998). The JNK pathway involves the sequential activation of a MAP kinase module (MEKK and MLKs, SEK, JNK) and sometimes also a MAP4K upstream of MEKK, which can be organized in a multi-protein complex via the scaffold protein JIP-1 (Whitmarsh *et al.*, 1998). The upstream regulators of the JNK/SAPK and p38 pathways are Rac, Rho and Cdc42 monomeric G-proteins, and their effectors are MAP4Ks. Ligand receptors implicated in JNK/SAPK and p38 pathways include cytokine receptors, G-protein-coupled receptors and tyrosine kinase receptors. The protein phosphatase PP2Cα negatively regulates the p38 pathway by dephosphorylating and inactivating the MAP2K MKK6 and the MAPK p38, through a direct interaction (Takekawa *et al.*, 1998). A high number of stress-related elements has been characterized and the link between these elements inside signalling cascades seem to be complex. It is speculated that this complexity of interaction may be related to cell specificity and regulation of subcellular localization via multi-protein complexes leading to a regulation of MAP kinase responses in localized regions of the cell and depending upon the different upstream stimuli (Fanger *et al.*, 1997).

Within minutes, cell stimulation by growth factors or various stresses transiently activates MAPK, but the kinetics of the related MAPK activation differs significantly. In mammalian cells, SAPK activity increases rapidly (within a few minutes) and transiently, in response to cellular stresses such as heat and metabolic poisons (Woodgett *et al.*, 1996). The mammalian ERK phosphorylation during mitogen-dependent cell proliferation is different. As a general rule, activation of ERK is biphasic, comprising a rapid, transient peak of activity within a few minutes, preceding a longer sustained period of activation during which activity falls back to basal levels within a few hours (Ferrel, 1996). This is an obligatory event for growth-factor-induced cell cycle progression (Meloche *et al.*, 1992). The sustained period of activation is accompanied by translocation of ERK to the nucleus (Lenormand *et al.*, 1993) which occurs rapidly (5–30 min) and typically endures for several hours. Several MKPs are excluded from the nucleus and thus are regulators of cytoplasmic MAPK, whereas others localized within the nucleus thus provide a mechanism for the inactivation of nuclear MAPKs (Lewis *et al.*, 1998). It can be hypothesized that a cytoplasmic ERK phosphatase first decreases ERK activity within the cytoplasm during ERK translocation to the nucleus. Thereafter, inside the nucleus, a nuclear ERK phosphatase transcription is activated, and as a consequence the ERK kinase activity falls back.

C. REGULATION THROUGH SCAFFOLDING, ANCHORING AND ADAPTOR PROTEINS

A view emerging from the studies of model MAP kinase pathways is the formation of protein complexes through the assembly of various signalling molecules. The role of scaffolding, anchoring and adaptor proteins in the specificity and selectivity of signalling has been detailed by Pawson and Scott (1997) and Whitmarsh and Davis (1998). The specificity of interaction between the various elements is provided by recognition and binding-sequence signatures, located both within and outside the kinase catalytic domain. Such highly conserved interaction motifs are present in elements of the MAP kinase cascade (e.g. SH2, SH3 and PH domains).

We have detailed the Ste5 complex, composed of numerous elements constituting the pheromone pathway in yeast. It has been suggested that Ste5 serves as an anchor for the MAP2K and permits the recruitment of unphosphorylated MAPK. In mammals a scaffold protein, named JIP1, has recently been identified, and shown (Dickens *et al.*, 1997; Whitmarsh *et al.*, 1998) to interact with components (i.e. MLK3, MKK7 and JNK) of the JNK pathway. This scaffold protein has different isoforms, characterized by different tissue specificities, which may share different protein specificities

(Kim *et al.*, 1999). Recently, the MP1 protein (MEK partner 1) was shown to be bound to MEK1 and to ERK1, and revealed to participate in their activation (see Schaeffer *et al.*, 1998). Using genetic studies, the authors have demonstrated that the above scaffold proteins are essential for the localization of signalling proteins, a critical step for the signalling machinery. Moreover, these proteins possess protein interaction domains susceptible to the building of large multi-protein complexes. Furthermore, the association of PBS2, a MAP2K, to other elements of the yeast HOG pathway has been detailed by Posas and Saito (1997). The sequential MEKK1–JNKK1 and JNKK1–JNK interactions, via the JNKK1 amino-terminal extension, activate the JNK module (Xia *et al.*, 1998) in a PBS2-like process.

The use of adaptor proteins may provide additional docking sites for MAP kinase signalling elements. For example, leucine-rich repeats (LRR) are characterized by a 23- or 24-residue consensus sequence, composed of leucines at invariant positions. These repeats have been shown to exist in plants, yeast and animals. It has been proposed that LRR participates in protein–protein interactions (Kobe and Deisenhofer, 1994). Based on the results obtained by Sieburth *et al.* (1998), LRR's adaptor proteins probably establish a link within the proteins involved in MAP kinase signalling, and are part of a protein complex. Potential interactions with other adaptors, such as the 14-3-3 proteins, help in building a signalling network. 14-3-3 dimers have been shown to associate with Raf-like proteins such as CTR1 (Solano and Ecker, 1998) and other protein kinases (Tzivion *et al.*, 1998). Hence, 14-3-3 proteins regulate protein–protein interactions and determine their localization and activation (Pawson and Scott, 1997). It has been suggested that the 14-3-3 proteins detected in plants may provide a structural link between signal transduction proteins, by acting as putative anchor proteins, through a chaperone-like action (Aitken, 1996; Muslin *et al.*, 1996). These mechanisms are used to link a MAP kinase module element to its target or to increase its affinity to an upstream or downstream partner. As a general rule, adapter proteins would enhance efficiency and specificity of the MAP kinase pathway.

The assembly of signalling molecules involves adaptor proteins that contain modules mediating association between proteins. Many proteins, including MLK, regulated by receptor tyrosine kinases contain, in addition to the src-homology-2 (SH2) domain, a SH3 domain. SH2 domains recognize short peptide motifs bearing phosphotyrosine residues, while SH3 domains bind to the proline-rich motifs of the target proteins (Pawson, 1995). Proline-rich sequences within the carboxy-terminal tail of MAP4Ks have been demonstrated to bind to proteins with SH3 domains (Anafi *et al.*, 1997). RTK tyrosine residues serve as docking sites for SH2 domains of adaptor proteins (Watanabe and Arai, 1996). A variety of adaptor proteins has been shown to possess SH2 and SH3 domains. Several adaptor proteins possess kinase activities, for example the Src family tyrosine kinase. The other adaptor proteins only consist of SH2 and SH3 domains, namely Nck, Grb2, Sos and

Sem5. This may link a protein to two or more partners, and may recruit kinases into an active complex (Pawson and Scott, 1997). The prominent part of such domains has been detailed earlier through the mammalian pathways (see Section II.B).

An emerging idea is that multi-protein complexes could be involved in subcellular compartmentalization of signalling elements, thus enabling efficient signal propagation. The sequence of particular domains or docking sites will select the molecule containing the target domain, thus determining the activated signalling pathway. The subcellular distribution of several kinases is restricted by association with targeting proteins or scaffold proteins (Mochly-Rosen, 1995; Faux and Scott, 1996). Immunolocalization of MEKK1, MEKK2, MEKK3 and MEKK4 proteins, shown to be able to phosphorylate the same MAP2Ks, revealed different subcellular distributions, probably linked to the formation of different multi-protein complexes (Fanger *et al.*, 1997). Differential MEKK subcellular distributions and interactions with small GTP-binding proteins provide a mechanism susceptible to regulate either MAP kinase responses in localized regions of the cell or different upstream stimuli. Recent results also demonstrate that kinases and phosphatases are physically joined (Camps *et al.*, 1998; Hafen, 1998). The concept of protein complexes involved in MAP kinase pathways seems to be of major importance for signal transmission in various organisms, including plants.

III. WHAT ABOUT MAP KINASE MODULE ELEMENTS IN PLANTS?

The proteins that act during MAP kinase signalling have been highly conserved during evolution, and are shared by vertebrates and invertebrates, but also by unicellular eukaryotes and plants. This allowed the application of genetic tools to the characterization of plant signal transduction pathways in different species. For example, genes encoding receptors with structural features similar to prokaryotic two-component regulators, and putative receptors that phosphorylate Tyr, Ser and Thr have recently been cloned in plants (Hanks and Hunter, 1995). A number of genes identified by positional cloning has been recognized by their sequences as likely MAP kinase signal transduction elements. In addition, an increasing number of plant MAP kinase module components has been isolated either by polymerase chain reaction (PCR) methods, by using expressed sequence tags (EST) or in systematic sequencing programmes. Genes encoding putative MAP kinase cascade components have been cloned from various species, namely alfalfa, *Arabidopsis*, *Citrus*, sweet potato, oat, parsley, pea, petunia, *Prunus*, rape, rice, *Ricinus*, tobacco and tomato (see reviews by Jonak *et al.*, 1994; Nishihama *et al.*, 1995; Tregear *et al.*, 1996; Hirt, 1997; Machida *et al.*, 1997; Mizoguchi *et al.*, 1997; Jouannic *et al.*, 1999b). Therefore, it seems likely that the basic MAP kinase module, as described above (Section

II), is adapted to a number of different functions in plants, as has already been shown to be the case in yeast and animals. This also suggests that the multiple MAP kinase pathways play a central role in plant cell regulation.

A. WHY STUDY MAP KINASE CASCADES IN PLANTS?

Signalling events involved in mechanisms protecting plant cells against damage caused by abiotic and biotic stress conditions, exerted by high salt concentrations, heavy metal ions, radiation, extreme pH, heat, wounding, drought and pathogen attack have received attention from plant biologists (Suzuki and Shinshi, 1995; Usami *et al.*, 1995; Bögre *et al.*, 1996, 1997; Sheen, 1996; Shinozaki and Yamaguchi-Shinozaki, 1996; Hirt, 1997; Mizoguchi *et al.*, 1997). Resistance of plants to stress conditions is particularly important for agriculture and is highly relevant for ecology. Moreover, to understand the mechanisms controlling plant development, highly dependent on environmental conditions, it will be important to elucidate the intracellular signalling events controlling cell fate, cell proliferation, cell differentiation and cell growth.

Studies on signalling pathways from model organisms provided evidence for the role of MAP kinase cascades in mediating various signals (see section above). Analogy-based studies will permit us to determine the signals mediated by MAP kinase signalling pathways, and lead to the understanding of the molecular mechanisms controlling plant development and plant environmental responses. Plant gene and cyclic DNA (cDNA) cloning showed that numerous MAP kinase kinase kinase kinases, MAP kinase kinase kinases, MAP kinase kinases and MAP kinases homologues do exist in plants (see Table II). An increasing body of evidence suggests that MAP kinase modules play important signalling roles in plants, particularly during the activation of various stress-associated responses and mechanosensing (Jonak *et al.*, 1994; Nishihama *et al.*, 1995; Bögre *et al.*, 1997; Hirt, 1997; Mizoguchi *et al.*, 1997; Zhang and Klessig, 1997), in cell cycle progression (Jonak *et al.*, 1993; Calderini *et al.*, 1998; Nakashima *et al.*, 1998; Bögre *et al.*, 1999) and in their response to hormones (Kieber *et al.*, 1993; Kovtun *et al.*, 1998). A large number of MAP kinase genes has been identified through the *Arabidopsis thaliana* sequencing programmes. The cloning of many genes in a plant model species has opened the way for examining crucial questions such as the perception and transduction of rapidly changing external and internal conditions. Many exciting new insights are likely to come from the plant model species *Arabidopsis*.

B. PLANT MAP KINASE CASCADE COMPONENTS

Many cDNA and genes encoding putative MAPKs, MAP2Ks, MAP3Ks and MAP4Ks have been identified in various plant species (see Table II). Several

TABLE II

The list of plant MAPK, MAP2K, MAP3K and MAP4K. The sequences have been classified according to Hunter and Plowman (1997)

	Name	Species	Accession number	Molecular mass	Yeast homologue	Regulation			References
						Tr	PTL	PTP	
CMGC									
PERK group									
PERKα subgroup	MMK1[1]	*Ms*	X66469 = L07042	44.4		M			Duerr *et al.* (1993); Jonak *et al.* (1993)
	AtMPK6	*At*	D21842	45.1		+			Mizoguchi *et al.* (1993)
	PsMAPK[2]	*Ps*	X70703	45.1	*hog 1*				Stafstrom *et al.* (1993); Pöpping *et al.* (1996)
	NTF4	*Nt*	X83880	45.1		+	H-M		Wilson *et al.* (1995, 1997)
	p48SIPK[3]	*Nt*	U94192	45.2		W	Sa-E	Sa	Zhang and Klessig (1997, 1998); Zhang *et al.* (1998)
	MAPK5	*Zm*	AB016802	44.9					Unpublished
PERKβ subgroup	AtMPK3	*At*	D21839	42.7		C-S-T			Mizoguchi *et al.* (1993)
	AsPK9[4]	*As*	X7993	42.9		+			Huttly and Phillips (1995)
	DS22[5]	*Nt*	D61377	42.8		W		+	Seo *et al.* (1995)
	MMK4[6]	*Ms*	X82270	43		D-C-W	D-C-W		Jonak *et al.* (1999)
	ERMK	*Pc*	Y12785	43			E		Ligterink *et al.* (1997)
	WCK	*Te*	AF079318	42.9					Unpublished
	MAPK4	*Zm*	AB016801	46.7					Unpublished
	IbMAPK	*Ib*	AF149424	41.7					Unpublished
PERKγ subgroup	AtMPK4	*At*	D21840	42.9					Mizoguchi *et al.* (1993)
	AtMPK5	*At*	D21841	43.1					Mizoguchi *et al.* (1993)
	MMK2	*Ms*	X82268	42.8	*mpk1*				Jonak *et al.* (1995)

TABLE II *Continued*

	Name	Species	Accession number	Molecular mass	Yeast homologue	Regulation			References
						Tr	PTL	PTP	
	NTF6[7]	*Nt*	X83879	42.7			Cy		Wilson *et al.* (1995); Calderini *et al.* (1998)
	MMK3	*Ms*	AJ224336	42.9			Cy		Bögre *et al.* (1999)
	AtMPK	*At*	AC005397	46.1					Unpublished
PERKδ subgroup	NTF3[8]	*Nt*	X69971	42.8		+			Wilson *et al.* (1993, 1995)
	AtMPK7	*At*	D21843	42.4					Mizoguchi *et al.* (1993)
	AtMPK1[9]	*At*	D14713	42.7					Mizoguchi *et al.* (1994)
	AtMPK2[9]	*At*	D14714	43.1			St		Mizoguchi *et al.* (1994)
	PMEK1	*Ph*	X83440	44.4		+			Decrooq-Ferrant *et al.* (1995)
	AtMAPK7	*At*	Z99708	42.0					Unpublished
	PaMAPK	*Pa*	AF134730	42.4					Unpublished
?	AtMPK8	*At*		?					Unpublished
	AtMPK9	*At*		?					Unpublished
	x	*Le*		?			W-E		Stratmann and Ryan (1997)
STE7/MEK									
PMEK subfamily									
PMEKα group	MAP2Kα[10]	*At*	Y07694	38.3					Jouannic *et al.* (1996)
	AtMKK4	*At*	AB015315	40.1					Ichimura *et al.* (1998b)
PMEKβ group	MAP2Kβ[11]	*At*	AJ006871	40					Hamal *et al.* (1999)
	AtMEK1[12]	*At*	AF000977	39.2	*pbs2*	W-EtDet			Morris *et al.* (1997); Mizoguchi *et al.* (1998)
	LeMEK1	*Le*	AJ000728	39.7					Unpublished

TABLE II *Continued*

	Name	Species	Accession number	Molecular mass	Yeast homologue	Regulation			References
						Tr	PTL	PTP	
	ZmMEK1	*Zm*	U83625	39.9		C			Hardin and Wolniak (1998)
PMEKγ group	NPK2	*Nt*	D31964	57.5					Shibata *et al.* (1995)
	AtMKK3	*At*	AB015314	57.5					Ichimura *et al.* (1998b)
STE11/STE20									
MEKK/STE11 family									
PMEKK subfamily									
MAP3Kα group	MAP3Kα	*At*	AJ010090	66.5					Jouannic *et al.* (1999a)
	MAP3Kα1	*Bn*	AJ010091	64.5					Jouannic *et al.* (1999a)
MAP3Kβ group	AtMEKK1	*At*	D50468	65.9	*ste11*	C-S-To			Mizoguchi *et al.* (1996)
	AtARAKIN	*At*	L43125	Truncated	*ste6-ste11-ste20*				Covic and Lew (1996)
	MAP3Kβ3	*At*	AJ010092	62.4					Jouannic *et al.* (1999a)
	MAP3Kβ4	*At*	AF076275	84.9					Jouannic *et al.* (1999a)
	MAP3Kβ5	*At*	AL049638	?					Unpublished
	MAP3Kβ1	*Bn*	AJ010093	62.6					Jouannic *et al.* (1999a)
MAP3Kγ group	MAP3Kγ	*At*	Y14316	58.9					Jouannic *et al.* (1999a)
MAP3Kζ group	NPK1	*Nt*	D26601	76.2	*bck1,pck1*	M	NA		Banno *et al.* (1993); Kovtun *et al.* (1998)
	ANPIL	*At*	AB000796	72.8	*ste11*				Nishihama *et al.* (1995)
	ANP1S	*At*	AB000797	41.4	*ste11*				Nishihama *et al.* (1995)
	ANP2	*At*	AB000798	70.8					Nishihama *et al.* (1995)
	ANP3	*At*	AB000799	71.7					Nishihama *et al.* (1995)
	MAP3Kζ1	*Cs*	C21846	Truncated					Unpublished

TABLE II *Continued*

	Name	Species	Accession number	Molecular mass	Yeast homologue	Regulation			References
						Tr	PTL	PTP	
RAF family									
PRAF subfamily									
MAP3Kδ group	CTR1	*At*	L08789	90.3			ER		Kieber *et al.* (1993)
	LeCTR1	*Le*	Y13273	92.0					Wang and Li (1997)
	LeCTR2	*Le*	AJ005077	107.3					Unpublished
	MAP3Kδ1	*At*	Y14199	Truncated					Jouannic *et al.* (1999a)
	MAP3Kδ1	*Os*	D47273	Truncated					Unpublished
	MAP3Kδ2	*At*	R29949	Truncated					Jouannic *et al.* (1999a)
	MAP3Kδ3	*At*	AC003981	104.9					Jouannic *et al.* (1999a)
	MAP3Kδ4	*At*	AL031018	82.3					Unpublished
	MAP3Kδ5	*At*	AL078637	108					Unpublished
MAP3Kη group	MAP3Kη1	*At*	H36649	Truncated					Jouannic *et al.* (1999a)
	MAP3Kη2	*At*	Ac003981	Truncated					Unpublished
MAP3Kθ group	MAP3Kθ1	*At*	AC004669	Truncated					Jouannic *et al.* (1999a)
	MAP3Kθ1	*Rc*	T14949	Truncated					Unpublished
	MAP3Kθ1	*Os*	D41138	Truncated					Unpublished
	MAP3Kθ2	*At*	AC007087	40.2					Unpublished
CDC7 family									
PCDC7 subfamily	MAP3Kϵ1	*At*	AJ224982	151.2			M		Authors' unpublished results
	MAP3Kϵ1	*Bn*	AJ238845	143.7					Authors' unpublished results

TABLE II *Continued*

	Name	Species	Accession number	Molecular mass	Yeast homologue	Regulation			References
						Tr	PTL	PTP	
MAP4K family									
GCK/SPS1 subfamily	MAP4Kα1	*Bn*	AJ009608	75.1			M		Leprince *et al*. (1999)
	MAP4Kα2	*Bn*	AJ009609	74.6			M		Leprince *et al*. (1999)
	SIK1	*At*	U96613	92.7					Unpublished
	T19F6.10	*At*	AC002343	63.5					Unpublished
	F17O7.3	*At*	Z97336	62.3					Unpublished

1. MsERK1 = MMK1 = MsK7; 2. PsMAPK = D5; 3. p48SIPK = SIPK = NtSAPK (stress-activated protein kinase); 4. Aspk9 = AsMAP1; 5. DS22 = WIPK; 6. MMK4 = SAMK; 7. NTF6 = p43; 8. NTF3 = NtERK; 9. A *Xenopus* MAPkinase kinase phosphorylates ATMPK1 and ATMPK2; 10. AtMAP2Kα = AtMKK5; 11. AtMAP2Kβ = AtMKK2; 12. AtMEK1 = AtMKK1.

Species: *Ms* = *Medicago sativa; Nt* = *Nicotiana tabacum; As* = *Avena sativa; Ps* = *Pisum sativum; Ph* = *Petunia hybrida; At* = *Arabidopsis thaliana; Pc* = *Petroselinum crispum; Bn* = *Brassica napus; Le* = *Lycopersicum esculentum; Os* = *Oryza sativa; Rc* = *Ricinus communis; Cs* = *Citrus sinensis; Zm* = *Zea mays; Te* = *Triticum aestivum; Ib* = *Ipomoea batatas; Pa* = *Prunus armeniaca*.

Regulation: Tr = transcriptional; PTL = post-translational; PTP = protein tyrosine phosphatase; Sa = salicylic acid; D = drought (= dehydration); C = cold; W = wounding; E = elicitor; ER = ethylene response; T = touch; S = salinity; H = hydration; EtDet = etiolation–De-etiolation; M = mitosis; Cy = cytokinesis; NA = negative regulator of auxin signalling; St = stress.

groups have reported biochemical activities of plant protein kinase putatively related to MAPK; however, their molecular characterization requires further investigations (Suzuki and Shinshi, 1995; Usami *et al.*, 1995; Stratmann and Ryan, 1997). Research in this area has rapidly expanded, and additional members have been identified from systematic sequencing of the genome from the model species (*A. thaliana* and *Oryza sativa*). As a consequence of the *S. cerevisiae* genome sequencing programmes (Mewes *et al.*, 1997), we know precisely that the budding yeast genome contains 118 conventional protein kinase genes, constituting about 2% of the total number of genes identified (Hunter and Plowman, 1997). More than 60% of these protein kinases have either known or suspected functions; the remainder are novel and still await a functional analysis. There are at least six MAPKs, four MAP2Ks, four MAP3Ks and six MAP4Ks (Hunter and Plowman, 1997). By linear extrapolation, we might estimate the putative number of protein kinase genes in *A. thaliana*, whose genome is about four times larger than that of yeast. Based on this and on the fact that the gene density is lower than in yeast, a total of about 400–500 protein kinases encoded by the *Arabidopsis* genome seems reasonable, and about 10–12 MAPKs, 5–10 MAP2Ks, 25–30 MAP3Ks and five MAP4Ks might be expected within its genome. At present (August 1999) about 65.5% of the whole genome has been sequenced. Only eight MAP2Ks (five from *A. thaliana*) and five MAP4Ks (three from *A. thaliana*), but 28 MAPKs (11 from *A. thaliana*) and 29 MAP3Ks (21 from *A. thaliana*) have been identified in plants (see Table II). The latter results suggest that MAPK cascades might have a wider range of functions in plants than in yeast.

The complexity and detailed regulation of specific signalling pathways may be markedly increased by the synthesis of multiple isoforms as a result of alternative splicing, as evidenced in *A. thaliana* (Nishihama *et al.*, 1997). This process may influence the activity of the protein, as suggested by the higher ability of the short isoform of AtANP1, to complement a yeast mutant compared to the longer isoform. Many vertebrate MAP kinase module elements with high sequence identities are thought to result from alternative splicing of a single gene. As an example, alternative processing of the transcripts from *JNK1, JNK2* and *JNK3* yields products containing different carboxy-termini, and thus proteins with different roles (Gupta *et al.*, 1996). However, this apparently occurs on a larger scale in animals than in plants.

In order to show that the putative MAP kinase genes encode functional protein kinases, their biochemical function was demonstrated by *in vitro* phosphorylation assays. Several MAPKs (Duerr *et al.*, 1993; Jonak *et al.*, 1995, 1996; Wilson *et al.*, 1995), the MAP2Ks NPK2 (Shibata *et al.*, 1995), ZmMEK1 (Hardin and Wolniak, 1998) and the MAP3Ks NPK1, constitutive triple response 1 (CTR1) (Kovtun *et al.*, 1998) and BnMAP3Kϵ1 (authors' unpublished results) exhibit autophosphorylation. Moreover, there are few examples of plant MAP kinase module members which are able to phosphorylate specific substrates *in vitro*. It has been shown that plant

MAPKs (Jonak *et al.*, 1995, 1996; Knetsch *et al.*, 1996), MAP2Ks (Hardin and Wolniak, 1998), MAP3Ks and MAP4Ks (authors' unpublished results) are able to phosphorylate the myelin basic protein (MBP). MBP phosphorylation is not restricted to MAPK, hence MBP is an unspecific substrate.

Some of the plant gene products have been shown to be related to members of the MAP kinase cascades of other organisms by functional complementation of yeast mutants (see Table II). These features indicate that there is a conservation of function as well as structural homology among the MAP kinase cascade components, through evolution.

Higher plants contain a large number of genes encoding various elements of a MAP kinase module. Although a whole pathway has not yet been reconstituted, recent results indicate that soon we will be able to do so in plants. Recently, Mizoguchi *et al.* (1998) pointed out, using the two-hybrid system, physical interactions between AtMEKK1, AtMEK1 and AtMPK4, suggesting that they may constitute a MAP kinase module in *A. thaliana*. Similarly, a MAP2K acting downstream of NPK1 and an activator for NPK1 has been identified, according to Nakashima *et al.* (1998). These results strongly suggest that MAP kinase cascades also exist in plants, but the physiological and biochemical relevance of these putative cascades remains to be evidenced. Moreover, plant MAP kinase activity after stress treatment is inhibited in the presence of a mammalian-specific MAP2K inhibitor (Romeis *et al.*, 1999). The transient overexpression of *NPK1* in mesophyl cells of maize (*Zea mays*) induced the phosphorylation of a protein immunodetected by a MAP kinase specific antibody (Kovtun *et al.*, 1998). Moreover, the NPK1-associated effect is abolished in the presence of a MAPK specific phosphatase. These results strongly suggest the existence of functional MAP kinase cascades in plant cells.

C. WHAT ABOUT THEIR BIOLOGICAL FUNCTION IN PLANTS?

The activity of the components of MAP kinase cascades must be regulated in a complex way. This is achieved at many different levels including protein synthesis, post-translational modification via protein phosphorylation–dephosphorylation and formation of multi-protein complexes. These processes confer specificity and selectivity to the signal transmission pathways. Several potential roles for MAP kinase signalling elements have been identified in higher plants on the basis of transcript accumulation, protein kinase activation or through the activation of phosphatases (see Table II). Experimental data suggest that plant MAP kinase pathways are involved in the responses to hormones and to stresses. They also suggest that plant MAP kinase modules may be involved in the control of the cell cycle. Moreover, several MAPKs are spatially and temporally regulated during development. *PMEK1*, *NTF3* and *NTF4* gene transcripts show a developmental regulation in either sporophytic or gametophytic cells and tissues (Wilson *et al.*, 1993, 1997; Ferrant *et al.*, 1994;

Decroocq-Ferrant *et al.*, 1995), indicating their involvement in particular during plant development; however, their biological functions remain unknown. To date, a single 'knock-out' mutant, corresponding to a MAPK element pathway element, has been described, showing that a MAP kinase pathway may be directly involved in the ethylene response pathway (Kieber *et al.*, 1993).

1. The Ethylene Pathway and Hormone Response

The simple gaseous hormone ethylene (C_2H_4) is involved in the regulation of a diverse array of mechanisms participating in plant development, physiological processes and stress responses (Ecker, 1995). Work with *A. thaliana* has contributed to a more complete understanding of the means by which plants recognize and transduce the ethylene signal. Genetic and molecular studies have defined a set of genes involved in the ethylene signal transduction pathway.

The *CTR1* gene was cloned using a T-DNA tagged allele (Kieber *et al.*, 1993). The recessive 'knock-out' mutant *ctr1* displays the constitutive triple response: inhibition of root and hypocotyl elongation, radial swelling of the hypocotyl and overstatement of the curvature of the apical hook (Kieber, 1997). *CTR1* encodes a protein possessing all the hallmarks of a serine/threonine protein kinase with significant similarity to the RAF family (Kieber *et al.*, 1993) known to be regulated by the small G-protein RAS (Daume *et al.*, 1994). Results obtained by Kieber *et al.* (1993) are consistent with the hypothesis that CTR1 is active in the absence of ethylene, and thus negatively regulates the downstream signal transduction pathway. CTR1 is inactivated in the presence of ethylene and thus permits the downstream elements to be activated (see Fig. 5).

The *ETR1* gene was isolated by positional cloning and its carboxy-terminal sequence was found to encode a protein showing similarity with the conserved domains (transmembrane segments, histidine kinase, receiver) of the prokaryotic two-component histidine kinase 'sensors' (Chang *et al.*, 1993; Chang and Meyerowitz, 1998). Moreover, Gamble *et al.* (1998) demonstrated that the *ETR1* gene encodes a functional histidine kinase, with His353 and Asp659 as the phosphorylation sites. The ETR1 protein contains a hydrophobic amino-terminus responsible for membrane localization (see Fig. 5). The first step in the ethylene perception is the binding of the molecule to the amino-terminal transmembrane domains of the receptor molecule ETR1, which results in the formation of membrane-associated disulfide-linked homodimers (Schaller and Bleeker, 1995). Biochemical studies of receptor binding to ethylene indicated that a copper (Cu) co-factor is required (Rodriguez *et al.*, 1999). ETR1 is active in the absence of ethylene and becomes inactivated upon binding to the hormone (Hua and Meyerowitz, 1998). Ethylene signalling involves protein–protein interactions between the receiver domain of the two-component receptor and the amino-terminal regulatory domain of CTR1 (Clark *et al.*,

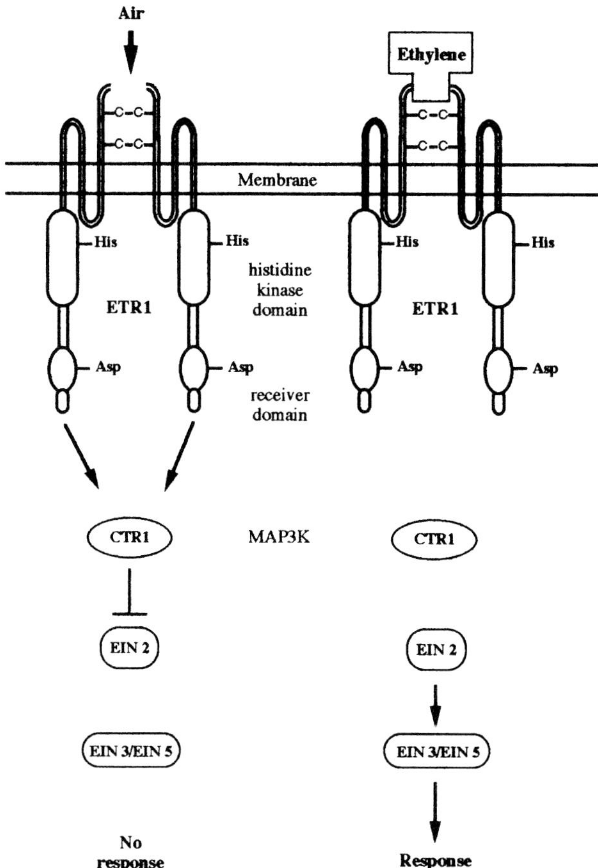

Fig. 5. The ethylene signalling cascade.

1998), suggesting that CTR1 might be part of an ethylene receptor complex in *Arabidopsis* (Clark *et al.*, 1998). Such a receptor complex may also involve a 14-3-3 protein (Solano and Ecker, 1998) which binds to the amino-terminal tail of CTR1. The identification of its physiological downstream elements remains to be completed even if CTR1 is able to phosphorylate *in vitro* the MAP2K AtMEK1 (Solano and Ecker, 1998). A model has been proposed in which ETR1 and CTR1 act in a multi-step signal transduction pathway through phosphorylation of proteins in a cascade that presumably couples the two-component system to *EREBP* genes encoding ethylene-response element binding proteins (Ohme-Takagi and Shinshi, 1995). This occurs through nuclear proteins acting sequentially in a cascade of transcriptional regulation (Solano *et al.*, 1998) downstream to the EIN2 predicted transmembrane protein (Alonso *et al.*, 1999).

Several *ETR1*-homologous genes and cDNAs encoding histidine kinases (ERS) which sometimes lack the C-terminal receiver domain have been identified in *Arabidopsis* (Ecker, 1995; Hua *et al.*, 1995, 1998; Sakai *et al.*, 1998), tobacco (Knoester *et al.*, 1997) and tomato (Payton *et al.*, 1996) (see Fig. 6). Mapping studies in tomato have revealed at least five independent loci with homology to *ETR1*, including the *never-ripe* (*NR*) mutant which is ethylene insensitive (Wilkinson *et al.*, 1995). Moreover, a tomato *CTR1*-like cDNA has been characterized recently (Wang and Li, 1997). This suggests both that the pathway for ethylene response is highly conserved in higher plants and that it is regulated by a receptor gene family whose members have overlapping functions (Hua and Meyerowitz, 1998; Hua *et al.*, 1998). Receptor multiplicity may serve as mechanisms which both regulate responses to ethylene under various conditions or concentrations, and achieve different sensitivities in different tissues. It appears (see Fig. 6) that the carboxy-terminal part of the ethylene receptors possesses a histidine kinase domain, fused to a receiver domain in ETR1, ETR2, EIN4 and LeETR1. On the contrary, these two domains remain separated in ERS1, ERS2 and NR (Lashbrook *et al.*, 1998).

Although the pathway for ethylene signalling in *Arabidopsis* shows similarities to the yeast HOG pathway, the plant RAF-like CTR1 and the prokaryotic-type ETR1 indicate that plants have merged evolutionary distinct components, one widely observed in prokaryotes (see Fig. 6) and the other associated with eukaryotes' signal transduction systems (Bleecker and Schaller, 1996). ETR, ETR-like, the putative cytokinin receptor CKI1 (Kakimoto, 1996) and its homologues (Brandstatter and Kieber, 1998) may represent members of an emerging family of eukaryotic histidine kinase sensors that have diverged to respond either to different plant hormones or to abiotic signals.

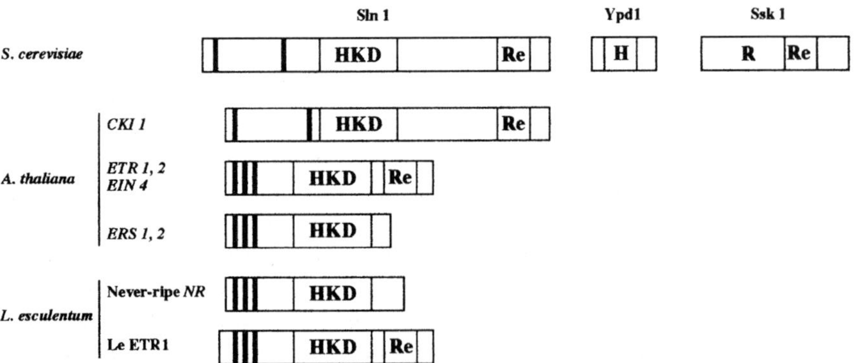

Fig. 6. Two-component signalling molecules. HKD is the histidine kinase domain (transmitter), Re the receiver domain, H the histidine phosphorelay, and R the response regulator. Black bars indicate transmembrane segments.

Other MAP kinase pathway elements are regulated by phytohormones, such as abscissic acid (ABA) and gibberellic acid (GA). These two phytohormones regulate stress responses and developmental processes, including seed dormancy, but the molecular mechanisms involved are still poorly understood. Nevertheless, in barley (*Hordeum vulgare*) a low concentration of ABA is able to induce MBP phosphorylation in aleurone protoplasts, suggesting the involvement of a MAPK pathway in the response to ABA (Knetsch *et al.*, 1996). This activation correlates with ABA-induced *rad16* gene expression and involves the action of a tyrosine phosphatase. Moreover, this protein kinase is not activated by GA. Huttly and Phillips (1995) reported a downregulation by GA of the expression of the *AsPK9* gene, encoding a MAPK, in cells from the aleurone layer from oat (*Avena sativa*). In contrast, no effect has been observed using the same tissue from barley. These results suggest a complex signalling network in response to these two hormones.

2. Plant MAP Kinase Pathways Mediate Abiotic and Biotic Stress Responses

Plants are subjected to a variety of abiotic and biotic stresses from their surronding environment (see Hirt, 1997). During their evolutionary history, plants have developed original strategies against pathogen (bacteria, fungi and insects) attacks, by using mechanisms involving specific gene interactions between plants and pathogens, referred to as 'gene-for-gene' interaction (see de Wit, 1997). This initiates a hypersensitive response inducing a local necrosis that may prevent the propagation of the pathogen. Moreover, a large set of genes is activated, locally or in a systemic way, mediated by salicylic acid, jasmonate acid and/or ethylene. The involvement of protein kinases in the plant defence mechanisms has been evidenced by a transient phosphorylation and the increase in kinase activity in elicitor-treated cells (Suzuki and Shinshi, 1995; Adam *et al.*, 1997). The signal transduction pathways involved seem to have incorporated MAP kinase cascades (Ligterink *et al.*, 1997; Stratmann and Ryan, 1997; Zhang and Klessig, 1997; Lebrun-Garcia *et al.*, 1998; Zhang *et al.*, 1998). Indeed, various proteins similar to MAP kinases, such as WIPK and SIPK from tobacco, are activated by elicitors in the AVR9/Cf9 system (see Fig. 7). These two MAPKs, activated locally or in a systemic way, do not participate in the synthesis of active oxygen species involved in the hypersensitive response. The data indicate that these two MAPKs are converging points of unspecific and specific interactions involved in a general stress response pathway to pathogens and wounding (Zhang and Klessig, 1997, 1998; Zhang *et al.*, 1998; Romeis *et al.*, 1999; Seo *et al.*, 1999). The relationship between these two kinases remains to be elucidated but we can hypothesize different functions for these two kinases in the signal transduction pathway. The WIPK protein may be involved in the regulation of jasmonic acid since an overexpression of this protein results in the overproduction of jasmonate (Seo *et al.*, 1999). The phospholipase A, which initiates the synthesis of jasmonate, could be a direct target for the MAPK. Moreover, the mechanisms of control

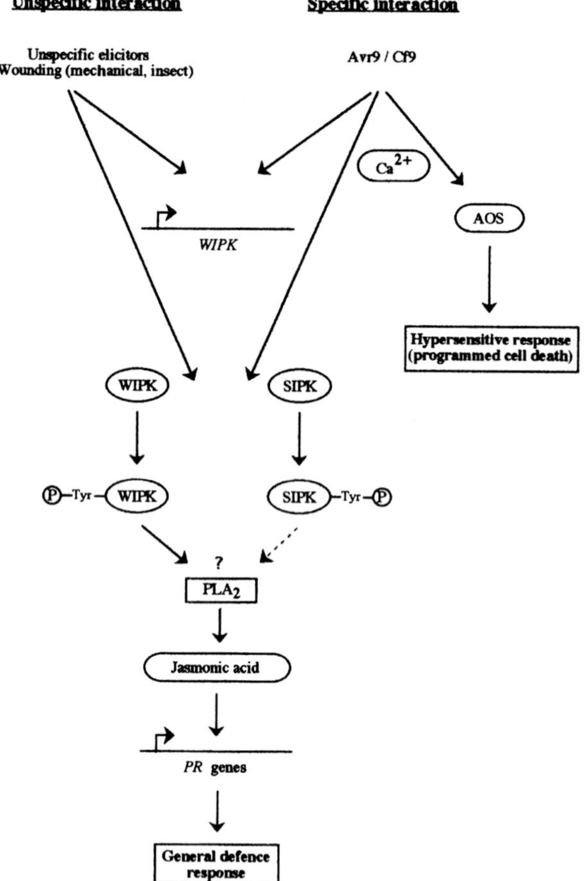

Fig. 7. MAP kinase involvement in general and specific defence response mechanisms. Adapted from Romeis *et al.* (1999) and Seo *et al.* (1999). AOS, active oxygen species; PLA, phospholipase A; PR, pathogen response; WIPK, wounding-induced protein kinase (MAPK); SIPK, salicylic-acid-induced protein kinase (MAPK); Tyr, Tyrosine.

of the two MAPKs share some differences. Indeed, *WIPK* gene expression, but not *SIPK*, is upregulated by elicitors. Data concerning PcERMK from parsley, a putative orthologue of WIPK, show that it is also activated by elicitors, and imported into the nucleus, suggesting that this protein may phosphorylate transcription factors playing an important function in the expression of defence genes (Ligterink *et al.*, 1997).

This general scheme for stress response signalling appears to be highly conserved in the plant kingdom. Using tobacco cell suspensions, hypo-osmotic stress has recently been demonstrated to activate both a phosphorylation-dependent oxidative burst and MAP kinases (Cazalé *et al.*, 1999). By contrast to defence response signal transmission pathways, these protein kinases may play a role in the activation of the oxidative burst. The same kinases are also activated by oligogalacturonides and salicylic acid, suggesting the importance of these MAP kinases as common components of different signalling pathways triggered by various extracellular stimuli (Cazalé *et al.*, 1999). The same complex feature is illustrated by the MMK4 MAP kinase from alfalfa, which shows 81% identity to WIPK from tobacco. *MMK4* gene expression and protein kinase activity are upregulated by drought, cold temperature and wounding (Jonack *et al.*, 1996; Bögre *et al.*, 1997), whereas the protein level remains stable. These observations suggest a regulation through turnover modulation. However, this protein is not activated in response to high temperature, high saline concentration or osmotic stress (Bögre *et al.*, 1997). The use of mastoparan and cholera toxin, which respectively activate seven transmembrane helice receptors and heterotrimeric G-proteins, affects activity of MMK4 (C. Jonak, personal communication), suggesting the involvement of a G-protein coupled receptor upstream of the putative MMK4 pathway. Serpentine receptors have recently been evidenced in plants (Josefsson and Rask, 1997; Plakydou-Dymock *et al.*, 1998) and studies have shown that in higher plants G-proteins are involved in signal transmission mechanisms for defence response (Legendre *et al.*, 1992) and auxin signalling (Millner, 1995). All the above data suggest the existence of a complex network of signalling transduction pathways either specific to particular stresses or responding to unspecific stresses.

The expression of several other MAP kinase pathway element genes is regulated by various stresses, namely enzymatic cell wall lysis (Tréhin *et al.*, 1998), high salt concentration, cold, dehydratation, touch and wounding (Mizoguchi *et al.*, 1996; Morris *et al.*, 1997; Hardin and Wolniak, 1998). The time course of activation of the corresponding proteins remains to be determined in order to evidence a direct link between the stress response and the proteins. Furthermore, results from various authors demonstrated that MAPK-like activities are transiently activated in response to several abiotic stresses (Usami *et al.*, 1995; Téna and Renaudin, 1998).

In mammalian systems, inactivation or downregulation of MAPKs is performed by dual-specificity protein phosphatases, which dephosphorylate both the phosphothreonine (or phosphoserine) and the phosphotyrosine residues of MAPKs (Groom *et al.*, 1996). Similarly, in plants, a transient SIPK activity is observed, which then drops below the initial level, in response to salicylic acid activation (Zhang *et al.*, 1998), suggesting that a MAPK phosphatase has been activated. The activation of the MAP kinase MMK4 depends on an activating post-translational mechanism triggered by drought, touch or wounding, and its subsequent inactivation requires the *de novo* synthesis of the MP2C protein phosphatase (Meskiene *et al.*, 1998). The recently evidenced protein tyrosine-specific phosphatase (PTP), which may function in response to a plant stress (Xu *et al.*, 1998), is likely to be a putative MAPK regulator. Evidence for the functional involvement of plant protein phosphatases in plant abscisic acid, pathogen and stress responses or development is overwhelming (see Luan, 1998). For example, treatment of plant cells with a tyrosine phosphatase (Adam *et al.*, 1997) or with a MAPK specific phosphatase (Gupta *et al.*, 1998; Kovtun *et al.*, 1998) eliminates MAPK activity.

3. *Cell Cycle Regulation*

As previously evidenced in yeasts and animals, the regulation of the plant cell cycle may involve various MAP kinase modules. These signalling pathways may be involved in different mechanisms at various steps of cell division, i.e. initiation, cytokinesis and in an anti-proliferative way. The MAP kinase Ntf4 from tobacco is activated during the re-entry in the cell cycle of cells released from a blockage due to phosphate starvation (Wilson *et al.*, 1998). Furthermore, Ntf4 is also activated in pollen rehydration (Wilson *et al.*, 1997). An unrelated gene, *PMEK1* from petunia, encoding a MAPK (Decroocq-Ferrant *et al.*, 1995), is upregulated by cytokinin during the G1 phase in relation to the entry of protoplast into the cell cycle (Tréhin *et al.*, 1998). A direct relationship between the expression of the stress-induced *PMEK1* gene and cell cycle-regulated proteins or genes remains to be demonstrated. These data suggest a possible link between stress response and cell cycle initiation in plant cells. Several other genes encoding MAPK- and MAP4K-related proteins have been demonstrated to possess a cell cycle-dependent expression (Jonack *et al.*, 1993; Leprince *et al.*, 1999) but studies of the encoded proteins remain to be performed.

The orthologues Ntf6 and MMK3, respectively from tobacco and alfalfa, are activated during late anaphase/early telophase. The activation results from a post-translational event (Calderini *et al.*, 1998; Bögre *et al.*, 1999). Furthermore, depolymerization of microtubules abolished MMK3 activation, indicating that intact microtubules are required for MMK3 activation. The Ntf6 and MMK3 proteins concentrate at the midplane of cell division in the phragmoplast and the forming cell plate, suggesting an involvement in cytokinesis (Calderini *et al.*, 1998; Bögre *et al.*, 1999). Although microtubules or other cytoskeletal elements seem to be necessary to localize the MMK3 protein to the cell plate during anaphase and early telophase, they are not clearly required to maintain MMK3 at this location during later stages of the cell cycle (Bögre *et al.*, 1999).

Furthermore, the expression of the *NPK1* gene, encoding a tobacco MAP3K-related protein from the PMEKK subfamily, is associated to proliferating cells (Banno *et al.*, 1993; Nakashima *et al.*, 1998). Both auxin and cytokinin induce *NPK1* expression before cell division. When an inhibitor of DNA synthesis is added during the germination or the induction of lateral roots by auxin, *NPK1* transcripts are detected, suggesting that *NPK1* expression precedes DNA replication (Nakashima *et al.*, 1998). According to

Machida *et al.* (1998), the NPK1 protein slightly accumulates from the S to M phases and disappears during the G1 phase. Kovtun *et al.* (1998) have shown that a transient expression of the NPK1 protein in mesophyl cells of maize is able to suppress auxin-induced gene expression via the activation of a MAPK-related protein. Moreover, tobacco transgenic plants overexpressing the NPK1 catalytic domain share defects in embryo and endosperm development (Kovtun *et al.*, 1998). As far as auxin acts in the initiation of the cell cycle, it is tempting to speculate that this protein may be involved in an anti-proliferative signal transduction pathway.

Similarly to their animal and yeast homologues, plant MAP kinase pathway elements have been shown to be involved in signalling processes associated with the cell cycle, development and differentiation, and also with environmental abiotic and biotic stress responses. Several plant MAPK, MAP2K and MAP3K have been shown to complement yeast mutants, suggesting both structural homology between the proteins and conservation of function. The genes or cDNAs recently detailed in *A. thaliana* show that its genome contains many more MAP3Ks than yeast, suggesting the existence of additional MAP kinase pathways and functions in plants. The plant elements having no counterparts in yeast can mostly be linked to a complexity in signalling that arises most probably from multicellularity, for example cell–cell interactions, cell differentiation and morphogenesis as suggested after a comparison of the whole proteosome of the unicellular yeast to the pluricellular nematode (Chervitz *et al.*, 1998). Indeed, RAF kinases have not been observed in yeast, indicating possible roles for the RAF-related plant proteins in phytohormone response and developmental processes. Another particular feature of plants is the fact that MAP kinase modules might be activated in response to various pathogens. Disease-resistance mechanisms may use regulatory pathways similar to animal developmental pathways. This suggests that plant developmental programmes differ significantly from animals, although they also involve MAP kinase module elements. The ethylene signalling pathway is the most studied MAP kinase pathway in plants. However, more insight will turn on other pathways in the near future.

IV. CONSERVED FEATURES OF THE MAP KINASE MODULE ELEMENTS

A. A RELATIONSHIP- AND EVOLUTIONARY-BASED CLASSIFICATION

Leonard *et al.* (1998) predicted the existence of an ancestral protein kinase gene prior to the divergence of eukaryotes. Gene duplication in the ancestor may have been the mechanism of origin for most kinase genes, suggesting that they are close relatives. Comparisons of the amino acid sequences of the catalytic domain sequences have been used to arrange kinases in relationship trees that cluster closely related members. The kinase superfamily has been classified by Hanks and Hunter (1995) into four main clusters, namely the AGC [protein kinase A (PKA), PKG, PKC], the CaMK (calcium and calmodulin-dependent protein kinases), the CMGC [cyclin-dependent protein kinase (CDK), MAPK, glycogen synthase kinase-3 (GSK3), casein kinase II (CKII)] and the PTK (protein tyrosine kinases). A complete analysis of the yeast protein kinases allowed Hunter and Plowman (1997) to designate two additional highly conserved new clusters in eukaryotes, namely the STE11/STE20 including the MAP3Ks and the MAP4Ks, and the STE7 including the MAP2Ks, the NIMA/NEKs and the NEK-like kinases. A relationships analysis of protein kinases

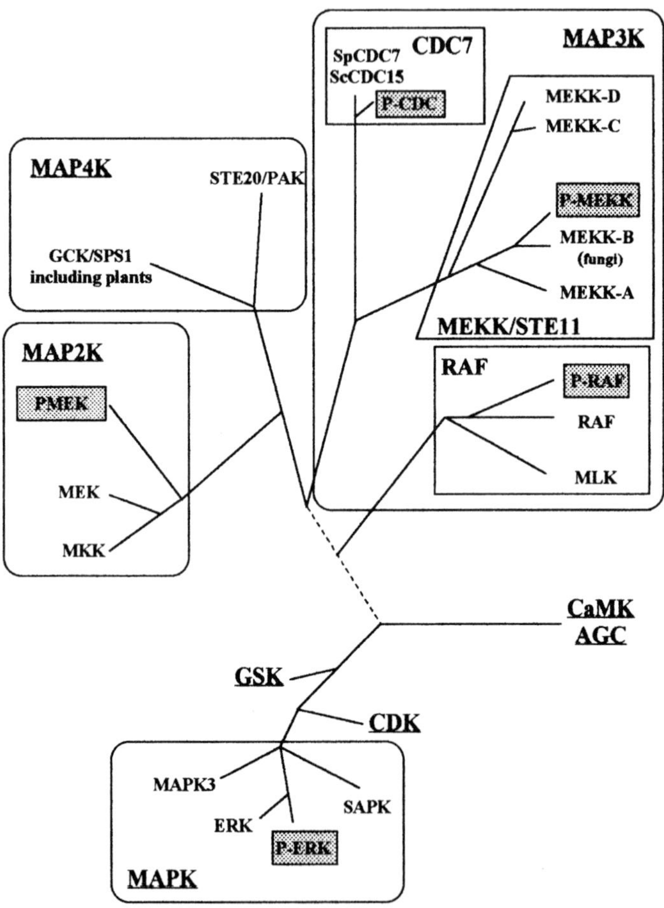

Fig. 8. Schematic view of the evolutionary relationships between MAPKs, MAP2Ks, MAP3Ks and MAP4Ks. Boxes containing plant kinases are shaded.

showed that the MAPKs are clearly distinct from the other MAP kinase module elements (see Fig. 8).

A detailed analysis of the similarities of the predicted amino acid sequences over the catalytic domains revealed a diversity of MAP kinase module elements. The MAPKs, the MAP2Ks and the MAP4Ks constitute three distinct but homogeneous clusters of proteins, referred to as families. This indicates that each molecular family diverged from a common ancestor before separation of the eukaryotic kingdoms. In contrast, the MAP3Ks, a biochemically and homology-based cluster of proteins, are characterized by a high sequence heterogeneity and have been resolved into five distinct families including the MEKK/STE11, the RAF, the MLK, the MOS and the CDC7 (see Jouannic *et al.*, 1999b). Studies on MAP3Ks (Jouannic *et al.*, 1999b) suggested that convergence of evolutionarily distinct protein kinases towards a biochemical function may have occurred during the evolution of higher eukaryotes. In each family, several subfamilies, groups and subgroups have been defined as discussed by Sells and Chernoff (1997), Kültz (1998), Hamal *et al.* (1999), Jouannic *et al.* (1999b) and Leprince *et al.* (1999). Inside a family or a subfamily, the plant elements form a cluster of proteins distinct from those of yeast and animals. Figure 8 shows that the PERK group is related to the ERK subfamily and the PMEK subfamily to the MAP2Ks family. Within the heterogeneous MAP3Ks, the PMEKK and PRAF, respectively from MEKK/STE11 and RAF families, are assigned to subfamilies separated from the non-plant members (Fig. 8). These observations illustrate the diversification of genes after the emergence of the plant kingdom. Inside a plant cluster of proteins, various groups or subgroups can be evidenced (see Fig. 9). Consistent with the fact that each family is homogeneous, signature sequences inside the catalytic domain have been identified for the MAPKs (Kültz, 1998), MAP2Ks, RAF and MAP4Ks families (authors' unpublished results), which permit discrimination of these proteins from other protein kinases (see Table III). In contrast, the high sequence heterogeneity of MAP3Ks does not allow the identification of a general MAP3K sequence signature. The presence of signature sequences inside the catalytic domain involved in substrate interaction, phosphotransfer reaction and phosphorylation sites suggests similar biochemical mechanisms for these proteins. As a consequence, we should speculate on a higher range of regulations and substrates within the MAP3Ks. Plant signature sequences were identified for the PERK group, PMEK and PRAF subfamilies.

The question flowing from the above sequence relationship-based classifications is the following: do these classifications have any predictive value as to the biological functions of the proteins? An emerging view from animal and yeast studies is the existence of two types of MAP kinase pathways: the first type involves ERK and mediates mitogenesis and cell differentiation signals, while the second type (SAPK/JNK) mediates stress responses (see Kültz, 1998). In mammals, a functional specialization is observed for MAPKs, MAP2Ks and MAP3Ks. The mitogen pathways are usually referred to as the ERK pathways and involve ERK, MEK and RAF. In contrast, the stress pathways, referred to as the SAPK pathways, involve other types of MAP kinase module elements, namely SAPK, MKK and MEKK/STE11 (or MLK). However, the situation may change according to the nature of the organism considered, as illustrated by the *Drosophila* SAPK-like pathway which was shown to be involved in an embryonic developmental process (Noselli, 1998). Furthermore, *S. cerevisiae*, possessing only one subfamily of MAP2Ks and one of MAP3Ks, each being involved in the mitogen as well as the stress pathway, possesses ERK- and SAPK-like proteins. In plants the situation seems to be rather

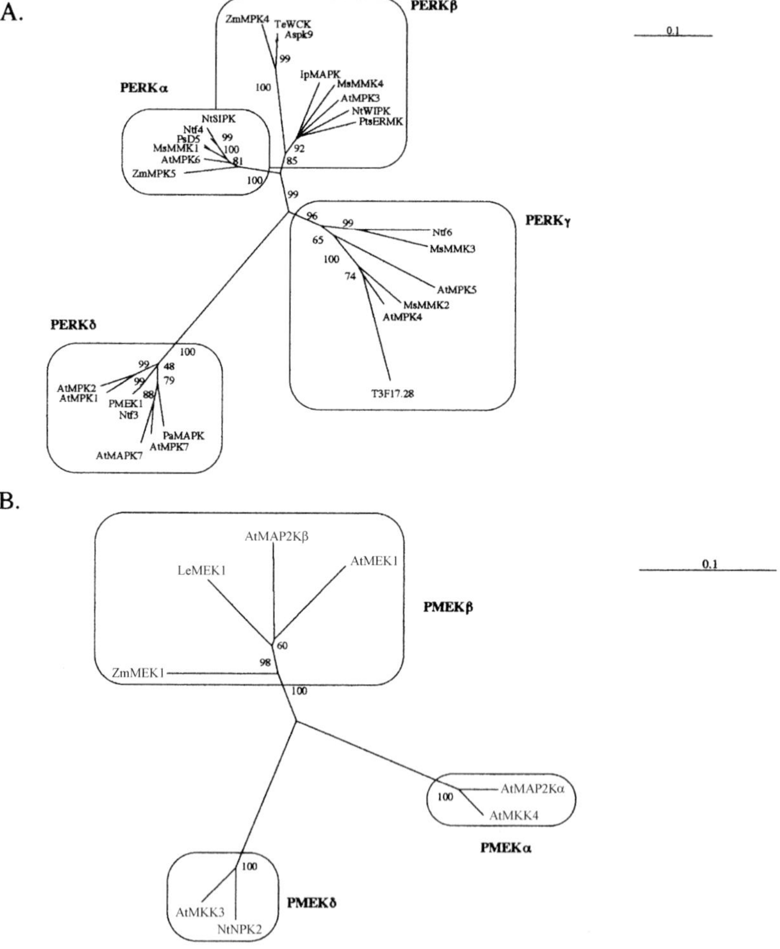

Fig. 9. *Caption opposite*

similar. In contrast to the metazoans, plants possess neither SAPK-like nor MKK-like proteins, but have MEKK/STE11 and RAF-like proteins. Various studies concerning the regulation of plant MAP kinases revealed that the PERKs and the PMEKs may indeed be involved in stress-response pathways (see Section III.C.3). Although plant MAP3K-related proteins seem to be involved in mitogen and stress pathways, currently available data do not allow us to relate directly sequence and biological function. Nevertheless, it is tempting to speculate that a functional specialization occurs inside plant proteins according to the groups or the subgroups. Current results suggest that different mechanisms of control may occur according to the subgroups of PERK (see Section III.C). Some proteins of the PERKα and PERKβ subgroups are activated after stress treatments but only the genes corresponding to the PERKβ subgroup are upregulated. Nevertheless, this mechanism of control has to be confirmed for all the genes and proteins of these subgroups.

The diversity of the catalytic domain sequences has been shown to be associated to a diversity of amino- and carboxy-terminal extensions of the MAP kinase module elements. These extensions are involved either in the regulation of the protein activity, or in the protein–protein interactions leading to the specificity of protein interaction and activation processes. For example, PERK and PMEK showed a high homogeneity in their extensions within each subgroup and group, respectively. On the contrary, this is not the case for the

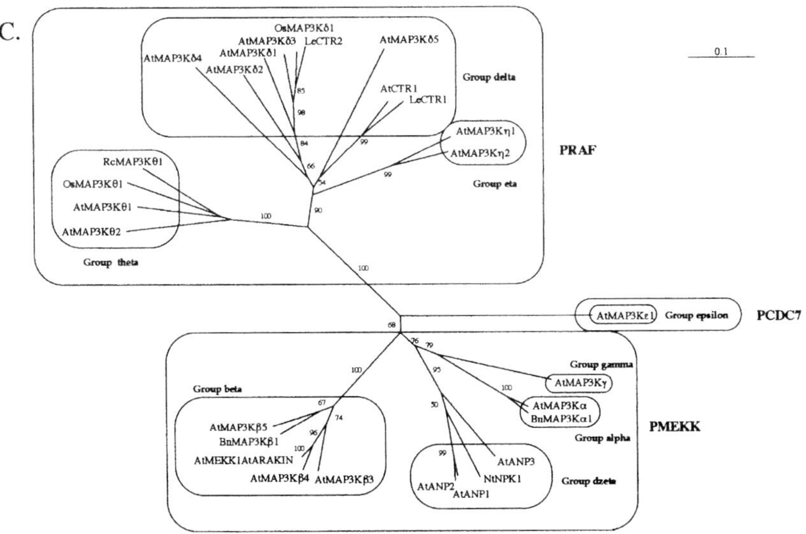

Fig. 9. Relationships between plant MAP kinases (A), MAP kinase kinases (B) and MAP kinase kinase kinases (C). The dendograms were obtained according to Jouannic *et al.* (1999b). The scale bar corresponds to 0.1 distance units.

TABLE III

Catalytic domain signature sequences of MAPKs, MAP2Ks, MAP3Ks and MAP4Ks

Family *Subfamily* Group	Sequences	Localization
MAPK	[LIVM][T*S*]X[Y*X][LIVM]XT[KR][WY]YRXPX[LIVM][LIVM]	VIII
PERK	ELMD[TS]DL[NHQ]XI[IL][KR]SXQ	V
	[QH][CS]X[FY][FL][LIVM][FY]Q[LIVM]LRGL[KR]Y[LIVM]HSAN	VIa
MAP2K	[AIVT][GA][TC]XX[YF]M[SAG]PER[IL]	VIII
	[YD]X[YNFVI]X[GSAC]D[IV]WS[LFTM]G[VIL][SVATM][VILMF][VILMFY][EGDQ][FC AVLM][GAYVL][LTIVNGS] [GQLA]XX[PV][YFWCIVL]	IX
PMEK	[SQY][LIV][ILAV]LE[YF]M[DN][GKQR]GSL[AE][DG][IFAL][L HIV][KIV]	V
	S[PQ][EQ][FL][CRS][SH]F[IV][SD][CATQS]C[LIV][QK][KRS][DEHN][PQ][PDNGK RA][KDASQ]R[LKPWMRS][ST]A	XI
RAF	[RK][IV]GXG[SF][FY]G[TE]VX[KRH][GA]X[WF][HFN]G	I
ARAF	AI[VI]TQWCEGSSLYXHXH[VI]	IV
	[NS]IIHRD[LM]K[ST]NNIFL[HM][ED][DG][LM]	VIb
	[TS]VKIGDFGLAT[VA]K[ST][RK]W[ST]	VII
PRAF	[LIM]X[SD]X[ST]X[AK]GTP[EQ]W	Upstream of VIII
MAP4K	[LM]W[IVML][VIAMC][MI]E[YFLH][CIML]X[AGS]G[SCA][LAVCI]XX[VIL][VI LMTY]	IV-V
	[GA]X[VI][KR][LVI][AVTG]DFG[VFY][CASG][AVG][QREHK][LIVN]	VII
	Y[GTND]X[KRVQL][VIACS]D[VIL]W[SA][LCTF]G[IV][MTS][LIATVMS][ILRY]E[MIL][VIALG][QEKHNT][GNLCI]X[PA]P	IX
STE20/PAK	[YIKFMLY][KEVNH][KIQ][AIL]G[QES]G[EASG][STA]G[S ADGTEV]V[YFE][LSTEA][AS]	I
	[KRQL][QEDK][LFIT][IV][IVMLF]NE[ILV][LTQGS][IVM][ML][KRN]	Upstream of III to III
	[NA][IV]V[NTQR][FY][ILF][DES][ATS][YFH][LIFV]X[DGS]	Between III and IV
	[IV]HRDIKSD[NS][IV]L[LI]	VIb
	[IV]K[LI][TA]DFG[FY][CAG][AVG][QRE][IL][TSGND]	VII
	[KR][VIA]D[VIL]WS[LF]GIM[LIAT][ILR]EM[LIVA]E[GN]EPP[YF][LIF][NRE][EDF]	Downstream of IX
	EPP[YF][LIF][NRE][EDF]X[PM][LIHF][RKQ]A[LIM][YKF][LR][IL][ATR]	IX to X

Signature sequences are shown in a Prosite format. The symbol X defines any amino acid. Letters between brackets define observed variations at a single position. The subdomain localizations of the signature sequences are indicated. Subdomain I: phosphatidyl binding site; VIb: catalytic region; VII: activation loop; VIII: P + 1 pocket. The MAPK family specific signature sequence was defined by Kültz (1998).

plant MAP3Ks (Jouannic *et al.*, 1999a). The motif [RK][RK][NK][PK]X [GP][LI]X[LI][NPA][PK] located within the non-catalytic amino-terminus of several MEKs has been shown to act as a 'docking site' (Bardwell and Thorner, 1996). This docking site plays a role in maintaining the fidelity of signalling events. This motif was not observed in the MAP2K that functions in stress-induced MAP kinase cascades such as yeast Pbs2, Mkk1 and Mkk2, and the mammalian JNKK. The amino-terminal sequences of several plant MAP2Ks show a putative similar docking site in AtMEK1, LeMEK1, ZmMEK1, AtMAP2Kα and AtMKK4, which are elements of different groups of PMEKs (see Fig. 9).

B. THE PHOSPHORYLATION SITES

Within the different groups of the MAP kinase cascade elements, protein kinases share similarities within the region where they undergo their activating phosphorylations. This suggests that the phosphorylation-site sequences may constitute a signature for groups or subgroups.

1. The MAP Kinases

In 1991, Payne *et al.* determined inside the catalytic domain of the murine p42 MAP kinase, two sites of phosphorylation for the MAP kinase MEK1. The two sites (Thr183 and Tyr185) were shown to be located in the activation loop, and only separated by a single residue in the tripeptide motif TxY. The phosphorylation reactions are time-ordered and occur usually, but not invariably, first on the tyrosine residue, and then on the threonine (Haystead *et al.*, 1992; Ferrel and Bhatt, 1997). It has been demonstrated that MEK1 phosphorylates p42 by a two-collision mechanism, rather than a single one, indicating that MEK1 dissociates fully from p42 between the first and the second phosphorylation (Ferrel and Bhatt, 1997). It has been suggested that such a process increases the sensitivity of the response. The specificity of the MEK/MAPK phosphorylation results from multiple bindings, including sites located in both the amino- and carboxy-terminal domains (Wilsbacher *et al.*, 1999). In order to expose and enable preferential phosphorylation of the tyrosine residue, MEK binding induces a major conformational change in the ERK protein (Canagarajah *et al.*, 1997).

Nearly all known MAP kinases are activated by dual phosphorylation at specific Thr and Tyr residues. Recent studies by Ferrel (1996) showed that the MAPK family can be divided into at least 10 clusters of related kinases according to the sequence signature of the phosphorylation sites. The various signatures observed in MAPKs are present just upstream of the conserved [WY]YR[ASP]PE peptide sequence of subdomain VIII of the catalytic core, inside the L12 loop. The two phosphorylation motifs TEY (owing to the largest cluster, including the PERK group) and TNY have been detected in the proteins of the ERK subfamily and in protozoan MAPKs. In contrast, the SAPKs are characterized by the presence of TGY and TPY motifs within the activation site. Unusual sequences like TQY, TNY and TDY have been detailed, including two *A. thaliana* TDY MAP kinases (i.e AtMPK8 and AtMPK9) (Mizoguchi *et al.*, 1997). A few MAPKs showed, at the same position, the tripeptide sequences GEY, KGY, THE (the most distant from other MAPKs) or SEG (MAPK3 group). The latter MAPKs are phosphorylated at a single residue, suggesting that either tyrosine or serine/threonine can be phosphorylated when either the threonine/serine or the tyrosine have been lost by mutation. These kinases represent the most distantly related MAPKs.

Figure 10 shows a phylogeny of the motifs of phosphorylation from MAP kinases and suggests that within the MAPKs the different SAPK groups are derived from the ERKs. However, it is much more difficult to determine whether single phosphorylation site MAP kinases were the ancestor or whether they derived from MAP kinases containing a dual specific phosphorylation motif. Moreover, this phylogeny also suggests the existence of other MAPK phosphorylation motifs. Phosphorylation of the equivalent residue occurs in other kinases (Draetta, 1997), for example two yeast kinases which are closely related to the MAP kinase family from an evolutionary point of view (Hunter and Plowman, 1997), namely SSN3 (a member of the CDK, i.e. cyclin-

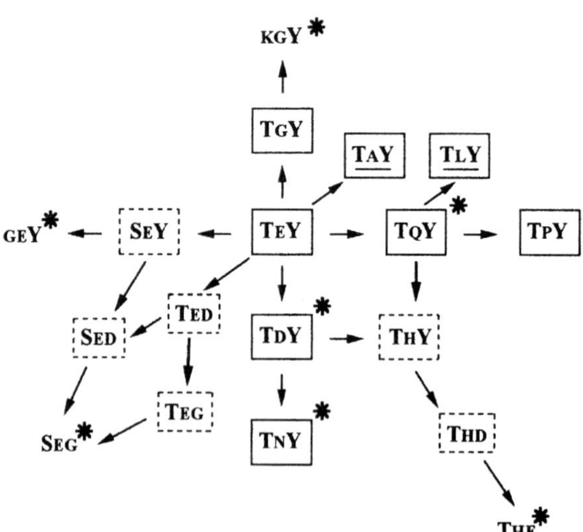

Fig. 10. Phylogeny of the highly conserved phosphorylation motif in MAP kinases. Dual specific motifs are boxed, putative motifs are indicated in discontinuous boxes, asterisks show rare motifs and CDK or CDK-like are underlined.

dependent kinase) family, and IME (a CDK-like). The latter two kinases possess, respectively, a TLY and a TAY motif (Ferrel, 1996).

2. The MAP Kinase Kinases

Several lines of evidence from mammals and yeast suggest that the phosphorylation of serine/threonine residues in the SxAxT/S motif within subdomains VII and VIII of the catalytic domain by upstream MAP3Ks is a general molecular mechanism for activation of members of the MAP2K family. Alessi *et al.* (1994) identified Ser218 and Ser222 as the two residues of a rabbit MAP2K (Ashworth *et al.*, 1992) that could be phosphorylated by an upstream MAP3K. Moreover, Zheng and Guan (1994) demonstrate that C-Raf activates human MEK1 by phosphorylating both serine 218 and 222 within the activation lip. The phosphorylation of either residue by B-Raf (Papin *et al.*, 1995) was shown to be sufficient for maintaining the enzyme in its active conformation. Similarly, equivalent phosphorylation sites were identified in STE7 at position 359 and 363 (Zheng and Guan, 1994). Furthermore, other MAP2Ks have been demonstrated to possess phosphorylated residues at the same site (Shieh *et al.*, 1998).

The alignment shown in Fig. 11 presents experimental and predicted phosphorylation sites of MAP2Ks including plant members. The structural features of *Candida albicans* HST7 (Clark *et al.*, 1995), *Leishmania chagasi* LPK (Li *et al.*, 1996), *D. discoideum* DdMEK1 (Ma *et al.*, 1997), and plant MAP2Ks (Shibata *et al.*, 1995; Jouannic *et al.*, 1996; Morris *et al.*, 1997; Ichimura *et al.*, 1998, Hamal *et al.*, 1999) do not fit within the consensus motif deduced from the other eukaryotes. The *HST7* gene of *C. albicans* (S*LAMADT*) complements disruption of the *Ste7* MAP2K from yeast (S*IADT*), suggesting that HST7 could be a Ste11 substrate. Thus, it is potentially phosphorylated on Ser402 and Thr408 although they are separated by two additional residues when compared to STE7 (Clark *et al.*, 1995). Similarly, the plant AtMKK2/AtMAP2Kα protein, which is able partially to complement the *pbs2* mutant of *S. cerevisiae*, does not complement the mutant when the two putative phosphorylation sites are mutated (Ichimura *et al.*, 1998b). This observation is consistent with the phosphorylation sites consensus sequence S/TxxxxxS/T for the plant MAP2Ks.

3. The MAP Kinase Kinase Kinases

Putative phosphorylation sites in the MAP3Ks are unclear and their consistent heterogeneity in the catalytic domain sequences is certainly related to various phosphorylation sites and mechanisms of regulation. Studies aiming at the determination of phosphorylation sites have mostly been performed on animal proteins of the RAF family. The RAF proteins act upstream of MAP2Ks in the mitogen signal transmission pathway (i.e the ERK pathway). Raf-1 activation is a highly complex process which involves both an interaction with Ras (or Src) and the 14-3-3 proteins, as well as phosphorylation events.

```
                 VII                      VIII
                            *      *
F35C8.3-Ce       DFGISGFMTDSM--AHSKSAGCPPYMAPE
K08A8.1-Ce       DFGIAGRLIES--RAHSKQAGCPLYMGPE
MKK7             DFGISGRLVDS--KAKTRSAGCAAYMAPE
HEP              DFGISGRLVDS--KANTR-AGCAAYMAPE
XlMPK2           DFGISGQLVDSIAK--TRDAGCRPYMAPE
MKK4             DFGISGQLVDSIAK--TRDAGCRPYMAPE
MKK3             DFGISGYLVDSVAK--TMDAGCKPYMAPE
MKK6             DFGISGYLVDSVAK--TIDAGCKPYMAPE
PBS2             DFGVSGNLVASLAK--T-NIGCOSYMAPE
WIS1             DFGVSGNLVASISK--T-NIGCOSYMAPE
MEK5             DFGVSTQLVNSIAK--TY-VGTNAYMAPE
MEK-2            DFGVSGMLIDSMA--NSF-VGTRSYMAPE
MEK2             DFGVSGQLIDSMA--NSF-VGTRSYMAPE
DSOR1            DFGVSGQLIDSMA--NSF-VGTRSYMSPE
MEK1             DFGVSGOLIDSMA--NSF-VGTRSYMSPE
STE7             DFGVSKKLINSIA--DTF-VGTSTYMSPE
HST7             DFGVSRELTNSLAMADTF-VGTSMYMSPE
FUZ7             DFGVSGELINSIA--DTF-VGTSTYMSPE
BYR1             DFGVSGELVNSVAQ--TF-VGTSTYMSPE
MKK1             DFGVSGEAVNSLA--TTF-TGTSFYMAPE
MKK2             DFGVSGEAVNSLAM--TF-TGTSFYMAPE
LPK              DFGVS-KLIQTLAVSSTY-VATMCFMAPE
DdMEK1           DFGVSGQLQHTLSKAVTW-VGTVTYMSPE
Physcomit        DFGVSAVLVHSLAQRDTF-VGTCTYMSPE
ZmMEK1           DFGVSAVLASSIGQRDTF-VGTYNYMAPE
AtMAP2Kβ         DFGVSTVMTNTAGLANTF-VGTYNYMSPE
LeMEK1           DFGVSAVLASTSGLANTF-VGTYNYMSPE
AtMEK1           DFGVSKILTSTSSLANSF-VGTYPYMSPE
AtMAP2Kα         DFGVSRILAQTMDPCNSS-VGTIAYMSPE
AtMKK4           DFGVSRILAQTMDPCNSS-VGTIAYMSPE
AtMKK3           DFGISAGLENSMAMCATF-VGTVTYMSPE
NtNPK2           DFGISAGLESSIAMCATF-VGTVTYMSPE

consensus        DFGVS-----S-----T---GT--YM-PE
                    IA     T-----S---AC--F
```

Fig. 11. Region of the catalytic domain highly conserved among the MAP2K family members. The protein sequence including the phosphorylated serine and threonine residues in MEK1, STE7, WIS1, PBS2 and AtMAP2Kβ is underlined. The asterisks indicate predicted phosphorylation sites. Plant MAP2K protein names are double underlined.

Phosphorylation of Raf-1 occurs *in vivo* at three major residues, namely Ser43, Ser259 and Ser621, whereas Raf-1 autophosphorylates at Thr268 and Thr269 (Morrison *et al.*, 1993; Morrison and Cutler, 1997). The residues flanking either Ser259 or Ser621 indicate that RSXS*XP may be the consensus sequence for the phosphorylation of the Raf family members by the upstream kinases (Morrison *et al.*, 1993). It has been shown that a phosphoserine (Ser621) participates in the interaction between Raf-1 and 14-3-3 (Muslin *et al.*, 1996), a critical step in the activation of Raf-1. A 14-3-3 dimer associates with the N- and C-terminus of Raf, which is thus maintained in an inactive state (Tzivion *et al.*, 1998). Upon Ras-GTP-mediated activation, 14-3-3 is displaced from the N-terminus, and Ras recruits the unfolded Raf molecule to the membrane for activation (Aitken, 1996). *In vitro*, Raf-1 activation also requires phosphorylation of Tyr340 and Tyr341, and probably Ser621 in order to enhance catalytic activity (Morrison and Cutler, 1997). A recent result from King *et al.* (1998) reveals that PAK-3, a MAP4K, phosphorylates Raf-1 on Ser338, within a QRDpSSYY motif. Results from Mason *et al.* (1999) show that activated Ras predominantly phosphorylates Raf-1 on Ser338, whereas activated Src gives predominantly Tyr341 phosphorylation.

V. CONCLUSION

Since the first MAP kinase was characterized by Ray and Sturgill in 1987, hundreds of MAPKs, MAP2Ks, MAP3Ks and MAP4Ks have been detailed in various organisms. During the 1990s, work in a number of laboratories has largely elucidated MAP kinase cascades in yeast, animals and plants. There are clearly additional components and pathways to be discovered, but progress in defining the elements not yet ascertained will undoubtedly be rapid. A large number of genes encoding MAP kinase cascade components has been isolated using molecular approaches. In addition to the homology-based approaches, genomic sequencing projects have contributed significantly to the discovery of new plant and animal MAP kinase cascade components.

Intracellular plant signal transduction mechanisms that involve MAP kinase modules bear significant homology to those in mammals, *Drosophila* and yeasts. The MAP kinase modules, likely to occur in plants, are both more numerous and much more complicated than could have been predicted even in 1995. Many conserved signalling molecules have been evidenced in plants, namely receptor-like kinases (Schmidt *et al.*, 1997), serpentine receptors (Josefsson and Rask, 1997; Plakydou-Dymock *et al.*, 1998), G-proteins (Ma, 1994), phospholipase C (Einsparh *et al.*, 1989), Ca^{2+}/calmodulin (Jena *et al.*, 1989), various 14-3-3 proteins (Aitken, 1996), and various kinases and phosphatases (Lin *et al.*, 1998; Meskiene *et al.*, 1998; Xu *et al.*, 1998). The MAP kinase cascade appears to be a general signalling pathway that regulates processes stimulated by either extracellular agents such as stressful physical and chemical stimuli or mitogenic agents. A topic now under investigation is the number, composition and regulation of the MAPK cascades in particular plant cells and/or tissues. By analogy to what is known in other organisms (Ambrosio *et al.*, 1989; Pang *et al.*, 1995) it is anticipated that the characterization of MAP kinase genes, expressed during development, will provide insights into the nature of the signalling pathways implicated in cell proliferation and in morphogenesis.

One major challenge for further studies will be the elucidation of the biological function of the signal transduction pathways identified in plants. The biochemical activity of the putative signalling proteins evidenced from the molecular approaches requires extensive studies. To address the function of a putative protein kinase as a signalling molecule, it is necessary to use biochemical, cytological and transgenic approaches, determine the biochemical activity and the targets, produce antibodies, carry out immunolocalization, and obtain knock-out and overexpression mutants. Since signalling proteins interact physically with upstream and downstream elements, complementary components will be evidenced using strategies based on protein interactions. The 'DNA arrays' will also enlarge our knowledge of the expression of related genes.

Many new plant kinases have been assigned to a particular group based on sequence similarity to previously defined elements. Although the classification of MAPK, MAP2K, MAP3K and MAP4K is based solely on catalytic domain sequences, members of a group defined by this means are usually also related in the regions extending outside the catalytic domain. In many cases, members of such groups have been shown to encode biochemical homologues (Hanks and Hunter, 1995). Chervitz *et al.* (1998) indicated that 'annotation of protein functions and activities will be reliably transferable between organisms as disparate as yeast and worm'. Nevertheless, the detailed biochemical and biological functions of each MAP kinase cascade element remain to be analysed, to assign the function unequivocally.

Reversible phosphorylation/dephosphorylation of target proteins can be viewed as a binary switch with the kinases and phosphatases pushing the target protein into the 'on' or 'off' stage. Significant advances in the understanding of the co-ordinated interplay of both kinase and phosphatase, which are responsible for the regulatory effects of reversible phosphorylation, will enable progress in the knowledge of cell signalling. An important issue will be to establish which phosphatase (PP or PTP) dephosphorylates a particular target phosphoprotein *in vivo*. A fundamental question to be addressed in further studies is whether the localization of signalling proteins to particular areas is a prerequisite for efficient and specific signalling. Such a prerequisite is an attractive explanation for why specificity of interaction between signalling molecules is often lost when proteins are reconstituted *in vitro*. Recent results also suggest that the members of several MAP kinase cascades, including their associated phosphatases, form a protein complex capable of entering into direct physical interactions (Faux and Scott, 1996; Paravicini and Friedli, 1996). Scaffolds and adaptor proteins, such as 14-3-3, organize kinases and phosphatases and co-ordinate their actions. A further challenge will be the understanding of the molecular mechanisms by which specificity is conferred to the plant MAP kinase pathways (Tan and Kim, 1999).

Recent papers reveal a greater sharing of components than previously appreciated. The identification of a whole plant MAP kinase module associated to a precise biological function remains to be achieved. Direction of keen interest also includes the investigation of both signalling events upstream of the MAP kinase cascade, and downstream targets of MAP kinases. It is anticipated that signalling pathways interact with one another. These interactions result in complex networks that may possess new properties.

The question of cross-talk between the plant MAP kinase pathways is of major interest in order to determine how this can modulate the cellular response. Mechanisms for cross-talk between plant MAP kinase cascades require extensive studies. The network of signalling pathways will reveal more details about the control of cell organization, proliferation and differentiation in the near future. Such networks may regulate complex processes such as the plasticity of plant development. For example, work from Zhou *et al.* (1998) has evidenced an interplay between the ethylene and the glucose signalling pathways in regulating plant development. Currently, investigators studying plant MAP kinase signalling pathways benefit greatly from the advances made in other model organisms. Furthermore, research will lead to the identification of plant-specific sequences and mechanisms that will undoubtedly contribute to the elucidation of the complex picture presently emerging on how eukaryotic development is controlled.

ACKNOWLEDGEMENTS

The authors wish to thank Professor P. Gullick (University Concordia, Canada) for valuable advice. The authors gratefully acknowledge the information and comments provided by colleagues from the BDP laboratory. Thanks to everybody for valuable support, interesting discussions and interest in our work. Y. Henry also acknowledges GB, who helped greatly in providing the opportunity to start the MAP kinase story. Part of this work was initiated and supported by an EU grant (SIME project BIOTEC-RTD-CEE PL 960275). The authors also acknowledge the scientific and financial support from the MERT and the CNRS to UMR 8618.

REFERENCES

Adam, A. L., Pike, S., Hoyos, M. E., Stone, J. M., Walker, J. C. and Novacky, A. (1997). Rapid and transient activation of a myelin basic protein kinase in tobacco leaves treated with harpin from *Erwinia amylovora*. *Plant Physiology* **115**, 853–861.

Aitken, A. (1996). 14-3-3 proteins on the MAP. *Trends in Biochemical Science* **20**, 95–97.

Alessi, D. R., Saito, Y., Campbell, D. G., Cohen, P., Sithanandam, G., Rapp, U., Ashworth, A., Marshall, C. J. and Cowley, S. (1994). Identification of the sites in MAP kinase kinase-1 phosphorylated by p74raf-1. *EMBO Journal* **13**, 1610–1619.

Alonso, J. M., Hirayama, T., Roman, G., Nourizadeh, S. and Ecker, J. R. (1999). EIN2, a bifunctional transducer of ethylene and stress responses in *Arabidopsis*. *Science* **284**, 2148–2152.

Alvarez, E., Northwood, I. C., Gonzalez, F. A., Latour, D. A., Seth, A., Abate, C., Curran, T. and Davis, R. J. (1991). Pro-Leu-Set/thr-Pro is a primary sequence for substrate protein phosphorylation. Characterization of the phosphorylation of c-myc and c-jun proteins by an epidermal growth factor receptor threonine 669 protein kinase. *Journal of Biological Chemistry* **266**, 15277–15285.

Ambrosio, L., Mahowald, A. P. and Perrimon, N. (1989). Requirement of the *Drosophila* raf homologue for torso function. *Nature* **342**, 288–291.

Anafi, M., Kiefer, F., Gish, G. D., Mbamalu, G., Tscove, N. N. and Pawson, T. (1997). SH2/SH3 adaptor proteins can link tyrosine kinases to a Ste20-related protein kinase HPK1. *Journal of Biological Chemistry* **272**, 27804–27811.

Ashworth, A., Nakielny, S., Cohen, P. and Marshall, C. (1992) The amino acid sequence of a mammalian MAP kinase. *Oncogene* **7**, 2555–2556.

Banno, H., Hirano, K., Nakamura, T., Irie, K., Nomoto, S., Matsumoto, K. and Machida, Y. (1993). *NPK1*, a tobacco gene that encodes a protein with a domain homologous to yeast BCK1, STE11 and Byr2 protein kinases. *Molecular and Cellular Biology* **13**, 4745–4752.

Bardwell, L. and Thorner, J. (1996). A conserved motif at the amino termini of MEKs might mediate high-affinity interaction with the cognate MAPKs. *Trends in Biochemical Sciences* **21**, 373–374.

Binari, R. and Perrimon, N. (1994). Stripe-specific regulation of pair-rule genes by hopscotch, a putative JAK family tyrosine kinase in *Drosophila. Genes and Development* **8**, 300–312.

Bleeker, A. B. and Schaller, G. E. (1996) The mechanism of ethylene perception. *Plant Physiology* **111**, 653–660.

Bögre, L., Ligterink, W., Heberle-Bors, E. and Hirt, H. (1996) Mechanosensors in plants. *Nature* **383**, 489–490.

Bögre, L., Ligterink, W., Meskiene, I., Barker, P. J., Heberle-Bors, E., Huskisson, N. S. and Hirt, H. (1997). Wounding induce the rapid and transient activation of a specific MAP kinase pathway. *Plant Cell* **9**, 75–83.

Bögre, L., Calderini, O., Binarova, P., Mattauch, M., Till, S., Kiegerl, S., Jonak, C., Pollaschek, C., Barker, P., Huskisson, N. S., Hirt, H. and Heberle-Bors, E. (1999). A MAP kinase is activated late in plant mitosis and becomes localized to the plane of cell division. *Plant Cell* **11**, 101–114.

Boguslawski, G. (1992). *PBS2* a yeast gene encoding a putative protein kinase, interacts with the Ras2 pathway and affects osmotic sensitivity of *Saccharomyces cerevisiae. Journal of Genetic and Microbiology* **138**, 2425–2432.

Bonni, A., Sun, Y., Nadal-Vicens, M., Bhatt, A., Frank, D. A., Rozovsky, I., Stohl, N., Yancopoulos, G. D. and Greenberg, M. E. (1997). Regulation of gliogenesis in the central nervous system by the JAK-STAT signaling pathway. *Science* **278**, 477–483.

Boulton, T. G., Nye, S. H., Robbins, D. J., Ip, N. Y., Radziejwska, E., Morgenbesser, S. D., DePinho, R. A., Panayotatos, N., Cobb, M. H. and Yancopoulos, G. D. (1991). ERKs: a family of protein-serine/threonine kinases that are activated and tyrosine phosphorylated in response to insulin and NGF. *Cell* **65**, 663–675.

Brandstatter, I. and Kieber, J. J. (1998) Two genes with similarities to bacterial response regulators are rapidly and specifically induced by cytokinin in *Arabidopsis. Plant Cell* **10**, 1009–1019.

Brewster, J. L., de Valoir, T., Dwyer, N. D., Winter, E. and Gustin, M. C. (1993). An osmosensing signal transduction pathway in yeast. *Science* **259**, 1760–1763.

Butty, A.-C., Pryciak, P. M., Huang, L. S., Herskowitz, I. and Peter, M. (1998). The role of Far1p in linking the heterotrimeric G protein to polarity establishment proteins during yeast mating. *Science* **282**, 1511–1516.

Calderini, O., Bögre, L., Vicente, O., Binarova, P., Heberle-Bors, E. and Wilson, C. (1998). A cell cycle regulated MAP kinase with a possible role in cytokinesis in tobacco cells. *Journal of Cell Science* **111**, 3091–3100.

Campbell, S. L., Khosravi-Far, R., Rossman, K. L., Clark, G. L. and Der, C. J. (1998). Increasing complexity of Ras signalling. *Oncogene* **17**, 1395–1413.

Camps, M., Nichols, A., Gillieron, C., Antonsson, B., Muda, M., Chabert, C., Boschert, V. and Arkinstall, S. (1998) Catalytic activation of the phosphatase MKP-3 by ERK2 mitogen-activated protein kinase. *Science* **280**, 1262–1265.

Canagarajah, B. J., Khokhlatchev, A., Cobb, M. H. and Goldsmith, E. J. (1997). Activation mechanism of the MAP kinase ERK2 by dual phosphorylation. *Cell* **90**, 859–869.

Cazalé, A.-C., Droillard, M.-J., Wilson, C., Heberle-Bors, E., Barbier-Brygoo, H. and Lauriere, C. (1999). MAP kinase activation by hypo-osmotic stress of tobacco cell suspensions: toward the oxidative burst response? *Plant Journal* **19**, 297–307.

Chang, C. and Meyerowitz, E. M. (1998). The ethylene hormone response in *Arabidopsis*: a eukaryotic two-component signaling system. *Proceedings of the National Academy of Sciences of the United States of America* **92**, 4129–4133.

Chang, C., Kwok, S. F., Bleecker, A. B. and Meyerowitz, E. M. (1993). *Arabidopsis* ethylene response gene ETR1: similarity of product to two-component regulators. *Science* **262**, 539–544.

Chervitz, S. A., Aravind, L., Sherlock, G., Ball, C. A., Koonin, E. V., Dwight, S. S., Harris, M. A., Dolinski, K., Mohr, S., Smith, T., Weng, S., Cherry, J. M. and Bolstein, D. (1998). Comparison of the complete protein sets of worm and yeast: orthology and divergence. *Science* **282**, 2022–2028.

Choi, K.-Y., Satterberg, B., Lyons, D. M. and Elion, E. A. (1994). Ste5 tethers multiple protein kinases in the MAP kinase cascade required for mating in *S. cerevisiae*. *Cell* **78**, 499–512.

Clark, K. L., Feldmann, P. J., Dignard, D., Larocque, R., Brown, A. J., Lee, M. G., Thomas, D. Y. and Whiteway, M. (1995). Constitutive activation of the *Saccharomyces cerevisiae* mating response pathway by a MAP kinase kinase from *Candida albicans*. *Molecular and General Genetics* **249**, 609–621.

Clark, K. L., Larsen, P. B., Wang, X. and Chang, C. (1998). Association of the *Arabidopsis* CTR1 Raf-like kinase with the ETR1 and ERS ethylene receptors. *Proceedings of the National Academy of Sciences of the United States of America* **95**, 5401–5406.

Clark-Lewis, I., Sanghera, J. S. and Pelech, S. L. (1991). Definition of a consensus sequence for peptide substrate recognition by $p44^{mpk}$, the meiosis-activated myelin basic protein kinase. *Journal of Biological Chemistry* **266**, 15180–15184.

Covic, L. and Lew, R. R. (1996). *Arabidopsis thaliana* cDNA isolated by functional complementation shows homology to serine/threonine protein kinases. *Biochimica et Biophysica Acta* **1305**, 125–129.

Daume, G., Eisenman-Tappe, I., Friez, H.-W., Troppmair, J. and Rapp, U. R. (1994). The ins and outs of Raf kinases. *Trends in Biochemical Sciences* **19**, 474–480.

Decroocq-Ferrant, V., Decroocq, S., Van Went, J., Schmidt, E. and Kreis, M. (1995). A homologue of the MAP/ERK family of protein kinase genes is expressed in vegetative and in female reproductive organs of *Petunia hybrida*. *Plant Molecular Biology* **27**, 339–350.

Derijard, B., Hibi, M., Wu, I. H., Barrett, T., Su, B., Deng, T., Karin, M. and Davis, R. J. (1994). JNK1: a protein kinase stimulated by UV light and Ha-Ras that binds and phosphorylates the c-Jun activation domain. *Cell* **76**, 1025–1037.

de Wit, P. J. G. M. (1997). Pathogen avirulence and plant resistance: a key role for recognition. *Trends in Plant Science* **2**, 452–458.

Dickens, M., Rogers, J. S., Cavanagh, J., Raitano, A., Xia, Z., Halpern, J. R., Greenberg, M. E., Sawyers, C. L. and Davis, R. J. (1997). A cytoplasmic inhibitor of the JNK signal transduction pathway. *Science* **277**, 693–696.

Draetta, G. F. (1997). Cell cycle: will the real Cdk-activating kinase please stand up. *Current Biology* **7**, R50–R52.

Duerr, B., Gawienowski, M., Ropp, T. and Jacobs, T. (1993). MsERK1: a mitogen-activated protein kinase from a flowering plant. *Plant Cell* **5**, 87–96.

Ecker, J. R. (1995). The ethylene signal transduction pathway in plants. *Science* **268**, 667–675.

Einsparh, K. J., Peeler, T. C. and Thompson, G. A. (1989). Phosphatidylinositol 4,5-biphosphate phospholipase C and phosphomonoesterase in *Dunaliella salina* membranes. *Plant Physiology* **90**, 1115–1120.

Errede, B. and Levin, D. E. (1993). A conserved kinase cascade for MAP kinase activation in yeast. *Current Opinion in Cellular Biology* **5**, 254–260.

Errede, B. and Ge, Q. Y. (1996). Feedback regulation of MAP kinase signal pathways. *Philosophical Transactions of the Royal Society of London* **351**, 143–149.

Fanger, G. R., Johnson, N. L. and Johnson, G. L. (1997). MEK kinases are regulated by EGF and selectively interact with Rac/Cdc42. *EMBO Journal* **16**, 4961–4972.

Faux, M. C. and Scott, J. D. (1996). Molecular glue, kinase anchoring and scaffold proteins. *Cell* **85**, 9–12.

Ferrant, V., van Went, J. and Kreis, M. (1994). Ovule cDNA clones of *Petunia hybrida* encoding proteins homologous to MAP and shaggy/zeste-white 3 protein kinases. *In* "Molecular and Cellular Aspects of Plant Reproduction" (R. J. Scott and A. D. Stead, eds) pp. 159–172. Society for Experimental Biology, Cambridge University Press.

Ferrel, J. E. (1996). MAP kinases in mitogenesis and development. *Current Topics in Developmental Biology* **33**, 1–60.

Ferrel, J. E. and Bhatt, R. R. (1997). Mechanistic studies of the dual phosphorylation of mitogen-activated protein kinase. *Journal of Biological Chemistry* **272**, 19008–19016.

Ferrigno, P., Posas, F., Koepp, D., Saito, H. and Silver, P. A. (1998). Regulated nucleo/cytoplasmic exchange of HOG1 MAPK requires the importin β homologs NMD5 and XPO1. *EMBO Journal* **17**, 5606–5614.

Gamble, R. L., Coonfield, M. L. and Schaller, G. (1998). Histidine kinase activity of the ETR1 ethylene receptor from *Arabidopsis*. *Proceedings of the National Academy of Sciences of the United States of America* **95**, 7825–7829.

Groom, L. A., Sneddon, A. A., Alessi, D. R., Dowd, S. and Keyse, S. M. (1996). Differential regulation of the MAP, SAP and RK/p38 kinases by Pyst1, a novel cytosolic dual-specificity phosphatase. *EMBO Journal* **15**, 3621–3632.

Gupta, R., Huang, Y., Kieber, J. and Luan, S. (1998). Identification of a dual-specificity protein phosphatase that inactivates a MAP kinase from *Arabidopsis*. *Plant Journal* **16**, 581–589.

Gupta, S., Han, H., Wong, L. H., Ralph, S. and Schindler, C. (1996). The SH2 domains of Stat1 and Stat2 mediate multiple interactions in the transduction of IFN-α signals. *EMBO Journal* **15**, 1075–1084.

Gustin, M. C., Albertyn, J., Alexander, M. and Davenport, K. (1998). MAP kinase pathways in the yeast *Saccharomyces cerevisiae*. *Microbiology and Molecular Biology Review* **62**, 1264–1300.

Hafen, E. (1998). Kinases and phosphatase – a marriage is consummated. *Science* **280**, 1212–1213.

Hamal, A., Jouannic, S., Leprince, A.-S., Kreis, M. and Henry, Y. (1999). Molecular characterization and expression of an *Arabidopsis thaliana* L. MAP kinase kinase cDNA, *AtMAP2Kα*. *Plant Science* **140**, 41–52.

Hanks, S. K. and Quinn, A. M. (1991). Protein kinase catalytic domain: sequence database. *Methods in Enzymology* **200**, 38–62.

Hanks, S. K. and Hunter, T. (1995). The eukaryotic protein kinase superfamily: kinase (catalytic) domain structure and classification. *FASEB Journal* **9**, 576–596.

Hanks, S. K., Quinn, A. M. and Hunter, T. (1988). The protein kinase family: conserved features and deduced phylogeny of the catalytic domains. *Science* **241**, 42–52.

Hardin, S. C. and Wolniak, S. M. (1998). Molecular cloning and characterization of maize ZmMEK1, a protein kinase with a catalytic domain homologous to mitogen- and stress-activated protein kinase kinases. *Planta* **206**, 577–584.

Haystead, T. A. J., Dent, P., Wu, J., Haystead, C. M. M. and Sturgill, T. W. (1992). Ordered phosphorylation of $p42^{MAPK}$ by MAP kinase kinase. *FEBS Letters* **306**, 17–22.

He, T.-C., Jiang, N., Zhuang, H. and Wojchowski, D. M. (1995). Erythropoietin-induced recruitment of Shc via a receptor phosphotyrosine-independent, Jak2-associated pathway. *Journal of Biological Chemistry* **270**, 11055–11061.

Herskowitz, I. (1995). MAP kinase pathways in yeast: for mating and more. *Cell* **80**, 187–197.

Hirt, H. (1997). Multiple roles of MAP kinases in plant signal transduction. *Trends in Plant Science* **2**, 11–15.

Horvath, C. M. and Darnel, J. E., Jr (1997). The state of the STATs: recent developments in the study of signal transduction to the nucleus. *Current Opinion in Cell Biology* **9**, 233–239.

Hou, X. S., Melnick, M. B. and Perrimon, N. (1996). *Marelle* acts downstream of the *Drosophila* HOP/JAK kinase and encodes a protein similar to the mammalian STATs. *Cell* **84**, 411–419.

Hua, J. and Meyerowitz, E. M. (1998). Ethylene responses are negatively regulated by a receptor gene family in *Arabidopsis thaliana*. *Cell* **94**, 261–271.

Hua, J., Chang, C., Sun, Q. and Meyerowitz, E. M. (1995). Ethylene insensitivity conferred by *Arabidopsis* ERS gene. *Science* **269**, 1712–1714.

Hua, J., Sakai, H., Nourizadeh, S., Chen, Q. G., Bleeker, A. B., Ecker, J. R. and Meyerowitz, E. J. (1998). *EIN4* and *ERS2* are members of the putative ethylene receptor gene family in *Arabidopsis*. *Plant Cell* **10**, 1321–1332.

Huang, C.-Y. and Ferrel, J. E. Jr (1996). Ultrasensitivity in the mitogen-activated protein kinase cascade. *Proceedings of the National Academy of Sciences of the United States of America* **93**, 10078–10083.

Hunter, T. and Plowman, G. D. (1997). The protein kinases of budding yeast: six score and more. *Trends in Biochemical Sciences* **22**, 18–22.

Huttly, A. K. and Phillips, A. L. (1995). Gibberellin-regulated expression in oat aleurone cells of two kinases that show homology to MAP kinase and a ribosomal protein kinase. *Plant Molecular Biology* **27**, 1043–1052.

Ichimura, K., Mizoguchi, T., Irie, K., Morris, P., Giraudat, J., Matsumoto, K. and Shinozaki, K. (1998a). Isolation of ATMEKK1 (a MAP kinase kinase kinase)-interacting proteins and analysis of a MAP kinase cascade in *Arabidopsis*. *Biochemical and Biophysical Research Communications* **253**, 532–543.

Ichimura, K., Mizoguchi, T., Hayashida, N., Seki, M. and Shinozaki, K. (1998b). Molecular cloning and characterization of three cDNAs encoding putative mitogen-activated protein kinase kinase (MAPKKs) in *Arabidopsis thaliana*. *DNA Research* **31**, 341–348.

Ihle, J. N. (1996). STATs: signal transducers and activators of transcription. *Cell* **84**, 331–334.

Ihle, J. N. and Kerr, I. M. (1995). JaKs and Stats in signaling by the cytokine receptor superfamily. *Trends in Genetics* **11**, 69–74.

Jena, P. K., Reddy, A. S. N. and Poovaiah, B. W. (1989). Molecular cloning and sequencing of a cDNA for plant calmodulin: signal-induced changes in the expression of calmodulin. *Proceedings of the National Academy of Sciences of the United States of America* **86**, 3644–3648.

Jonak, C., Pay, A., Bögre, L., Hirt, H. and Heberle-Bors, E. (1993). The plant homologue of MAP kinase is expressed in a cell cycle-dependent and organ-specific manner. *Plant Journal* **3**, 611–617.

Jonak, C., Heberle-Bors, E. and Hirt, E. (1994). MAP kinases: universal multi-purpose signalling tool. *Plant Molecular Biology* **24**, 407–416.

Jonak, C., Kiegerl, S., Lloyd, C., Chan, J. and Hirt, H. (1995). MMK2, a novel alfalfa MAP kinase, specifically complements the yeast MPK1 function. *Molecular and General Genetics* **248**, 686–694.

Jonak, C., Kiegerl, S., Ligterink, W., Barker, P. J., Huskisson, N. S. and Hirt, H. (1996). Stress signaling in plants: a mitogen-activated protein kinase pathway is activated by cold and drought. *Proceedings of the National Academy of Sciences of the United States of America* **93**, 11274–11279.

Josefsson, L. G. and Rask, L. (1997). Cloning of a putative G-protein-coupled receptor from *Arabidopsis thaliana. European Journal of Biochemistry* **249**, 415–420.

Jouannic, S., Hamal, A., Kreis, M. and Henry, Y. (1996). Molecular cloning of an *Arabidopsis thaliana* MAP kinase kinase-related cDNA. *Plant Physiology* **112**, 1397.

Jouannic, S., Hamal, A., Leprince, A.-S., Tregear, J. W., Kreis M. and Henry, Y. (1999a). Characterization of novel plant genes encoding MEKK/STE11 and RAF-related proteins. *Gene* **229**, 171–181.

Jouannic, S., Hamal, A., Leprince, A.-S., Tregear, J. W., Kreis M. and Henry, Y. (1999b). Plant MAP kinase kinase kinases structure, classification and evolution – a mini-review. *Gene* **233**, 1–11.

Kakimoto, T. (1996). *CKI1*, a histidine kinase homolog implicated in cytokinin signal transduction. *Science* **274**, 982–985.

Ketela, T., Brown, J. L., Stewart, R. C. and Bussey, H. (1998). Yeast Skn7p activity is modulated by the Sln1p-Ypd1p osmosensor and contributes to regulation of the HOG pathway. *Molecular and General Genetics* **259**, 372–378.

Khokhlatchev, A. V., Canagarajah, B., Wilsbacher, J., Robinson, M., Atkinson, M., Goldsmith, E. and Cobb, M. H. (1998). Phosphorylation of the MAP kinase ERK2 promotes its homodimerization and nuclear translocation. *Cell* **93**, 605–615.

Kieber, J. (1997). The ethylene signal transduction pathway in *Arabidopsis. Journal of Experimental Botany* **48**, 211–218.

Kieber, J., Rothenberg, M., Roman, G., Feldmann, K. A. and Ecker, J. R. (1993). *CTR1*, a negative regulator of the ethylene response pathway in *Arabidopsis*, encodes a member of the Raf family of protein kinases. *Cell* **72**, 427–441.

Kim, I. J., Lee, K. W., Park, B. Y., Lee, J. K., Park, J., Choi, I. Y., Eom, S. J., Chang, T. S., Kim, M. J., Yeom, Y. I., Chang, S. K., Lee, Y. D., Choi, E. J. and Han, P. L. (1999). Molecular cloning of multiple splicing variants of JIP-1 preferentially expressed in brain. *Journal of Neurochemistry* **72**, 1335–1343.

King, A. J., Sun, H., Diaz, B., Barnard, D., Miao, W., Bagrodia, S. and Marshall, M. S. (1998). The protein kinase Pak3 positively regulates Raf-1 through phosphorylation on serine 338. *Science* **396**, 180–183.

Knetsch, M. L. W., Wang, M., Snaar-Jagalska, B. E. and Heimovaara-Dijkstra, S. (1996). Abscisic acid induces mitogen-activated protein kinase activation in barley aleurone protoplasts. *Plant Cell* **8**, 1061–1067.

Knoester, M., Hennig, J., van Loon, L. C., Bol, J. F. and Linthorst, H. J. M. (1997). Isolation and characterization of a tobacco cDNA encoding an ETR1 homolog (accession number AFO 22727; PGR 97-188). *Plant Physiology* **115**, 1731.

Kobe, B. and Deisenhofer, J. (1994). The leucine-rich repeat: a versatile binding motif. *Trends in Biochemical Sciences* **19**, 415–421.

Kovtun, Y, Chiu, W.-L., Zeng, W. and Sheen, J. (1998). Suppression of auxin signal transduction by a MAPK cascade in higher plants. *Nature* **395**, 716–720.

Kültz, D. (1998). Phylogenetic and functional classification of mitogen- and stress-activated protein kinases. *Journal of Molecular Evolution* **46**, 571–588.

Kyriakis, J. M. and Avruch, J. (1996). Protein kinase cascades activated by stress and inflammatory cytokines. *Bioessays* **18**, 567–577.

Lashbrook, C. C., Tieman, D. M. and Klee, H. J. (1998). Differential regulation of the tomato *ETR* gene family throughout plant development. *Plant Journal* **15**, 243–252.

Leberer, E., Dignard, D., Harcus, D., Thomas, D. Y. and Whiteway, M. (1992). The protein kinase homologue Ste20p is required to link the yeast pheromone response G-protein $\beta\gamma$ subunits to downstream signalling components. *EMBO Journal* **11**, 4815–4824.

Lebrun-Garcia, A., Ouaked, F., Chiltz, A. and Pugin, A. (1998). Activation of MAPK homologues by elicitors in tobacco cells. *Plant Journal* **15**, 773–781.

Leeuw, T., Fourest-Lieuvin, A., Wu, C., Chenevert, J., Clark, K., Whiteway, M., Thomas, D. Y. and Leberer, E. (1995). Pheromone response in yeast: association of Bemp1 with proteins of the MAP kinase cascade and actin. *Science* **270**, 1210–1213.

Leeuw, T., Wu, C., Schrag, J. D., Whiteway, M., Thomas, D. Y. and Leberer, E. (1998). Interaction of a G-protein β-subunit with a conserved sequence in Ste20/PAK family protein kinase. *Nature* **311**, 191–195.

Legendre, L., Heinstein, P. F. and Low, P. S. (1992). Evidence for participation of GTP-binding proteins in elicitation of the rapid oxidative burst in cultured soybean cells. *Journal of Biological Chemistry* **267**, 20140–20147.

Lenormand, P., Sardet, C., Pages, G., L'Allemain, G., Brunet, A. and Pouyssegur, J. (1993). Growth factor induce nuclear translocation of MAP kinases ($p42^{mapk}$ and $p44^{mapk}$) but not of their activator MAP kinase ($p45^{mapkk}$) in fibroblasts. *Journal of Cellular Biology* **1222**, 1079–1088.

Leonard, C. J., Aravind, B. and Koonin, E. V. (1998). Novel families of putative protein kinases in bacteria and Archaea: evolution of the 'Eukaryotic' protein kinase superfamily. *Genome Research* **8**, 1038–1047.

Leprince, A.-S., Jouannic, S., Hamal, A., Kreis, M. and Henry, Y. (1999). Molecular characterisation of plant cDNAs *BnMAP4Kα1* and *BnMAP4Kα2* belonging to the GCK/SPS1 subfamily of MAP kinase kinase kinase kinase. *Biochimica et Biophysica Acta* **1444**, 1–13.

Lewis, T. S., Shapiro, P. S. and Ahn, N. C. (1998). Signal transduction through MAP kinase cascades. *Advances in Cancer Research* **74**, 49–139.

Li, S., Wilson, M. E. and Donelson, J. E. (1996). *Leishmania chagasi*: a gene encoding a protein kinase with a catalytic domain structurally related to MAP kinase kinase. *Experimental Parasitology* **82**, 87–96.

Ligterink, W., Kroj, T., Nieden, U., Hirt, H. and Scheel, D. (1997). Receptor-mediated activation of a MAP kinase in pathogen defense in plants. *Science* **276**, 2054–2057.

Lin, Q., Li, J., Smith, R. D. and Walker, J. C. (1998). Molecular cloning and chromosomal mapping of type one serine/threonine protein phosphatases in *Arabidopsis thaliana*. *Plant Molecular Biology* **37**, 471–481.

Lopez-Ilasaca, M., Crespo, P., Pellici, P. G., Gutkin, J. S. and Wetzker, R. (1997). Linkage of G protein-coupled receptors to the MAP kinase signaling pathway through PI3-kinase γ. *Science* **275**, 394–397.

Luan, S. (1998). Protein phosphatases and signaling cascades in higher plants. *Trends in Plant Science* **3**, 271–275.

Ma, H. (1994). GTP-binding proteins in plants: new members of an old family. *Plant Molecular Biology* **26**, 1611–1636.

Ma, H., Gamper, M., Parent, C. and Firtel, R. A. (1997). The *Dictyostelium* MAP kinase kinase DdMEK1 regulates chemotaxis and is essential for chemoattractant-mediated activation of guanylyl cyclase. *EMBO Journal* **16**, 4317–4332.

Machida, Y., Nishihama, R. and Kitakura, S. (1997). Progress in studies of plant homologs of mitogen-activated protein (MAP) kinases and potential upstream components in kinase cascades. *Critical Review in Plant Sciences* **16**, 481–496.

Machida, Y., Nakashima, M., Morikiyo, K., Banno, H., Ishikawa, M., Soyano, T. and Nishihama R. (1998). MAPKKK related protein kinase NPK1: regulation of the M phase of plant cell cycle. *Journal of Plant Research* **111**, 243–246.

Madhani, H. D. and Fink, G. R. (1998). The riddle of MAP kinase signaling specificity. *Trends in Genetics* **14**, 151–155.

Madhani, H. D., Styles, C. A. and Fink, G. R. (1997). MAP kinases with distinct inhibitory functions import signaling specificity during yeast differentiation. *Cell* **91**, 673–684.

Maeda, T., Wurgler-Murphy, S. M. and Saito, H. (1994). A two-component system that regulates an osmosensing MAP kinase cascade in yeast. *Nature* **396**, 242–245.

Maeda, T., Takekawa, M. and Saito, H. (1995). Activation of yeast PBS2 MAPKK by MAPKKKs or by binding of an SH3-containing osmosensor. *Science* **269**, 554–558.

Marshall, C. J. (1994). MAP kinase kinase kinase, MAP kinase kinase and MAP kinase. *Current Biology* **4**, 82–89.

Mason, C. S., Springer, C. J., Cooper, R. G., Superti-Furga, G., Marshall, C. J. and Marais, R. (1999). Serine and tyrosine phosphorylations cooperate in Raf-1, but no B-Raf activation. *EMBO Journal* **18**, 2137–2148.

Meloche, S., Seuwen, K., Pages, G. and Pouyssegur, J. (1992). Biphasic and synergistic activation of p44mapk (ERK1) by growth factors: correlation between late phase activation and mitogenicity. *Molecular Endocrinology* **6**, 845–854.

Meskiene, I., Bögre L., Glaser, W., Balog, J., Brandstötter, M., Zwerger, K., Ammerer, G. and Hirt, H. (1998). MP2C, a plant protein phosphatase 2C, functions as a negative regulator of mitogen-activated protein kinase pathways in yeast and plants. *Proceedings of the National Academy of Sciences of the United States of America* **95**, 1938–1943.

Mewes, H. W., Albermann, K., Bähr, M., Frishman, D., Gleissner, A., Hani, J., Heumann, K., Kleine, K., Maieri, A., Oliver, S. G., Pfeiffer, F. and Zollner, A. (1997). Overview of the yeast genome. *Nature* **387**, 7–73.

Millner, P. A. (1995). The auxin signal. *Current Opinion in Cell Biology* **7**, 224–231.

Mizoguchi, T., Hayashida, N., Yamaguchi-Shinozaki, K., Kamada, H. and Shinozaki, K. (1993). *ATMPKs*: a gene family of plant MAP kinases in *Arabidopsis thaliana. FEBS Letters* **336**, 440–444.

Mizoguchi, T., Gotoh, Y., Nishida, E., Yamaguchi-Shinozaki, K., Hayashida, N., Iwasaki, T., Kamada, H. and Shinozaki, K. (1994). Characterization of two cDNAs that encode MAP kinase homologues in *Arabidopsis thaliana* and analysis of the possible role of auxin in activating such kinase activities in cultured cells. *Plant Journal* **5**, 111–122.

Mizoguchi, T., Irie, K., Hirayama, T., Hayashida, N., Yamagushi-Shinozaki, K., Matsumoto, K. Shinozaki, K. (1996). A gene encoding a mitogen-activated protein kinase kinase kinase is induced simultaneously with genes for a mitogen-activated protein kinase and an S6 ribosomal protein kinase by touch, cold and water stress in *Arabidopsis thaliana. Proceedings of the National Academy of Sciences of the United States of America* **93**, 765–769.

Mizoguchi, T., Ichimura, K. and Shinozaki, K. (1997). Environmental stress response in plants: the role of mitogen-activated protein kinases. *Trends in Biotechnology* **15**, 15–19.

Mizoguchi, T., Ichimura, K., Irie, K., Morris, P., Giraudat, J., Matsumoto, K. and Shinozaki, K. (1998). Identification of a possible MAP kinase cascade in *Arabidopsis thaliana* based on pairwise yeast two-hybrid analysis and functional complementation tests of yeast mutants. *FEBS Letters* **437**, 56–60.

Mochly-Rosen, D. (1995). Localization of protein kinases by anchoring proteins: a theme in signal transduction. *Science* **268**, 247–251.

Moriguchi, T., Gotoh, Y. and Nishida, E. (1996). Roles of the MAP kinase cascade in vertebrates. *In* "Intracellular Signal Transduction" (H. Hidaka and A. C. Nairn, eds). *Advances in Pharmacology* **36**, 121–137.

Morris, P. C., Guerrier, D., Leung, J. and Giraudat, J. (1997). Cloning and characterization of *MEK1*, an *Arabidopsis* gene encoding a homologue of MAP kinase kinase. *Plant Molecular Biology* **35**, 1057–1064.

Morrison, D. K. and Cutler, R. E. Jr (1997). The complexity of Raf-1 regulation. *Current Opinion in Cell Biology* **9**, 174–179.

Morrison, D. K., Heidecker, G., Rapp, U. R. and Copeland, T. D. (1993). Identification of the major phosphorylation sites of the Raf-1 kinase. *Journal of Biological Chemistry* **268**, 17309–17316.

Mösch, H.-U., Roberts, R. and Fink, G. R. (1996). Ras 2 signals via the Cdc 42/Ste 20/mitogen-activated protein kinase module to induce filamentous growth in *Saccharomyces cerevisiae*. *Proceedings of the National Academy of Sciences of the United States of America* **93**, 5352–5356.

Moutoussamy, S., Kelly, P. A. and Finidori, J. (1998). Growth-hormone-receptor and cytokine-receptor-family signalling. *European Journal of Biochemistry* **255**, 1–11.

Muslin, A. J., Tanner, J. W., Allen, P. M. and Shaw, A. S. (1996). Interaction of 14-3-3 with signalling proteins is mediated by the recognition of phosphoserine. *Cell* **84**, 889–897.

Musti, A. M., Treier, M. and Bohmann, D. (1997). Reduced ubiquitin-dependent degradation of c-Jun after phosphorylation by MAP kinases. *Science* **275**, 400–402.

Nakashima, A., Hirano, K., Nakashima, S., Banno, H., Nishihama, R. and Machida, Y. (1998). The expression pattern for the gene for NPK1 protein kinase related to mitogen-activated protein kinase kinase kinase (MAPKKK) in a tobacco plant: correlation with cell proliferation. *Plant and Cell Physiology* **39**, 690–700.

Nern, A. and Arkowitz, R. A. (1998). A GTP-exchange factor required for cell orientation. *Nature* **391**, 195–198.

Nishihama, R., Banno, H., Shibata, W., Hirano, K., Nakashima, M., Usami, S. and Machida, Y. (1995). Plant homologues of components of MAPK (mitogen-activated protein kinase) signal pathways in yeast and animal cells. *Plant and Cell Physiology* **36**, 749–757.

Nishihama, R., Banno, H., Kawahara, E., Irie, K. and Machida, Y. (1997). Possible involvement of differential splicing in regulation of the activity of *Arabidopsis* ANP1 that is related to mitogen-activated protein kinase kinase kinase (MAPKKKs). *Plant Journal* **12**, 39–48.

Noselli, S. (1998). JNK signaling and morphogenesis in *Drosophila*. *Trends in Genetics* **14**, 33–38.

Ohme-Takagi, M. and Shinshi, H. (1995). Ethylene-inducible DNA binding proteins that interact with an ethylene-responsive element. *Plant Cell* **7**, 173–182.

Ota, I. and Varshawsky, A. (1993). A yeast protein similar to bacterial two-component regulators. *Science* **262**, 566–569.

Pang, L., Sawada, T., Decker, S. J. and Saltiel, A. R. (1995). Inhibition of MAP kinase kinase blocks the differentiation of PC-12 cells induced by nerve growth factor. *Journal of Biological Chemistry* **270**, 13585–13588.

Papin, C., Eychene, A., Brunet, A., Pages, G., Pouyssegur, J., Calothy, G. and Vianney-Barnier, J. (1995). B-Raf protein isoforms interact with and phosphorylate Mek-1 on serine residues 218 and 222. *Oncogenes* **10**, 1647–1651.

Paravicini, G. and Friedli, L. (1996). Protein–protein interactions in the yeast *PKC1* pathway: Pkc1p interacts with a component of the MAP kinase cascade. *Molecular and General Genetics* **251**, 682–691.

Pawson, T. (1995). Protein modules and signalling networks. *Nature* **373**, 573–580.

Pawson, T. and Scott, J. D. (1997). Signaling through scaffold, anchoring and adaptor proteins. *Science* **278**, 2075–2085.

Payne, D., Rossomando, A., Martino, P., Erickson, A., Her, J.-H., Shabanowitz, J., Hunt, D., Weber, W. and Sturgill, T. (1991). Identification of the regulatory phosphorylation sites in PP42/mitogen-activated protein kinase (MAP kinase). *EMBO Journal* **10**, 885–892.

Payton, S., Fray, R. G., Brown, S. and Griessen, D. (1996). Ethylene receptor expression is regulated during fruit ripening, flower senescence and abscission. *Plant Molecular Biology* **31**, 1227–1231.

Pelech, S. L. and Sanghera, J. S. (1992). Mitogen-activated protein kinases: versatile transducers for cell signaling. *Trends in Biochemical Sciences* **17**, 233–238.

Plakydou-Dymock, S., Dymock, D. and Hooley, R. (1998). A higher plant seven-transmembrane receptor that influences sensitivity to cytokinins. *Current Biology* **8**, 315–324.

Pöpping, B., Gibbons, T. and Watson, M. D. (1996). The *Pisum sativum* MAP kinase homologue (PsMAPK) rescues the *Saccharomyces cerevisiae hog1* deletion mutant under conditions of high osmotic stress. *Plant Molecular Biology* **31**, 355–363.

Posas, F. and Saito, H. (1997). Osmotic activation of the HOG MAPK pathway via Ste11p MAP3K: scaffold role of Pbs2p MAPKK. *Science* **276**, 1702–1705.

Posas, F. and Saito, H. (1998). Activation of the yeast SSK2 MAP kinase kinase kinase by the SSK1 two-component response regulator. *EMBO Journal* **17**, 1385–1394.

Pryciak, P. M. and Huntress, F. A. (1998). Membrane recruitment of the kinase cascade scaffold protein STE5 by the $G\beta\gamma$ complex underlies activation of the yeast pheromone response pathway. *Genes and Development* **12**, 2684–2697.

Pulverer, B. J., Kyriakis, J. M., Avruch, J., Nikolakaki, E., Woodgett, J. R. (1991). Phosphorylation of c-jun mediated by MAP kinases. *Nature* **353**, 670–674.

Ray, L. B. and Sturgill, T. W. (1987). Rapid stimulation by insulin of a serine/threonine kinase in 3T3-L1 adipocytes that phosphorylates microtubule-associated protein-2 *in vitro*. *Proceedings of the National Academy of Sciences of the United States of America* **84**, 1502–1506.

Robinson, M. J. and Cobb, M. H. (1997). Mitogen-activated protein kinase pathways. *Current Opinion in Cell Biology* **9**, 180–186.

Rodriguez, I., Esch, J. J., Hall, A. E., Binder, B. M., Schaller, G. E. and Bleeker, A. B. (1999). A copper cofactor for the ethylene receptor ETR1 from *Arabidopsis*. *Science* **283**, 996–998.

Romeis, T., Piedras, P., Zhang, S., Klessig, D. F., Hirt, H. and Jones, J. D. G. (1999). Rapid Avr-9 and Cf-9-dependent activation of MAP Kinases in tobacco cell cultures and leaves: convergence of resistance gene, elicitor, wound, and salicylate responses. *Plant Cell* **11**, 273–287.

Rossomando, A. J., Payne, D. M., Weber, M. J. and Sturgill, T. W. (1989). Evidence that pp42, a major tyrosine kinase target protein, is a mitogen-activated serine/

threonine protein kinase. *Proceedings of the National Academy of Sciences of the United States of America* **86**, 6940–6943.

Rouse, J., Cohen, P., Trigon, S., Morange, M., Alonso-Llamazares, A., Zamanillo, D., Hunt, T. and Nebrada, A. R. (1994). A novel kinase cascade triggered by stress and heat shock that stimulates MAPKAP kinase-2 and phosphorylation of the small heat shock proteins. *Cell* **78**, 1027–1037.

Sakai, H., Hua, J., Chen, Q. G., Chang, C., Bleeker, A. B. and Meyerowitz, E. M. (1998). *ETR2* is an *ETR1*-like gene involved in ethylene signaling in *Arabidopsis*. *Proceedings of the National Academy of Sciences of the United States of America* **95**, 5812–5817.

Schaeffer, H. J., Catling, A. D., Eblen, S. T., Collier, L. S., Krauss, A. and Weber, M. J. (1998). MP1: a MEK binding partner that enhances enzymatic activation of the MAP kinase cascades. *Science* **281**, 1668–1671.

Schaller, G. E. and Bleeker, A. B. (1995). Ethylene-binding sites generated in yeast expressing the *Arabidopsis ETR1* gene. *Science* **270**, 1809–1811.

Schmidt, E. D. L., Guzzo, F., Toonen, M. A. J. and De Vries, S. C. (1997). A leucine-rich repeat containing receptor-like kinase mark somatic plant cells competent to form embryos. *Development* **124**, 2049–2062.

Schüller, C., Brewster, J. L., Alexander, M. R., Gustin, F. C. and Ruis, H. (1994). The HOG pathway controls osmotic regulation of transcription via the stress response element (STRE) of the *Saccharomyces cerevisiae CTT1* gene. *EMBO Journal* **13**, 4382–4389.

Schultz, J., Ferguson, B. and Sprague, J. F. (1995). Signal transduction and growth control in yeast. *Current Biology* **5**, 31–37.

Sells, M. A. and Chernoff, J. (1997). Emerging from the Pak: the p21-activated protein kinase family. *Trends in Cell Biology* **7**, 162–167.

Seo, S., Okamoto, M., Seto, H., Ishizuka, K., Sano, H. and Ohashi, Y. (1995). Tobacco MAP kinase: a possible mediator in wound signal transduction pathways. *Science* **270**, 1988–1992.

Seo, S., Sano, H. and Ohashi, Y. (1999). Jasmonate-based wound signal transduction requires activation of WIPK, a tobacco mitogen-activated protein kinase. *Plant Cell* **11**, 288–298.

Seth, A., Alvarez, E., Gupta, S. and Davis, R. J. (1991). A phosphorylation site located at the NH2-terminal domain of c-Myc increases transactivation of gene expression. *Journal of Biological Chemistry* **266**, 23521–23524.

Sheen, J. (1996). Ca^{2+}-dependent protein kinases and stress signal transduction in plants. *Science* **274**, 1900–1902.

Shibata, W., Banno, H., Hirano, Y. I. K., Irie, K., Machida, S. U. C. and Machida, Y. (1995). A tobacco protein kinase, NPK2, has a domain homologous to a domain found in activators of mitogen-activated protein kinases (MAPKKs). *Molecular and General Genetic* **246**, 401–410.

Shieh, J.-C., Martin, H. and Millar, J. B. A. (1998). Evidence for a novel MAPKKK-independent pathway controlling the stress activated Sty1/Spc1 MAP kinase in fission yeast. *Journal of Cell Science* **111**, 2799–2807.

Shinozaki, K. and Yamaguchi-Shinozaki, K. (1996). Molecular responses to drought and cold stress. *Current Opinion in Biotechnology* **7**, 161–167.

Sieburth, D. S., Sun, Q. and Han, M. (1998). SUR-8, a conserved Ras-binding protein with leucine-rich repeats, positively regulates Ras-mediated signaling in *C. elegans*. *Cell* **94**, 119–130.

Smith, C. M., Shindyalov, I. N. and Veretnik, S. (1997). The Protein Kinase Resource. *Trends in Biochemical Sciences* **22**, 444–446.

Solano, R. and Ecker, J. R. (1998). Ethylene gas: perception, signaling and response. *Current Opinion in Plant Biology* **1**, 393–398.

Solano, R., Stepanova, A. and Ecker, J. R. (1998). Nuclear events in ethylene signaling: a transcriptional cascade mediated by ETHYLENE-INSENSITIVE3 and ETHYLENE-RESPONSE-FACTOR1. *Genes and Development* **12**, 3703–3714.

Stafstrom, J. P., Altschuller, M. and Anderson, D. H. (1993). Molecular cloning and expression of a MAP kinase homologue from pea. *Plant Molecular Biology* **22**, 83–90.

Stratmann, J. W. and Ryan, C. A. (1997). Myelin basic protein kinase activity in tomato leaves is induced systemically by wounding and increases in response to systemin and oligosaccharide elicitors. *Proceedings of the National Academy of Sciences of the United States of America* **94**, 11085–11089.

Suzuki, K. and Shinshi, H. (1995). Transient activation and tyrosine phosphorylation of a protein kinase in tobacco cells treated with a fungal elicitor. *Plant Cell* **7**, 639–647.

Takekawa, M., Maeda, T. and Saito, H. (1998). Protein phosphatase 2Cα inhibits the human stress-responsive p38 and JNK MAPK pathways. *EMBO Journal* **17**, 4744–4752.

Tan, P. B. O. and Kim, S. K. (1999). Signaling specificity: the RTK/RAS/MAP kinase pathway in metazoans. *Trends in Genetics* **15**, 145–149.

Tedford, K., Kim, S., Sa, D., Stevens, K. and Tyers, M. (1997). Regulation of the mating pheromone and invasive growth responses in yeast by two MAP kinase substrates. *Current Biology* **7**, 228–238.

Téna, G. and Renaudin, J.-P. (1998). Cytosolic acidification but not auxin at physiological concentration is an activator of MAP kinases in tobacco cells. *Plant Journal* **16**, 173–182.

Tregear, J. W., Decroocq-Ferrant, V., Jouannic, S. and Kreis, M. (1996). Protein kinase genes expressed during reproductive plant development. *In* "Embryogenesis, the Generation of a Plant" (T. L. Wang and A. C. Cuming, eds) pp. 77–88. BIOS Scientific, Oxford.

Tréhin, C., Planchais, S., Glab, N., Perennes, C., Tregear, J. W. and Bergounioux, C. (1998). Cell cycle regulation by plant growth regulators: involvement of auxin and cytokinin in the re-entry of *Petunia* protoplasts into the cell cycle. *Planta* **206**, 215–224.

Tzivion, G., Luo, Z. and Avruch, J. (1998). A dimeric 14-3-3 protein is an essential cofactor for Raf kinase activity. *Nature* **394**, 88–92.

Usami, S., Banno, H., Ito, Y., Nishihama, R. and Machida, Y. (1995). Cutting activates a 46-kilodalton protein kinase in plants. *Proceedings of the National Academy of Sciences of the United States of America* **92**, 8660–8664.

VanderKuur, J., Allevato, G., Billestrup, N., Norstedt, G. and Carter-Su, C. (1995). Growth-hormone-promoted tyrosyl phosphorylation of SHC proteins and SHC association with Grb2. *Journal of Biological Chemistry* **270**, 7587–7593.

Wan, Y., Kurosaki, T. and Huang, X. Y. (1996). Tyrosine kinases in inactivation of the MAP kinase cascade by G-protein-coupled receptors. *Nature* **380**, 541–544.

Wang, Y. and Li, N. (1997). A cDNA sequence isolated from the ripening tomato fruit encodes a putative protein kinase (PGR97-096). *Plant Physiology* **114**, 1135.

Watanabe, S. and Arai, K.-C. (1996). Roles of the JAK-STAT system in signal transduction via cytokine receptors. *Current Opinion in Genetics and Development* **6**, 587–596.

Whitmarsh, A. J. and Davis, R. J. (1998). Structural organization of MAP-kinase signaling modules by scaffold proteins in yeast and mammals. *Trends in Biochemical Sciences* **23**, 481–485.

Whitmarsh, A. J., Cavanagh, J., Tournier, C., Yasuda, J. and Davis, R. J. (1998). A mammalian scaffold complex that selectively mediates MAP kinase activation. *Science* **281**, 1671–1674.

Whiteway, M. S., Wu, C., Leeuw, T., Clark, K., Fourest-Lieuvin, A., Thomas, D. Y. and Leberer, E. (1995). Association of the yeast pheromone response G protein $\beta\gamma$ subunits with the MAP kinase scaffold Ste5p. *Science* **269**, 1572–1575.

Wilkinson, J. Q., Lanahan, M. B., Yen, H.-C., Giovannoni, J. J. and Klee, H. J. (1995). An ethylene-inducible component of signal transduction encoded by *Never-ripe*. *Science* **270**, 1807–1809.

Wilsbacher, J. L., Goldsmith, E. J. and Cobb, M. H. (1999). Phosphorylation of MAP kinases by MAP/ERK involves multiple regions of MAP kinases. *Journal of Biological Chemistry* **274**, 16988–16994.

Wilson, C., Eller, N., Gartner, A., Vicente, O. and Heberle-Bors, E. (1993). Isolation and characterization of a tobacco cDNA encoding a putative MAP kinase. *Plant Molecular Biology* **23**, 543–551.

Wilson, C., Anglmayer, R., Vicente, O. and Heberle-Bors, E. (1995). Molecular cloning, functional expression in *E. coli*, and characterization of multiple mitogen-activated protein kinases from tobacco. *European Journal of Biochemistry* **233**, 249–257.

Wilson, C., Voronin, V., Touraev, A., Vicente, O. and Heberle-Bors, E. (1997). A developmentally regulated MAP kinase activated by hydration in tobacco pollen. *Plant Cell* **9**, 2093–2100.

Wilson, C., Pfosser, M., Jonak, C., Hirt, H., Heberle-Bors, E. and Vicente, O. (1998). Evidence for the activation of a MAP kinase upon phosphate-induced cell cycle re-entry in tobacco cells. *Physiologia Plantarum* **102**, 532–538.

Winston, L. A. and Hunter, T. (1995). JAK2, Ras, and Raf are required for activation of extracellular signal-regulated kinase/mitogen-activated protein kinase by growth hormone. *Journal of Biological Chemistry* **270**, 30837–30840.

Wittenberg, C. and Reed, S. I. (1996). Plugging it in: signaling circuits and the yeast cell cycle. *Current Opinion in Cellular Biology* **8**, 223–230.

Woodgett, J. R., Kyriakis, J. M., Avruch, J., Zon, L. I., Zanke, B. and Templeton, D. J. (1996). Reconstitution of novel signaling cascades responding to cellular stresses. *Philosophical Transactions of the Royal Society of London B* **351**, 135–142.

Xia, Y., Wu, Z., Su, B., Murray, B. and Karin, M. (1998). JNKK1 organizes a MAP kinase module through specific and sequential interactions with upstream and downstream components mediated by its amino-terminal extension. *Genes and Development* **12**, 3369–3381.

Xu, Q., Fu, H. H., Gupta, R. and Luan, S. (1998). Molecular characterization of a tyrosine-specific protein phosphatase encoded by a stress-responsive gene in *Arabidopsis*. *Plant Cell* **10**, 849–857.

Yablonski, D., Marbach, I. and Levitzki, A. (1996). Dimerization of Ste5, a mitogen-activated protein kinase cascade scaffold protein, is required for signal transduction. *Proceedings of the National Academy of Sciences of the United States of America* **93**, 13863–13869.

Yan, R., Small, S., Desplan, C., Dearolf, C. R. and Darnell, J. E., Jr (1996). Identification of a *Stat* gene that function in *Drosophila* development. *Cell* **84**, 421–430.

Zhang, S. and Klessig, D. F. (1997). Salicylic acid activates a 48-kD MAP kinase in tobacco. *Plant Cell* **9**, 809–824.

Zhang, S. and Klessig, D. F. (1998). The tobacco wounding-activated mitogen-activated protein kinase is encoded by *SIPK*. *Proceedings of the National Academy of Sciences of the United States of America* **95**, 7225–7230.

Zhang, S., Du, H. and Klessig, D. F. (1998). Activation of the tobacco SIP kinase by both a cell wall-derived carbohydrate elicitor and purified proteinaceous elicitins from *Phytophtora* sp. *Plant Cell* **10**, 435–449.

Zheng, C.-F. and Guan, K. L. (1994). Activation of MEK family kinases requires phosphorylation of two conserved Ser/Thr residues. *EMBO Journal* **13**, 1123–1131.

Zhou, L., Jang, J. C., Jones, T. L. and Sheen, J. (1998). Glucose and ethylene signal transduction crosstalk revealed by an *Arabidopsis* glucose-insensitive mutant. *Proceedings of the National Academy of Sciences of the United States of America* **95**, 10294–10299.

Protein Phosphorylation and Dephosphorylation in Environmental Stress Responses in Plants

K. ICHIMURA, T. MIZOGUCHI, R. YOSHIDA, T. YUASA and K. SHINOZAKI

Laboratory of Plant Molecular Biology, Tsukuba Life Science Center, The Institute of Physical and Chemical Research (RIKEN), 3-1-1, Koyadai, Tsukuba, Ibaraki 305–0074, Japan

I. INTRODUCTION

In contrast to animals that can escape an environmental stress, plants are unable to move to avoid the stress; instead they respond and adapt themselves to the environmental stresses to survive. Growth of land plants is greatly affected by a variety of environmental stresses. Therefore, plants have unique

Advances in Botanical Research Vol. 32
incorporating Advances in Plant Pathology
ISBN 0-12-005932-0

systems to sense and respond to a variety of environmental stresses, such as dehydration, low temperature, heat, mechanical perturbation (touch), mechanical wounding and pathogen infection. Environmental stresses induce various biochemical, physiological and molecular responses including gene expression in plants (for reviews see Dixon and Lamb, 1990; Ryan 1990; Grillo and Leone, 1996; Ingram and Bartels, 1996; Shinozaki and Yamaguchi-Shinozaki, 1996, 1997; Braam *et al.*, 1997; Bray, 1997; Lamb and Dixon, 1997; Yang *et al.*, 1997; Satoh and Murata, 1998; Shinozaki *et al.*, 1998). These responses are considered to be promoted as a result of a perception of environmental stimuli and signal transduction in plant cells, but the mechanisms by which plants sense and transduce signals in response to environmental stresses are largely unknown.

Protein phosphorylation/dephosphorylation has been shown to regulate diverse cellular processes in eukaryotes (Boyer and Krebs, 1986; Hunter, 1995). Thousands of protein kinases have been reported and classified into several groups based on their structures, substrate specificities and regulatory ligands (Hanks *et al.*, 1988; Hanks and Quinn, 1991; Smith *et al.*, 1997). Budding yeast, *Saccharomyces cerevisiae*, has 113 conventional protein kinases, corresponding to approximately 2% of the total genes (Hunter and Plowman, 1997). This fact implicates that protein kinases are essential for cellular activities. Genetic, biochemical and pharmacological analyses have shown that protein kinases and protein phosphatases play important roles in environmental stress responses including responsiveness to stress-induced phytohormones in plants. Recently, enzymatic activity of MAPKs (mitogen-activated protein kinases) and MAPK-like protein kinases has been shown to be modulated in response to biotic and abiotic stresses and phytohormones in higher plants (Table I and Fig. 1). MAPK cascades are modules of signal transduction from the cell surface to the nucleus and play a pivotal role in the regulation of biochemical and physiological changes associated with extracellular stimuli, such as growth regulators, differentiation signals and environmental stimuli (Cobb *et al.*, 1991; Pelech and Sanghera, 1992; Nishida and Gotoh, 1993; Cano and Mahadevan, 1995; Herskowitz, 1995; Levin and Errede, 1995) (Fig. 1). A novel MAPK cascade has also been identified in *Arabidopsis*. In this review, we summarize the biological roles of protein phosphorylation and dephosphorylation in environmental stress responses, and mainly discuss a role for MAP kinase in environmental stress signal transduction in higher plants.

II. MECHANICAL WOUNDING AND TOUCH

A. INVOLVEMENT OF PHOSPHORYLATION AND DEPHOSPHORYLATION IN GENE EXPRESSION RESPONSIVE TO MECHANICAL WOUNDING STRESS

Mechanical wounding, one of the most severe environmental stresses to plants, is mostly caused by weather conditions (strong wind and hail) or herbivore

TABLE I
Activation of MAPK and MAPK-like protein kinase by environmental stresses

Source	Protein kinase	Molecular mass	Signal(s)	References
Tobacco		47 kDa	Elicitor, touch	Suzuki and Shinshi (1995)
Tobacco	PMSAPK	46 kDa	Wounding, elicitor, SA	Usami *et al.* (1995)
Tobacco	WIPK[a]	46 kDa	Wounding, TMV infection, elicitor	Seo *et al.* (1995, 1999); Zhang and Klessig (1998b); Romeis *et al.* (1999)
Tobacco	SIPK[a]	48 kDa	SA, elicitor, wounding, TMV infection	Zhang and Klessig (1997, 1998a, b); Zhang *et al.* (1998); Romeis *et al.* (1999)
Tobacco	HAPK	49 kDa	Elicitor	Ádám *et al.* (1997)
Tobacco		50 kDa, 46 kDa	Elicitor	Lebrun-Garcia *et al.* (1998)
Alfalfa	MMK4[a] (SAMK)	44 kDa	Cold, drought, touch, wounding	Bögre *et al.* (1996, 1997); Jonak *et al.* (1996)
Parsley	ERMK[a]	45 kDa	Elicitor	Ligterink *et al.* (1997)
Tomato	MBPK	48 kDa	Systemin, wounding	Stratmann and Ryan (1997)
Arabidopsis	ATMPK4[a]	43 kDa	Cold, drought, touch, wounding	Ichimura (1999)

[a]Corresponding genes have been cloned.

attack by animals and insects. Mechanical wounding activates the production of phytohormones such as jasmonic acid (JA), ethylene, abscisic acid (ABA) (Hildmann *et al.*, 1992), and the 18 amino acid polypeptide, systemin (Pearce *et al.*, 1991), and finally causes induction of wound-responsive gene expression (Peña-Cortés *et al.*, 1988; Farmer and Ryan, 1990; Mason and Mullet, 1990; Titarenko *et al.*, 1997). Titarenko *et al.* (1997) isolated wound- and JA-inducible genes by the differential display. A comparable study of the wound response in wild-type and JA-insensitive *coi1* mutant plants indicated that *Arabidopsis* wound signals are transmitted via at least two different pathways. One of the pathways, which does not involve JA as a mediator, is preferentially responsible for gene activation in the vicinity of the wound site, whereas another requires JA perception and activation of gene expression throughout the aerial part of the plant. Pharmacological approaches using protein kinase

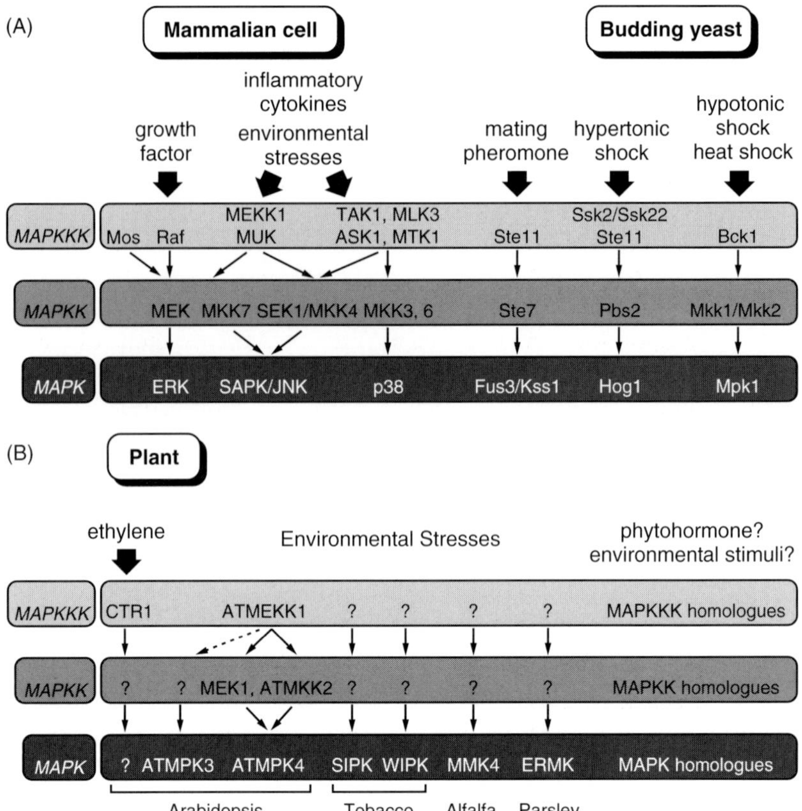

Fig. 1. Comparison of conserved mammalian, budding yeast (A) and plant (B) MAP kinase cascade. Each level in the cascade is colour-coded. Arrows indicate substrate specificity of MAPKKKs or MAPKKs. Question marks indicate a particular component that has not been identified. In the MAPK cascades, MAPK is activated by phosphorylation catalysed by MAPKK (MAPK kinase). MAPKKs are also regulated through the phosphorylation by several different groups of MAPKK kinases (MAPKKK). Several MAP kinase cascades, responsive to growth signals, differentiation signals, inflammatory cytokines and environmental stresses, have been identified in both the animal cells and budding yeast cell. In higher plants, several environmental stress-activated MAPKs have already been identified (described elsewhere in this volume). However, the stimulatory signal of a number of other MAPKs, MAPKKs, and MAPKKKs still remains unclear.

inhibitors or protein phosphatase inhibitors showed that JA-induced gene expression was negatively regulated by protein kinase sensitive to staurosporine, and protein phosphatases (Rojo *et al.*, 1998). This JA-induced gene expression was blocked in the *Arabidopsis* JA-insensitive mutants *jin1*, *jin4* and *coi1* (Rojo *et al.*, 1998). Wound-induced activation of JA-responsive genes is also regulated by this protein phosphorylation step. An alternative wound signaling pathway, independent of JA, has also been identified, leading to the transcriptional activation of a different set of genes. This JA-independent pathway is also regulated by a protein phosphorylation switch, in which the protein kinase positively regulated the pathway while the protein phosphatase negatively regulated it.

The *proteinase inhibitor II* (*pin2*) gene family of potato and tomato (Graham *et al.*, 1985; Sánchez-Serrano *et al.*, 1986) is widely used to monitor release and perception of wound signals. Not only ABA (Peña-Cortés *et al.*, 1989) and JA (Farmer and Ryan, 1990) but also systemin (Pearce *et al.*, 1991), a potent inducer of proteinase inhibitor accumulation, has been shown to be involved in *pin2* activation upon wounding. Systemin action is ABA dependent because it does not induce *pin2* expression in ABA-deficient plants (Peña-Cortés *et al.*, 1996). Dammann *et al.* (1997) studied a reversible phosphorylation involved in ABA- or JA-induced expression of *pin2* gene. A protein kinase inhibitor, staurosporine, blocked ABA-induced transcriptional activation of *pin2* gene in tomato leaves. By contrast, a protein phosphatase inhibitor, okadaic acid, did not inhibit it. Interestingly, okadaic acid prevented JA-induced gene expression of *pin2* gene while staurosporine did not inhibit it. These observations suggest that hormonal induction of *pin2* gene in leaves involves reversible phosphorylation and is differently regulated by ABA and JA, respectively. In roots, okadaic acid did not inhibit the JA-dependent transcriptional activation of *pin2* gene. The ABA-induced expression of *pin2* gene was abolished in roots (Dammann *et al.*, 1997). These findings suggest the presence of a complex, organ-specific transduction network for regulation of the effects of the phytohormones ABA and JA on gene expression upon wounding.

B. ENZYMATIC ACTIVATION OF PROTEIN KINASES BY MECHANICAL WOUNDING

Usami *et al.* (1995) reported the cutting-activated 46-kDa protein kinase in tobacco leaves using an in-gel phosphorylation assay. The in-gel phosphorylation is an efficient method for detecting stimuli-responsive protein kinases, because we can monitor the fluctuation and molecular size of the protein kinases at the same time. They termed the 46-kDa protein kinase as a plant multi-signal-activated protein (PMSAP) kinase, because this activity was activated not only by wounding but also by water-soluble chitosan, a tobacco cell-wall digest, and salicylic acid (Table I). The PMSAP kinase was activated within 1 min after cutting. Maximum activity was observed at 2–5 min, and the activity dropped at 30 min. They also found that (1) the PMSAP kinase did not respond to cutting at 30 min after the first cutting, and (2) cycloheximide, an inhibitor of protein synthesis, led to constitutive activation of PMSAP kinase without wounding. These results suggested that newly synthesized protein phosphatase is involved in inactivation of PMSAP kinase. Phosphorylation of serine (Ser)/threonine (Thr) and tyrosine (Tyr) residues is possibly required for activation of the PMSAP kinase.

A tobacco protein kinase, WIPK (wound-induced protein kinase), is a member of the MAP kinase family (Seo *et al.*, 1995). The *WIPK* gene was initially isolated by differential screening. Transcripts of the *WIPK* gene rapidly accumulated within 1 min after wounding. Transgenic tobacco with co-suppressed *WIPK* gene exhibited (1) an increased salicylic acid (SA) level and induction of pathogenesis-related genes and (2) inhibition of wound-induced JA accumulation and gene expression (Seo *et al.*, 1995). JA inhibits biosynthesis of SA, while SA inhibits JA biosynthesis (Sano and Ohashi, 1995). Seo *et al.* (1995) reported that WIPK is involved in the initial response of mechanical wounding, and possibly regulates wound-induced JA biosynthesis. Recently, using a specific antibody, Seo *et al.* (1999) found that WIPK was rapidly and transiently activated by wounding (Fig. 2 and Table I). Systemic activation of WIPK was also detected in the adjacent upper leaves by cutting the lower stem. Transgenic tobacco plants carrying cauliflower mosaic virus 35S-promoter-driven *WIPK* gene (not co-suppressed) showed constitutive expression of the wound-inducible *PI-II* gene in healthy leaves. The activity of WIPK and the levels of JA and Me-JA were correlated in the transgenic tobacco over-expressing the *WIPK* gene and *PI-II* gene.

Tobacco SIPK (salicylic-acid-induced protein kinase) was initially identified as a salicylic-acid-activated protein kinase in tobacco suspension cells (Zhang

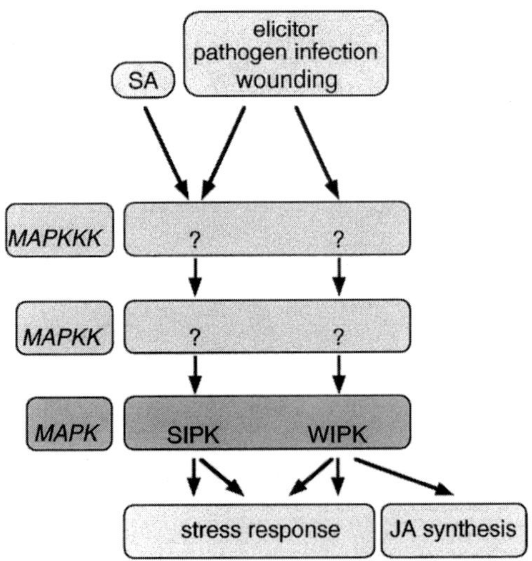

Fig. 2. Two MAPKs, SIPK and WIPK, commonly activated by environmental stresses. Elicitor, pathogen infection and mechanical wounding activate both SIPK and WIPK. Salicylic acid activates SIPK. Study using transgenic tobacco which, over-expressing the *WIPK* gene, suggests that WIPK regulates JA biosynthesis.

and Klessig, 1997). SIPK is a 48-kDa kinase and a member of the plant MAP kinase family (Table I and Fig. 2). Using a specific antibody, Zhang and Klessig (1998a) found that infiltration of water alone into tobacco leaves activated the SIPK. Not only infiltration, but also various mechanical stresses including cutting and wounding activated the SIPK (Zhang and Klessig, 1998a). This indicates that mechanical wounding subsequent to water infiltration causes activation of SIPK. Activation of SIPK after wounding was associated with phosphorylation of its tyrosine residue. Unlike *WIPK* and *MMK4*, expression of the *SIPK* gene did not respond to wounding (Zhang and Klessig, 1998a). The protein level of SIPK was also constant.

Bögre *et al.* (1997) showed wound-induced rapid and transient activation of $p44^{MMK4}$, an alfalfa MAP kinase, using a specific antibody raised against a synthetic peptide encoding the C-terminal 10 amino acids of the alfalfa MMK4 kinase (Table I). Bögre *et al.* (1997) suggested that the inactivation machinery for the $p44^{MMK4}$-mediating MAP kinase pathway was activated around 25 min after wounding. The inactivation of $p44^{MMK4}$ required *de novo* protein synthesis because pretreatment of leaves with α-amanitin or cycloheximide, inhibitors of transcription or translation, resulted in sustained activity of $p44^{MMK4}$ after wounding. These results agree with the observations by Usami *et al.* (1995). An alfalfa protein phosphatase MP2C (Medicago phosphatase 2C) is a candidate component for inactivation machinery of MMK4 and PMSAPK, because wounding-induced expression of the *MP2C* gene and recombinant MP2C protein dephosphorylated and inactivated wound-activated MMK4 *in vitro* (Meskiene *et al.*, 1998). Transcription of the alfalfa *MMK4* gene is upregulated by wounding, like *WIPK*. The transcript amount increased markedly at 20 min after wounding and decreased to the basal levels of the non-wounding leaves by 4 h, but the protein levels of alfalfa MMK4 did not increase upon wounding. This observation suggests the presence of some mechanism allowing the $p44^{MMK4}$ to stay in a constitutive protein level. Recently, Hirt (1999) presented an idea that the activated MAP kinase was degraded and the transient transcriptional upregulation of the MAP kinase gene was to compensate for the loss of the MAP kinase protein.

Tomato plants respond to insect herbivory and mechanical wounding damage by inducing a localized and systemic synthesis of several defence-related proteins (Bergey *et al.*, 1996). An 18 amino acid polypeptide, called systemin, isolated from tomato leaves, appears to be a primary wound signal (Pearce *et al.*, 1991). Wounding tomato leaves rapidly activated 48-kDa myelin basic protein (MBP) kinase as seen in tobacco and alfalfa leaves (Stratmann and Ryan, 1997) (Table I). This activity was recovered by immunoprecipitation with anti-phosphotyrosine antibody. Application of systemin through cut stems also induced activation of 48-kDa MBP kinase. The Ala17-systemin, which is an inactive analogue of systemin, did not activate the 48-kDa MBP kinase. Moreover, systemin-induced activation of the 48-kDa MBP kinase was inhibited by the presence of an excess amount of Ala17-systemin. This fact shows that systemin specifically activated the 48-kDa MBP kinase. Wounding-induced activation of the 48-kDa MBP kinase was detected in the tomato *def1* (*defenseless1*) mutant that harbours a mutation in the octadecanoid pathway for JA synthesis (Stratmann and Ryan, 1997). This indicates that the MBP kinase functions between the perception of primary signals and the *DEF1* gene product.

C. ENZYMATIC ACTIVATION OF PROTEIN KINASES BY MECHANICAL PERTURBATION (TOUCH)

In response to mechanical stimuli such as wind and touch, plants undergo physiological and developmental changes that enhance resistance to subsequent mechanical stress. In general, plants are grown in windy environments or are exposed to repetitive touch stimulation. They become shorter and stockier, and often have altered flexibility. These changes in development in response to mechanostimulation are collectively known as thigmomorphogenesis (Mitchell and Myers, 1995; Ennos, 1997). In plants, wind or touch stimulation results in the enhancement of expression of several genes encoding calmodulin, a calmodulin-related protein, xyloglucan endotransglycosylase, ACC synthase and protein kinases (Braam and Davis, 1990; Botella *et al.*, 1995, 1996; Mizoguchi *et al.*, 1995; Arteca and Arteca, 1999). Bögre *et al.* (1996) reported that mechanical stimulation rapidly and transiently activated an alfalfa MAP kinase MMK4 (Table I). In tobacco, simple transfer of the cell culture activated the 47-kDa protein kinase (Suzuki and Shinshi, 1995). The 47-kDa protein kinase was originally identified as an elicitor-activated protein kinase (Suzuki and Shinshi, 1995). This implicates that the 47-kDa protein kinase was also activated by mechanical stimuli. The magnitude and duration of the transfer-induced activation of 47-kDa protein kinase were less than those of the activation of this kinase by elicitor treatment.

III. PATHOGEN INFECTION AND ELICITORS

A. INVOLVEMENT OF PHOSPHORYLATION AND DEPHOSPHORYLATION IN GENE EXPRESSION RESPONSIVE TO PATHOGEN INFECTION AND ELICITOR TREATMENT

Plant cells exposed to elicitors, whether crude fungal cell wall fragments or defined molecules, such as purified proteins and avirulence gene products, respond with a battery of cellular changes (Dixon and Lamb, 1990; Yu, 1995; Hammond-Kosack and Jones, 1997; Yang *et al.*, 1997). Some of these responses, such as changes in ion fluxes and the generation of reactive oxygen species, occur very rapidly and may involve events that occur primarily at the post-translational level (Levine *et al.*, 1994; Mehdy, 1994; Viard *et al.*, 1994; Jabs *et al.*, 1997). Other responses, such as the accumulation of phytoalexins and synthesis of chitinases, glucanases and other pathogenesis-related proteins, involve induction of gene expression (Dixon and Lamb, 1990; Yang *et al.*, 1997).

Several lines of evidence have implicated the involvement of protein phosphorylation in the regulation of the elicitor-stimulated defence responses, including those induced by elicitins. Elicitins are a family of small extracellular proteins produced by the pathogenic fungal genus *Phytophthora* (Yu, 1995). In tobacco, elicitins induce a hypersensitive-like response and confer protection against subsequent infection by microbial pathogens. Treatment of plant cells with either purified elicitins or cell wall-derived carbohydrate elicitors has been shown to cause rapid changes in phosphoprotein profiles (Dietrich *et al.*, 1990; Felix *et al.*, 1991, 1994; Viard *et al.*, 1994). Inhibition of protein phosphorylation by K-252a and staurosporine correlates with the blockage of medium alkalization, reactive oxygen species generation and defence gene activation (Grosskopf *et al.*, 1990; Felix *et al.*, 1991; Viard *et al.*, 1994; Suzuki *et al.*, 1995). By contrast, protein phosphatase inhibitors can mimic the effects of fungal elicitors, presumably by elevating the protein phosphorylating state (Felix *et al.*, 1994).

Potential substrates for protein kinases responding to fungal elicitors include DNA-binding proteins, such as KAP-1 and KAP-2. Both proteins recognize the H-box element of the elicitor-responsive chalcone synthase promoter. Dephosphorylation of KAP-1 and KAP-2 alters the mobility of their protein–DNA complexes (Yu *et al.*, 1993). Phosphorylation of G/HBF-1, a soybean basic leucine zipper (b-ZIP) transcription factor, by a bacterial pathogen-induced serine kinase, enhances its ability to bind to the chalcone synthase *chs15* promoter (Dröge-Laser *et al.*, 1997). Similarly, elicitor-induced phosphorylation of the nuclear factor PBF-1 is required for effective binding and activation of the potato *PR-10a* gene (Després *et al.*, 1995). A kinase with characteristics of mammalian protein kinase C is implicated in this process (Subramaniam *et al.*, 1997).

B. ENZYMATIC ACTIVATION OF PROTEIN KINASES BY PATHOGEN INFECTION AND ELICITORS

Exposure of plants to elicitors induces changes in the phosphorylation status of proteins (Grab *et al.*, 1989; Dietrich *et al.*, 1990; Grosskopf *et al.*, 1990; Felix *et al.*, 1991), and inhibition of such changes by protein kinase inhibitors has been demonstrated (Grosskopf *et al.*, 1990; Conrath *et al.*, 1991; Felix *et al.*, 1991). Suzuki and Shinshi (1995) reported biochemical analysis of the PiE (elicitor from *Phytophthora infestans* cell wall)-induced transient activation of a 47-kDa protein kinase in tobacco suspension cells (Table I). The TvX (a proteineous elicitor, xylanase from *Trichoderma viride*) also activated the 47-kDa protein kinase with delayed and prolonged kinetics, and elicitor-induced cell death (Suzuki *et al.*, 1999). Staurosporine, a protein kinase inhibitor, and Gd^{3+} ion, a plasma membrane Ca^{2+} channel blocker, inhibited the PiE and TvX-responsive activation of the 47-kDa protein kinase and the TvX-induced cell death (Suzuki and Shinshi, 1995; Suzuki *et al.*, 1999). Treatment with cells with calyculin A, an inhibitor of protein phosphatase type 1 and 2A, or cycloheximide, an inhibitor of protein synthesis, resulted in sustained activation of the 47-kDa protein kinase (Suzuki and Shinshi, 1995). Treatment of calyculin A as well as TvX induced hypersensitive cell death (Suzuki *et al.*, 1999). These results showed that both Ca^{2+} influx through the Ca^{2+} channel on the plasma membrane and protein phosphorylation are involved in the elicitor-induced activation of the 47-kDa protein kinase and hypersensitive cell death in tobacco cells. Furthermore, inactivation of the 47-kDa protein kinase requires *de novo* synthesized factor(s), possibly protein phosphatase(s). Elicitor-induced phosphorylation of tyrosine residue of 47-kDa protein kinase correlated with the activation of the protein kinase. These characteristics are very similar to those of MAP kinases in animals and yeasts. Harpin, a bacterial protein elicitor, also rapidly and transiently activated harpin-activated protein kinase (HAPK), a MAP kinase-like protein kinase, in tobacco cells (Ádám *et al.*, 1997). Lebrun-Garcia *et al.* (1998) presented lines of evidence that cryptogein, a low-molecular-weight secreted proteinous elicitor, activates two MAP kinase homologues (50 kDa and 48 kDa) in tobacco cells. They also showed that oligogalacturonides (OGs) activated the MAP kinase homologues. These lines of evidence strongly support the involvement of MAP kinase pathway in the elicitor-induced defence response in tobacco cells.

Zhang *et al.* (1998) found that two purified proteinaceous fungal elicitors, parasiticein and cryptogein, as well as fungal cell wall-derived carbohydrate elicitor(s), rapidly activated a 48-kDa protein kinase in tobacco suspension cells (Table I and Fig. 2). The 48-kDa kinase was immunoprecipitated with a specific antibody raised against SIPK. SIPK was initially identified as salicylic-acid-activated protein kinase in tobacco suspension cells (Zhang and Klessig, 1997). These results showed that not only SA, but also elicitors activated the 48-kDa SIPK. Both infiltration of SA and fungal cell wall elicitors into leaves also caused activation of SIPK with different kinetics (Zhang and Klessig, 1998a). The transcript level of the *SIPK* gene as well as the protein level remained constant during the activation of SIPK by the treatment with cell wall elicitor and SA. The activation of SIPK by cell wall elicitors and SA was associated with phosphorylation of tyrosine residues. These results suggest that SIPK is activated by tyrosyl phosphorylation. The maximum activation of this kinase paralleled or preceded medium alkalization and activation of the defence gene phenylalanine ammonia-lyase (PAL). Inhibition of SIPK activation by protein kinase inhibitors correlated with the suppression of cell wall elicitor-induced medium alkalization and *PAL* gene activation, suggesting that SIPK has a role in these defence responses. Activation of 48-kDa and 44-kDa protein kinases was detected in tobacco mosaic virus (TMV)-infected tobacco carrying resistance gene *N* (Zhang and Klessig, 1998b). Using specific antibodies against SIPK and WIPK, Zhang and Klessig showed that the 48-kDa and 44-kDa protein kinases were encoded by the *SIPK* and *WIPK* genes, respectively. The two MAP kinases were activated within 4 h post-shifting (hps) TMV-infected plants (Fig. 2). Both messenger RNA (mRNA) and protein levels of WIPK increased in response to TMV infection, which induced an enzymatic activation of WIPK. TMV-induced elevation of mRNA level of the *WIPK* gene was systemic. Its activation at the mRNA, protein and enzyme activity levels was independent of SA. They suggested that regulation of WIPK at multiple levels by an *N*-gene-mediated signal might play an important role in the signal transduction pathway leading to local and systemic resistance to TMV.

Cf-9 gene from tomato confers resistance to races of the leaf mould fungus *Cladosporium fulvum* expressing the complementary *Avr9* gene. *Cf-9* has been cloned and appeared to encode an extracytoplasmic membrane-anchored glycoprotein encompassing 27 leucine-rich repeats (Jones *et al.*, 1994). Avr9 protein was purified based on its capacity to elicit a Cf-9-dependent necrotic response in tomato leaves (De Wit and Spikman, 1982). The isolation of the corresponding gene revealed that *Avr9* encodes a precursor protein of 63 amino acids that becomes processed by fungal and plant proteases to a 28 amino acid peptide when secreted (Van den Ackerveken *et al.*, 1992, 1993). Hammond-Kosak *et al.* (1994, 1998) showed, in accordance with the gene-for-gene model, that Cf-9 and Avr9 retain their specificity in other plants: functional apoplastically expressed, the Avr9 peptide that harbours the *Avr9* transgene under the control of the cauliflower mosaic virus 35S promoter induced necrosis in transgenic *Cf-9* tobacco and potato. Romeis *et al.* (1999) showed that two protein kinases of 46 and 48 kDa were rapidly and transiently activated by Avr9 in a gene-for-gene manner. Studies using pharmacological inhibitors and effectors suggested that protein phosphorylation, Ca^{2+} influx, calmodulin and phospholipase A_2 were required for Avr9-induced activation of the 46- and 48-kDa protein kinases. Activation of the two protein kinases is independent of oxidative burst. Using antibodies which specifically recognize the tobacco MAP kinases SIPK and WIPK, the 48- and 46-kDa protein kinases were shown to be related to SIPK and WIPK, respectively (Fig. 2). The *WIPK* mRNA was induced by addition of Avr9 in the plants and cell cultures containing Cf-9. These data indicated that (1) the R/Avr-mediated induction of plant defence is accomplished via several parallel signalling mechanisms, and (2) R/Avr-dependent signal transduction pathways were interlinked at MAP kinases with responses of plants, not only to non-race-specific elicitors, but also to abiotic stimuli, such as wounding and mechanical stresses.

An extracellular 42-kDa fungal glycoprotein has been identified as the principal elicitor of the multi-component defence response through sequential activation of ion channels and an oxidative burst in parsley cells (Parker *et al.*, 1991). An oligopeptide fragment of 13 amino acids in length (Pep13) within this glycoprotein is necessary and sufficient to induce the same reactions as the intact glycoprotein (Nürnberger *et al.*, 1994; Sacks *et al.*, 1995). Pep13 specifically interacts with a plasma membrane target site in the plant and initiates a signal transduction cascade leading to the transient activation of plant defence genes and the accumulation of phytoalexines (Nürnberger *et al.*, 1994). The Pep13 and Pep25 elicitors rapidly and transiently activated 43-kDa ERMK (elicitor-responsive MAP kinase) (Ligterink *et al.*, 1997) (Table I). Several compounds, such as an ion channel blocker (anthracene-9-carboxylate), an agonist of ion fluxes (Amphotericin B) and inhibitor of NADH or NADPH oxidase (diphenylene iodonium), are known to inhibit or mimic elicitor-induced ion fluxes, production of reactive oxygen species, induction of phytoalexin and defence gene activation. An analysis using these compounds showed that the elicitor-induced ion channel activation is necessary for activation of ERMK, and the ERMK plays a role either upstream or independently of the oxidative burst. The subcellular location of ERMK was determined by immunofluoresence microscopy before and after treatment of parsley cells with the elicitor. Immediately after Pep25 treatment, ERMK was translocated into the nucleus (Ligterink *et al.*, 1997).

IV. LOW-TEMPERATURE AND OSMOTIC STRESS

A. INVOLVEMENT OF PHOSPHORYLATION AND DEPHOSPHORYLATION IN LOW-TEMPERATURE AND OSMOTIC-STRESS-INDUCIBLE GENE EXPRESSIONS

In response to low but non-freezing temperatures, plants activate transcription of many genes such as *cor* (cold-regulated), *lti* (low-temperature-induced) and *cas* (cold-acclimation-specific) genes (see reviews by Grillo and Leone, 1996; Ingram and Bartels, 1996; Shinozaki and Yamaguchi-Shinozaki, 1996, 1997; Bray, 1997; Shinozaki *et al.*, 1998). Using pharmacological inhibitors and effectors, protein phosphorylation and Ca^{2+} have been shown to be required for cold-acclimation-specific gene expression of the *cas* genes (Monroy and Dhindsa, 1995; Monroy *et al.*, 1998). Low-temperature-induced expression of alfalfa *cas15* gene (cold-acclimation-specific gene encoding a 14.5-kDa protein) was prevented by a protein kinase inhibitor, staurosporine (Monroy *et al.*, 1998). By contrast, treatment with okadaic acid, a protein phosphatase inhibitor, induced expression of the *cas15* gene at room temperature (Monroy and Dhindsa, 1995). These results suggested that the activity of staurosporine-sensitive protein kinases and inactivation of protein phosphatases are required for the low-temperature signal transduction leading to the expression of the *cas15* gene. The activity of protein phosphatases in the total cellular extract of alfalfa plants subjected to cold acclimation was decreased to the steady level (70% of room temperature) within 30 min. Okadaic acid preferentially inhibits protein phosphatase type 2A (PP2A) rather than protein phosphatase type 1 (PP1) (Cohen *et al.*, 1990). This fact raises the possibility that cold-induced inactivation of PP2A regulates induction of the *cas15* gene at cold acclimation.

Mizoguchi *et al.* (1996) showed that expression of the *Arabidopsis ATMPK3* gene encoding a MAPK is upregulated by drought, salt and cold stresses. An

Arabidopsis MAPKKK gene, *ATMEKK1*, was also induced by the same environmental stresses. These results suggest that ATMPK3 and ATMEKK1 possibly have a role in the environmental stress response in *Arabidopsis* (Fig. 1). The transcription levels of *ATPK19* and *ATPK6* (*atpk1*) genes, which encode homologues of ribosomal-protein S6 kinase, known as a substrate for animal MAPKs, were also rapidly and markedly increased in response to cold and salt stresses (Mizoguchi *et al.*, 1995). An alfalfa MAPK gene, *MMK4*, is also induced by low-temperature, dehydration and mechanical wounding stresses (Jonak *et al.*, 1996). Calcium-dependent protein kinases (CDPKs) appear to represent a large protein kinase gene family in plants. Gene expression of two *Arabidopsis* CDPKs, ATCDPK1 and ATCDPK2, was up-regulated by drought and salt stresses but not by cold and heat stresses (Urao *et al.*, 1994). Thus these protein kinases appear to be involved in low-temperature and osmotic stress responses in higher plants.

B. ENZYMATIC ACTIVATION OF PROTEIN KINASES BY LOW-TEMPERATURE AND OSMOTIC STRESS

An earlier study showed that the *in vivo* protein phosphorylation level of several proteins was changed by cold acclimation in comparison to non-acclimation (Monroy *et al.*, 1993). This suggests the presence of low-temperature-induced protein kinase(s) in plants. Recently, Jonak *et al.* (1996) identified a cold-activated protein kinase and its gene. An alfalfa MAP kinase, MMK4, was shown to be activated by not only low temperature but also dehydration, based on biochemical analysis using a specific antibody (Table I). Application of ABA to alfalfa plants did not activate MMK4, suggesting that MMK4 was activated by an ABA-independent pathway. In *Arabidopsis*, ATMPK4 is activated by cold, low humidity, touch and wounding, as shown in the next section (Ichimura, 1999). Accumulation of mRNA of the *MMK4* gene was detected under cold and drought stresses, but the protein levels of MMK4 remained constant. Recently, Hirt (1999) proposed an idea that the activated MAP kinase was degraded and the transient transcriptional upregulation of the MAP kinase gene was to compensate for the loss of the MAP kinase protein. Seo *et al.* (1999) showed a constant protein level of WIPK during wound-induced transcriptional activation of its gene, but Zhang and Klessig (1998b) showed TMV-induced accumulation of *WIPK* transcript and WIPK protein. In such cases, protein levels of MAPKs may be regulated in a stress-dependent manner.

V. POSSIBLE INVOLVEMENT OF AN *ARABIDOPSIS* MAPK CASCADE IN ABIOTIC ENVIRONMENTAL STRESS SIGNALLING

Mizoguchi *et al.* (1996) showed that the mRNA level of the *ATMEKK1* (a MAP kinase kinase kinase) gene increased markedly in response to water stress, cold and touch. This suggests that the signal transduction pathway through ATMEKK1 functions in the presence of environmental stress. Functional analyses using both the yeast two-hybrid system and the complementation of yeast mutants suggest that MEK1/ATMKK2 (MAPKKs) and ATMPK4 (a MAPK) constitute a MAPK cascade with ATMEKK1 and function in the downstream of ATMEKK1 (Figs 1 and 3) (Mizoguchi *et al.*, 1993, 1996, 1997, 1998; Morris *et al.*, 1997; Ichimura *et al.*, 1998). ATMEKK1 also possibly functions as a scaffold protein because ATMEKK1 can specifically interact with all the members of the putative MAP kinase cascade at different regions (Ichimura *et al.*, 1998) (Fig. 3). ATMEKK1 interacts with ATMPK4 in the N-terminal non-catalytic region, and also interacts with both MEK1 and ATMKK2 in the C-terminal catalytic region (Ichimura *et al.*, 1998; Mizoguchi *et al.*, 1998). In the budding yeast, Pbs2 protein kinase plays the role of MAPKK in the MAPK pathway responsive to hyperosmotic shock. Pbs2 specifically interacts with both its activator MAPKKKs and its MAPK, and functions as a scaffold protein of this MAPK cascade (Posas and Saito, 1997).

ATMEKK1 may have a similar role in the environmental stress-responsive MAPK cascade in *Arabidopsis*. In order to examine whether these components of the putative MAPK cascade (ATMEKK1 → MEK1/ATMKK2 → ATMPK4) are involved in environmental stress signal transduction, we raised a polyclonal antibody (named ATMPK4CT antibody) against the C-terminal

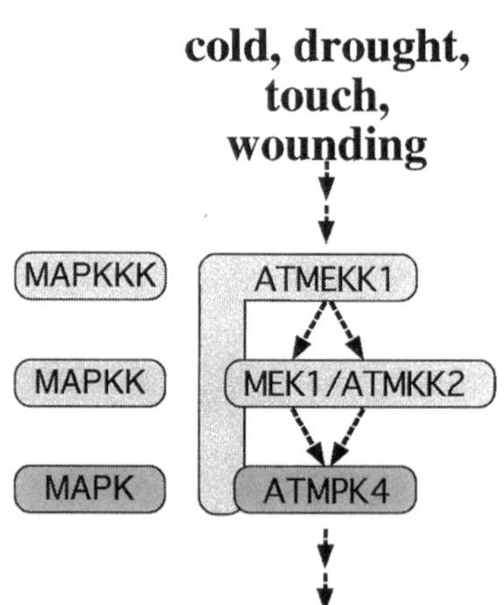

Fig. 3. Schematic model of a novel MAP kinase cascade – ATMEKK1 (a MAPKKK), MEK1 and ATMKK2 (MAPKKs) and ATMPK4 (a MAPK) – in *Arabidopsis*. Environmental signals such as cold, drought, touch and wounding may be transmitted to ATMEKK1. ATMEKK1 may activate both MEK1 and ATMKK2. Activated MEK1 and ATMKK2 phosphorylate and activate ATMPK4. ATMEKK1 may also function as a scaffold protein since ATMEKK1 can specifically interact with all the members of the putative MAP kinase cascade at different regions.

16 amino acid polypeptide of the ATMPK4. The ATMPK4CT antibody specifically cross-reacted to the ATMPK4 protein but not to other *Arabidopsis* MAP kinases. Using this antibody, we performed immunoprecipitation with crude protein extracts prepared from detached leaves of stress-treated plants. The recovered immunoprecipitates were subjected to the in-gel phosphorylation assay for analysing protein kinase activity. We found transient activation of ATMPK4 in response to dehumidification (as a model of dehydration), cold, touch and wounding stress treatments (Fig. 3) (Ichimura, 1999). By contrast, neither heat stress nor abscisic acid led to activation of ATMPK4. Neither transcript nor protein levels of the ATMPK4 were affected by all the stress conditions examined (Ichimura, 1999). These results suggest that ATMPK4 is involved in the signal transduction of dehydration, low-temperature, touch and wounding stresses. Alfalfa MMK4 is also activated by dehydration, low-temperature, touch and wounding stresses. However, ATMPK4 is distantly related to MMK4 and SIPK.

In tobacco, a specific pathogen, Avr9, leads to activation of at least two known MAP kinases, SIPK and WIPK (Romeis *et al.*, 1999). Moreover, two proteinaceous elicitins, parasiticein and cryptogein, activated at least three MBP kinases (Zhang *et al.*, 1998). Our preliminary results showed that a number of MAPK-like protein kinases was activated by dehydration, low-temperature, touch and wounding stresses. Taken together, a number of MAP kinases is possibly activated by various environmental stresses. This indicates the involvement of redundant MAPK cascades in stress signals.

VI. CONCLUSIONS AND PERSPECTIVES

In mammalian cells, various cellular stresses activate two types of MAPK homologues, stress-activated protein kinase (SAPK)/c-Jun N-terminal kinase (JNK) and p38 groups (Fig. 1). These groups have several isoforms. Taken together, different MAPK homologues are activated by cellular stresses in mammalian cells. It seems that a redundant mode of signal pathway also exists in higher plants. In tobacco, wounding and elicitors activate at least two known MAP kinases (Zhang and Klessig, 1998a, b; Romeis *et al.*, 1999; Seo *et al.*, 1999) (Fig. 2). Moreover, two proteinaceous elicitins, parasiticein and cryptogein, activated at least three MAPK-like protein kinases (Zhang *et al.*, 1998). Our preliminary results showed that a number of MAPK-like protein kinases (at least three) was activated by dehydration, low-temperature, touch and wounding stresses in *Arabidopsis*. Taken together, these results implicate the existence of redundant environmental stress-responsive MAPK cascades in plants. An alfalfa MAPK, MMK4 is activated by multiple stresses such as dehydration, low-temperature, touch and wounding stresses. Although *Arabidopsis* ATMPK4 is also activated by the same stresses, ATMPK4 is distantly related to MMK4. This indicates not only involvement of redundant

MAPK cascades in environmental stress signal transduction but also different stress signals transmitted through the same pathway.

How can different stimuli be transmitted through the same pathway and how is the specificity of the stress response determined? Two ideas arise about the multi-stress-induced activation of MAPKs. The first, a common signal such as calcium might be mobilized by different environmental stresses. The second, cross-talk of stress signal might occur at the level of MAPKKK. In the mammalian MAPK cascade, MAPKKKs activate various types of MAPKK. The stress response might be defined by duration and/or amplitude of several MAPK cascades like the animal MAPK cascade (Marshall, 1995; Chen *et al.*, 1996). Another idea is that a multi-protein complex formation related to the MAPK cascade may determine the signal specificity like the budding yeast pheromone responsive pathway and high osmolarity responsive pathway. The last idea is that an alternative signal pathway independent of the MAPK cascade may be simultaneously activated and might be concerned with the determination of response specificity. So far, no mutants of *MAPK* genes have been isolated using forward genetics. This may be due to redundancy of the MAPK cascade in a specific signal transduction. *Arabidopsis* ATMPK4 interacts with two MAPKKs, MEK1 and ATMPK2, and may constitute a redundant cascade. There may be a complicated mechanism of signal transduction in the MAPK cascade in higher plants.

Recently, forward and reverse genetic approaches have led to isolation of genes encoding protein kinases and protein phosphatases that play important roles in stress signalling in plants. In most cases, however, their direct upstream or downstream factors remain to be identified. The yeast two-hybrid system and the functional complementation tests with yeast mutants will be powerful tools for the isolation of regulators and substrates of the protein kinases and phosphatases. Biochemical analysis using specific antibodies will help to elucidate their activity and complex formation *in vivo*. Once we have the regulators or interaction factors of the key protein kinases and phosphatases, isolation of insertional mutants of the genes which contain transfer DNA (T-DNA) or transposons will be useful for analysing the biological function of the regulators or interaction factors for the protein kinases and phosphatases. Analysis of transgenic plants that overexpress or repress genes for the kinases or phosphatases will help to elucidate the molecular mechanisms of the stress responses to environmental changes in higher plants. Screening for mutants which have altered sensitivity to protein kinase or phosphatase inhibitors will also enable us to identify novel factors involved in the stress signalling pathway.

REFERENCES

Ádám, A. L., Pike, S., Hoyos, M. E., Stone, J. M., Walker, J. C. and Navacky, A. (1997). Rapid and transient activation of myelin basic protein kinase in tobacco leaves treated with harpin from *Erwinia amylovora*. *Plant Physiology* **115**, 853–861.

Arteca, J. M. and Arteca, R. N. (1999). A multi-responsive gene encoding 1-aminocyclopropane-1-carboxylate synthase (*ACS6*) in mature *Arabidopsis* leaves. *Plant Molecular Biology* **39**, 209–219.

Bergey, D. R., Howe, G. A. and Ryan, C. A. (1996). Polypeptide signaling for plant defensive genes exhibits analogies to defense signaling in animals. *Proceedings of the National Academy of Sciences of the United States of America* **93**, 12053–12058.

Bögre, L., Ligterink, W., Heberle-Bors, E. and Hirt, H. (1996). Mechanosensors in plants. *Nature* **383**, 489–450.

Bögre, L., Ligterink, W., Meskiene, I., Barker, P. J., Heberle-Bors, E., Huskisson, N. S. and Hirt, H. (1997). Wounding induces the rapid and transient activation of a specific MAP kinase pathway. *Plant Cell* **9**, 75–83.

Botella, J. R., Arteca, J. M., Schlagnhaufer, C. D., Arteca, R. N. and Phillips A. T. (1995). A mechanical strain-induced ACC synthase gene. *Proceedings of the National Academy of Sciences of the United States of America* **92**, 1595–1598.

Botella, J. R., Arteca, J. M., Somodevilla, M. and Arteca, R. N. (1996). Calcium-dependent protein kinase gene expression in response to physical and chemical stimuli in mungbean (*Vigna radiata*). *Plant Molecular Biology* **30**, 1129–1137.

Boyer, P. D. and Krebs, E. G. (eds) (1986). "The Enzymes", Vol. 17. Academic Press, New York.

Braam, J. and Davis, R. W. (1990). Rain-, wind-, and touch-induced expression of calmodulin and calmodulin-related genes in *Arabidopsis*. *Cell* **60**, 357–364.

Braam, J., Sistrunk, M. L., Polisensky, D. H., Xu, W., Purugganan, M. M., Antosiewicz, D. M., Campbell, P. and Johnson, K. A. (1997). Plant responses to environmental stress: regulation and functions of the *Arabidopsis TCH* genes. *Planta* **203**, S35–S41.

Bray, E. A. (1997). Plant responses to water deficit. *Trends in Plant Science* **2**, 48–54.

Cano, E. and Mahadevan, L. C. (1995). Parallel signal processing among mammalian MAPKs. *Trends in Biochemical Sciences* **20**, 117–122.

Chen, Y.-R. Meyer, C. F. and Tan, T.-H. (1996). Persistent activation of c-Jun N-terminal kinase 1 (JNK1) in γ radiation-induced apoptosis. *Journal of Biological Chemistry* **271**, 631–634.

Cobb, M. H., Boulton, T. G. and Robbins, D. J. (1991). Extracellular signal-related kinases: ERKs in progress. *Cell Regulation* **2**, 965–978.

Cohen, P., Holomes, C. F. B. and Tsukitani, Y. (1990). Okadaic acid: a new probe for the study of cellular regulation. *Trends in Biochemical Sciences* **15**, 98–102.

Conrath, U., Jeblick, W. and Kauss, H. (1991). The protein kinase inhibitor, K-252a, decreases elicitor-induced Ca^{2+} uptake and K^+ release, and increases coumarin synthesis in parsley cells. *FEBS Letters* **279**, 141–144.

Damman, C., Rojo, E. and Sanchez-Serrano, J. J. (1997). Abscisic acid and jasmonic acid activate wound-inducible gene in potato through separate, organ-specific signal transduction pathways. *Plant Journal* **11**, 773–782.

Després, C., Subramaniam, R., Matton, D. P. and Brisson, N. (1995). The activation of the potato *PR-10a* gene requires the phosphorylation of the nuclear factor PBF-1. *Plant Cell* **7**, 589–598.

De Wit, P. J. G. M. and Spikman, G. (1982). Evidence for the occurrence of race and cultivar-specific elicitors of necrosis in intercellular fluids of compatible interactions of *Cladosporium fulvum* and tomato. *Physiological Plant Pathology* **21**, 1–11.

Dietrich, A., Mayer, J. E. and Hahlbrock, K. (1990). Fungal elicitor triggers rapid, transient, and specific protein phosphorylation in parsley cell suspension cultures. *Journal of Biological Chemistry* **265**, 6360–6368.

Dixon, R. A. and Lamb, C. J. (1990). Molecular communication in interactions between plants and microbial pathogens. *Annual Review of Plant Physiology and Plant Molecular Biology* **41**, 339–367.

Dröge-Laser, W., Kaiser, A., Lindsay, W. P., Halkier, B. A., Loake, G. J., Doerner, P., Dixon, R. A. and Lamb, C. (1997). Rapid stimulation of a soybean protein serine kinase which phosphorylates a novel bZIP DNA-binding protein, G/HBF-1, during the induction of early transcription-dependent defenses. *EMBO Journal* **16**, 726–738.

Ennos, A. R. (1997). Wind as an ecological factor. *Trends in Ecology and Evolution* **12**, 108–111.

Farmer, E. E. and Ryan, C. A. (1990). Interplant communication: airborne methyl jasmonate induces synthesis of proteinase inhibitors in plant leaves. *Proceedings of the National Academy of Sciences of the United States of America* **87**, 7713–7716.

Felix, G., Grosskopf, D. G., Regenass, M. and Boller, T. (1991). Rapid changes of protein phosphorylation are involved in transduction of the elicitor signal in plant cells. *Proceedings of the National Academy of Sciences of the United States of America* **88**, 8831–8834.

Felix, G., Regenass, M., Spanu, P. and Boller, T. (1994). The protein phosphatase inhibitor calyculin A mimics elicitor action in plant cells and induced rapid hyperphosphorylation of specific protein as revealed by pulse labeling with [^{33}P] phosphate. *Proceedings of the National Academy of Sciences of the United States of America* **91**, 952–956.

Grab, D., Feger, M. and Ebel, J. (1989). An endogenous factor from soybean (*Glycine max L.*) cell cultures activates phosphorylation of a protein which is dephosphorylation *in vivo* in elicitor-challenged cells. *Planta* **179**, 430–348.

Graham, J. S., Pearce, G., Merryweather, J., Titani, K., Ericsson, L. H. and Ryan, C. A. (1985). Wound-induced proteinase inhibitors from tomato leaves. II. The cDNA-deduced primary structure of pre-inhibitor II. *Journal of Biological Chemistry* **260**, 6561–6564.

Grillo, S. and Leone, A. (eds) (1996). "Physical Stresses in Plants, Genes and their Products for Tolerance". Springer, Berlin.

Grosskopf, D. G., Felix, G. and Boller, T. (1990). K-252a inhibits the response of tomato cells to fungal elicitors *in vivo* and their microsomal protein kinase in vitro. *FEBS Letters* **275**, 177–180.

Hammond-Kosack, K. E. and Jones, J. D. G. (1997). Plant disease resistance genes. *Annual Review of Plant Physiology and Plant Molecular Biology* **48**, 575–607.

Hammond-Kosack, K. E., Harrison, K. and Jones, J. D. G. (1994). Developmentally regulated cell death on expression of the fungal avirulence gene *Avr9* in tomato seedlings carrying the disease-resistance gene *Cf-9*. *Proceedings of the National Academy of Sciences of the United States of America* **91**, 10445–10449.

Hammond-Kosack, K. E., Tang, S., Harrison, K. and Jones, J. D. G. (1998). The tomato *Cf-9* disease resistance gene functions in tobacco and potato to confer responsiveness to the fungal avirulence gene product Avr9. *Plant Cell* **10**, 1251–1266.

Hanks, S. K. and Quinn, A. M. (1991). Protein kinase classification. *Methods in Enzymology* **200**, 38–62.

Hanks, S. K., Quinn, A. M. and Hunter, T. (1988). The protein kinase family: conserved features and deduced phylogeny of the catalytic domains. *Science* **241**, 42–52.

Herskowitz, I. (1995). MAP kinase pathways in yeast: for mating and more. *Cell* **80**, 187–197.

Hildmann, T., Ebneth, M., Peña-Cortés, H., Sánchez-Serrano. J. J., Willmitzer, L. and Prat, S. (1992). General roles of abscisic and jasmonic acids in gene activation as a result of mechanical wounding. *Plant Cell* **4**, 1157–1170.

Hirt, H. (1999). Transcriptional upregulation of signaling pathways: more complex than anticipated? *Trends in Plant Science* **4**, 7–8.

Hunter, T. (1995). Protein kinases and phosphatases: the yin and yang of protein phosphorylation and signaling. *Cell* **80**, 225–236.

Hunter, T. and Plowman, G. D. (1997). The protein kinases of budding yeast: six score and more. *Trends in Biochemical Sciences* **22**, 18–21.

Ichimura, K. (1999). Molecular analysis of a MAP kinase cascade involved in environmental stress response in *Arabidopsis thaliana*. PhD Thesis, University of Tsukuba, Tsukuba, Ibaraki, Japan.

Ichimura, K., Mizoguchi, T., Irie, K., Morris, P., Giraudat, J., Matsumoto, K. and Shinozaki, K. (1998). Isolation of ATMEKK1 (a MAP kinase kinase)-interacting proteins and analysis of a MAP kinase cascade in *Arabidopsis. Biochemical and Biophysical Research Communications* **253**, 532–543.

Ingram, J. and Bartels, D. (1996). The molecular basis of dehydration tolerance in plants. *Annual Review of Plant Physiology and Plant Molecular Biology* **47**, 377–403.

Jabs, T., Tschöpe, M., Colling, C., Hahlbrock, K. and Scheel, D. (1997). Elicitor-stimulated ion fluxes and O_2 from the oxidative burst are essential components in triggering defense gene activation and phytoalexin synthesis in parsley. *Proceedings of the National Academy of Sciences of the United States of America* **94**, 4800–4805.

Jonak, C., Kiegerl, S., Ligterink, W., Barker, P. J., Huskisson, N. S. and Hirt, H. (1996). Stress signaling in plants: a mitogen-activated protein kinase pathway is activated by cold and drought. *Proceedings of the National Academy of Sciences of the United States of America* **93**, 11274–11279.

Jones, D. A., Thomas, C. M., Hammond-Kosack, K. E., Balint-Kurti, P. J. and Jones, J. D. G. (1994). Isolation of the tomato *Cf-9* gene for resistance to *Cladosporium fulvum* by transposon tagging. *Science* **266**, 789–793.

Lamb, C. J. and Dixon, R. A. (1997). The oxidative burst in plant disease resistance. *Annual Review of Plant Physiology and Plant Molecular Biology* **48**, 251–275.

Lebrun-Garcia, A., Ouaked, F., Chiltz, A. and Pugin, A. (1998). Activation of MAPK homologues by elicitor in tobacco cells. *Plant Journal* **15**, 773–781.

Levin, D. E. and Errede, B. (1995). The proliferation of MAP kinase signaling pathways in yeast. *Current Opinion in Cell Biology* **7**, 197–202.

Levine, A., Tenhaken, R., Dixon, R. and Lamb, C. (1994). H_2O_2 from the oxidative burst orchestrates the plant hypersensitive disease resistance response. *Cell* **79**, 583–593.

Ligterink, W., Kroj, T., Nieden, U., Hirt, H. and Scheel, D. (1997). Receptor-mediated activation of a MAP kinase in pathogene defense of plants. *Science* **276**, 2504–2507.

Marshall, C. J. (1995). Specificity of receptor tyrosine kinase signaling: transient versus sustained extracellular signal-regulated kinase activation. *Cell* **80**, 179–185.

Mason, H. S. and Mullet, J. E. (1990). Expression of two soybean vegetative storage protein genes during development and in response to water deficit, wounding, and jasmonic acid. *Plant Cell* **2**, 569–579.

Mehdy, M. C. (1994). Active oxygen species in plant defense against pathogens. *Plant Physiology* **105**, 467–472.

Meskiene, I., Bögre, L., Glaser, W., Balog, J., Brandstötter, M., Zwerger, K., Ammerer, G. and Hirt, H. (1998). MP2C, a plant protein phosphatase 2C, functions as a negative regulator of mitogen-activated protein kinase pathways in yeast and plants. *Proceedings of the National Academy of Sciences of the United States of America* **95**, 1938–1943.

Mitchell, C. A. and Myers, P. N. (1995). Mechanical stress regulation of plant growth and development. *Horticultural Reviews* **17**, 1–42.

Mizoguchi, T., Hayashida, N., Yamaguchi-Shinozaki, K., Kamada, H. and Shinozaki, K. (1993). ATMPKs: a gene family of plant MAP kinases in *Arabidopsis thaliana*. *FEBS Letters* **336**, 440–444.

Mizoguchi, T., Hayashida, N., Yamaguchi-Shinozaki, K., Kamada, H. and Shinozaki, K. (1995). Two genes that encode ribosomal-protein S6 kinase homologs are induced by cold or salinity stress in *Arabidopsis thaliana*. *FEBS Letters* **358**, 199–204.

Mizoguchi, T., Irie, K., Hirayama, T., Hayashida, N., Yamaguchi-Shinozaki, K., Matsumoto, K. and Shinozaki, K. (1996). A gene encoding a mitogen-activated protein kinase kinase kinase is induced simultaneously with genes for a mitogen-activated protein kinase and an S6 ribosomal protein kinase by touch, cold and water stress in *Arabidopsis thaliana*. *Proceedings of the National Academy of Science of the United States of America* **93**, 765–769.

Mizoguchi, T., Ichimura, K. and Shinozaki, K. (1997). Environmental stress response in plants: the role of mitogen-activated protein kinases. *Trends in Biotechnology* **15**, 15–19.

Mizoguchi, T., Ichimura K., Irie K., Morris, P., Giraudat, J., Matsumoto, K. and Shinozaki K. (1998). Identification of a possible MAP kinase cascade in *Arabidopsis thaliana* based on pairwise yeast two-hybrid analysis and functional complementation tests of yeast mutants. *FEBS Letters* **437**, 56–60.

Monroy, A. F., and Dhindsa, R. S. (1995). Low-temperature signal transduction: induction of cold acclimation-specific genes of alfalfa by calcium at 25°C. *Plant Cell* **7**, 321–331.

Monroy, A. F., Castonguay, Y., Laberge, S., Sarhan, F., Vezina, L. P. and Dhindsa, R. S. (1993). A new cold-induced alfalfa gene is associated with enhanced hardening at subzero temperature. *Plant Physiology* **102**, 873–879.

Monroy, A. F., Sangwan, V. and Dhindsa, R. S. (1998). Low temperature signal transduction during cold acclimation: protein phosphatase 2A as an early target for cold-inactivation. *Plant Journal* **13**, 653–660.

Morris, P. C., Guerroer, D., Leung, J. and Giraudat, J. (1997). Cloning and characterisation of *MEK1*, an *Arabidopsis* gene encoding a homologue of MAP kinase kinase. *Plant Molecular Biology* **35**, 1057–1064.

Nishida, E. and Gotoh, Y. (1993). The MAP kinase cascade is essential for diverse signal transduction pathways. *Trends in Biochemical Sciences* **18**, 128–131.

Nürnberger, T., Nennstiel, D., Jabs, T., Sacks, W. R., Hahlbrock, K. and Scheel, D. (1994). High affinity binding of a fungal oligopeptide elicitor to parsley plasma membranes triggers multiple defense responses. *Cell* **78**, 449–460.

Parker, J. E., Schulte, W., Hahlbrock, K. and Scheel, D. (1991). An extracellular glycoprotein from *Phytophthora megasperma f.* sp. glycinea elicits phytoalexin synthesis in cultured parsley cells and protoplasts. *Molecular Plant–Microbe Interactions* **4**, 19–27.

Pearce, G., Strydom, D., Johnson, S. and Ryan, C. A. (1991). A polypeptide from tomato leaves induces wound-inducible proteinase inhibitor proteins. *Science* **253**, 895–898.

Pelech, S. L. and Sanghera, J. S. (1992). Mitogen-activated protein kinases: versatile transducers for cell signaling. *Trends in Biochemical Sciences* **17**, 233–238.

Peña-Cortés, H., Liu, X., Sanchez-Serrano, J. J., Schmidt, R. and Willmitzer, L. (1988). Systemic induction of proteinase inhibitor II gene expression in potato plants by wounding. *Planta* **186**, 495–502.

Peña-Cortés, H., Sánchez-Serrano, J. J., Mertens, R., Willmitzer, L. and Prat, S. (1989). Abscisic acid is involved in the wound-induced expression of the proteinase inhibitor II gene in potato and tomato. *Proceedings of the National Academy of Sciences of the United States of America* **86**, 9851–9855.

Peña-Cortés, H., Prat, S., Atzorn, R., Wasternack, C. and Willmitzer, L. (1996). Abscisic acid-deficient plants do not accumulate proteinase inhibitor II following systemin treatment. *Planta* **198**, 447–451.

Posas, F. and Saito, H. (1997). Osmotic activation of the HOG MAPK pathway via Ste11p MAPKKK: scaffold role of Pbs2p MAPKK. *Science* **276**, 1702–1705.

Rojo, E., Titarenko, E., Leon, J., Berger, S., Vancanneyt, G. and Sánchez-Serrano, J. J. (1998). Reversible protein phosphorylation regulates jasmonic acid-dependent and -independent wound signal transduction pathways in *Arabidopsis thaliana*. *Plant Journal* **13**, 153–165.

Romeis, T., Piedras, P., Zhang, S., Klessig, D. F., Hirt, H. and Jones, J. D. G. (1999). Rapid Avr 9- and Cf9-dependent activation of MAP kinases in tobacco cell cultures and leaves: convergence of resistance gene, elicitor, wound, and salicylate responses. *Plant Cell* **11**, 273–288.

Ryan, C. A. (1990). Proteinase inhibitors in plants: genes for improving defenses against insects and pathogens. *Annual Review of Phytopathology* **28**, 425–449.

Sacks, W. R., Nürnberger, T., Hahlbrock, K. and Scheel, D. (1995). Molecular characterization of nucleotide sequences encoding the extracellular glycoprotein elicitor from *Phytophthora megasperma* *Molecular and General Genetics* **246**, 45 55.

Sánchez-Serrano, J. J., Schmidt, R., Schell, J. and Willmitzer, L. (1986). Nucleotide sequence of proteinase inhibitor II encoding complementary DNA of potato solanum-tuberosum and its mode of expression. *Molecular and General Genetics* **203**, 15–20.

Sano, H. and Ohashi, Y. (1995). Involvement of small GTP-binding proteins in defense signal-transduction pathways of higher plants. *Proceedings of the National Academy of Science of the United States of America* **92**, 4138–4144.

Satoh, K. and Murata, N. (eds) (1998). "Stress Responses of Photosynthetic Organisms, Molecular Mechanisms and Molecular Regulations." Elsevier Science, Amsterdam.

Seo, S., Okamoto, M., Ishizuka, K., Sano, K. and Ohashi, Y. (1995). Tobacco MAP kinase: a possible mediator in wound signal transduction pathways. *Science* **270**, 1988–1992.

Seo, S., Sano, H. and Ohashi, Y. (1999). Jasmonate-based wound signal transduction requires activation of WIPK, a tobacco mitogen-activated protein kinase. *Plant Cell* **11**, 289–298.

Shinozaki, K. and Yamaguchi-Shinozaki, K. (1996). Molecular responses to drought and cold stress. *Current Opinion in Plant Biology* **7**, 161–167.

Shinozaki, K. and Yamaguchi-Shinozaki, K. (1997). Gene expression and signal transduction in water-stress response. *Plant Physiology* **115**, 327–334.

Shinozaki, K., Yamaguchi-Shinozaki, K., Mizoguchi, T., Urao, T., Katagiri, T., Nakashima, K., Abe, H., Ichimura, K., Liu, Q., Nanjyo, T., Uno, Y., Iuchi, S., Seki, M., Ito, T., Hirayama, T. and Mikami, K. (1998). Molecular responses to water stress in *Arabidopsis thaliana*. *Journal of Plant Research* **111**, 345–351.

Smith, C. M., Shindyalov, I. N., Veretnik, S., Gribskov, M., Taylor, S. S., Ten Eyck, L. F. and Bourne, P. E. (1997). The protein kinase resource. *Trends in Biochemical Sciences* **22**, 444–446.

Stratmann, J. W. and Ryan, C. A. (1997). Myelin basic protein kinase activity in tomato leaves is induced systemically by wounding and increases in response to systemin and oligosaccharide elicitors. *Proceedings of the National Academy of Sciences of the United States of America* **94**, 11085–11089.

Subramaniam, R., Després, C. and Brisson, N. (1997). A functional homolog of mammalian protein kinase C participates in the elicitor-induced defense response in potato. *Plant Cell* **9**, 653–664.

Suzuki, K. and Shinshi, H. (1995). Transient activation and tyrosine phosphorylation of a protein kinase in tobacco cells treated with a fungal elicitor. *Plant Cell* **7**, 639–647.

Suzuki, K., Fukuda, Y. and Shinshi, H. (1995). Studies on elicitor-signal transduction leading to differential expression of defense genes in cultured tobacco cells. *Plant Cell and Physiology* **36**, 281–289.

Suzuki, K., Yano, Y. and Shinshi, S. (1999). Slow and prolonged activation of the p47 protein kinase during hypersensitive cell death in a culture of tobacco cells. *Plant Physiology* **119**, 1465–1472.

Titarenko, E., Rojo, E., Leon, J. and Sanchez-Serrano J. J. (1997). Jasmonic acid-dependent and -independent signaling pathways control wound-induced gene activation in *Arabidopsis thaliana. Plant Physiology* **115**, 817–826.

Urao, T., Katagiri, T., Mizoguchi, T., Yamaguchi-Shinozaki, K., Nobuaki, H. and Shinozaki, K. (1994). Two genes that encode Ca^{2+}-dependent protein kinases are induced by drought and high-salt stresses in *Arabidopsis thaliana. Molecular General Genetics* **244**, 331–340.

Usami, S., Banno, H., Ito, Y., Nishihama, R. and Machida, Y. (1995). Cutting activates a 46-kilodalton protein kinase in plants. *Proceedings of the National Academy of Sciences of the United States of America* **92**, 8660–8664.

Van den Ackerveken, G. F. J. M., Van Kan, J. A. L. and De Wit, P. J. G. M. (1992). Molecular analysis of the avirulence gene *avr9* of the fungal tomato pathogen *Cladosporium fulvum* fully supports the gene-for-gene hypothesis. *Plant Journal* **2**, 359–366.

Van den Ackerveken, G. F. J. M., Vossen, P. and De Wit, P. J. G. M. (1993). The AVR9 race-specific elicitor of *Cladosporium fulvum* is processed by endogenous and plant proteases. *Plant Physiology* **103**, 91–96.

Viard, M.-P., Martin, F., Pugin, A., Ricci, P. and Blein, J.-P. (1994). Protein phosphorylation is induced in tobacco cells by the elicitor cryptogein. *Plant Physiology* **104**, 1245–1249.

Yang, Y., Shah, J. and Klessig, D. F. (1997). Signal perception and transduction in plant defense responses. *Genes and Development* **11**, 1621–1639.

Yu, L. M. (1995). Elicitins from *Phytophthora* and basic resistance in tobacco. *Proceedings of the National Academy of Sciences of the United States of America* **92**, 4088–4094.

Yu, L. M., Lamb, C. J., and Dixon, R. A. (1993). Purification and biochemical characterization of proteins which bind to the H-box cis-element implicated in transcriptional activation of plant defense genes. *Plant Journal* **3**, 805–816.

Zhang, S. and Klessig, D. F. (1997). Salicylic acid activates a 48-kD MAP kinase in tobacco. *Plant Cell* **9**, 809–824.

Zhang, S. and Klessig, D. F. (1998a). The tobacco wounding-activated mitogen-activated protein kinase is encoded by *SIPK. Proceedings of the National Academy of Sciences of the United States of America* **95**, 7225–7230.

Zhang, S. and Klessig, D. F. (1998b). Resistance gene *N*-mediated *de novo* synthesis and activation of a tobacco mitogen-activated protein kinase by tobacco mosaic virus infection. *Proceedings of the National Academy of Sciences of the United States of America* **95**, 7433–7438.

Zhang, S., Du, H. and Klessig, D. F. (1998). Activation of the tobacco SIP kinase by both a cell-wall-derived carbohydrate elicitor and purified proteinaceous elicitins from *Phytophthora* spp. *Plant Cell* **10**, 435–450.

Protein Kinases in the Plant Defense Response

GUIDO SESSA and GREGORY B. MARTIN

Boyce Thompson Institute for Plant Research and Department of Plant Pathology, Cornell University, Ithaca, NY 14853, USA

I. PLANT DEFENSE MECHANISMS AND REVERSIBLE PROTEIN PHOSPHORYLATION

A. PLANT DEFENSE STRATEGIES

The ability to sense and rapidly respond to external stimuli allows plants to resist pathogen attack and colonization. Despite the large variety of phytopathogens present in nature, disease is rare and resistance is prevalent as an outcome of plant–pathogen interaction. Plant disease resistance is mostly dependent on the genetic background of both host and invading agent, and often relies on complex mechanisms of molecular recognition and cellular

Advances in Botanical Research Vol. 32
incorporating Advances in Plant Pathology
ISBN 0-12-005932-0

signal transduction. In general, resistance occurs in the following circumstances:

1. The pathogen fails to infect the plant, because the plant belongs to a taxonomic group outside its host range (nonhost resistance).
2. The plant contains preformed physical and chemical barriers, which impede pathogen penetration and spreading.
3. The plant recognizes the presence of a pathogen and rapidly triggers a battery of defense mechanisms, which efficiently arrest further invasion.

In the latter type of resistance, early recognition of the pathogen at the level of single cells is essential to mount successfully efficient defense responses. Plant cells are able to sense pathogen invasion by recognizing either endogenous signal molecules derived by degradation of plant cell wall components, or exogenous molecules synthesized by the invading pathogen. Endogenous and exogenous signal compounds are termed elicitors, and include proteins, glycoproteins, oligosaccharides and lipids. In several instances, elicitors are sufficient to induce a complete set of plant defense responses, and they presumably interact with specific plant receptors (Benhamou, 1996). Elicitors are defined as race specific and nonrace specific, according to the range of plants in which they are able to induce defense responses. Race-specific elicitors are often products of avirulence (*avr*) genes encoded by pathogens, and are specifically recognized by products of plant resistance (*R*) genes. This recognition event between the products of a pathogen *avr* gene and a plant *R* gene is referred to as 'gene-for-gene' interaction, and represents the molecular basis of race/cultivar-specific host resistance (Baker *et al.*, 1997). Nonrace-specific elicitors are able to activate defense responses by mechanisms independent of plant *R* genes. Their recognition in the host plant is probably mediated by high-affinity receptors present in the plasma membrane (Yang *et al.*, 1997).

Successful pathogen recognition triggers the activation of several and diverse defense responses. In many plant–pathogen interactions resistance is manifested at the macroscopic level by the appearance of necrotic lesions at the site of infection. This is the result of a rapid localized cell death, termed hypersensitive response (HR), which is thought to limit pathogen growth and spreading throughout the infected plant. Early and local molecular responses associated with the hypersensitive response include the production of reactive oxygen species, transient opening of ion channels, cell wall fortifications, production of antimicrobial phytoalexins, and synthesis of pathogenesis-related (PR) proteins, such as glucanases and chitinases (Hammond-Kosack and Jones, 1996; Somssich and Hahlbrock, 1998). In addition to responses localized at the site of pathogen infection, plants often induce defense mechanisms in noninfected areas. Defense responses at such secondary sites are collectively referred to as systemic acquired resistance (SAR). SAR often confers a long-lasting and nonspecific resistance to subsequent infection of a broad variety of avirulent and normally virulent pathogens (Ryals *et al.*, 1996).

B. PROTEIN PHOSPHORYLATION AND DEFENSE RESPONSES

A remarkable difference between resistant and susceptible plants is in the timing with which they recognize pathogen invasion and activate defense responses. The ability quickly to induce defense mechanisms is a characteristic of incompatible (resistant) plant–pathogen interactions. Resistant plants are equipped with a molecular alert system which allows them not only to recognize pathogen intrusion, but also to amplify very efficiently the initial alarm signal and to activate self-defense. In recent years it has become evident that reversible protein phosphorylation plays a cardinal role in transducing signals leading to disease resistance in plants. Protein kinases and phosphatases in plants, as well as in animals, are implicated as key components in signaling mechanisms critical for responses to environmental stresses and attack by pathogens. The use of protein phosphorylation for signal transduction is particularly suited for defense, because it allows efficient amplification of the original signal, negative feedback for desensitization, cross-talk and branching for the activation of diverse lines of defense.

Early evidence for the involvement of protein phosphorylation in defense mechanisms was provided by the analysis of changes in the profile of phosphorylated proteins upon elicitor application to cell-suspension cultures. Treatment of tomato, parsley, soybean or tobacco cell cultures with different types of elicitors, including oligosaccharides, chitin oligomers, yeast extracts, xylanase from *Trichoderma viride* and cryptogein from *Phytophtora cryptogea*, caused rapid (within minutes) induction of protein phosphorylation (Grab *et al.*, 1989; Dietrich *et al.*, 1990; Felix *et al.*, 1991, 1994; Viard *et al.*, 1994). In addition, pharmacological studies have shown that protein kinase inhibitors, such as K-252a and staurosporine, interfere with many different defense reactions, including extracellular alkalization (Felix *et al.*, 1993), ion fluxes (Kauss *et al.*, 1992), oxidative burst (Schwacke and Hager, 1992; Levine *et al.*, 1994; Chandra and Low, 1995), production of pathogenesis-related proteins (Raz and Fluhr, 1993; Suzuki and Shinshi, 1995), and accumulation of secondary metabolites (Conrath *et al.*, 1991).

In cellular signal transduction pathways, phosphorylation by protein kinases is antagonized by dephosphorylation activity of protein phosphatases (Hunter, 1995). The balance between these two activities in the plant cell also appears to regulate the induction of defense responses. Specific phosphatase inhibitors, in contrast to kinase inhibitors, induce defense responses in the absence of elicitors. For instance, in tomato cell cultures calyculin A induced medium alkalization and protein hyperphosphorylation (Felix *et al.*, 1994), while in soybean cell cultures okadaic acid induced accumulation of phytoalexins (MacKintosh *et al.*, 1994; Gianfagna and Lawton, 1995). In addition, okadaic acid elicited accumulation of pathogenesis-related proteins in tobacco plants (Raz and Fluhr, 1993), and initiated an oxidative burst in cultured soybean cells (Chandra and Low, 1995).

A major target of signal transduction pathways originated by pathogen invasion is the cell nucleus. Terminal signals activate transcription of a variety of genes, which encode products with antimicrobial activity and contribute to pathogen defeat (Rushton and Somssich, 1998). Activation or deactivation of transcription factors is an additional level of regulation where protein phosphorylation appears to play an important role. For instance, a DNA–protein complex, between a G-box motif of the pathogenesis-related *PRB-1b* promoter and nuclear factor(s), was found to be phosphorylated in tobacco plants (Sessa *et al.*, 1995). Two bean transcription factors, KAP-1 and KAP-2, which recognize H-boxes in the elicitor- and stress-induced chalcone synthase *chs15* promoter, were also proposed to be regulated by phosphorylation (Yu *et al.*, 1993). Dephosphorylation of KAP-1 and KAP-2 was observed to alter the mobility of the complexes they form with the H-box element.

Direct phosphorylation of the G/HBF-1 transcription factor was found to occur in soybean suspension cells in response to elicitor treatment (Dröge-Laser *et al.*, 1997). G/HBF-1, encodes a basic leucine zipper (bZIP) protein and binds both a G-box and an H-box of the chalcone synthase *chs15* promoter. Interestingly, in elicitor-treated soybean suspension cultures, G/HBF-1 was rapidly and transiently phosphorylated by a cytosolic serine protein kinase. Phosphorylation of G/HBF-1 occurred at multiple sites and enhanced its DNA-binding activity (Dröge-Laser *et al.*, 1997). Similarly, phosphorylation enhanced binding of the potato PBF-1 nuclear factor to the stress-induced *PR10a* promoter (Després *et al.*, 1995). Moreover, the phosphorylation state of PBF-1 correlated with *PR10a* gene activation. Pharmacological and biochemical analysis suggests that PBF-1 phosphorylating activity originates from a potato protein kinase related to the mammalian protein kinase C family (Subramaniam *et al.*, 1997).

The recent isolation and characterization of protein kinases, either directly involved in disease resistance or associated with defense mechanisms, has confirmed the central role of protein phosphorylation in plant defense responses. In the next sections we will focus our attention on protein kinases that appear to be directly involved in signal transduction pathways leading to defense responses. These include products of disease-resistance genes, receptor-like protein kinases, and mitogen-activated protein (MAP) kinases. We will review their known characteristics related to activation, function in signaling, interaction with other proteins and cellular localization.

II. PHOSPHORYLATION CASCADES MEDIATED BY PLANT RESISTANCE (*R*) GENES

A. RESISTANCE GENES AND PROTEIN PHOSPHORYLATION

In race/cultivar-specific disease resistance, rapid activation of the defense responses is mediated by a specific recognition event, which involves the product of an avirulent (*avr*) gene in the pathogen and the corresponding resistance (*R*) gene in the plant (Flor, 1971). In the last few years, many *R* genes have been isolated that confer resistance to different pathogens including viruses, bacteria, fungi and nematodes (Baker *et al.*, 1997; Hammond-Kosack and Jones, 1997). Based on their structural characteristics, *R* gene products can be grouped into five different classes: (I) intracellular protein kinases (e.g. Pto from tomato); (II) transmembrane receptor-like proteins with extracellular leucine-rich repeats (LRR) and cytoplasmic protein kinase domain (e.g. Xa21 from rice); (III) intracellular receptor-like proteins with LRR domains and nucleotide binding sites (NBS) (e.g. RPS2 and RPM1 from *Arabidopsis*, and Prf from tomato); (IV) intracellular receptor-like proteins with LRR domains, NBS and a region of homology to the *Drosophila* Toll and to the mammalian interleukin-1 receptor (e.g. N from tobacco and RPP5 from *Arabidopsis*); and (V) transmembrane receptor-like proteins with extracellular LRR domains (e.g. Cf2 and Cf9 from tomato).

Several lines of evidence convincingly indicate that kinase signaling cascades may originate from the specific recognition of the products of a plant *R* gene and the corresponding pathogen *avr* gene. First, the tomato *Pto* and the rice *Xa21* resistance genes encode a serine/threonine protein kinase and a receptor kinase-like protein, respectively (Martin *et al.*, 1993; Song *et al.*, 1995). Second, a class of plant resistance genes shares homologous domains with the interleukin-1 receptor in mammals and Toll in *Drosophila*, which transduce extracellular signals by activating downstream phosphorylation cascades (O'Neil and Greene, 1998). Third, in tobacco plants two mitogen-activated protein kinases, a wound-induced protein kinase (WIPK) and a salicylic-acid-induced protein kinase (SIPK), are rapidly and transiently activated in a gene-for-gene manner, which is dependent on the expression of the plant *Cf-9* and the fungal *Avr9* genes (Romeis *et al.*, 1999). The protein kinase WIPK was also shown to be activated by tobacco mosaic virus (TMV) only in plants expressing the TMV resistance gene *N* (Zhang and Klessig, 1998a).

B. THE Pto KINASE AND SIGNAL RECOGNITION

In tomato, the Pto protein kinase confers resistance to bacterial speck disease, which represents a model system for a phosphorylation cascade activating defense responses (Gu and Martin, 1998). Pto-mediated resistance conforms to a gene-for-gene interaction and is observed in plants expressing the *Pto* gene upon infection of *Pseudomonas syringae* pv. tomato strains, which express the avirulence gene *avrPto*. The *Pto* gene was isolated by map-based cloning (Martin *et al.*, 1993), and its introduction into a susceptible tomato cultivar resulted in a marked increase in resistance to *avrPto*-expressing strains of *P. syringae*. The ability of the *Pto* gene product to trigger defense mechanisms was further confirmed by the finding that overexpression of *Pto* in transgenic tomato plants conferred broad resistance, and resulted in the activation of various defense responses (Tang *et al.*, 1998). To date, several different components of the Pto-mediated signal transduction pathway have been cloned and characterized. These include the serine/threonine protein kinases Pto and Pti1 (Martin *et al.*, 1993; Zhou *et al.*, 1995), the putative transcription factors Pti4, Pti5 and Pti6 (Zhou *et al.*, 1997), and the leucine zipper, LRR- and NBS-containing protein Prf (Salmeron *et al.*, 1996).

The *Pto* gene was originally derived from the wild tomato species *Lycopersicon pimpinellifolium*, and introgressed into cultivated tomato species. It belongs to a small gene family, which consists of five members clustered on tomato chromosome five. One member of this family is the *Fen* gene, which encodes a functional serine/threonine protein kinase (Martin *et al.*, 1994; Loh and Martin, 1995). Interestingly, *Fen* mediates a hypersensitive-like response in tomato plants treated with the organophosphorous insecticide fenthion. A *Pto* orthologous gene with high similarity to Pto has been recently isolated and characterized from tomato lines susceptible to speck disease (Jia *et al.*, 1997).

The *Pto* gene encodes a functional serine/threonine protein kinase (Loh and Martin, 1995), which autophosphorylates *in vitro* at multiple sites (Sessa *et al.*, 1998), and shares significant homology with the mammalian interleukin-1 receptor-associated kinase (IRAK) and the *Drosophila* Pelle protein kinases (O'Neill and Greene, 1998). The IRAK and Pelle proteins have also been implicated in signal mechanisms activated by environmental stresses and pathogen attack. An intracellular localization has been proposed for the Pto kinase based on the observation that it does not contain obvious extracellular or membrane-spanning domains. A putative myristoylation motif is present at the N-terminus of Pto, which may be utilized by the cell to recruit Pto to the plasma membrane. However, the insertion of a mutation in the invariant glycine of the Pto myristoylation site did not affect Pto-mediated speck-disease resistance (Loh *et al.*, 1998).

The molecular basis of gene-for-gene specificity is thought to reside in the recognition between a receptor encoded by the plant *R* gene and an elicitor molecule produced by the pathogen. In support of this receptor–ligand model, a physical interaction was detected between the Pto kinase and the AvrPto protein, by using the yeast two-hybrid system (Scofield *et al.*, 1996; Tang *et al.*, 1996). The specificity of this interaction was substantiated by the observation that the closely related kinase Fen and the product of the *pto* susceptible allele were unable to interact with AvrPto in the same system. Additional evidence indicates that the interaction between Pto and the small and hydrophilic AvrPto molecule also occurs *in vivo*, and takes place inside the plant cell. Transient expression of AvrPto in plant cells via *Agrobacterium*, in fact, activated a hypersensitive response only in plants carrying the *Pto* gene. This is consistent with reports indicating that some bacterial avirulence proteins are only active if present in intracellular compartments, where they are probably delivered by the bacterial type III secretion system (Gopalan *et al.*, 1996;

Leister *et al.*, 1996; Van den Ackerveken *et al.*, 1996; Mudgett and Staskawicz, 1998). The Pto–AvrPto interaction detected in yeast was strictly correlated with the onset of disease resistance in plant. Deletion of AvrPto sequences, required for the interaction with Pto, impaired the ability of AvrPto to elicit defense responses. Conversely, removal of AvrPto portions dispensable for Pto interaction did not affect resistance (Tang *et al.*, 1996).

A domain-swapping analysis between Pto and the closely related Fen protein kinase has identified a region in the Pto kinase activation domain that is a determinant for Pto–AvrPto interaction and specificity (Scofield *et al.*, 1996; Tang *et al.*, 1996). Within this region, threonine-204 (Thr204) was required for the specific recognition of AvrPto, as tested in the yeast two-hybrid system, and for the elicitation of the hypersensitive response in plants (Frederick *et al.*, 1998). Moreover, introduction of a threonine at the amino acid location corresponding to Pto Thr204 conferred to Fen the ability to interact with AvrPto, and to induce a hypersensitive response. It is interesting to note that several protein kinases related to Pto, from plants and other organisms, have a threonine conserved at the position corresponding to Pto Thr204. Among these kinases are the rice disease resistance protein Xa21 (GeneBank accession U37133), the *Arabidopsis* ERECTA (U47029) and the maize CRINKLY proteins (U67422), the *Drosophila* Pelle kinase (108476), and the human interleukin-1 receptor-associated kinase (IRAK; L76191).

Pto autophosphorylation activity appears to be a prerequisite for the Pto–AvrPto interaction. In fact, forms of Pto mutated in residues essential for kinase activity did not interact with AvrPto in the yeast two-hybrid system (Scofield *et al.*, 1996; Tang *et al.*, 1996). Pto Thr204 may represent a target for autophosphorylation or for phosphorylation by a different protein kinase. Interestingly, the requirement of a phosphorylated residue for the interaction between a protein kinase and a regulatory subunit has been observed for the cyclic adenosine monophosphate (AMP)-dependent protein kinase (cAPK; Levin and Zoller, 1990). cAPK requires phosphorylation of a conserved threonine residue for the interaction with an associated subunit, which regulates its activity. Similarly, the *Arabidopsis* serine/threonine receptor kinase, receptor-like kinase 5 (RLK5), interacts *in vitro* with the kinase-associated type 2C protein phosphate KAPP type 2C protein phosphatase only in its autophosphorylated form (Stone *et al.*, 1994).

The physical interaction between the Pto protein kinase and AvrPto provides a molecular explanation for gene-for-gene specificity in plant disease resistance. However, it remains to be elucidated how the formation of the AvrPto–Pto complex results in the activation of downstream effectors. A model involving Pto activation through reciprocal phosphorylation of two Pto molecules, which are bridged by an AvrPto molecule, is unlikely. Pto autophosphorylation, in fact, was found to proceed *in vitro* via an intramolecular mechanism (Sessa *et al.*, 1998). Alternative modes of activation include the possibility that interaction with AvrPto causes conformational changes in Pto, that possibly expose certain domains, which were not previously available, to autophosphorylation or to phosphorylation by an additional protein kinase. A similar mechanism occurs during activation of cyclin-dependent protein kinases (CDK) by cyclins (Morgan, 1995). The activity of CDK2, for example, which is involved in regulation of events in the eukaryotic cell cycle, is stimulated by a two-step mechanism of activation. First, the regulatory subunit cyclin A associates with CDK2, causing conformational changes in the kinase catalytic sites that make Thr160 more accessible for phosphorylation by the CDK-activating kinase (CAK). Second, CAK phosphorylation of Thr160 determines full activation of CDK2 (Jeffrey *et al.*, 1995).

An additional possible mode of Pto activation is that interaction with AvrPto allows it to take part in a multi-subunit receptor complex with other plant molecules. The effect of complex formation could be either activation of Pto kinase activity or its recruitment to a target cellular compartment. A possible candidate for the AvrPto-mediated complex formation with Pto is the Prf molecule, which is required for bacterial speck-disease resistance and the related phenotype of fenthion sensitivity (Salmeron *et al.*, 1996). However, despite its involvement in resistance and its similarity to other disease-resistance genes, the function of Prf in the Pto-mediated signal transduction pathway remains unclear.

C. Pto DOWNSTREAM EFFECTORS: THE Pti1 KINASE AND THE TRANSCRIPTION FACTORS Pti4, Pti5 AND Pti6

The Pto–AvrPto recognition event is postulated to activate the Pto kinase and induce phosphorylation of downstream components in signaling pathways leading to defense responses. A model for the signal transduction pathway mediated by the Pto protein kinase is shown in Fig. 1. Downstream effectors, which physically interact with Pto and represent putative targets for its phosphorylation, were isolated by using a yeast two-hybrid system (Zhou *et al.*, 1995, 1997). Among them and of particular significance are the protein kinase Pti1, and the putative transcription factors Pti4, Pti5 and Pti6.

Pti1 is a cytoplasmic serine/threonine protein kinase, and its role in Pto-mediated disease resistance was tested in transgenic tobacco plants (Zhou *et al.*, 1995). In these plants, overexpression of Pti1 enhanced the hypersensitive response in leaves challenged with *Pseudomonas syringae* pv. tabaci expressing the *avrPto* gene. *In vitro* analysis of glutathione *S*-transferase (GST)-Pti1 fusion protein revealed that Pti1 is able to autophosphorylate mainly on threonine residues, and its autophosphorylation proceeded via an intramolecular mechanism (Zhou *et al.*, 1995; Sessa *et al.*, 1998). In addition, in cross-phosphorylation experiments *in vitro* Pto phosphorylated Pti1, but Pti1 failed to phosphorylate Pto. The specificity of Pti1 phosphorylation by Pto was corroborated by the inability of the closely related protein kinase Fen to phosphorylate Pti1. Taken together, these results suggest that Pti1 is a specific

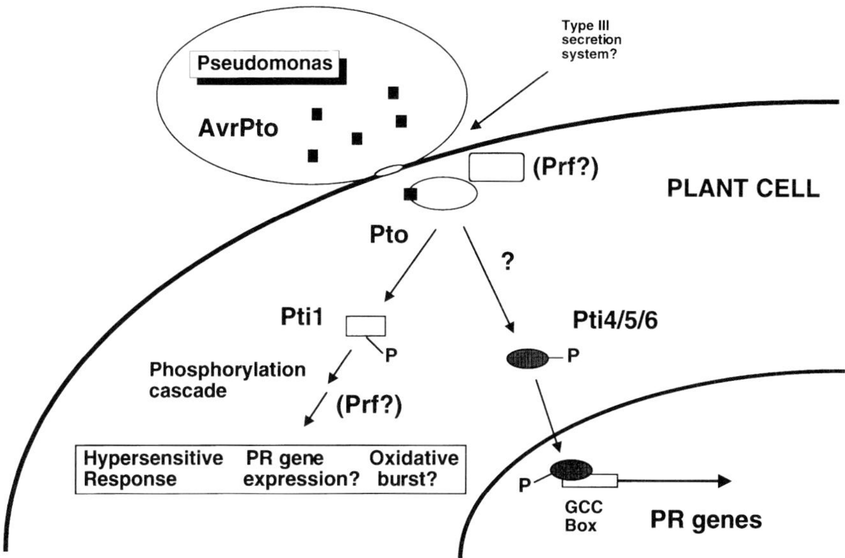

Fig. 1. A model for the signal transduction pathway mediated by the Pto protein kinase.

substrate for Pto and it acts downstream of Pto in the phosphorylation cascade that triggers the hypersensitive response.

Phosphorylation of a protein kinase by an upstream regulatory kinase is a common strategy used in transduction pathways for signal amplification (Johnson *et al.*, 1996). The Pto–Pti1 interaction is the first of this kind to be identified in plants, and it is therefore of particular interest to understand its involvement in the molecular strategies leading to defense responses. A biochemical approach was used to characterize further the molecular mechanisms involved in the interaction between the Pto and Pti1 protein kinases (Sessa *et al.*, 1998, 2000). Kinetic analysis of Pti1 phosphorylation by Pto revealed that this reaction takes place with a K_m of 4.1 μM and a V_{max} of 0.55 nmol min^{-1} mg^{-1}. In addition, Pto phosphorylated Pti1 at one major and several minor sites, as observed by phosphopeptide mapping. Similar sites were also phosphorylated during Pti1 autophosphorylation. By a combination of different biochemical techniques, the major site phosphorylated by Pto in Pti1 was identified as threonine-233 (Thr233; Sessa *et al.*, 2000). Mutational analysis revealed that this site is also the major Pti1 autophosphorylation site. Thr233 is located in the Pti1 kinase activation domain, which is defined as the region between the DFG and PE conserved motifs of kinase subdomains VII and VIII (Hanks and Quinn, 1991). Phosphorylation of residues in this domain is responsible for activation of several kinases from different organisms, probably by stabilizing the kinase in an active conformation (Johnson *et al.*,

1996). Interestingly, a group of plant kinases involved in different physiological processes, and the human interleukin-1 receptor-associated protein kinase IRAK, all contain a threonine at the corresponding position of Pti1 Thr233 (Sessa *et al.*, 1999).

Because Pti1 Thr233 appears to be the main site phosphorylated *in vitro* both by Pto and by Pti1 autophosphorylation, this raises the question of how Pti1 regulation is achieved *in vivo*. Assuming that the phosphorylation state of Pti1 determines its activity, other components in addition to Pto are thus required to regulate the extent of Pti1 phosphorylation. A type 2A protein phosphatase has been found to interact with Pti1 in a yeast two-hybrid interaction screen (X.-Y. Tang and G. B. Martin, unpublished results), and it may take part in a mechanism that negatively regulates Pti1 activity in the absence of bacterial challenge.

Phosphorylation was found to be involved in the physical interaction between Pto and Pti1, as observed in the yeast two-hybrid system (Zhou *et al.*, 1995; Sessa *et al.*, 1998). Pto kinase activity was required for this interaction, while Pti1 kinase activity was dispensable. Mutational analysis of Pti1 Thr233 to alanine abolished the Pto–Pti1 physical interaction. However, substitution of Thr233 with a negatively charged residue to mimic phosphorylation restored Pto–Pti1 interaction. These observations suggest that Pto interacts in yeast with the phosphorylated form of Pti1. The dynamics of Pto–Pti1 physical interaction and phosphorylation *in vivo* remain to be established. Pto and Pti1 could form a transient complex that is released upon Pti1 phosphorylation, or more likely a durable complex stabilized by phosphorylation. It will be important also to assess the physiological relevance of Pti1 phosphorylation by Pto *in vivo*, and how phosphorylation of Pti1 Thr233 affects Pti1 kinase activity in the onset of speck-disease resistance. In addition, physiological substrates for Pti1 phosphorylation have yet to be identified.

A major target of signal transduction pathways leading to defense responses is the transcriptional activation of defense-related genes (Rushton and Somssich, 1998). The rapid and efficient induction of defense genes appears to be important for the onset of resistance and incompatibility. In speck-disease resistance of tomato and tobacco, for example, a set of defense genes encoding pathogenesis-related proteins was observed to accumulate during incompatible interactions earlier than in compatible interactions (Zhou *et al.*, 1997; Jia and Martin, 1999). In the Pto signaling pathway, a possible bridge linking signal recognition to activation of defense genes is represented by the interaction between Pto and the transcription factors Pti4, Pti5 and Pti6 (Pti4/5/6; Zhou *et al.*, 1997). In a yeast two-hybrid interaction screen, the Pti4/5/6 proteins were found to interact physically with Pto, but not with the product of the *pto* recessive allele or with the closely related Fen protein kinase. Analysis of the Pti4/5/6 amino acid sequence revealed the presence of characteristics typical of transcription factors. All three proteins contain a central DNA-binding domain rich in basic residues, a region of acidic residues possibly involved in transcriptional activation, and nuclear localization sequences. Interestingly, Pti4/5/6 share significant homology with the ethylene-responsive element-binding proteins from tobacco (EREBPs; Ohme-Takagi and Shinshi, 1995). EREBPs were shown in a gel-shift assay to bind a *cis*-acting element required for ethylene responsiveness and present in basic-type pathogenesis-related gene promoters (GCC-box; Ohme-Takagi and Shinshi, 1995). Pti5 and Pti6 were also able to bind to a GCC-box element from the β-1,3-glucanase *gln2* promoter gene. Taken together, the interaction between Pto and Pti4/5/6, and the binding of Pti5 and Pti6 to the GCC-box, suggest that Pto may regulate PR gene expression by the direct activation of transcription factors. Moreover, the binding of both Pto-interactors and EREBPs to the same *cis*-acting element hints to a convergence of the Pto disease-resistance pathway and the ethylene pathway at the level of transcriptional activation. The ethylene signal is also recognized and amplified by a phosphorylation cascade, although the kinases involved are of different nature (Fluhr, 1998).

The mechanisms that allow Pto to interact with and possibly activate Pti4/5/6 by Pto are still unclear. Recently, Pti4 has been found to be phosphorylated at multiple sites by Pto in a kinase assay *in vitro* (Y.-Q. Gu and G.B. Martin, unpublished results). A phosphorylation event may result in the activation of the transcription factors Pti4/5/6, by regulating either their binding activity or transactivation potential, as observed in mammalian cells (Hunter and Karin, 1992). In addition, phosphorylation may play a role in localizing Pti4/5/6 to the nucleus as reported for other proteins containing nuclear localization signals (Jean and Hubner, 1996). To distinguish between these different mechanisms of activation, it will be necessary to elucidate *in vivo* the effect of phosphorylation on Pti4/5/6 activity and localization during pathogen challenge.

D. THE RICE Xa21 RECEPTOR-LIKE KINASE

The *Xa21* gene, which confers resistance to bacterial leaf blight, encodes a protein kinase with characteristics of receptor-like kinases (RLKs; Song *et al.*, 1995). Structural characteristics of RLKs include an extracellular domain, a transmembrane sequence, and an intracellular kinase domain. RLKs are widely utilized in animal and plant systems to transduce extracellular signals into the cell (Fantl *et al.*, 1993; Braun and Walker, 1996; Lease *et al.*, 1998). In plants, RLKs are involved in diverse physiological processes, including development, self-incompatibility, and disease resistance (Lease *et al.*, 1998). The rice *Xa21* gene encodes an RLK with serine/threonine specificity and confers resistance to the bacterial pathogen *Xanthomonas oryzae* pv. oryzae race 6 (Song *et al.*, 1995). The Xa21 predicted amino acid sequence contains regions with similarity to known protein domains. The N-terminus of Xa21 consists of a predicted extracellular domain with a hydrophobic stretch of amino acids characteristic of a signal peptide, and 23 imperfect copies of a leucine-rich repeat (LRR) with similarity to the *Cf9* resistance gene product. Next to the extracellular domain lies a hydrophobic region, which is likely to form a membrane-spanning helix. The C-terminal part of the protein contains an intracellular kinase catalytic domain with similarity to the tomato Pto. As observed for other resistance genes, *Xa21* belongs to a multi-gene family, which is clustered on rice chromosome 11 (Song *et al.*, 1997). Isolation and sequence analysis of seven family members defined two classes of *Xa21* genes based on their degree of similarity. The sequence identity of different members within the same class is more than 93%, compared to 69% overall identity between classes. Sequence comparison revealed that transposition, recombination, duplication and point mutations have played an important role in evolution of the *Xa21* gene family. In fact, only one member from each class encodes a complete receptor kinase-like open reading frame, while the other five members encode truncated open reading frames.

Little is known about cellular signaling mechanisms mediated by the product of the *Xa21* gene, and the current model resembles the mode of action of animal RLKs (Ronald, 1997). Leaf-blight-disease resistance is proposed to be mediated by extracellular signal recognition, as opposed to the intracellular Pto–AvrPto interaction, which triggers speck-disease resistance. Based on observations related to animal LRR-containing RLKs (Braun *et al.*, 1991), ligand binding and recognition specificity of Xa21 is proposed to be mediated by the LRR motif, which is located in the presumed extracellular portion of the protein. In support of this concept, the *Xa21D* gene, which encodes an LRR but lacks the transmembrane and kinase domains, was able to transduce a partial resistance response in a race-specific manner (Wang *et al.*, 1998). However, the isolation from *X. oryzae* pv. oryzae of an avirulence determinant corresponding to the Xa21 protein will be necessary to investigate directly its interaction with Xa21. In animal RLKs, ligand binding to extracellular domains causes receptor dimerization, activation of the cytoplasmic kinase domain by intermolecular phosphorylation, and transduction of the signal to downstream effectors (Ullrich and Schlessinger, 1990; Ten Dijke *et al.*, 1996). In line with a similar mechanism in rice, the ability of the *Xa21D* gene to induce partial resistance to normally susceptible lines, despite its encoding a truncated RLK form, was ascribed to a putative dimerization step with an endogenous RLK (Wang *et al.*, 1998). Detailed molecular and biochemical analysis will be necessary to establish whether a chain of events, similar to those mediated by animal RLKs, also takes place in rice in the defense pathway(s) mediated by the *Xa21* and *Xa21D* gene products.

E. CONSERVATION OF DEFENSE PATHWAYS IN PLANTS, MAMMALS AND INSECTS

One class of resistance genes includes the tobacco *N*, the *Arabidopsis RPP5*, and the flax *L6* (Hammond-Kosack and Jones, 1997). These genes encode putative cytoplasmic proteins with nucleotide binding site (NBS) domains, leucine-rich repeats and an intriguing region of homology to the cytoplasmic domains of the mammalian interleukin-1 receptor (IL-1R) protein and of the *Drosophila* Toll protein. IL-1R plays a central role in the immune and inflammatory responses (O'Neill and Greene, 1998). Binding of interleukin-1 to IL-1R triggers phosphorylation events, which involve the protein kinase IRAK and determines activation of the transcription factor NF-κB, by releasing it from a cytoplasmically localized complex with the inhibitor protein IκB. The *Drosophila* Toll protein is involved in the determination of dorsoventral polarity and in the production of antimicrobial peptides in the adult fly (Morisato and Anderson, 1995; Lemaitre *et al*., 1996). The binding of the extracytoplasmic ligand Spätzle to Toll leads to activation of the protein kinase Pelle by recruiting it to the plasma membrane. Pelle, which is the functional homolog of the mammalian IRAK, phosphorylates the inhibitory protein Cactus, which is complexed in the cytosol to the transcription factor Dorsal. Phosphorylation of Cactus releases Dorsal and allows it to relocalize to the nucleus, where it activates transcription of its target genes. The sequence similarity of the N, RPP5 and L6 gene products with Toll and IL-1R, and the evidence that all these proteins are involved in response to infection, suggest that plant resistance gene products, Toll and IL-1R, may activate similar downstream signaling pathways. This is further supported by the similarity between the protein kinase encoded by the *Pto* resistance gene and the IRAK and Pelle kinases. The Pto protein kinase shares significant sequence similarity with IRAK and Pelle, and among its different functions, it is proposed to phosphorylate and activate transcription factors. Transcription factor activation is also the result of IRAK and Pelle activities. It is still unclear whether protein kinases, functionally similar to Pto, are also acting downstream of N, RPP5 and L6 gene products. If this is indeed the case, different plant-pathogen recognition events could rapidly converge into one or a few common phosphorylation pathways that co-ordinate the onset of the defense response.

III. RECEPTOR-LIKE KINASES RELATED TO DEFENSE RESPONSES

A number of proteins with structural characteristics of receptor-like kinases has been associated with defense responses, although their biological function in defense mechanisms has yet to be demonstrated. Among them is the product of the *LRK10* gene from wheat, which was isolated in a screen of near isogenic lines carrying different leaf rust resistance genes, by using as probe a serine/

threonine kinase sequence (Feuillet *et al.*, 1997). The *LRK10* gene belongs to the *Lr10* locus conferring race-specific resistance to leaf rust, but its involvement in disease resistance, if any, is still unknown. An additional RLK potentially involved in pathogen recognition is the *Arabidopsis* PR5K (Wang *et al.*, 1996). PR5K displays an high degree of similarity in its extracellular domain to the PR5 family of pathogenesis-related proteins. This is reminiscent of the homology between the LRRs present in the product of *Cf9* resistance gene, which confers resistance to races of *Cladosporium fulvum* expressing the *avr9* gene, and the LRRs of the antimicrobial polygalacturonase inhibitor protein (PGIP; De Lorenzo *et al.*, 1994). Based on the similarity of PR5K and PR5 proteins, it has been proposed that the interaction of PR5K with a similar target of PR5 allows microbial recognition, and activation of defense mechanisms. The intracellular kinase catalytic domain of PRK5 was shown to be functional in a kinase assay *in vitro*. In addition, *PRK5* transcripts were found to accumulate in roots and inflorescence stems, although they were not induced by viral infection or salicylic acid treatment (Wang *et al.*, 1996). Inducibility at the transcriptional level in response to pathogen infection was observed for the *S* family receptor2 (SFR2) and cell wall-associated kinase1 (WAK1) RLKs (Pastuglia *et al.*, 1997; He *et al.*, 1998). The SFR2 protein is encoded by a member of the *S*-gene family of *Brassica oleracea*. Accumulation of its transcript was most abundant in flower buds and was rapidly induced in leaves by bacterial infection, wounding, and salicylic acid. The WAK1 RLK is an *Arabidopsis* cell wall-associated protein and its expression is induced by pathogen infection, salicylic acid and its analog 2,2-dichloroisonicotinic acid (INA; He *et al.*, 1998). Interestingly, overexpression of WAK1 provided resistance to otherwise lethal levels of salicylic acid, and *Arabidopsis* plants transformed with truncated or antisense forms or WAK1 were unable to survive INA stimulation. The results from this transgenic approach support a role for WAK1 in protecting the plant from detrimental side-effects derived from the response of plants to pathogens.

IV. MAP KINASES AND DEFENSE SIGNAL TRANSDUCTION

Mitogen-activated protein kinase (MAPK) signaling cascades are utilized in yeast and mammals to transduce extracellular stimuli into intracellular responses (Herskowitz, 1995; Seger and Krebs, 1995). MAP kinases are activated by MAP kinase kinase (MKK), through dual phosphorylation of threonine and tyrosine residues of a TXY motif located between subdomain VII and VIII of the kinase catalytic domain. In turn, MKK are activated by phosphorylation of a MAP kinase kinase kinase (MKKK). Recent studies provided evidence that MAP kinases are also involved in plant signaling pathways, particularly during the activation of stress-associated responses (Hirt, 1997; Mizoguchi *et al.*, 1997; Somssich, 1997).

The first evidence for the involvement of MAP kinases in defense responses was provided by the analysis of kinase activities from plant cell extracts that were able to phosphorylate myelin basic protein (MBP). Tyrosine phosphorylation was found to be associated with MBP kinases, supporting the theory that these activities originated from a MAP kinase. For instance, a rapid and transient activation of a 47-kDa and 49-kDa MBP kinase was observed in cell cultures and tobacco leaves treated with fungal or bacterial elicitors, respectively (Suzuki and Shinshi, 1995; Adam *et al.*, 1997; Lebrun-Garcia *et al.*, 1998). MBP kinase activities were also induced in tobacco by additional signals, which activate defense responses, such as mechanical wounding, cutting, and salicylic acid (Seo *et al.*, 1995; Usami *et al.*, 1995; Zhang and Klessig, 1997). Activation was found to be systemic for an MBP kinase of 48 kDa induced in tomato in response to wounding, systemin and oligosaccharide elicitors (Stratmann and Ryan, 1997). More recently, isolation of MAP kinase genes and generation of specific antibodies against their protein products have contributed to the characterization of several MAP kinase activities involved in defense mechanisms. Of particular interest are the elicitor-activated ERM kinases from parsley, a group of highly homologous wound-induced MAP kinases, and the salicylic-acid-induced SIP kinase, which appears to be a convergence point for different defense pathways.

A. THE ELICITOR-ACTIVATED ERM KINASE

In parsley cells, an oligopeptide elicitor derived from *Phytophtora sojae* binds to a plasma membrane receptor and activates a broad array of defense responses (Nürnberger *et al.*, 1994). The same elicitor was shown to determine the rapid and transient activation of the MAP kinase ERMK (Ligterink *et al.*, 1997). ERMK was immunoprecipitated from parsley cells by using antibodies raised against a peptide sequence from the alfalfa MAP kinase homologue MMK4 (Jonak *et al.*, 1996). Remarkably, only oligopeptides that were able to bind the plasma membrane receptor and triggered defense responses also induced ERMK kinase activity (Ligterink *et al.*, 1997). This strongly suggested that ERMK is involved in the signaling pathway activated by the specific recognition between the elicitor and a putative plasma-membrane receptor. Upon elicitor binding, activation of ion channels preceded an oxidative burst, and activation of defense genes (Nürnberger *et al.*, 1994). Pharmacological studies positioned ERMK activation downstream of ion-channel activation, and upstream or independent of the oxidative burst (Ligterink *et al.*, 1997). Moreover, the ERM kinase was translocated from the cytoplasm to the nucleus within 10 min after elicitor treatment. Translocation to the nucleus was shown to allow yeast and mammalian MAP kinases to phosphorylate, and induce the transactivation potential of target transcription factors (Seger and Krebs, 1995). Similarly, translocation of ERMK to the nucleus upon elicitation may determine phosphorylation of transcription factors involved in the regulation of defense genes. An *ERMK* complementary DNA (cDNA) clone was isolated from parsley cells and found to be homologous to a group of wound-inducible MAP kinase genes. The products of these genes are the MAP kinases MMK4 from alfalfa (Jonak *et al.*, 1996), WIPK from tobacco (Seo *et al.*, 1995), and ATMPK3 from *Arabidopsis* (Mizoguchi *et al.*, 1993). Their degree of homology to ERMK is shown in Table I, and their characteristics are discussed below.

B. WOUNDING-ACTIVATED MAP KINASES

The alfalfa ortholog of the parsley ERMK is the MMK4 MAP kinase (Jonak *et al.*, 1996). Specific antibodies were used to study the inducibility of MMK4 by external stimuli in alfalfa plants and cell cultures. MMK4 kinase activity was induced within minutes by mechanical stimulation (Bögre *et al.*, 1996), wounding (Bögre *et al.*, 1997), cold and drought (Jonak *et al.*, 1996). All these treatments also induced accumulation of *MMK4* transcript, although activation of the MMK4 kinase was achieved through post-translational mechanism. Inducibility at the transcriptional level appears to be a distinctive characteristic of wounding-induced MAP kinases (Seo *et al.*, 1995; Mizoguchi *et al.*, 1996; Bögre *et al.*, 1997), but is not entirely uncommon among protein kinases. Inactivation of protein kinases in several instances involves their degradation. Transcriptional activation of protein kinase genes and the subsequent protein synthesis may represent a strategy to re-establish the original protein level in the cell. Interestingly, messenger RNAs (mRNAs) of MAPKKK and a MAPKK from *Arabidopsis* plants were also found to accumulate in response to wounding (Mizoguchi *et al.*, 1996; Morris *et al.*, 1997).

Following wound-induced activation, the MMK4 MAP kinase was found to be subject to a refractory period, which is initiated by an inactivation mechanism dependent on *de novo* transcription and translation (Bögre *et al.*, 1997). A protein phosphatase 2C, MP2C, has been recently proposed as a negative regulator of MMK4 activity (Meskiene *et al.*, 1998). *MP2C* transcript

TABLE I
MAP kinases related to the parsley ERM kinase

MAP kinase	Plant	Homology[a]	Reference
ERMK	Parsley	100%	Ligterink *et al.* (1997)
WIPK	Tobacco	83%	Seo *et al.* (1995)
ATMPK3	*Arabidopsis*	83%	Mizoguchi *et al.* (1993)
MMK4	Alfalfa	81%	Jonak *et al.* (1996)

[a]The homology is as reported in Ligterink *et al.* (1997).

accumulation is transiently induced by wounding and its induction correlated with the MMK4 refractory period. In addition, when a bacterial expressed MP2C–glutathione *S*-transferase fusion was added to a wound-induced leaf extract, MMK4 activity was specifically inhibited.

An additional component of the group of plant kinases activated by wounding is the MAP kinase WIPK (Seo *et al.*, 1995). WIPK was shown first to be activated by wounding at the transcriptional level. Recently, local and systemic induction of WIPK kinase activity has been demonstrated as well (Seo *et al.*, 1999). A transgenic approach was utilized to assess the involvement of WIPK in the wound signal transduction pathway (Seo *et al.*, 1995, 1999). Introduction of the *WIPK* gene into transgenic plants resulted either in its overexpression or in inactivation of the endogenous gene. Interestingly, in both cases, defense responses induced by wounding were altered. Upon wounding, silencing of the endogenous gene resulted in abnormal accumulation of PR proteins, increased accumulation of salicylic acid and lack of jasmonic acid accumulation. Conversely, plants with constitutively elevated levels of WIPK activity accumulated a high amount of jasmonic acid in the absence of wound induction. These results indicate that the MAP kinase pathway mediated by WIPK is upstream of jasmonic acid. In addition this suggests, by analogy to prostaglandin synthesis in mammalian cells, that a possible target of the WIPK pathway may be a phospholipase involved in jasmonic acid synthesis.

Recently, the WIPK MAP kinase has been shown to be activated also by biotic stress in signaling pathways mediated by resistance genes. WIPK kinase activity was induced by tobacco mosaic virus infection only in plants expressing the resistance gene *N* (Zhang and Klessig, 1998a). In addition, an increase in WIPK kinase activity was detected in tobacco when plants or tissue cultures expressing the tomato *Cf9*-resistance gene were elicited with the *Cladosporium fulvum* avirulence protein Avr9 (Romeis *et al.*, 1999). In both cases induction of WIPK kinase activity was dependent on the expression of a resistance gene and was associated with tyrosine phosphorylation. However, the kinetics of activation were different in the two systems. Induction by TMV started about 2–3 h after treatment and lasted for several hours. Induction by Avr9 was very rapid (within minutes) and WIPK kinase activity returned to basal levels during the subsequent 2–3 h. Moreover, an accumulation of *WIPK* mRNA was observed after both TMV and Avr9 treatments.

Taken together, the emerging evidence related to the induction of WIPK by several biotic and abiotic signals, and via distinct activation mechanisms, suggests a role for the MAP kinase WIPK in the integration of different pathways leading to similar defense responses. Alternatively, WIPK may be involved in one of multiple pathways activated by different signals, or it may participate in distinct pathways and lead to different outputs as determined by the specific association with additional molecular components.

C. THE MULTI-STRESS-RESPONSIVE SIP KINASE

An additional plant MAP kinase associated with defense responses is the tobacco salicylic-acid-induced protein (SIP) kinase. SIPK was first identified as a salicylic acid (SA)-activated MAP kinase and purified to homogeneity from SA-treated tobacco suspension cells (Zhang and Klessig, 1997). By biochemical analysis of the purified protein, SIPK was found to have an absolute requirement for Mg^{2+} ions, and K_m and V_{max} values for Mg^{2+}-adenosine triphosphate (ATP) of 24 μM and 0.39 μmol min^{-1} mg^{-1}, respectively. SIPK was phosphorylated on tyrosine residue(s), and with MBP as phosphate acceptor it had a specific activity of 340 nmol min^{-1} mg^{-1}, which is higher than that of previously characterized animal MAP kinases. In addition, both tyrosine-specific protein phosphatase and serine/threonine-specific protein phosphatase treatments inactivated its kinase activity, suggesting that phosphorylation of these residues was required for SIP kinase activation. Oligonucleotides were designed according to partial sequences obtained from SIPK tryptic peptides, and used to isolate the *SIPK* gene. Analysis of its deduced amino acid sequence revealed that SIPK encodes a MAP kinase, with similarity to Ntf4 from tobacco (Wilson *et al.*, 1993), MMK1 from alfalfa (Duerr *et al.*, 1993), AtMPK6 from *Arabidopsis* (Mizoguchi *et al.*, 1993), and PsMAPK from pea (*Pisum sativum*) (Pöpping *et al.*, 1996; Table II).

SIPK kinase activity was shown to be induced by various biotic and abiotic elicitors in both host and nonhost resistance mechanisms. Rapid and transient activation of the SIP kinase was observed after nonspecific elicitation in tobacco suspension cells by salicylic acid (Zhang and Klessig, 1997) and fungal elicitors (Zhang *et al.*, 1998), and in leaves by salicylic acid and wounding (Zhang and Klessig, 1998b). In tobacco plants, SIPK elicitation also occurred in an *N*-gene-dependent manner after TMV infection, and in a *Cf9*-dependent manner after inoculation with the avirulence gene product Avr9 (Romeis *et al.*, 1999). In all these treatments SIPK activation was a result of post-translational event(s) and was not associated with an increase in mRNA or protein levels. As

TABLE II
MAP kinases related to the tobacco SIP kinase

MAP kinase	Plant	Homology[a]	Reference
SIPK	Tobacco	100%	Zhang and Klessig (1997)
Ntf4	Tobacco	93.6%	Wilson *et al.* (1993)
MMK1	Alfalfa	87.9%	Duerr *et al.* (1993)
PsMAPK	Pea	87.3%	Pöpping *et al.* (1996)
AtMPK6	*Arabidopsis*	86%	Mizoguchi *et al.* (1993)

[a]The homology is as reported in Zhang and Klessig, 1997.

for the WIP kinase, the induction of SIP kinase activity by an array of inducers suggests that signaling pathways derived from different resistance genes, and from biotic and abiotic stimuli may be connected at the level of certain MAP kinases. This raises the question of how different stimuli, which are transmitted via identical signaling components, result in different responses. Based on observations in animal systems, it has been proposed that the duration and activation level of SIPK and WIPK MAP kinases may be involved in the determination of the final output (Romeis *et al.*, 1999). In addition, the association with different components available in the cell only under specific circumstances may be important for the activation of specific responses.

V. PERSPECTIVES

In the last few years it has been clearly established that protein phosphorylation plays a central role in signaling pathways linking the perception of pathogen attack and the elaboration of defense responses. Protein kinases of different families appear to be involved in defense mechanisms at different cellular levels, including elicitor recognition, as extracellular or intracellular receptors, signal transduction, and induction of transcriptional activation. However, only a few kinases participating in defense signaling have been isolated to date, and their molecular characterization is still very limited. A major future challenge will be to dissect defense pathways into single components, and to elucidate their mutual interactions at the molecular and cellular levels.

A great deal remains to be learned about the mechanisms that allow receptor protein kinases to recognize elicitors and propagate warning signals into the plant cell. In addition, little is known about the biochemical properties of plant protein kinases, their mode of activation, and the role of their phosphorylation activities in signal transduction. Valuable tools for future research are protein kinases already associated with defense. These may be utilized to identify physical interactors from plant or pathogen, and to isolate physiological substrates and cellular targets for their activities. Furthermore, characterization of several signaling components will shed light on whether different elicitors activate distinct defense pathways, or different plant-pathogen recognition events rapidly converge into one or a few common pathways that co-ordinate the overall defense response. Finally, elucidation of mechanisms regulating defense responses will lead to new strategies for the improvement of plant disease resistance.

ACKNOWLEDGEMENTS

The authors wish to thank Daniel F. Klessig for sharing prepublication material. Research in the laboratory is supported by National Science

Foundation grant MCB-96-30635/MCB-9896308, and a David and Lucile Packard Foundation Fellowship. Guido Sessa is supported by Postdoctoral Award No. FI-248-97 from BARD, the United States–Israel Binational Agricultural Research and Development Fund.

REFERENCES

Adam, A. L., Pike, S., Hoyos, M. E., Stone, J. M., Walker, J. C. and Novacky, A. (1997). Rapid and transient activation of a myelin basic protein kinase in tobacco leaves treated with harpin from *Erwinia amylovora*. *Plant Physiology* **115**, 853–861.

Baker, B., Zambryski, P., Staskawicz, B. and Dinesh-Kumar, S. P. (1997). Signaling in plant–microbe interactions. *Science* **276**, 726–733.

Benhamou, N. (1996). Elicitor-induced plant defence pathways. *Trends in Plant Science* **1**, 233–240.

Bögre, L., Ligterink, W., Heberle-Bors, E. and Hirt, H. (1996). Mechanosensors in plants. *Nature* **383**, 489–490.

Bögre, L., Ligterink, W., Meskiene, I., Barker, P. J., Heberle-Bors, E., Huskisson, N. S. and Hirt, H. (1997). Wounding induces the rapid and transient activation of a specific MAP kinase pathway. *Plant Cell* **9**, 75–83.

Braun, D. M. and Walker, J. C. (1996). Plant transmembrane receptors: new pieces in the signaling puzzle. *Trends in Biochemical Sciences* **21**, 70–73.

Braun, T., Schofield, P. R. and Sprengel, R. (1991). Amino-terminal leucine-rich repeats in gonadotropin receptors determine hormone selectivity. *EMBO Journal* **10**, 1885–1890.

Chandra, S. and Low, P. S. (1995). Role of phosphorylation in elicitation of the oxidative burst in cultured soybean cells. *Proceedings of the National Academy of Sciences of the United States of America* **92**, 4120–4123.

Conrath, U., Jeblick, W. and Kauss, H. (1991). The protein kinase inhibitor, K-252a, decreases elicitor-induced Ca^{2+} uptake and K^+ release, and increases coumarin synthesis in parsley cells. *FEBS Letters* **279**, 141–144.

De Lorenzo, G., Cervone, F., Bellincampi, D., Caprari, C., Clark, A. J., Desiderio, A., Devoto, A., Forrest, F., Leckie, F., Nuss, L. and Salvi, G. (1994). Polygalacturonase, PGIP and oligogalacturonides in cell–cell communication. *Biochemical Society Transactions* **22**, 394–397.

Després, C., Subramaniam, R., Matton, D. P. and Brisson, N. (1995). The activation of the potato *PR-10a* gene requires phosphorylation of the nuclear factor PBF-1. *Plant Cell* **7**, 589–598.

Dietrich, A., Mayer, J. E. and Hahlbrock, K. (1990). Fungal elicitor triggers rapid, transient, and specific protein phosphorylation in parsley cell suspension cultures. *Journal of Biological Chemistry* **265**, 6360–6368.

Dröge-Laser, W., Kaiser, A., Lindsay, W. P., Halkier, B. A., Loake, G. J., Doerner, P., Dixon, R. A. and Lamb, C. (1997). Rapid stimulation of a soybean protein–serine kinase that phosphorylates a novel bZIP DNA-binding protein, G/HBF-1, during the induction of early transcription-dependent defenses. *EMBO Journal* **16**, 726–738.

Duerr, B., Gawienowski, M., Ropp, T. and Jacobs, T. (1993). MsERK1: a mitogen-activated protein kinase from a flowering plant. *Plant Cell* **5**, 87–96.

Fantl, W. J., Johnson, D. E. and Williams, L. T. (1993). Signalling by receptor tyrosine kinases. *Annual Review of Biochemistry* **62**, 453–481.

Felix, G., Grosskopf, D. G., Regenass, M. and Boller, T. (1991). Rapid changes of protein phosphorylation are involved in transduction of the elicitor signal in plant cells. *Proceedings of the National Academy of Sciences of the United States of America* **88**, 8831–8834.

Felix, G., Regenass, M. and Boller, T. (1993). Specific perception of subnanomolar concentrations of chitin fragments by tomato cells: induction of extracellular alkalinization, changes in protein phosphorylation, and establishment of a refractory state. *Plant Journal* **4**, 307–316.

Felix, G., Regenass, M., Spanu, P. and Boller, T. (1994). The protein phosphatase inhibitor calyculin A mimics elicitor action in plant cells and induces rapid hyperphosphorylation of specific proteins as revealed by pulse labeling with [^{33}P]phosphate. *Proceedings of the National Academy of Sciences of the United States of America* **91**, 952–956.

Feuillet, C., Schachermayr, G. and Keller, B. (1997). Molecular cloning of a new receptor-like kinase gene encoded at the *Lr10* locus of wheat. *Plant Journal* **11**, 45–52.

Flor, H. H. (1971). Current status of the gene-for-gene concept. *Annual Review of Phytopathology* **9**, 275–296.

Fluhr, R. (1998). Ethylene perception: from two-component signal transducers to gene induction. *Trends in Plant Science* **3**, 141–146.

Frederick, R. D., Thilmony, R. L., Sessa, G. and Martin, G. B. (1998). Recognition specificity for the bacterial avirulence protein AvrPto is determined by Thr-204 in the activation loop of the tomato Pto kinase. *Molecular Cell* **2**, 241–245.

Gianfagna, T. J. and Lawton, M. A. (1995). Specific activation of soybean defense genes by the phosphoprotein phosphatase inhibitor okadaic acid. *Plant Science* **109**, 165–170.

Gopalan, S., Bauer, D. W., Alfano, J. R., Loniello, A. O., He, S. Y. and Collmer, A. (1996). Expression of the *Pseudomonas syringae* avirulence protein AvrB in plant cells alleviates its dependence in the hypersensitive response and pathogenicity (Hrp) secretion system in eliciting genotype-specific hypersensitive cell death. *Plant Cell* **8**, 1095–1105.

Grab, D., Feger, M. and Ebel, J. (1989). An endogenous factor from soybean (*Glycine max* L.) cells activates phosphorylation of a protein which is dephosphorylated *in vivo* in elicitor-challenged cells. *Planta* **179**, 340–348.

Gu, Y. Q. and Martin, G. B. (1998). Molecular mechanisms involved in bacterial speck disease resistance of tomato. *Philosophical Transactions of the Royal Society of London B* **353**, 1455–1461.

Hammond-Kosack, K. E. and Jones, J. D. G. (1996). Resistance gene-dependent plant defense responses. *Plant Cell* **8**, 1773–1791.

Hammond-Kosack, K. E. and Jones, J. D. G. (1997). Plant disease resistance genes. *Annual Review of Plant Physiology and Plant Molecular Biology* **48**, 575–607.

Hanks, S. K. and Quinn, M. (1991). Protein kinase catalytic domain sequence database: identification of conserved features of primary structure and classification of family members. *Methods in Enzymology* **200**, 38–62.

He, Z.-H., He, D. and Kohorn, B. D. (1998). Requirement for the induced expression of a cell wall associated receptor kinase for survival during the pathogen response. *Plant Journal* **14**, 55–63.

Herskowitz, I. (1995). MAP kinase pathways in yeast: for mating and more. *Cell* **80**, 187–197.

Hirt, H. (1997). Multiple roles of MAP kinases in plant signal transduction. *Trends in Plant Science* **2**, 11–15.

Hunter, T. (1995). Protein kinases and phosphatases: the yin and yang of protein phosphorylation and signaling. *Cell* **80**, 225–236.

Hunter, T. and Karin, M. (1992). The regulation of transcription by phosphorylation. *Cell* **70**, 375–378.

Jean, D. A. and Hubner, S. (1996). Regulation of protein transport to the nucleus: central role to phosphorylation. *Physiological Reviews* **76**, 651–685.

Jeffrey, P. D., Russo, A. A., Polyak, K., Gibbs, E., Hurwitz, J., Massague, J. and Pavletich, N. P. (1995). Mechanism of CDK activation revealed by the structure of a cyclinA-CDK2 complex. *Nature* **376**, 313–320.

Jia, Y. and Martin, G. B. (1999). Rapid transcript accumulation of pathogenesis-related genes during an incompatible interaction in bacterial speck-disease resistant tomato plants. *Plant Molecular Biology* **40**, 455–465.

Jia, Y., Loh, Y.-T., Zhou, J. and Martin, G. B. (1997). Alleles of *Pto* and *Fen* occur in bacterial speck-susceptible and fenthion-insensitive tomato cultivars and encode active protein kinases. *Plant Cell* **9**, 61–73.

Johnson, L. N., Noble, M. E. M. and Owen, D. J. (1996). Active and inactive protein kinases: structural basis for regulation. *Cell* **85**, 149–158.

Jonak, C., Kiegerl, S., Ligtering, W., Barker, P. J., Huskisson, N. S. and Hirt, H. (1996). Stress signaling in plants: a mitogen-activated protein kinase pathway is activated by cold and drought. *Proceedings of the National Academy of Sciences of the United States of America* **93**, 11274–11279.

Kauss, H., Jeblick, W. and Conrath, U. (1992). Protein kinase inhibitor K-252a and fusicoccin induce similar initial changes in ion transport of parsley suspension cells. *Physiologia Plantarum* **85**, 483–488.

Lease, K., Ingham, E. and Walker, J. C. (1998). Challenges in understanding RLK function. *Current Opinion in Plant Biology* **1**, 388–392.

Lebrun-Garcia, A., Ouaked, F., Chiltz, A. and Pugin, A. (1998). Activation of MAPK homologues by elicitors in tobacco cells. *Plant Journal* **15**, 773–781.

Leister, R. T., Ausubel, F. M. and Katagiri, F. (1996). Molecular recognition of pathogen attack occurs inside plant cells in plant disease resistance specified by the *Arabidopsis* genes *RPS2* and *RPM1*. *Proceedings of the National Academy of Sciences of the United States of America* **93**, 15497–15502.

Lemaitre, B., Nicolas, E., Michaut, L., Reichart, J-M. and Hoffmann, J. A. (1996). The dorsoventral regulatory gene cassette *spatzle/Toll/cactus* controls the potent antifungal response in *Drosophila* adults. *Cell* **86**, 973–983.

Levin, L. R. and Zoller, M. J. (1990). Association of catalytic and regulatory subunits of cyclic AMP-dependent protein kinase requires a negatively charged side group at a conserved threonine. *Molecular and Cellular Biology* **10**, 1066–1075.

Levine, A., Tenhaken, R., Dixon, R. and Lamb, C. (1994). H_2O_2 from the oxidative burst orchestrates the plant hypersensitive disease resistance response. *Cell* **79**, 583–593.

Ligterink, W., Kroj, T., zur Nieden, U., Hirt, H. and Scheel, D. (1997). Receptor-mediated activation of a MAP kinase in pathogen defense of plants. *Science* **276**, 2054–2057.

Loh, Y.-T. and Martin, G. B. (1995). The *Pto* bacterial resistance gene and the *Fen* insecticide sensitivity gene encode functional protein kinases with serine/threonine specificity. *Plant Physiology* **108**, 1735–1739.

Loh, Y.-T., Zhou, J. and Martin, G. B. (1998). The myristoylation motif of Pto is not required for disease resistance. *Molecular Plant–Microbe Interactions* **11**, 572–576.

MacKintosh, C., Lyon, G. D. and MacKintosh, R. W. (1994). Protein phosphatase inhibitors activate anti-fungal defence responses of soybean cotyledons and cell cultures. *Plant Journal* **5**, 137–147.

Martin, G. B., Brommonschenkel, S., Chunwongse, J., Frary, A., Ganal, M. W., Spivey, R., Wu, T., Earle, E. D. and Tanksley, S. D. (1993). Map-based cloning of a protein kinase gene conferring disease resistance in tomato. *Science* **262**, 1432–1436.

Martin, G. B., Frary, A., Wu, T., Brommonschenkel, S., Chunwongse, J., Earle, E. D. and Tanksley, S. D. (1994). A member of the tomato *Pto* gene family confers sensitivity to fenthion resulting in rapid cell death. *Plant Cell* **6**, 1543–1552.

Meskiene, I., Bögre, L., Glaser, W., Balog, J., Brandstötter, M., Zwerger, K., Ammerer, G. and Hirt, H. (1998). MP2C, a plant protein phosphatase 2C, functions as a negative regulator of mitogen-activated protein kinase pathways in yeast and plants. *Proceedings of the National Academy of Sciences of the United States of America* **95**, 1938–1943.

Mizoguchi, T., Hayashida, N., Yamaguchi-Shinozaki, K., Kamada, H. and Shinozaki, K. (1993). ATMPKs: a gene family of plant MAP kinases in *Arabidopsis thaliana*. *FEBS Letters* **336**, 440–444.

Mizoguchi, T., Irie, K., Hirayama, T., Hayashida, N., Yamaguchi-Shinozaki, K., Matsumoto, K. and Shinozaki, K. (1996). A gene encoding a mitogen-activated protein kinase kinase kinase is induced simultaneously with genes for a mitogen-activated protein kinase and an S6 ribosomal protein kinase by touch, cold, and water stress in *Arabidopsis thaliana*. *Proceedings of the National Academy of Sciences of the United States of America* **93**, 765–769.

Mizoguchi, T., Ichimura, K. and Shinozaki, K. (1997). Environmental stress response in plants: the role of mitogen-activated protein kinases. *Trends in Biotechnology* **15**, 15–19.

Morgan, D. O. (1995). Principles of CDK regulation. *Nature* **374**, 131–134.

Morisato, D. and Anderson, K. V. (1995). Signaling pathways that establish the dorsal-ventral pattern of the *Drosophila* embryo. *Annual Review in Genetics* **29**, 371–399.

Morris, P. C., Guerrier, D., Leung, J. and Giraudat, J. (1997). Cloning and characterisation of MAK1, an *Arabidopsis* gene encoding a homologue of MAP kinase kinase. *Plant Molecular Biology* **35**, 1057–1064.

Mudgett, M. B. and Staskawicz, B. J. (1998). Protein signaling via type III secretion pathways in phytopathogenic bacteria. *Current Opinion in Microbiology* **1**, 109–114.

Nürnberger, T., Nennstiel, T. J., Sacks, W. R., Hahlbrock, K. and Scheel, D. (1994). High affinity binding of a fungal oligopeptide elicitor to parsley plasma membranes triggers multiple defense responses. *Cell* **78**, 449–460.

O'Neill, L. A. J. and Greene, C. (1998). Signal transduction pathways activated by the IL-1 receptor family: ancient signaling machinery in mammals, insects, and plants. *Journal of Leukocyte Biology* **63**, 650–657.

Ohme-Takagi, M. and Shinshi, H. (1995). Ethylene-inducible DNA binding proteins that interact with an ethylene-responsive element. *Plant Cell* **7**, 173–182.

Pastuglia, M., Roby, D., Dumas, C. and Cock, J. M. (1997). Rapid induction by wounding and bacterial infection of an *S* gene family receptor-like kinase gene in *Brassica oleracea*. *Plant Cell* **9**, 49–60.

Pöpping, B., Gibbons, T. and Watson, M. D. (1996). The *Pisum sativum* MAP kinase homologue (PsMAPK) rescues the *Saccharomyces cerevisiae hog1* deletion mutant under conditions of high osmotic stress. *Plant Molecular Biology* **31**, 355–363.

Raz, V. and Fluhr, R. (1993). Ethylene signal is transduced via protein phosphorylation events in plants. *Plant Cell* **5**, 523–530.

Romeis, T., Piedras, P., Zhang, S., Klessig, D. F., Hirt, H. and Jones, J. D. G. (1999). Rapid *Avr9*- and *Cf9*-dependent activation of MAP kinases in tobacco cell cultures and leaves; convergence of resistance gene, elicitor, wound, and salicylate responses. *Plant Cell* **11**, 273–287.

Ronald, P. C. (1997). The molecular basis of disease resistance in rice. *Plant Molecular Biology* **35**, 179–186.

Rushton, P. J. and Somssich, I. E. (1998). Transcriptional control of plant genes responsive to pathogens. *Current Opinion in Plant Biology* **1**, 311–315.

Ryals, J. A., Neuenschwander, U. H., Willits, M. G., Molina, A., Steiner, H.-Y. and Hunt, M. D. (1996). Systemic acquired resistance. *Plant Cell* **8**, 1809–1819.

Salmeron, J. M., Oldroyd, G. E. D., Rommens, C. M. T., Scofield, S. R., Kim, H.-S., Lavelle, D. T., Dahlbeck, D. and Staskawicz, B. J. (1996). Tomato *Prf* is a member of the leucine-rich-repeat class of plant disease resistance genes and lies embedded within the *Pto* kinase gene cluster. *Cell* **86**, 123–133.

Schwacke, R. and Hager, A. (1992). Fungal elicitors induce a transient release of active oxygen species from cultured spruce cells that is dependent on Ca^{2+} and protein-kinase activity. *Planta* **187**, 136–141.

Scofield, S. R., Tobias, C. M., Rathjen, J., Chang, J. H., Lavelle, D. T., Michelmore, R. W. and Staskawicz, B. J. (1996). Molecular basis of gene-for-gene specificity in bacterial speck disease of tomato. *Science* **274**, 2063–2065.

Seger, R. and Krebs, E. G. (1995). The MAPK signaling cascade. *FASEB Journal* **9**, 726–735.

Seo, S., Okamoto, M., Seto, H., Ishizuka, K., Sano, H. and Ohashi, Y. (1995). Tobacco MAP kinase: a possible mediator in wound signal transduction pathways. *Science* **270**, 1988–1992.

Seo, S., Sano, H. and Ohashi, Y. (1999). Jasmonate-based wound signal transduction requires activation of WIPK, a tobacco mitogen-activated protein kinase. *Plant Cell* **11**, 289–298.

Sessa, G., Meller, Y. and Fluhr, R. (1995). A GCC element and a G-box motif participate in ethylene induced expression of the *PRB-1b* gene. *Plant Molecular Biology* **28**, 145–153.

Sessa, G., D'Ascenzo, M., Loh, Y.-T. and Martin, G. B. (1998). Biochemical properties of two protein kinases involved in disease resistance signaling in tomato. *Journal of Biological Chemistry* **273**, 15860–15865.

Sessa, G., D'Ascenzo, M. and Martin, G. B. (2000). The major site of the Pti1 kinase phosphorylated by the Pto kinase is located in the activation domain and is required for the Pto–Pti1 physical interaction. *European Journal of Biochemistry* **267**, 171–178.

Somssich, I. E. (1997). MAP kinases and plant defence. *Trends in Plant Science* **2**, 406–408.

Somssich, I. E. and Hahlbrock, K. (1998). Pathogen defence in plants – a paradigm of biological complexity. *Trends in Plant Science* **3**, 86–90.

Song, W.-Y., Wang, G.-L., Chen, L.-L., Kim, H.-S., Pi, L.-Y., Holsten, T., Gardner, J., Wang, B., Zhai, W.-X., Zhu, L.-H., Fauquet, C. and Ronald, P. (1995). A receptor kinase-like protein encoded by the rice disease resistance gene *Xa21*. *Science* **270**, 1804–1806.

Song, W.-Y., Pi, L.-Y., Wang, G.-L., Gardner, J., Holsten, T. and Ronald, P. C. (1997). Evolution of the rice *Xa21* disease resistance gene family. *Plant Cell* **9**, 1279–1287.

Stone, J. M., Collinge, M. A., Smith, R. D., Horn, M. A. and Walker, J. C. (1994). Interaction of a protein phosphatase with an *Arabidopsis* serine–threonine receptor kinase. *Science* **266**, 793–795.

Stratmann, J. W. and Ryan, C. A. (1997). Myelin basic protein kinase activity in tomato leaves is induced systemically by wounding and increases in response to systemin and oligosaccharide elicitors. *Proceedings of the National Academy of Sciences of the United States of America* **94**, 11085–11089.

Subramaniam, R., Després, C. and Brisson, N. (1997). A functional homolog of mammalian protein kinase C participates in the elicitor-induced defense response in potato. *Plant Cell* **9**, 653–664.

Suzuki, K. and Shinshi, H. (1995). Transient activation and tyrosine phosphorylation of a protein kinase in tobacco cells treated with a fungal elicitor. *Plant Cell* **7**, 639–647.

Tang, X., Frederick, R. D., Zhou, J., Halterman, D. A., Jia, Y. and Martin, G. B. (1996). Initiation of plant disease resistance by physical interaction of AvrPto and Pto kinase. *Science* **274**, 2060–2063.

Tang, X., Xie, M., Kim, Y. J., Zhou, J., Klessig, D. F. and Martin, G. B. (1998). Overexpression of Pto activates defense responses and confers broad resistance. *Plant Cell* **11**, 15–29.

Ten Dijke, P., Miyazono, K. and Heldin, C.-H. (1996). Signaling via hetero-oligomeric complex of type I and type II serine/threonine kinase receptors. *Current Opinion in Cell Biology* **8**, 139–145.

Ullrich, A. and Schlessinger, J. (1990). Signal transduction by receptors with tyrosine kinase activity. *Cell* **61**, 203–212.

Usami, S., Banno, H., Ito, Y., Nishihama, R. and Machida, Y. (1995). Cutting activates a 46-kilodalton protein kinase in plants. *Proceedings of the National Academy of Sciences of the United States of America* **92**, 8660–8664.

Van den Ackerveken, G., Marois, E. and Bonas, U. (1996). Recognition of the avirulence protein AvrBs3 occurs inside the host plant cell. *Cell* **87**, 1307–1316.

Viard, M-P., Martin, F., Pugin, A., Ricci, P. and Blein, J.-P. (1994). Protein phosphorylation is induced in tobacco cells by the elicitor cryptogein. *Plant Physiology* **104**, 1245–1249.

Wang, G.-L., Ruan, D. L., Song, W.-Y., Sideris, S., Chen, L., Pi, L.-Y., Zhang, S., Zhang, Z., Fauquet, C., Gaut, B. S., Whalen, M. C. and Ronald, P. C. (1998). *Xa21D* encodes a receptor-like molecule with a leucine-rich repeat domain that determines race-specific recognition and is subject to adaptive evolution. *Plant Cell* **10**, 765–779.

Wang, X., Zafian, P., Choudhary, M. and Lawton, M. (1996). The PR5K receptor protein kinase from *Arabidopsis thaliana* is structurally related to a family of plant defense proteins. *Proceedings of the National Academy of Sciences of the United States of America* **93**, 2598–2602.

Wilson, C., Eller, N., Gartner, A., Vicente, O. and Heberle-Bors, E. (1993). Isolation and characterization of a tobacco cDNA clone encoding a putative MAP kinase. *Plant Molecular Biology* **23**, 543–551.

Yang, Y., Shah, J. and Klessig, D. F. (1997). Signal perception and transduction in plant defense responses. *Genes and Development* **11**, 1621–1639.

Yu, L. M., Lamb, C. J. and Dixon, R. A. (1993). Purification and biochemical characterization of proteins which bind to the H-box cis-element implicated in transcriptional activation of plant defense genes. *Plant Journal* **3**, 805–816.

Zhang, S. and Klessig, D. F. (1997). Salicylic acid activates a 48-kD MAP kinase in tobacco. *Plant Cell* **9**, 809–824.

Zhang, S. and Klessig, D. F. (1998a). Resistance gene *N*-mediated *de novo* synthesis and activation of a tobacco mitogen-activated protein kinase by tobacco mosaic virus infection. *Proceedings of the National Academy of Sciences of the United States of America* **95**, 7433–7438.

Zhang, S. and Klessig, D. F. (1998b). The tobacco wounding-activated mitogen-activated protein kinase is encoded by SIPK. *Proceedings of the National Academy of Sciences of the United States of America* **95**, 7225–7230.

Zhang, S., Du, H. and Klessig, D. F. (1998). Activation of the tobacco SIP kinase by both a cell-wall-derived carbohydrate elicitor and purified proteinaceous elicitins from *Phytophtora* spp. *Plant Cell* **10**, 435–449.

Zhou, J., Loh, Y.-T., Bressan, R. A. and Martin, G. B. (1995). The tomato gene Pti1 encodes a serine/threonine kinase that is phosphorylated by Pto and is involved in the plant hypersensitive response. *Cell* **83**, 925–935.

Zhou, J., Tang, X. and Martin, G. B. (1997). The Pto kinase conferring resistance to tomato bacterial speck disease interacts with proteins that bind a *cis*-element of pathogenesis-related genes. *EMBO Journal* **16**, 3207–3218.

SNF1-Related Protein Kinases (SnRKs) – Regulators at the Heart of the Control of Carbon Metabolism and Partitioning

N. G. HALFORD[1], J.-P. BOULY[2] and M. THOMAS[2]

[1]IACR-Long Ashton Research Station, Department of Agricultural Sciences, University of Bristol, Long Ashton, Bristol BS41 9AF, UK
[2]Laboratoire de Biologie du Développement des Plantes, Institut de Biotechnologie des Plantes, UMR CNRS 8618, Bâtiment 630 Université de Paris-Sud, 91405 Orsay Cedex, France

I. INTRODUCTION

SnRKs (SNF1-related protein kinases) are a family of plant protein kinases with catalytic domains similar to that of SNF1 (sucrose non-fermenting-1) of yeast and AMPK (adenosine monophosphate-activated protein kinase) of

Advances in Botanical Research Vol. 32
incorporating Advances in Plant Pathology
ISBN 0-12-005932-0

animals. We regard them as a distinct family within the protein kinase superfamily, but they are closely related to the 'CaMK' group described by Hanks and Hunter (1995), which includes the animal calmodulin-dependent protein kinases (CaMKs) and the plant calmodulin domain protein kinases (CDPKs).

SNF1 and AMPK will be discussed in detail later in the chapter; both can be described as global regulators of carbon metabolism in their respective systems. The plant SnRK family comprises at least three subfamilies (SnRK1, 2 and 3). Relatively little is known about the functions of SnRK2s and SnRK3s, but it is already clear that SnRK1s play an important role in the regulation of carbon metabolism and in the cross-talk between metabolic and other signalling pathways in plants. This role has some similarities with the roles of SNF1 and AMPK, but also has some intriguing differences. We will begin by describing the structure and organization of the genes and gene families that encode this important family of plant proteins.

II. THE PLANT SNF1-RELATED PROTEIN KINASE (SnRK) GENE FAMILY

A. THE SnRK1 SUBFAMILY

The first plant *SNF1*-related sequence to be reported was *RKIN1*, a complementary DNA (cDNA) isolated from a rye endosperm cDNA library (Alderson *et al.*, 1991). *RKIN1* encodes a 57 710 Da protein of 502 amino acid residues showing 48% amino acid sequence identity with SNF1 and AMPK. The domain structure of the RKINI1 protein kinase is also similar to that of SNF1 and AMPK, with the protein kinase catalytic domain, as defined by Hanks *et al.* (1988) in the N-terminal half (Fig. 1). The protein kinase catalytic domains of the three proteins show approximately 62–64% amino acid sequence identity. This degree of similarity between protein kinases is usually indicative of similar substrate specificity and function. AMPK and SNF1 are both larger than RKIN1 at 63 kDa and 72 kDa, respectively. The size variation is due almost entirely to differences in the C-terminal regions that, however, still retain 29–34% amino acid sequence identity. SNF1 also has a slightly longer N-terminal region, prior to the catalytic domain, which contains a characteristic stretch of 13 histidine residues, the function of which is not known.

Genomic clones, cDNAs and polymerase chain reaction (PCR) products very similar to *RKIN1* have now been cloned from *Arabidopsis* (*AKIN10, AKIN20, AKIN30* and *ATSKIN1*), barley (*BKIN2* and *BKIN12*), oat (*ASPK1–3*), potato (*PKIN1*), rice (*RSK1*), sugar beet (*SBKIN154*) and tobacco (*NPK5*). They all encode very similar protein kinases, the lowest degree of amino acid sequence identity between any two being 68% and the size varying only from 57 710 Da (RKIN1) to 58 910 Da (AKIN10). This subfamily of plant protein kinases has been given the name SnRK1 (SNF1-related protein kinase-1). The

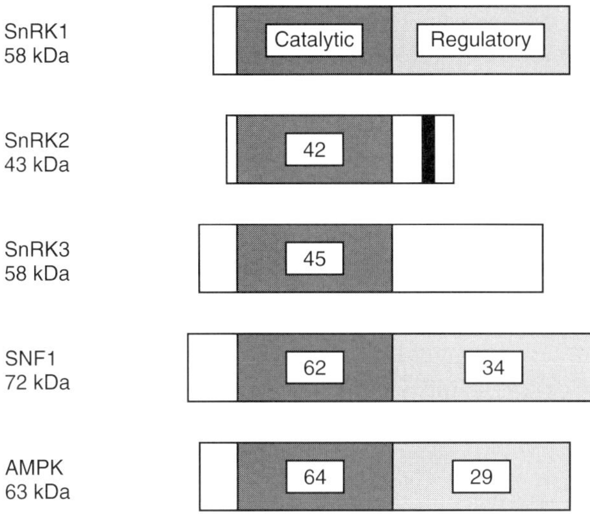

Fig. 1. Domain structure of members of the SNF1 family of protein kinases. The N-terminal protein kinase catalytic domain is shaded dark grey and the C-terminal regulatory region light grey. The short acidic patch that is characteristic of SnRK2s is shaded black. Figures within the N- and C-terminal domains give the degree of amino acid sequence identity, where significant, with SnRK1. The sizes given for SnRK1, SnRK2 and SnRK3 are typical of each subfamily, but there is some size variation between different subfamily members.

European Molecular Biology Laboratory (EMBL) accession numbers of several of its members are given in Table I.

The *SnRK1* gene subfamily of barley was the first to be studied in detail. Southern blot analyses showed that it compromised approximately 10–20 members per haploid genome (Halford *et al.*, 1992) and several related but different cDNAs, PCR products and genomic clones have been isolated (Halford *et al.*, 1992; Hannappel *et al.*, 1995). They fall into two groups, *SnRK1a* and *SnRK1b*, on the basis of amino acid sequence similarity and expression patterns (Halford and Hardie, 1998). The former, typified by gene *BKIN2*, is expressed throughout the plant, whereas the latter, typified by *BKIN12*, is expressed only in the seed. The BKIN2 and BKIN12 proteins have approximately 68% amino acid sequence identity with each other, with most divergence in the C-terminal domain. This region may play a role in protein–protein interactions. The structure of the *BKIN12* gene is shown in Fig. 2.

The *Arabidopsis* and potato *SnRK1* gene subfamilies have also been studied in detail. Three *SnRK1* genes have been identified in *Arabidopsis*, namely *AKIN10, AKIN20 and AKIN30* (Le Guen *et al.*, 1992; Lessard, 1997). Both *AKIN10* and *AKIN20* are located on chromosome 3, whereas *AKIN30* has recently been identified by genome sequencing on chromosome 5–83 (Genbank

TABLE I

Details of members of the SNF1 family of protein kinases. The plant family members (SnRKs) are divided into subfamilies based on degrees of amino acid sequence identity (see also Figs 1 and 2). Accession numbers are from the EMBL, GENBANK or PIR2 databases

		Name	Accession no.	Species
SnRK1	SnRK1a	AKIN10	M93023	*Arabidopsis*
		ASPK2	S56722	Oat
		ASPK3	S56723	Oat
		ATSKIN1	X94755	*Arabidopsis*
		BKIN2	X82548	Barley
		NPK5	D26602	Tobacco
		PKIN1	X95996-96000	Potato
		SBKIN154	AJ000475	Sugar beet
		–	U83797	Potato
		OSK1	D82039	Rice
	SnRK1b	ASPK1	S56721	Oat
		BKIN12	X65606	Barley
		RKIN1	M74113	Rye
		RSK1	U55768	Rice
		OSK2-5	D82035-8	Rice
SnRK2	SnRK2a	MCPK9	Z26846	Ice plant
		PKABA1	M94726	Wheat
		–	S56718	*Arabidopsis*
		–	S71172	*Arabidopsis*
	SnRK2b	ASK1	M91548	*Arabidopsis*
SnRK3		ASK2	Z12120	*Arabidopsis*
		BSK1	L12393	Oilseed rape
		BSK2	L12394	Oilseed rape
		SPK1	L01453	Soy bean
		SPK2	L19360	Soy bean
		SPK3	L19361	Soy bean
		WPK4	D21204	Wheat
		–	Z97336	*Arabidopsis*
		SNFL1	Y12464	Sorghum
		SNF1	M13971	Budding yeast
		NIM1	P07334	Fission yeast
		NIK1	U65921	Budding yeast
		AMPK	Z29486	Rat

accession number AB009054). The structure of the three genes is very similar, the positions of the introns being conserved (Fig. 2), but they differ in their leader sequences, with the presence of an extra intron in the 5′ untranslated regions of *AKIN10* and *AKIN20* genes. The role of the leader intron in the regulation of *AKIN10* expression has been investigated by promoter/reporter gene expression analysis. Its removal was found to lower the level of GUS

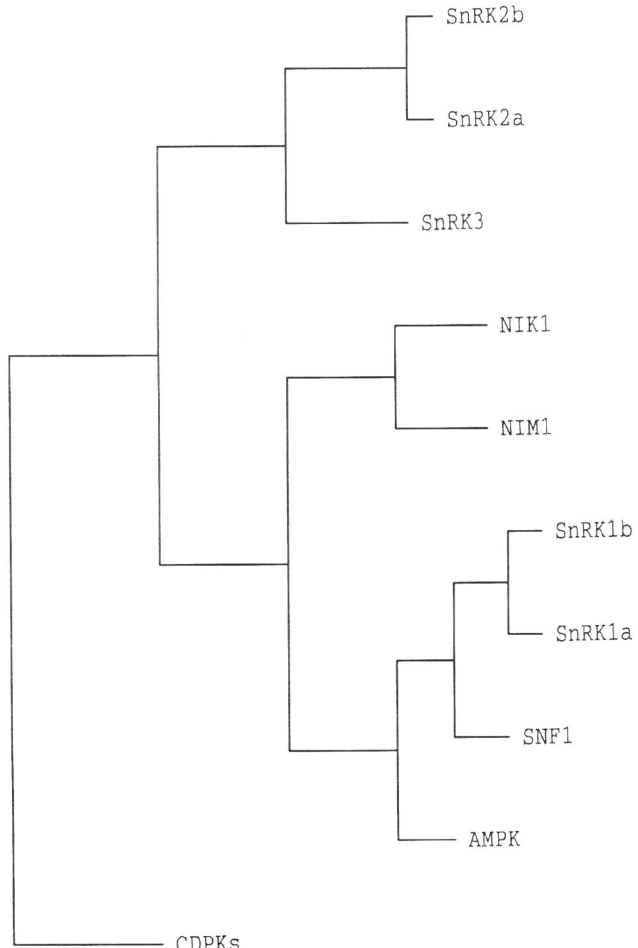

Fig. 2. Dendrogram showing relative evolutionary distances within the SNF1 family and between the SNF1-type kinases and CDPKs. Alignments were produced using the PILEUP program (Genetics Computer Group, 1994) with a gap creation penalty of 3.0 and a gap extension penalty of 1.0. Evolutionary distances were calculated using the DISTANCES program, correcting for multiple substitutions at a single site by the method of Kimura (1980), and displayed by the unweighted pair group method using arithmetic averages with the GROWTREE program. Refer to Table I for accession numbers.

activity by 10-fold in transgenic plants, leading to a marked decrease of expression in vegetative tissues (Bouly and Thomas, unpublished data). Two types of cDNAs, *AKINα1* and *AKINα2*, corresponding to *AKIN10* and *AKIN20* genes, respectively, have been isolated. The deduced amino acid sequences of *AKINα1* and *AKINα2* contain very few differences in their N-terminal kinase catalytic domains (90% identity), but show more sequence divergence within the C-terminal regulator domains (69% identity). As yet, no cDNA has been identified for *AKIN30*. The predicted amino acid sequence of AKINα3 is 20 residues shorter than those of the AKINα1 and AKINα2 proteins. Its catalytic domain is 78% identical with AKINα1 and AKINα2, but the three proteins diverge considerably in the C-terminal region (47% identity). Similarities between *AKIN10* and *AKIN20* genes extend to their patterns of expression, which were shown to be very similar during development and in different organs of *Arabidopsis*. Low levels of transcripts were detected in all tissues (Lessard and Thomas, unpublished data). At the moment, we have been unable to detect *AKIN30* transcripts, suggesting that *AKIN30* could be either a pseudogene or a gene whose expression is very low or restricted to specific cells.

Southern blot analysis of potato genomic DNA also showed the presence of a small family of *SnRK1* genes, comprising 5–10 members (Man *et al.*, 1997). In the same study, one cDNA and five genomic clones were isolated. The nucleotide sequences of a portion of all five genomic clones were shown to be identical and the name *PKIN1* was assigned to all. The cDNA contained only 15 substitutions compared with the genomic sequence, two of which resulted in changes in the derived amino acid sequence. However, a different potato SnRK1, StuSNF1, has been reported by another group (Lakatos and Banfalvi, 1997). Both PKIN1 and StuSNF1 are in the SnRK1a group and, although Northern blot analysis revealed some tissue-specific differences in *PKIN1* transcript levels, the lowest being detected in leaves and the highest in stolons, transcripts were detectable in all tissues.

The system of dividing the SnRK1s into SnRK1a and b groups holds up well in a plot of the evolutionary distances between them (Fig. 3). Barley SnRK1a (BKIN2) clusters with ASPK2 and ASPK3 from oat and all of the SnRK1s from dicotyledonous plants, whereas the barley SnRK1b (BKIN12), clusters with ASPK1 (oat), RKIN1 (rye) and RSK1 (rice). *RKIN1* appears to show a similar expression pattern to *BKIN12*, being expressed in endosperms but not in leaves (Alderson *et al.*, 1991). Similarly, rice contains *SnRK1a* and *SnRK1b*-type genes, and they show similar expression patterns to those of barley, with the *SnRK1b* type being expressed at high levels in developing seeds (Takano *et al.*, 1998). This suggests that the *SnRK1a* and *SnRK1b* groups differ on the basis of expression patterns as well as amino acid sequence alignments. As yet, the *SnRK1b* group appears to be present only in cereals whereas the *SnRK1a* group appears to be present in both dicotyledonous and monocotyledonous plants.

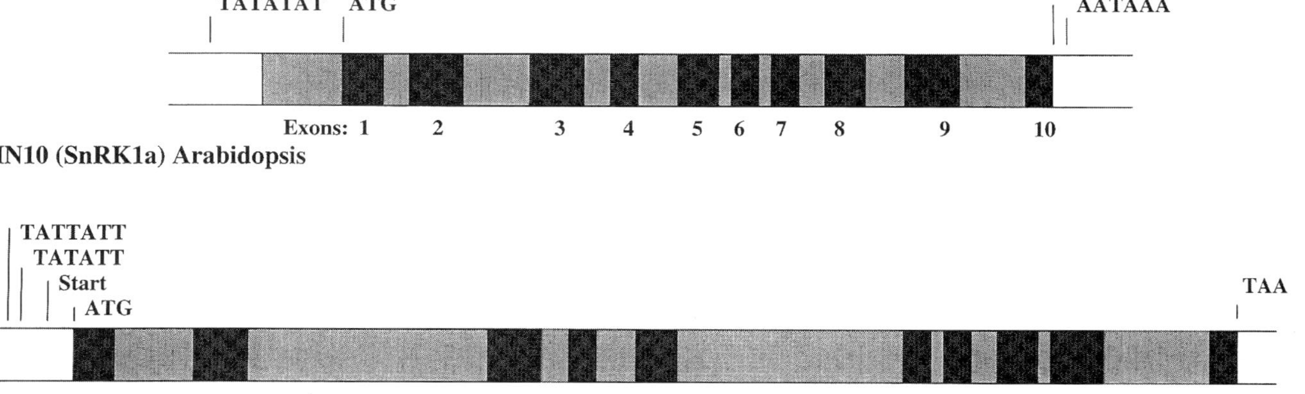

Fig. 3. Diagrammatic representation of the structures of the *SnRK1a* and *SnRK1b* genes *AKIN10* (*Arabidopsis*) and *BKIN12* (barley). Exons and introns are shaded dark and light grey, respectively. The positions of introns within the coding region are matched exactly in all of the *SnRK1* genes characterized so far.

B. THE SnRK2 AND SnRK3 SUBFAMILIES

Plants contain two other subfamilies of protein kinases containing catalytic domains with sequences that place them clearly within the SNF1 family. They have 42–45% amino acid sequence identity with the SnRK1s, SNF1 and AMPK in this region; they are, therefore, significantly less similar to SNF1 and AMPK than the SnRK1s are (Fig. 1). One of these subfamilies, known as SnRK2, comprises proteins with a molecular weight of approximately 40 kDa. They have relatively short C-terminal domains compared with SnRK1s, characterized by the presence of a short acidic 'patch' (Fig. 1) that is present in all of those identified so far with the exception of SPK1 and SPK2 from soy bean. SnRK2s divide further into two groups and these have been called SnRK2a and SnRK2b (Halford and Hardie, 1998). The C-terminal acidic patch in the SnRK2a group is rich in aspartic acid residues, while that of the SnRK2b group is rich in glutamic acid residues.

The other subfamily of plant protein kinases related to the SnRK1s currently comprises three members, WPK4 (Sano and Youssefian, 1994), which was isolated from wheat, SNFL1 from sorghum (Annen and Stockhaus, 1998) and a hitherto un-named gene from *Arabidopsis* (Table I). This subfamily has been called SnRK3. The WPK4 protein kinase has a molecular weight of 58 kDa, which means that it is the same size as members of the SnRK1 subfamily. However, it is clearly a member of a distinct subfamily, with no similarity in the C-terminal domain (Figs 1 and 3). The SNFL1 catalytic domain presents approximately 35% identity with that of SNF1, but SNFL1 is unable to complement yeast *snf1* mutants.

The SnRK2 and SnRK3 gene subfamilies appear to be unique to plants. Although they have diverged further from SNF1 and AMPK than the SnRK1s have, they retain sufficient similarity with them to suggest that they may have similar substrate specificity and functions. However, any functional overlap with each other and with the other members of the SNF1 family remains to be determined. The most comprehensive functional analysis of any of these proteins has been performed on PKABA1, a SnRK2a from wheat that is upregulated at the level of gene expression by the phytohormone abscisic acid (ABA; Anderberg and Walker-Simmons, 1992). PKABA1 has been shown to be involved in mediating ABA-induced suppression of α-amylase gene expression in barley aleurone layers (Gómez-Cadenas *et al.*, 1999).

III. IMMUNODETECTION OF SnRK1s AND MEASUREMENTS OF SnRK ACTIVITY IN CRUDE EXTRACTS FROM PLANTS

Independent antisera were raised to the RKIN1 protein expressed as a fusion with maltose-binding protein, and to a peptide (the NIP peptide) with the amino acid sequence Pro Phe Asp Asp/Glu Asn Ile Pro Asn Leu Phe Lys Lys

Ile Lys (Barker *et al.*, 1996). The NIP peptide sequence corresponds to residues 214–228 of RKIN1, which lie between subdomains IX and X of the kinase catalytic domain (Hanks *et al.*, 1988). It is highly conserved in the SnRK1 protein kinases and in SNF1 and AMPK, but not in the SnRK2s and SnRK3s.

The maltose binding protein (MBP)-RKIN1 fusion protein antibody recognizes polypeptides of approximately 60 kDa in crude extracts from barley (Barker *et al.*, 1996) and sugar beet (Monger *et al.*, 1997). In the former, a doublet of proteins, probably corresponding to BKIN2 and BKIN12, was recognized in endosperm extracts, while a single protein, probably BKIN2, was recognized in root extracts.

These antisera were also used to confirm the identity of protein kinases that were being studied independently at the biochemical level in the laboratory of Professor Grahame Hardie, University of Dundee. These protein kinases phosphorylated the SAMS peptide (His Met Arg Ser Ala Met Ser Gly Leu His Leu Val Lys Arg Arg), a synthetic peptide based on the sequence around the primary phosphorylation site for AMPK on rat acetyl-CoA carboxylase (Davies *et al.*, 1989). SAMS peptide kinase activity was detected in several plant extracts (MacKintosh *et al.*, 1992) and one of these activities was purified from cauliflower (Ball *et al.*, 1994). Although this protein kinase was not activated by AMP, it was found to have similar biochemical properties to those of mammalian AMPK in other respects (MacKintosh *et al.*, 1992; Ball *et al.*, 1994). It was shown to phosphorylate and inactivate a bacterially expressed *Arabidopsis* HMG-CoA (3-hydroxy-3-methylglutamyl-coenzyme A) reductase and was therefore given the name HMG-CoA reductase kinase-A (HRK-A), the suffix (-A) distinguishing it from another, minor SAMS peptide kinase activity (HRK-B) (Ball *et al.*, 1994).

As well as having similar biochemical properties to AMPK (MacKintosh *et al.*, 1992; Ball *et al.*, 1994; Dale *et al.*, 1995b), HRK-A was found to have a molecular mass of 58 kDa, exactly the size predicted for the plant homologues of AMPK, the SnRK1s. HRK-A was confirmed as a SnRK1 when cauliflower HRK-A and a similar protein kinase activity purified from barley endosperm were found to cross-react with the NIP and MBP-RKIN1 antisera (Ball *et al.*, 1995; Barker *et al.*, 1996).

More recently, biochemical studies of SnRK1s in Hardie's laboratory have utilized spinach leaf and at least four SAMS peptide kinases, named HRK-A through to -D, have been resolved from this tissue (Sugden *et al.*, 1999b). Both HRK-A and HRK-C contain polypeptides of approximately 60 kDa that cross-react with the NIP peptide antiserum. They probably represent isoforms of SnRK1, although it is also possible that they represent the same SnRK1 catalytic subunit complexed with alternative regulatory subunits (see Section V). There is no direct evidence that HRK-B is a SnRK, but cauliflower and spinach HRK-B are 40–45 kDa in size, approximately the size predicted for the SnRK2s. HRK-D remains uncharacterized.

IV. REGULATION OF SnRK1 ACTIVITY

SnRK1 transcripts are detectable in all of the plant tissues that we have analysed but the expression levels are not uniform. In barley, for example, there is more *SnRK1* transcript in the endosperm and aleurone than elsewhere in the plant. This is brought about in the main by activity of the *SnRK1b* gene, which is expressed seed-specifically. Even in dicotyledonous plants, however, which only contain the *SnRK1a*-type gene, there is differential regulation of gene expression. In *Arabidopsis*, for example, *AKIN10* and *AKIN20* are expressed in leaves throughout development but the amount of transcripts varies. *In situ* hybridization experiments and promoter studies have shown that *AKIN*α*1* transcripts are more abundant in vascular tissues of shoots and roots and in the basal tissues of axillary buds (Bouly and Thomas, unpublished data). In potato, highest levels of expression occur in stolons as they begin to develop into tubers (Man *et al.*, 1997). Expression gradually declines in maturing tubers, but is lowest in leaves. High levels of expression can be found in the developing storage root of sugar beet as well (Monger *et al.*, 1997) and may be a characteristic of developing storage tissues.

There is also evidence for post-transcriptional regulation of SnRK1 activity. Experiments undertaken in Hardie's laboratory have shown that cauliflower SnRK1, like AMPK, is regulated *in vitro* by phosphorylation; it can be inactivated by treatment with protein phosphatases and will reactivate in a time-dependent manner in the presence of Mg-adenosine triphosphate (ATP) and phosphatase inhibitors. This reactivation is due to the action of an endogenous upstream kinase, rather than autophosphorylation, because it is lost if the kinase is purified further (MacKintosh *et al.*, 1992). However, this kinase kinase has so far proved elusive (G. Hardie, C. Sugden and N. G. Halford, unpublished data) and has not been purified. Further experiments showed that spinach SnRK1 is regulated by phosphorylation on a threonine residue within the so-called T-loop that is conserved in many protein kinases (Sugden *et al.*, 1999a). A comparison of SAMS peptide kinase activities and *SnRK1* transcript levels in potato suggested that SnRK1 activity is regulated post-transcriptionally *in vivo* (Man *et al.*, 1997). The SAMS peptide kinase activity in mini-tubers and callus cultures was measured at 0.84 and 1.66 $\text{nmoles}^{-1}\text{min}^{-1}\text{mg}^{-1}$, respectively, whereas the activity in mature tubers was measured at 0.04 $\text{nmoles}^{-1}\ \text{min}^{-1}\ \text{mg}^{-1}$, even though the amounts of transcript in each sample were approximately the same.

The exact nature of the signal that brings about changes in SnRK1 gene expression or activation state is not known. SNF1 activity changes in response to glucose availability, and AMPK is activated in response to a high AMP:ATP ratio, a characteristic of animal cells under a variety of stresses, including metabolic stress. However, a clear change in SnRK1 activity in response to the presence/absence of a sugar, or to starvation or any other stimulus has not yet been reported. Intriguingly, dephosphorylation and inactivation of spinach SnRK1 was found to be inhibited by low concentrations of 5′-AMP (Sugden *et al.*, 1999a).

V. IDENTIFICATION OF SnRK1-INTERACTING PROTEINS

A. THE SNF1 COMPLEX

In *Saccharomyces cerevisiae*, SNF1 and SNF4, both defined by mutations which prevent growth on sucrose, are required for the derepression of genes repressed by glucose. Genetic studies indicated that SNF4 is a positive regulator of SNF1 and is essential for maximal kinase activity (Celenza and Carlson, 1989). These proteins have been shown to form a complex *in vivo* and *in vitro* (Celenza and Carlson, 1989; Fields and Song, 1989).

Another family of proteins (SIP1, SIP2 and GAL83) has been identified as associated proteins in the SNF1/SNF4 complex. The *SIP1* and *SIP2* genes (SNF1-interacting protein 1 and 2) were identified in a two-hybrid screen for genes encoding proteins that interact with SNF1, while *GAL83* was found in a genetic screen (Yang *et al.*, 1992, 1994). The function of these proteins is still unclear. The triple mutant (*sip1*Δ, *sip2*Δ, *gal83*Δ) is defective in sporulation but has no *snf* phenotype. Moreover, these subunits are not essential for the activity of the yeast enzyme; complexes formed by SNF1 and SNF4 can respond to the absence of glucose and give a normal phenotype. After co-immunoprecipitation, SIP1, SIP2 and GAL83 were not found in stochiometric amounts with SNF1 and SNF4, suggesting that the purified enzyme is a mixture of different combinations and that these proteins could participate in alternative forms of the SNF1 complex.

These proteins can interact directly and independently with SNF1 and SNF4. Carlson's group has shown by immunoprecipitation that the carboxy-terminal region (80 amino acids) of the SIP1 protein is sufficient for the interaction with the SNF1 complex *via* SNF4. This region was named the ASC domain (association with SNF1 complex) (Yang *et al.*, 1994; Jiang and Carlson, 1997). A second conserved domain was also defined. All three proteins present a conserved internal region named KIS (kinase interacting sequence) responsible for the interaction with the regulatory domain of SNF1 (Jiang and Carlson, 1997). It has been shown that the extreme C-terminal sequence of SNF1 (residues 515–633) is necessary and sufficient for the interaction with SIP2. The similarity between the three SIP proteins is much higher in the ASC and KIS domains than in the amino-terminal regions. These data suggest that one role of these proteins is to anchor SNF1 and SNF4 *via* the conserved ASC and KIS regions. Nevertheless, strains overexpressing SIP1 or GAL83 present distinct phenotypes with respect to *SUC* and *GAL* gene expression, suggesting specific functions for each of these proteins in addition to this common one. These proteins may serve as adaptors to direct the complex to specific targets by binding to the target or by directing the kinase to specific cellular locations. The high variability of the amino-terminal regions of SIP1, SIP2 and GAL83 suggests that these sequences could be involved in these specific functions.

The activity of the SNF1 protein kinase is regulated by glucose with an increase of activity in the absence of glucose in the medium (Wilson *et al.*, 1996). Co-immunoprecipitation experiments demonstrated that the components of the complex remain associated in the presence of high or low glucose (Yang *et al.*, 1992, 1994; Jiang and Carlson, 1997). However, it has been shown using the two-hybrid system that some inter- and intra-subunit interactions are regulated by the level of glucose. For example, the C-terminal regulatory region of SNF1 interacts with the N-terminal kinase catalytic domain in high glucose concentrations, while in low glucose concentrations it interacts with SNF4. The regions of SNF1 involved in these interactions overlap with each other and it has been suggested that interaction with SNF4 under low glucose conditions could counteract autoinhibition of SNF1 kinase activity (Jiang and Carlson, 1996). This hypothesis is in accordance with data obtained previously by Celenza and Carlson (1989) showing that a truncated SNF1 lacking the regulatory region was constitutively active in yeast cells.

B. THE MAMMALIAN AMPK COMPLEX

In mammals, the purification of AMPK revealed the existence of three prominent polypeptides of 63, 38 and 35 kDa (Carling *et al.*, 1989; Davies *et al.*, 1994; Mitchelhill *et al.*, 1994). The 63-kDa polypeptide was identified as the catalytic subunit and the 38-kDa and 35-kDa polypeptides were shown to co-migrate with the 63-kDa peptide, suggesting that they were subunits rather than contaminants. It is now established that these proteins represent the components of a heterotrimeric complex and the subunits have been named AMPKα (63 kDa), AMPKβ (38 kDa) and AMPKγ (35 kDa). The corresponding cDNAs have been cloned and shown to be homologues of SNF1 (AMPKα), SNF4 (AMPKγ) and SIP1/SIP2/GAL83 (AMPKβ) (Carling *et al.*, 1994; Gao *et al.*, 1995; Woods *et al.*, 1996). The C-terminal region of AMPKα has been shown to be involved in interactions with the other two subunits (Dyck *et al.*, 1996).

The function of the β and γ subunits remains unclear but, in contrast to the yeast SNF1 complex, co-expression of all three subunits is essential for significant kinase activity (Dyck *et al.*, 1996; Woods *et al.*, 1996). Moreover, α and γ subunits interact directly with β subunits, while no stable interaction is obtained between α and γ in CCL13 cells (Woods *et al.*, 1996). Although $\alpha\gamma$ complexes could be detected in COS cells, these findings suggest that the heterotrimeric complex is mediated by the β subunits (Dyck *et al.*, 1996). These results are consistent with those of the two-hybrid analysis of subunit interactions in the SNF1 system and it seems likely that the holoenzyme $\alpha\beta\gamma$ is more stable than the partial complexes $\alpha\beta$, $\alpha\gamma$ or $\beta\gamma$.

C. EVIDENCE FOR THE EXISTENCE OF A SnRK1 COMPLEX

Several lines of evidence suggest that SnRK1 also interacts with non-catalytic subunits. These are:

1. Complementation of *snf1* mutants by SnRK1 suggests that it interacts with yeast non-catalytic subunits to form an active complex (Alderson *et al.*, 1991; Muranaka *et al.*, 1994; Takano *et al.*, 1998; Bouly *et al.*, 1999).
2. Tobacco NPK5 was shown to interact with SNF4, SIP1 and SIP2 in two-hybrid experiments and the SNF4 subunit is required for the complementation of *snf1* by NPK5 (Muranaka *et al.*, 1994; Jiang and Carlson, 1997).
3. Purified cauliflower SnRK1 (HRK-A) was found to have a molecular mass of about 200 kDa (Ball *et al.*, 1995).

At the moment, non-catalytic additional subunits have not been purified from plant extracts. However, proteins interacting with the SnRK1 kinases and cDNAs encoding proteins showing some similarity with yeast SNF4 and SIP1/SIP2/GAL83 proteins have been characterized.

1. SNF4-Related Proteins in Plants

Two-hybrid system experiments indicated that both AKINα1 and NPK5 can interact with yeast SNF4 (Jiang and Carlson, 1997; Bouly *et al.*, 1999). Moreover, SNF4 is needed for complementation of *snf1* mutants by NPK5 (Muranaka *et al.*, 1994). This indicates that SnRK1 kinases are probably able to respond to the glucose-signalling pathway controlling the SNF1–SNF4 interaction in yeast. In addition, reported gene assays monitoring the interaction of an *Arabidopsis* SnRK1 with SNF4 revealed 40-fold higher activation in cultures growing under glucose limitation compared with cultures growing on glucose (Bhalerao *et al.*, 1999). These results suggest that plants are likely to contain a SNF4-like activating subunit. However, SNF4 and AMPKγ subunits share 46% amino acid sequence identity with each other when compared pair-wise using the GAP algorithm, and no plant protein has been identified so far that shows a comparable degree of similarity.

The apparent lack of a clear SNF4/AMPKγ homologue in plants is a puzzle, but there are three families of plant proteins that may be related to SNF4. The first is a family of proteins that comprises PV42 from bean (*Phaseolus vulgaris*) (Abe *et al.*, 1995), and two *Arabidopsis* open reading frames (ORFs; accession numbers AC007591 and AB017066) that could be identified in BLAST database searches using either PV42 or yeast SNF4 peptide sequences. PV42 shows 21% amino acid sequence identity with SNF4. The AC007591 and

AB017066 products show 63.3% and 27.8% amino acid sequence identity, respectively, with PV42, and 25.8% and 21.8% amino acid sequence identity, respectively, with SNF4. The second SNF4-like plant protein is AKINγ, an *Arabidopsis* protein that has been shown to interact with *Arabidopsis* SnRK1 in a two-hybrid assay (Bouly *et al.*, 1999). AKINγ shows 23% and 25% amino acid sequence identity with AMPKγ and SNF4, respectively. The third type of SNF4-like plant protein is a family that comprises BSnIP1 from barley (Slocombe, Bertini, Beaudoin, Dickinson and Halford, unpublished data) (accession numbers AJ007837 and AJ007838), and proteins encoded by sequences from maize (accession number AI691404), *Arabidopsis* (accession number AB016886) and poplar (accession number AI166543). BSnIP1 shares 19.6% amino acid sequence identity with SNF4 and 20.7% with AMPKγ.

These three types of putative SNF4-related proteins show no significant similarity with each other, and none of them is as closely related to SNF4 and AMPKγ as the yeast and animal proteins are to each other. However, they do all contain a conserved hydrophobic region of 17 residues with the motif: Hyd–Xxx–Aci–Xxx–Hyd–Xxx–Xxx–Hyd–Xxx–Bas–Xxx–Hyd–Xxx–Xxx–Xxx–Hyd–Hyd (where Hyd represents a hydrophobic residue, Xxx any residue, Aci an acidic residue and Bas a basic residue). This motif (the SnIP motif) includes a sequence that matches the consensus for the SnRK1 substrate recognition site (Halford and Hardie, 1998), with the exception of the lack of a target serine, and it may act as a pseudosubstrate site (Slocombe, Bertini, Beaudoin, Dickinson and Halford, unpublished data).

AKINγ also contains four repeat modules known as cystathionine-β-synthase domains (CBS) and previously described in SNF4 and AMPKγ. The function of these domains still remains unclear, but Bateman (1997) suggested that they might associate to form β-barrel structures.

2. *SIP1/SIP2/GAL83-Related Proteins in Plants*

Three cDNAs related to the SIP1/SIP2/GAL83/AMPKβ proteins have been characterized in plants; two *Arabidopsis* cDNAs, *AKINβ1* and *AKINβ2* (Bouly *et al.*, 1999), and one potato cDNA named *StubGAL83* (Lakatos *et al.*, 1999). The deduced proteins show similarity with SIP1/SIP2/GAL83/AMPKβ proteins restricted to the KIS and ASC domains. AKINβ1 and AKINβ2 interact with AKINα1 and AKINγ in the two-hybrid system but also with the yeast SNF1 and SNF4 proteins (Bouly *et al.*, 1999). Potato *StubGAL83* was isolated by screening of a yeast two-hybrid cDNA library with *StubSNF1*, a potato *SnRK1* cDNA (Lakatos *et al.*, 1999). This interaction has been confirmed by *in vitro* binding experiments. Interestingly, it has been shown that StubGAL83 is also able to interact with yeast SNF4, while StubSNF1 does not interact independently with SNF4, suggesting that StubGAL83 is required for StubSNF1/SNF4 interaction (Lakatos *et al.*, 1999). These data suggest that the ASC- and KIS-related domains of the plant proteins could have the same anchoring function as their yeast and animal counterparts. Moreover, additional sequences related to the SIP family have been found in *Arabidopsis* (AKINβ3 and AKINβ4) (Gissot and Thomas, unpublished results) and reported as expressed sequence tags in databases from other species (tomato, poplar, rice and bean).

At the present time (1999), we do not have any biochemical evidence that these are regulatory subunits and the physiologically role of the different subunits and isoforms has not been determined. However, we have shown that the *AKINβ1* and *AKINβ2* genes are differentially expressed during development. Moreover, they show contrasting responses to light. A high accumulation of AKINβ1 transcripts is detectable in plants 30 min after transfer from light to dark, whereas there is a decrease in AKINβ2 transcripts during the same treatment. These findings suggest transcriptional regulation of these subunits and indirectly of the SnRK1 complex. It is possible that AKINβ subunits direct the kinase activity to different targets, or have a role in the response to signals such as light (Bouly *et al.*, 1999).

3. *Prl1, Another Protein that Interacts with SnRK1*

Recently, AKINα1 and AKINα2 have been isolated in the course of a two-hybrid screen using PRL1 as bait (Bhalerao *et al.*, 1999). PRL1 encodes a protein of 54 kDa that carries WD-40 repeats similar to those of the TUP1 and WD-repressor proteins (Nemeth *et al.*, 1998). This is intriguing because one of the substrates of SNF1 is a transcription factor, MIG1, that acts as a repressor of glucose repressible genes by recruiting the repressor complex TUP1/SSN6. In *Arabidopsis*, the mutation of *PRL1* has pleiotropic effects in sugar-, light- and stress-regulated gene expression, as well as in starch and sugar accumulation and in response to several phytohormones. These effects are detectable essentially in light-grown plants, suggesting light dependence of PRL1 regulatory functions. The PRL1 protein has also been shown to interact with an α-importin nuclear import receptor (Nemeth *et al.*, 1998).

The binding of PRL1 with AKINα proteins was found to be enhanced in the absence of glucose and binding was shown to inhibit autophosphorylation of the kinase expressed in *Escherichia coli*. Activity in immunoprecipitated SnRK1 preparations from *Arabidopsis* was also slightly higher in a *PRL1* mutant than in wild type (Bhalerao *et al.*, 1999). However, in both cases the kinase activity that was measured was several orders of magnitude lower than the activities measured previously in plant extracts and purified SnRK1 preparations, and the significance of the PRL1/AKINα, therefore, still requires further investigation.

VI. FUNCTIONS OF SnRK1s

A combination of *in vitro* biochemical studies and analyses of transgenic plants in which SnRK1 activity has been reduced is beginning to provide information on the functions of SnRK1s in plants. In order to put the results of these studies into context, it is worthwhile summarizing briefly what is known about the functions of SNF1 and AMPK in their respective systems.

A. THE YEAST SNF1 SYSTEM

The preferred carbon source for budding yeast (*Saccharomyces cerevisiae*) is glucose, and if adequate glucose is available, the expression of a large battery of genes is repressed, including those required for growth on sucrose, galactose and maltose, and those required for growth on non-fermentable carbon sources such as glycerol and ethanol (Ronne, 1995; Gancedo, 1998). This process is called glucose repression and its reversal when the yeast is deprived of glucose is known as glucose derepression. SNF1 (Celenza and Carlson, 1986) is required for derepression of essentially all glucose-repressed genes. It also directly modulates the phosphorylation state of a number of metabolic enzymes, including acetyl-CoA carboxylase (Woods *et al.*, 1994) and glycogen synthase (Hardy *et al.*, 1994), and is required for the arrest of growth and the cell cycle under conditions of glucose deprivation (Thompson-Jaeger *et al.*, 1991). The cell cycle is also affected by disruption of the *SNF4* gene, confirming a role for the SNF1 complex in cell cycle control (Aon and Cortassa, 1999).

B. THE MAMMALIAN AMP-ACTIVATED PROTEIN KINASE SYSTEM

The AMP-activated protein kinase (AMPK) was originally detected in the form of crude protein fractions, which caused time- and ATP-dependent inactivation of HMG-CoA reductase (Beg *et al.*, 1973) or acetyl-CoA carboxylase (Carlson and Kim, 1973). It was not until the late 1980s that it was realized that these were functions of the same kinase (Carling *et al.*, 1987; Hardie *et al.*, 1989), and the early 1990s before AMPK was found to be related to SNF1 and the SnRKs (Carling *et al.*, 1994; Mitchelhill *et al.*, 1994). AMPK is activated by AMP (Carling *et al.*, 1987, 1989) and by phosphorylation by an upstream protein kinase [AMP-activated protein kinase kinase (AMPKK)] (Hawley *et al.*, 1996). Activation of AMPK by AMP is described in more detail elsewhere (Hardie and Carling, 1997; Halford and Hardie, 1998). It is antagonized by high (mM) concentrations of ATP. A high AMP:ATP ratio is symptomatic of low cellular energy levels and, for this reason, Hardie and Carling likened AMPK to a cellular fuel gauge (Hardie and Carling, 1997). When activated, it acts to conserve ATP by phosphorylating and inactivating regulatory enzymes of ATP-consuming, anabolic pathways such as acetyl-CoA carboxylase (fatty acid synthesis) (Davies *et al.*, 1990, 1992) and HMG-CoA reductase (sterol/isoprenoid synthesis) (Clarke and Hardie, 1990; Gillespie and Hardie, 1992). Evidence is also emerging to suggest that AMPK regulates gene expression, but this evidence has yet to be published.

C. IDENTIFICATION OF SUBSTRATES FOR SnRK1s

1. *Peptide Substrates*

A minimal recognition motif for SnRK1s was established in Hardie's laboratory using variants of the SAMS peptide (Weekes *et al.*, 1993). In order to act as good substrates, peptides must have bulky hydrophobic residues (Met, Leu, Ile, Phe or Val) at positions P − 5 and P + 4, where P is the phosphorylated serine. There must also be at least one basic residue (R > K > H) which could be at P − 3 or P − 4. The minimal recognition motif is, therefore, as shown in Fig. 4.

SnRK1s will phosphorylate threonine but phosphorylate serine much more efficiently. This is the only difference between the preferences for peptide substrates of SnRK1s and AMPK. The AMARA peptide (Ala Met Ala Arg Ala Ala Ser Ala Ala Ala Leu Ala Arg Arg Arg), in which the minimal recognition motif is retained but other residues are alanine, apart from the basic C-terminus, which is not essential, appears to be a better substrate than the SAMS peptide (Dale *et al.*, 1995b).

2. *Metabolic Enzyme Targets*

The first plant metabolic enzyme to be tested as a substrate for SnRK1 was HMG-CoA reductase. This enzyme catalyses the NADH-dependent reduction of HMG-CoA to mevalonic acid. This is the overall rate-limiting step for the whole isoprenoid biosynthetic pathway in animals and yeast, and the activity of HMG-CoA reductase in animals is regulated through transcription and translation (reviewed by Goldstein and Brown, 1990) and by turnover rates (Chun and Simoni, 1992; Correll and Edwards, 1994), as well as through

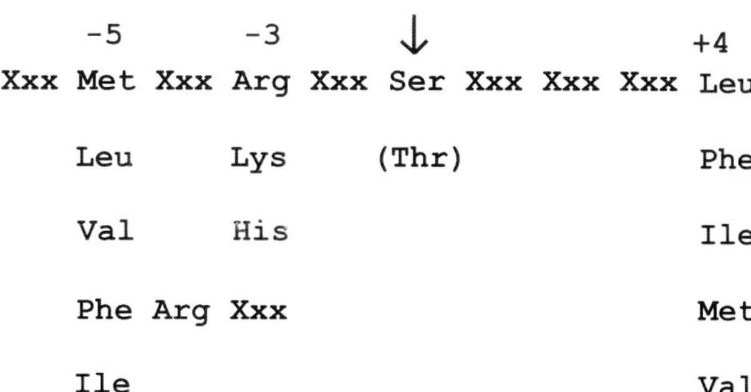

Fig. 4. Minimal recognition motif for SnRK1s.

phosphorylation by AMPK. The situation in plants is complicated by the existence of a second pathway for the synthesis of isopentenyl diphosphate (the precursor of all isoprenoids in eukaryotes) alongside the mevalonate pathway (Eisenrach *et al.*, 1996). However, mevalonate synthesis is the main pathway by which phytosterols and carotenoids are synthesized, and it is widely accepted that HMG-CoA reductase plays an important regulatory role in phytosterol accumulation (Bach, 1995). This view was supported by the results of experiments in which an animal HMG-CoA reductase gene was overexpressed in transgenic tobacco (Chappell *et al.*,1995). Isoprenoid synthesis was enhanced, although the nature of the isoprenoids that accumulated suggested that additional controls operate at later branch points in the pathway.

Hardie's group tested HMG-CoA reductase as a substrate because mammalian HMG-CoA reductase is a substrate for AMPK. Cauliflower SnRK1 (HRK-A) was found to phosphorylate mammalian HMG-CoA reductase at the same site as AMPK (MacKintosh *et al.*, 1992), and subsequent experiments with the catalytic domain of *Arabidopsis* HMG-CoA reductase expressed in *E. coli* showed that it, too, would act as a substrate and was inactivated by the phosphorylation. The phosphorylation site was serine-577 (Ser577) (Dale *et al.*, 1995a), equivalent to the single site for AMPK at Ser871 on the Chinese hamster enzyme (Clarke and Hardie, 1990). The sequence around this site fits the minimal recognition motif for SnRK1s (Section VI.C.1) and is conserved in almost all plant HMG-CoA reductases (Halford and Hardie, 1998), so it seems likely that HMG-CoA reductase is a physiological target for the SnRK1 subfamily.

Two other very important biosynthetic enzymes in the cytoplasm of plants are sucrose phosphate synthase (SPS) and nitrate reductase (NR). SPS catalyses a key step in sucrose biosynthesis in source tissues, prior to export of the sugar to sink tissues such as roots, storage organs such as tubers and seeds, and growing meristems, whereas NR catalyses the first step in the assimilation of nitrogen from nitrate into nitrogen-containing organic compounds such as amino acids. Both enzymes have been purified from spinach leaf and shown to be regulated by phosphorylation by multiple protein kinases from the same source (McMichael *et al.*, 1995), but until recently these kinases were not characterized at the molecular level.

SPS is phosphorylated at Ser158 and NR at Ser543 (Douglas *et al.*, 1995; Bachmann *et al.*, 1996b; Su *et al.*, 1996). In both cases phosphorylation results in inactivation of the enzyme, although the inactivation of NR also requires the binding of a 14-3-3 protein to the phosphorylation site (Bachmann *et al.*, 1996a; Moorhead *et al.*, 1996). The sequences around both sites conform to the SnRK1 consensus recognition motif. Immunological evidence that one of three protein kinases from spinach leaf shown to phosphorylate NR was a SnRK1 was obtained by Douglas *et al.* (1997) and recently two spinach leaf SnRK1s, HRK-A and -C, were shown to inactivate, by phosphorylation, both SPS and NR (Sugden *et al.*, 1995b).

D. SnRK1s REGULATE GENE EXPRESSION

As we have described in Section VI.A, one of the functions of SNF1 in yeast is the transcriptional regulation of genes encoding enzymes of carbohydrate metabolism and a key question regarding SnRK1s is whether or not they play an analogous role in plants. Expression of the rye *SnRK1* cDNA, *RKIN1*, in a yeast *snf1* mutant restored SNF1 function to the extent that the yeast could utilize non-fermentable carbon sources such as ethanol and glycerol (Alderson *et al.*, 1991). A similar experiment has been performed with the tobacco *SnRK1*, *NPK5*, using sucrose utilization as the selectable marker (Muranaka *et al.*, 1994). The *snf1* mutant yeast expressing the NPK5 protein kinase was able to grow on the sucrose medium and was shown to contain an invertase activity. These complementation experiments showed that plant SnRK1s can substitute for SNF1 in the sugar-sensing signalling pathway in yeast, suggesting that an analogous signalling system might exist in plant cells. Further studies with NPK5 showed that it rescued *snf1* mutants in a glucose-regulated manner, indicating that the regulation of the yeast and plant systems may be very similar (Jiang and Carlson, 1996).

This has been investigated further by expressing an antisense potato *SnRK1* (*PKIN1*) sequence in the tubers and leaves of transgenic potato (Halford *et al.*, 1994; Purcell *et al.*, 1998). This resulted in a reduction of up to 79% in SAMS peptide kinase activity in the tubers, confirming that PKIN1 is responsible for most if not all of the SAMS peptide kinase activity in this tissue. Since SNF1 controls the expression of the invertase gene, *SUC2*, in yeast (Neigeborn and Carlson, 1984), it made sense to look for effects of antisense *PKIN1* expression on invertase activity and the activity of sucrose synthase (since sucrose synthase also catalyses the conversion of sucrose to hexoses). Fructokinase and glucokinase activities were also measured, since these might be affected by changes in hexose concentrations. No change in invertase, glucokinase or fructokinase activity was observed in the transgenic plants, but the activity of sucrose synthase was found to have decreased in one of the lines expressing the antisense sequence in the tubers. On further investigation, sucrose synthase *gene expression* was shown to have decreased dramatically in transgenic tubers expressing the antisense sequence (Fig. 5). Sucrose synthase gene expression was also found to be uninducible by sucrose in excised leaves expressing the antisense sequence. In wild-type plants, the sucrose synthase gene, *Sus4*, is expressed in tubers and is induced in excised leaves by incubation with sucrose (Fu and Park, 1995).

These experiments showed clearly that SnRK1 activity is required for normal sucrose synthase gene expression in potato. Given that SNF1 directly regulates gene expression in yeast, and that sucrose synthase activity in the transgenic tubers was not reduced sufficiently to affect tuber development or composition (making it less likely that the reduction in sucrose synthase gene expression resulted from a change in the concentration of a metabolite), the

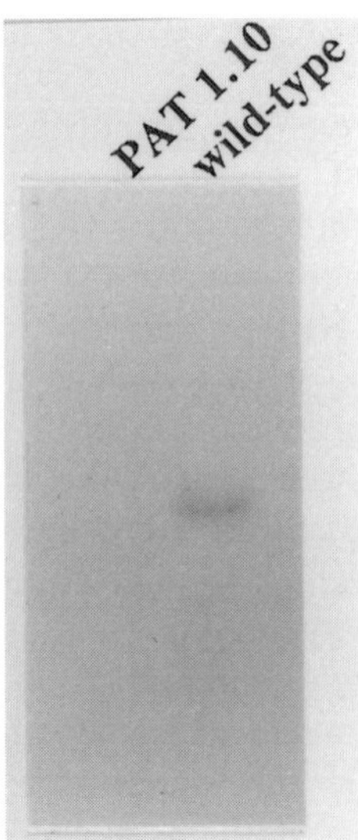

Fig. 5. Northern analysis of total RNA from mature tubers of wild-type potato and transgenic line PAT 1.10 in which SnRK1 activity had been reduced by approximately 80% by antisense expression of a *SnRK1* sequence (Purcell *et al.*, 1998). The blot was hybridized with a radiolabelled probe prepared from a potato sucrose synthase PCR product.

results support the hypothesis that SnRK1s play a role in the control of carbohydrate metabolism through the direct regulation of sucrose synthase gene expression. Sucrose synthase is the most important sucrose cleavage enzyme in potato tubers (Morrell and ap Rees, 1986; Sung *et al.*, 1989) and its activity is closely correlated with sink strength and starch accumulation (Zrenner *et al.*, 1995). It still has to be determined whether SnRK1s regulate the expression of other genes. It is quite possible that invertase gene expression, for example, is not under SnRK1 control in the tissues that were analysed in these experiments, but is affected by SnRK1 activity elsewhere in the plant or at different developmental stages. Changes in invertase gene expression have been observed in transgenic *Arabidopsis* plants overexpressing AKINα1 (Lessard and Thomas, unpublished results) but this is still under investigation.

E. SnRK1s AND SUGAR SENSING IN PLANT CELLS

The results of the experiments described in Section VI.D implicate SnRK1s in the *sucrose*-sensing signalling pathway that controls sucrose synthase gene expression in potato leaves and tubers. Sugar sensing clearly plays an important role in plant cell metabolism and the expression of a large number of genes, including many involved in photosynthesis and carbohydrate metabolism, is affected by sugars. The hypothesis that SnRK1s might be involved in sugar sensing in plants was first proposed when the rye *SnRK1, RKIN1*, was shown to complement the *snf1* mutation in yeast. However, there are some intriguing questions still to be answered.

In yeast, the repression of genes required for growth on carbon sources other than glucose and their depression when glucose is removed is essentially a response to the availability of glucose (Gancedo, 1998). The intracellular signals that mediate repression and derepression of genes in response to glucose in yeast remain unknown, but the fact that a functional *SNF1* gene is essential for derepression of all glucose-repressed genes is not disputed. This places SNF1 in the *glucose*-sensing system of yeast.

One clear difference between yeast and plant cells is the importance of sucrose as a carbon source. In plants it is the major transported sugar, linking carbon source organs (mature leaves) with carbon sinks (developing leaves, roots and storage organs such as seeds and tubers). Concentrations of sucrose and hexoses may be linked in plant cells, depending on the sucrose-metabolizing enzymes present. It is not surprising, therefore, that sucrose and glucose treatments have been shown to affect in a similar manner β-amylase gene expression, for example (Mita *et al.*, 1995). However, it is clear that sucrose itself can act as an effector in plant cells, generating a signal independent of that generated by glucose. The regulation of sucrose synthase gene expression in potato is a good example, since it is induced by sucrose but not by glucose (Salanoubat and Belliard, 1989; Fu and Park, 1995), and clearly, this sucrose induction requires SnRK1 activity. The evidence available so far, therefore, places SnRK1 in a *sucrose*- rather than *glucose*-sensing signalling pathway in plant cells, and it is not clear exactly how this relates to the glucose-sensing system of yeast. This is discussed in more detail by Halford *et al.* (1999).

F. EVIDENCE THAT SnRK1s ARE A CONDUIT FOR CROSS-TALK BETWEEN METABOLIC AND CELL CYCLE SIGNALLING

Several yeast protein kinases related to SNF1 are involved in cell cycle control. They include NIM1 from fission yeast (*Schizosaccharomyces pombe*) and its budding yeast homologue NIK1 (nim1-like kinase). NIM1 is a mitotic inducer in fission yeast that phosphorylates and thereby inactivates the cell cycle inhibitory protein kinase, WEE1 (Russell and Nurse, 1987). NIK1 is a negative regulator of SWE1 (*Saccharomyces* homologue of wee1) that is involved in calcium-dependent control of mitosis (Tanaka and Nojima, 1996). NIK1 interacts with the CDC28 complex and functions not only at the G2/M transition but also at other points in the cell cycle (Tanaka and Nojima, 1996).

As we have described above, SnRK1s have approximately 67% amino acid sequence identity with SNF1 in the catalytic domain. They also show striking similarity with NIM1 and NIK1 (Fig. 3), with approximately 50% amino acid sequence identity in the catalytic domain (Dickinson *et al.*, 1999). In addition, SNF1 itself has been implicated in cell cycle control, since *snf1* mutants fail to arrest the cell cycle when they are in a state of nutrient deprivation, in other words they do not enter a normal stationary phase (Thompson-Jaeger *et al.*, 1991). Other data suggest that SNF1 is also involved in the control of the onset of meiosis (Honigberg and Lee, 1998).

Dickinson *et al.* (1999) looked for cell cycle effects of RKIN1 overexpression in yeast and found that it resulted in a dramatic reduction in yeast cell size, suggesting that the yeast cells were completing their cell cycles too early. One possible interpretation of these results is that RKIN1 was behaving like NIM1 or NIK1 in the yeast cells. Certainly, overexpression of NIK1 has been shown to have a similar effect (Tanaka and Nojima, 1996). Another interpretation is that acting as a conduit for cross-talk between metabolic and cell cycle signalling is part of the normal activity of SnRK1 and SNF1, and that this system was perturbed in the yeast cells by overexpression of the SnRK1 protein.

VII. CONCLUSION – SnRK1s ARE GLOBAL REGULATORS OF CARBON METABOLISM IN PLANTS AND ARE AT THE HEART OF CARBON PARTITIONING

The identification of physiological substrates for SnRK1s is continuing but, as described in Section VI.C.2, three important enzymes of plant metabolism have already been shown to act as substrates *in vitro*. These are HMG-CoA reductase, sucrose phosphate synthase and nitrate reductase. Although the nature of stimulus that leads to induction of SnRK1 activity has not been established, our hypothesis is that SnRK1s are activated in response to changes in the carbon status of the cell. They act to conserve carbon by inactivating biosynthetic enzymes. It may seem strange that nitrate reductase activity could be regulated in part by such a system (it is also regulated in response to nitrogen availability and other factors), but incorporation of nitrogen into organic compounds does require carbon. Antisense expression of a *SnRK1* sequence in potato demonstrated clearly that, like SNF1, SnRK1s also regulate carbon metabolism through the control of gene expression. A decrease in sucrose synthase *gene expression* occurred in transgenic tubers and in excised transgenic leaves treated with sucrose.

The known functions of SnRK1s are summarized in Fig. 6 and we can already propose that SnRK1 protein kinases are global regulators of carbon metabolism in plants and that they are at the heart of the control of carbon partitioning. The manipulation of carbon partitioning in the harvested organs (such as seeds, tubers and fruits) of crop plants is an important biotechnological target (reviewed by Halford, 1999). Carbon assimilate entering these sink organs is partitioned between compounds for general and storage use, and between different storage compounds (sugars, starch, storage proteins and oils) and micro-nutrients (Fig. 7). The relative amount of each of these compounds differs greatly between different crops and varieties, and is influenced by environmental conditions and farming practice. It has great bearing on the processing properties and nutritional quality of a crop and, therefore, its value; plant breeders and farmers have long sought to control it. The genetic modification of SnRK1 activity and function may be one way of improving this important trait in crop plants.

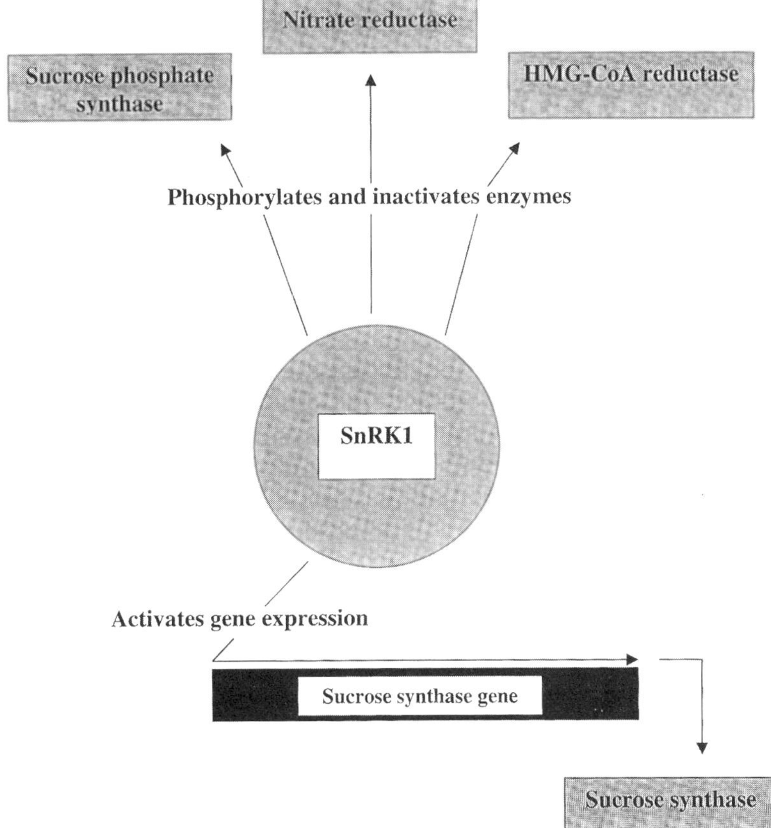

Fig. 6. The known targets for SnRK1. Sucrose phosphate synthase, nitrate reductase and HMG-CoA reductase are regulated post-translationally, while sucrose synthase is regulated at the level of gene expression. These activities put SnRK1 at the heart of the control of carbon partitioning in plants.

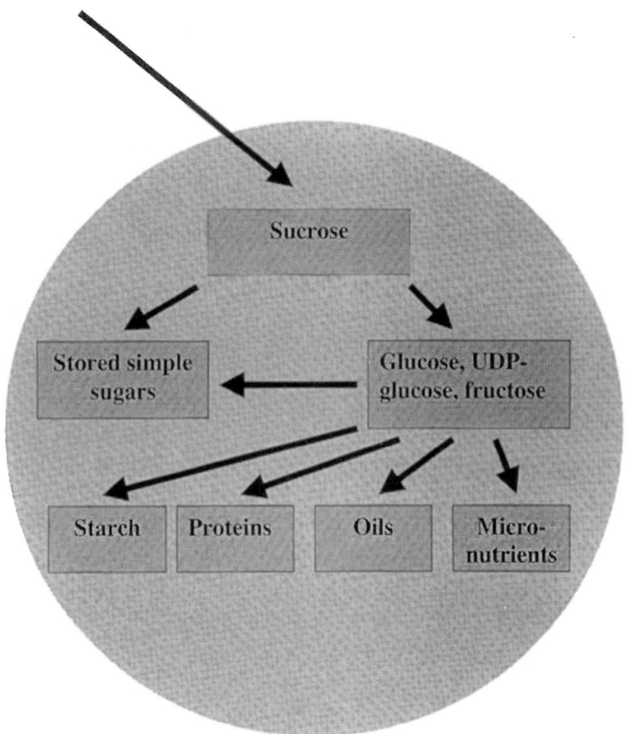

Fig. 7. Carbon partitioning in a sink organ such as a seed or tuber. Carbon is imported as sucrose from photosynthesizing leaves and used to make a range of products. The relative amounts of these products at harvest influence processing and nutritional properties.

ACKNOWLEDGEMENTS

IACR receives grant-aided support from the Biotechnology and Biological Sciences Research Council of the United Kingdom. M. Thomas thanks the EU for their support (B104-CT96-0311). J. P. Bouly was supported by the Ministère de l'Education et de la Recherche, France.

REFERENCES

Abe, H., Kamiya, Y. and Sakurai, A. A. (1995). cDNA clone encoding yeast SNF4-like protein from *Phaseolus vulgaris*. *Plant Physiology* **110**, 336.

Alderson, A., Sabelli, P. A., Dickinson, J. R., Cole, D., Richardson, M., Kreis, M., Shewry, P. R. and Halford, N. G. (1991). Complementation of *snf1*, a mutation affecting global regulation of carbon metabolism in yeast, by a plant protein kinase cDNA. *Proceedings of the National Academy of Sciences of the United States of America* **88**, 8602–8605.

Anderberg, R. J. and Walker-Simmons, M. K. (1992). Isolation of a wheat cDNA clone for an abscisic acid-inducible transcript with homology to protein kinases. *Proceedings of the National Academy of Sciences of the United States of America* **89**, 10183–10187.

Annen, F. and Stockhaus, J. (1998). Characterization of a *Sorghum bicolor* gene family encoding putative protein kinases with a high similarity to the yeast SNF1 protein kinase. *Plant Molecular Biology* **36**, 529–539.

Aon, M. A. and Cortassa, S. (1999). Quantitation of the effects of disruption of catabolite (de)repression genes on the cell cycle behavior of *Saccharomyces cerevisiae*. *Current Microbiology* **38**, 57–60.

Bach, T. J. (1995). Some new aspects of isoprenoid biosynthesis in plants – a review. *Lipids* **30**, 191–202.

Bachmann, M., Huber, J. L., Liao, P. C., Gage, D. A. and Huber, S. C. (1996a). The inhibitor protein of phosphorylated nitrate reductase from spinach (*Spinaea oleracea*) leaves is a 14-3-3 protein. *FEBS Letters* **387**, 127–131.

Bachmann, M., Shiraishi, N., Campbell, W. H., Yoo, B. C., Harmon, A. C. and Huber, S. C. (1996b). Identification of Ser-543 as the major regulatory phosphorylation site in spinach leaf nitrate reductase. *Plant Cell* **8**, 505–517.

Ball, K. L., Dale, S., Weekes, J. and Hardie, D. G. (1994). Biochemical characterization of two forms of 3-hydroxy-3-methylglutaryl-CoA reductase kinase from cauliflower (*Brassica oleracea*). *European Journal of Biochemistry* **219**, 743–750.

Ball, K. L., Barker, J., Halford, N. G. and Hardie, D. G. (1995). Immunological evidence that HMG-CoA reductase kinase-A is the cauliflower homologue of the RKIN1 subfamily of plant protein kinases. *FEBS Letters* **377**, 189–192.

Barker, J. H. A., Slocombe, S. P., Ball, K. L., Hardie, D. G., Shewry, P. R. and Halford, N. G. (1996). Evidence that barley 3-hydroxy-3-methylglutaryl-coenzyme A reductase kinase is a member of the sucrose nonfermenting-1-related protein kinase family. *Plant Physiology* **112**, 1141–1149.

Bateman, A. (1997). The structure of a domain common to archaebacteria and the homocystinuria disease protein. *Trends in Biochemical Sciences* **22**, 12–13.

Beg, Z. H., Allmann, D. W. and Gibson, D. M. (1973). Modulation of 3-hydroxy-3-methylglutaryl coenzyme A reductase activity with cAMP and with protein fractions of rat liver cytosol. *Biochemical and Biophysical Research Communications* **54**, 1362–1369.

Bhalerao, R. P., Salchert, K., Bako, L., Okresz, L., Szabados, L., Muranaka, T., Machida, Y., Schell, J. and Koncz, C. (1999). Regulatory interaction of PRL1 WD protein with *Arabidopsis* SNF1-like protein kinases. *Proceedings of the National Academy of Sciences of the United States of America* **96**, 5322–5327.

Bouly, J. P., Gissot, L., Lessard, P., Kreis, M. and Thomas, M. (1999). *Arabidopsis thaliana* homologues of the yeast SIP and SNF4 proteins interact with AKINα1, a SNF1-like protein kinase. *Plant Journal* **18**, 541–550.

Carling, D., Zammit, V. A. and Hardie, D. G. (1987). A common bicyclic protein kinase cascade inactivates the regulatory enzymes of fatty acid and cholesterol biosynthesis. *FEBS Letters* **223**, 217–222.

Carling, D., Clarke, P. R., Zammit, V. A. and Hardie, D. G. (1989). Purification and characterization of the AMP-activated protein kinase – co-purification of acetyl-CoA carboxylase and 3-hydroxy-3-methylglutaryl CoA reductase kinase activity. *European Journal of Biochemistry* **186**, 129–136.

Carling, D., Aguan, K., Woods, A., Verhoeven, A. J., Beri, R. K., Brennan, C. H., Sidebottom, C., Davison, M. D. and Scott, J. (1994). Mammalian AMP-activated protein kinase is homologous to yeast and plant protein kinases involved in the regulation of carbon metabolism. *Journal of Biological Chemistry* **269**, 11442–11448.

Carlson, C. A. and Kim, K. H. (1973). Regulation of hepatic acetyl coenzyme A carboxylase by phosphorylation and dephosphorylation. *Journal of Biological Chemistry* **248**, 378–380.

Celenza, J. L. and Carlson, M. (1986). A yeast gene that is essential for release from glucose repression encodes a protein kinase. *Science* **233**, 1175–1180.

Celenza, J. L. and Carlson, M. (1989). Mutational analysis of the *Saccharomyces cerevisiae* SNF1 protein kinase and evidence for functional interaction with the SNF4 protein. *Molecular and Cellular Biology* **9**, 5034–5044.

Chappell, J., Wolf, F., Proulx, J., Cuellar, R. and Saunders, C. (1995). Is the reaction catalyzed by 3-hydroxy-3-methylglutaryl coenzyme A reductase a rate-limiting step for isoprenoid biosynthesis in plants? *Plant Physiology* **109**, 1337–1343.

Chun, K. T. and Simoni, R. D. (1992). The role of the membrane domain in the regulated degradation of 3-hydroxy-3-methylglutaryl coenzyme A reductase. *Journal of Biological Chemistry* **267**, 4236–4246.

Clarke, P. R. and Hardie, D. G. (1990). Regulation of HMG-CoA reductase: identification of the site phosphorylated by the AMP-activated protein kinase *in vitro* and in intact rat liver. *EMBO Journal* **9**, 2439–2446.

Correll, C. C. and Edwards, P. A. (1994). Mevalonic acid-dependent degradation of 3-hydroxy-3-methylglutaryl-coenzyme A reductase *in vivo* and *in vitro*. *Journal of Biological Chemistry* **269**, 633–638.

Dale, S., Arró, M., Becerra, B., Morrice, N. G., Boronat, A., Hardie, D. G. and Ferrer, A. (1995a). Bacterial expression of the catalytic domain of 3-hydroxy-3-methylglutaryl CoA reductase (isoform HMGR1) from *Arabidopsis*, and its inactivation by phosphorylation at serine-577 by *Brassica oleracea* 3-hydroxy-3-methylglutaryl CoA reductase kinase. *European Journal of Biochemistry* **233**, 506–513.

Dale, S., Wilson, W. A., Edelman, A. M. and Hardie, D. G. (1995b). Similar substrate recognition motifs for mammalian AMP-activated protein kinase, higher plant HMG-CoA reductase kinase-A, yeast SNF1, and mammalian calmodulin-dependent protein kinase I. *FEBS Letters* **361**, 191–195.

Davies, S. P., Carling, D. and Hardie, D. G. (1989). Tissue distribution of the AMP-activated protein kinase, and lack of activation by cyclic AMP-dependent protein kinase, studies using a specific and sensitive peptide assay. *European Journal of Biochemistry* **186**, 123–128.

Davies, S. P., Sim, A. T. R. and Hardie, D. G. (1990). Location and function of three sites phosphorylated on rat acetyl-CoA carboxylase by the AMP-activated protein kinase. *European Journal of Biochemistry* **187**, 183–190.

Davies, S. P., Carling, D., Munday, M. R. and Hardie, D. G. (1992). Diurnal rhythm of phosphorylation of rat-liver acetyl-CoA carboxylase by the AMP-activated protein kinase, demonstrated using freeze-clamping – effects of high-fat diets. *European Journal of Biochemistry* **203**, 615–623.

Davies, S. P., Hawley, S. A., Woods, A., Carling, D., Haystead, T. A. and Hardie D. G. (1994). Purification of the AMP-activated protein kinase on ATP-gammasepharose and analysis of its subunit structure. *European Journal of Biochemistry* **223**, 351–357.

Dickinson, J. R., Cole, D. and Halford, N. G. (1999). A cell cycle role for a plant sucrose non-fermenting-1 related protein kinase (SnRK1) is indicated by expression in yeast. *Plant Growth Regulation* **28**, 169–174.

Douglas, P., Morrice, N. and MacKintosh, C. (1995). Identification of a regulatory phosphorylation site in the hinge 1 region of nitrate reductase from spinach (*Spinacea oleracea*) leaves. *FEBS Letters* **377**, 113–117.

Douglas, P., Pigaglio, E., Ferrer, A., Halford, N. G. and MacKintosh, C. (1997). Three spinach leaf nitrate reductase/3-hydroxy-3-methylglutaryl-CoA reductase kinases that are regulated by reversible phosphorylation and/or Ca^{2+} ions. *Biochemical Journal* **325**, 101–109.

Dyck, J. R. B., Gao, G., Widmer, J., Stapleton, D., Fernandez, C. S., Kemp, B. E. and Witters, L. A. (1996). Regulation of 5′-AMP-activated protein kinase activity by the noncatalytic beta and gamma subunits. *Journal of Biological Chemistry* **271**, 17798–17803.

Eisenrach, W., Menhard, B., Hylands, P. J., Zenk, M. H. and Bacher, A. (1996). Studies on the biosynthesis of taxol: the taxane carbon skeleton is not of mevalonoid origin. *Proceedings of the National Academy of Sciences of the United States of America* **93**, 6431–6436.

Fields, S. and Song, O. (1989). A novel genetic system to detect protein–protein interactions. *Nature* **340**, 245–246.

Fu, H. and Park, W. D. (1995). Sink- and vascular-associated sucrose synthase functions are encoded by different gene classes in potato. *Plant Cell* **7**, 1369–1385.

Gancedo, J. M. (1998). Yeast carbon catabolite repression. *Microbiology and Molecular Biology Reviews* **62**, 334–361.

Gao, G., Widmer, J., Stapleton, D., Teh, T., Cox, T., Kemp, B. E. and Witters, L. A. (1995). Catalytic subunits of the porcine and rat 5′-AMP-activated protein kinases are members of the SNF1 protein kinase family. *Biochimica et Biophysica Acta* **1266**, 73–82.

Genetics Computer Group (1994). "Program Manual for the Wisconsin Package, Version 8". Genetics Computer Group, Madison, Wisconsin.

Gillespie, J. G. and Hardie, D. G. (1992). Phosphorylation and inactivation of HMG-CoA reductase at the AMP-activated protein kinase site in response to fructose treatment of isolated rat hepatocytes. *FEBS Letters* **306**, 59–62.

Goldstein, J. L. and Brown, M. S. (1990). Regulation of the mevalonate pathway. *Nature* **343**, 425–430.

Gómez-Cadenas, A., Verhey, S. D., Holappa, L. D., Shen, Q., Ho, T.-H. D. and Walker-Simmons, M. K. (1999). An abscisic acid-induced protein kinase, PKABA1, mediates abscisic acid-suppressed gene expression in barley aleurone layers. *Proceedings of the National Academy of Sciences of the United States of America* **96**, 1767–1772.

Halford, N. G. (1999). Metabolic signaling and the partitioning of resources in plant storage organs. *Journal of Agricultural Science* **133**, 243–249.

Halford, N. G. and Hardie, D. G. (1998). SNF1-related protein kinases: global regulators of carbon metabolism in plants? *Plant Molecular Biology* **37**, 735–748.

Halford, N. G., Vicente-Carbajosa, J., Shewry, P. R., Hannappel, U. and Kreis, M. (1992). Molecular analyses of a barley multigene family homologous to the yeast protein kinase gene SNF1. *Plant Journal* **2**, 791–797.

Halford, N. G., Man, A. L., Barker, J. H. A., Monger, W., Shewry, P. R., Smith, A. M. and Purcell, P. C. (1994). Investigating the role of plant SNF1-related protein kinases. *Biochemical Society Transactions* **22**, 953–957.

Halford, N. G., Purcell, P. C. and Hardie, D. G. (1999). Is hexokinase really a sugar sensor in plants? *Trends in Plant Science* **4**, 117–120.

Hanks, S. K. and Hunter, T. (1995). The eukaryotic protein kinase superfamily. *In* "The Protein Kinase Factsbook", Vol. I (D. G. Hardie and S. Hanks, eds) pp. 7–47. Academic Press, London.

Hanks, S. K., Quinn, A. M. and Hunter, T. (1988). The protein kinase family: conserved features and deduced phylogeny of the catalytic domains. *Science* **241**, 42–52.

Hannappel, U., Vicente-Carbajosa, J., Barker, J. H. A., Shewry, P. R. and Halford, N. G. (1995). Differential expression of two barley *SNF1*-related protein kinase genes. *Plant Molecular Biology* **27**, 1235–1240.

Hardie, D. G. and Carling, D. (1997). The AMP-activated protein kinase: fuel gauge of the mammalian cell? *European Journal of Biochemistry* **246**, 259–273.

Hardie, D. G., Carling, D. and Sim, A. T. R. (1989). The AMP-activated protein kinase – a multisubstrate regulator of lipid metabolism. *Trends in Biochemical Sciences* **14**, 20–23.

Hardy, T. A., Huang, D. and Roach, P. J. (1994). Interactions between cAMP-dependent and SNF1 protein kinases in the control of glycogen accumulation in *Saccharomyces cerevisiae. Journal of Biological Chemistry* **269**, 27907–27913.

Hawley, S. A., Davison, M., Woods, A., Davies, S. P., Beri, R. K., Carling, D. and Hardie, D. G. (1996). Characterization of the AMP-activated protein kinase kinase from rat liver, and identification of threonine-172 as the major site at which it phosphorylates and activates AMP-activated protein kinase. *Journal of Biological Chemistry* **271**, 27879–27887.

Honigberg, S. M. and Lee, R. H. (1998). SNF1 kinase connects nutritional pathways controlling meiosis in *Saccharomyces cerevisiae. Molecular and Cellular Biology* **18**, 4548–4555.

Jiang, R. and Carlson, M. (1996). Glucose regulates protein interactions within the yeast SNF1 protein kinase complex. *Genes and Development* **10**, 3105–3115.

Jiang, R. and Carlson, M. (1997) The SNF1 protein kinase and its activating subunit, SNF4, interact with distinct domains of the SIP1/SIP2/GAL83 component in the kinase complex. *Molecular and Cellular Biology* **17**, 2099–2106.

Kimura, M. (1980). "The Neutral Theory of Molecular Evolution". Cambridge University Press, Cambridge.

Lakatos, L. and Banfalvi, Z. (1997). Nucleotide sequence of a cDNA clone encoding an SNF1 protein kinase homologue (accession no. U83797) from *Solanum tuberosum. Plant Physiology* **113**, 1004.

Lakatos, L., Klein, M., Hofgen, R. and Banfalvi Z. (1999). Potato StubSNF1 interacts with StubGAL83: a plant protein kinase complex with yeast and mammalian counterparts. *Plant Journal* **17**, 569–574.

Le Guen, L., Thomas, M., Bianchi, M., Halford, N. G. and Kreis, M. (1992). Structure and expression of a gene from *Arabidopsis* encoding a protein related to SNF1 protein kinase. *Gene* **120**, 249–254.

Lessard, P. (1997). Caracterisation de la famille de genes codant les proteines kinases homologues a SNF1 chez *Arabidopsis*. Contribution a l'etude de leur fonction chez les plantes superieures. PhD Thesis, University of Paris XI.

Mackintosh, R. W., Davies, S. P., Clarke, P. R., Weekes, J., Gillespie, J. G., Gibb, B. J. and Hardie, D. G. (1992). Evidence for a protein kinase cascade in higher plants: 3-hydroxy-3-methylglutaryl-CoA reductase kinase. *European Journal of Biochemistry* **209**, 923–931.

Man, A. L., Purcell, P. C., Hannappel, U. and Halford, N. G. (1997). Potato SNF1-related protein kinase: molecular cloning, expression analysis and peptide kinase activity measurements. *Plant Molecular Biology* **34**, 31–43.

McMichael, R. W., Bachmann, M. and Huber, S. C. (1995). Spinach leaf sucrose phosphate synthase and nitrate reductase are phosphorylated by multiple protein kinases *in vitro*. *Plant Physiology* **108**, 1077–1082.

Mita, S., Suzuki-Fujii, K. and Nakamura, K. (1995). Sugar-inducible expression of a gene for β-amylase in *Arabidopsis*. *Plant Physiology* **107**, 895–904.

Mitchelhill, K. I., Stapleton, D., Gao, G., House, C., Michell, B., Katsis, F., Witters, L. A. and Kemp, B. E. (1994). Mammalian AMP-activated protein kinase shares structural and functional homology with the catalytic domain of yeast SNF1 protein kinase. *Journal of Biological Chemistry* **269**, 2361–2364.

Monger, W. A., Thomas, T. H., Purcell, P. C. and Halford, N. G. (1997). Identification of a sucrose nonfermenting-1-related protein kinase in sugar beet (*Beta vulgaris* L.). *Plant Growth Regulation* **22**, 181–188.

Moorhead, G., Douglas, P., Morrice, N., Scarabel, M., Aitken, A. and MacKintosh, C. (1996). Phosphorylated nitrate reductase from spinach leaves is inhibited by 14-3-3 proteins and activated by fusicoccin. *Current Biology* **6**, 1104–1113.

Morrell, S. and ap Rees, T. (1986). Sugar metabolism in developing tubers of *Solanum tuberosum*. *Phytochemistry* **25**, 1579–1585.

Muranaka, T., Banno, H. and Machida, Y. (1994). Characterization of the tobacco protein kinase NPK5, a homologue of *Saccharomyces cerevisiae SNF1* that constitutively activates expression of the glucose-repressible *SUC2* gene for a secreted invertase of *S. cerevisiae*. *Molecular and Cellular Biology* **14**, 2958–2965.

Neigeborn, L. and Carlson, M. (1984). Genes affecting the regulation of gene expression by glucose repression in *Saccharomyces cerevisiae*. *Genetics* **108**, 845–858.

Nemeth, K., Salchert, K., Putnoky, P., Bhalerao, R., Koncz-Kalman, Z., Stankovic-Stangeland, B., Bako, L., Mathur, J., Okresz, L., Stabel, S., Geigenberger, P., Stitt, M., Redei, G. P., Schell, J. and Koncz, C. (1998). Pleiotropic control of glucose and hormone responses by PRL1, a nuclear WD protein, in *Arabidopsis*. *Genes and Development* **12**, 3059–3073.

Purcell, P. C., Smith, A. M. and Halford, N. G. (1998). Antisense expression of a sucrose nonfermenting-1-related protein kinase sequence in potato results in decreased expression of sucrose synthase in tubers and loss of sucrose-inducibility of sucrose synthase transcripts in leaves. *Plant Journal* **14**, 195–202.

Ronne, H. (1995). Glucose repression in fungi. *Trends in Genetics* **11**, 12–17.

Russell, P. and Nurse, P. (1987). The mitotic inducer nim1$^+$ functions in a regulatory network of protein kinase homologs controlling the initiation of mitosis. *Cell* **49**, 569–576.

Salanoubat, M. and Belliard, G. (1989). The steady-state level of potato sucrose synthase mRNA is dependent on wounding, anaerobiosis and sucrose concentration. *Gene* **84**, 181–185.

Sano, H. and Youssefian, S. (1994). Light and nutritional regulation of transcript encoding a wheat protein kinase homolog is mediated by cytokinins. *Proceedings of the National Academy of Sciences of the United States of America* **91**, 2582–2586.

Su, W., Huber, S. C. and Crawford, N. M. (1996). Identification *in vitro* of a post-translational regulatory site in the hinge 1 region of *Arabidopsis* nitrate reductase. *Plant Cell* **8**, 519–527.

Sugden, C., Crawford, R. M., Halford, N. G. and Hardie, D. G. (1999a). Regulation of spinach SNF1-related (SnRK1) kinases by protein kinases and phosphatases is associated with phosphorylation of the T loop and is inhibited by 5′-AMP. *Plant Journal* **19**, 433–439.

Sugden, C., Donaghy, P., Halford, N. G. and Hardie, D. G. (1999b). Two SNF1-related protein kinases from spinach leaf phosphorylate and inactivate 3-hydoxy-3-

methylglutaryl-coenzyme A reductase, nitrate reductase and sucrose phosphate synthase *in vitro*. *Plant Physiology*, **120**, 257–274.

Sung, S.-J. S., Xu, D.-P. and Black, C. C. (1989). Identification of actively filling sucrose sinks. *Plant Physiology* **89**, 1117–1121.

Takano, M., Kajiya-Kanegae, H., Funatsuki, H. and Kikuchi, S. (1998). Rice has two distinct classes of protein kinase genes related to SNF1 of *Saccharomyces cerevisiae*, which are differently regulated in early seed development. *Molecular and General Genetics* **260**, 388–394.

Tanaka, S. and Nojima, H. (1996). Nik1, a Nim1-like protein kinase of *S. cerevisiae*, interacts with the Cdc28 complex and regulates cell cycle progression. *Genes to Cells* **1**, 905–921.

Thompson-Jaeger, S., Francois, J., Gaughran, J. P. and Tatchell, K. (1991). Deletion of *SNF1* affects the nutrient response of yeast and resembles mutations which activate the adenylate cyclase pathway. *Genetics* **12**, 697–706.

Weekes, J., Ball, K. L., Caudwell, F. B. and Hardie, D. G. (1993). Specificity determinants for the AMP-activated protein kinase and its plant homologue analyzed using synthetic peptides. *FEBS Letters* **334**, 335–339.

Wilson, W. A., Hawley, S. A. and Hardie, D. G. (1996). Glucose repression/derepression in budding yeast: SNF1 protein kinase is activated by phosphorylation under derepressing conditions, and this correlates with a high AMP:ATP ratio. *Current Biology* **6**, 1426–1434.

Woods, A., Munday, M. R., Scott, J., Yang, X., Carlson, M. and Carling, D. (1994). Yeast SNF1 is functionally related to mammalian AMP-activated protein kinase and regulates acetyl-CoA carboxylase *in vivo*. *Journal of Biological Chemistry* **269**, 19509–19515.

Woods, A., Cheung, P. C., Smith, F. C., Davison, M. D., Scott, J., Beri, R. K. and Carling, D. (1996). Characterization of AMP-activated protein kinase beta and gamma subunits. Assembly of the heterotrimeric complex *in vitro*. *Journal of Biological Chemistry* **271**, 10282–10290.

Yang, X., Hubbard, E. J. and Carlson, M. (1992). A protein kinase substrate identified by the two-hybrid system. *Science* **257**, 680–682.

Yang, X., Jiang, R. and Carlson, M. (1994). A family of proteins containing a conserved domain that mediates interaction with the yeast SNF1 protein kinase complex. *EMBO Journal* **13**, 5878–5886.

Zrenner, R., Salanoubat, M., Willmitzer, L. and Sonnewald, U. (1995). Evidence of the crucial role of sucrose synthase for sink strength using transgenic potato plants (*Solanum tuberosum L.*). *Plant Journal* **7**, 97–107.

Carbon and Nitrogen Metabolism and Reversible Protein Phosphorylation

D. TOROSER and S. C. HUBER

United States Department of Agriculture, Agricultural Research Service and Departments of Crop Science and Botany, North Carolina State University, Raleigh, NC 26795, USA

I. INTRODUCTION

Carbohydrate metabolism and N-assimilation are important processes that occur in both source and sink tissues. In a vegetative plant, leaves are the primary site of photosynthetic CO_2 fixation for carbohydrate biosynthesis and also for the reduction of nitrate to ammonia for amino acid biosynthesis. In many plant species, sucrose (Suc) is the primary sugar that is synthesized and subsequently translocated through the phloem to heterotrophic sink tissues. Many of the amino acids synthesized in leaves are also translocated along with Suc via the phloem. Sink tissues such as roots, developing tubers, and reproductive tissues are all actively engaged in carbohydrate utilization, although some (e.g. roots) are also capable of nitrate assimilation. In general, the pathways involved have been identified, but mechanisms for control and coordination between pathways are poorly understood.

Advances in Botanical Research Vol. 32
incorporating Advances in Plant Pathology
ISBN 0-12-005932-0

Co-ordination between carbohydrate metabolism and N-assimilation is essential in order to avoid direct competition and potentially involves control at several levels including gene expression, membrane transport, and enzyme activity. One important mechanism that may impact all three levels is reversible protein phosphorylation. In this review we will focus on the regulation of activity of several enzymes involved in C- and N-metabolism that are phosphorylated by either calmodulin-like domain protein kinases (CDPKs) (Harmon *et al.*, 1997) or sucrose nonfermenting-1 (SNF1)-related protein kinases (SnRKs) (Hardie and MacKintosh, 1992; Halford and Hardie, 1998). The enzymes considered are sucrose-phosphate synthase (SPS), sucrose synthase (SuSy), NADH:nitrate reductase (NR) and 3-hydroxy-3-methylglutaryl-coenzyme A reductase (HR). It should be noted that several other enzymes critical to plant metabolism are also known or thought to be regulated by phosphorylation, but are outside the scope of this review. Of particular importance is phosphoenolpyruvate carboxylase (PEPC), which is known to be phosphorylated and activated by a rather specific protein kinase (for review see Vidal and Chollet, 1997). Recent studies have also identified several other enzymes, including trehalose-6-phosphate synthase and glutamine synthetases, as phosphoproteins because they interact with 14-3-3 proteins in a phosphorylation-dependent manner (Moorhead *et al.*, 1999). These results are exciting but the significance of phosphorylation and 14-3-3 protein binding to these enzymes remains to be determined, and consequently will not be considered further here. Lastly, it is also relevant to note that several transport activities that may impact metabolism either directly or indirectly may also be controlled by phosphorylation. Of particular importance is the possible regulation by phosphorylation of ion and solute transport, e.g. the plasma membrane H^+-adenosine triphosphate (ATP)ase (Schaller and Sussman, 1988; Assmann and Haubrick, 1996), plasma membrane K^+ channel (Li *et al.*, 1998), and the sucrose transporter (Roblin *et al.*, 1998). These transport activities are important to metabolism but are also outside the scope of this short review and will not be considered further.

II. TARGET ENZYMES

A. SUCROSE-PHOSPHATE SYNTHASE

The biosynthesis of Suc is catalyzed by the sequential action of SPS and sucrose 6′-phosphate phosphatase (SPPase):

$$\text{UDP-Glc} + \text{Fru-6-P} \leftrightarrow \text{Suc-6}'\text{-P} + \text{UDP} + \text{H}^+$$
$$\text{Suc-6}'\text{-P} + \text{H}_2\text{O} \rightarrow \text{Suc} + \text{Pi}$$

Suc-6′-P is rapidly removed by a specific and high-activity phosphatase that displaces the reversible SPS reaction from equilibrium *in vivo* (Stitt *et al.*, 1987).

Thus, it is thought that regulation of SPS activity is important and contributes to a significant extent to the control of flux into Suc (Geigenberger *et al.*, 1995). In many tissues, including spinach leaves, SPS activity is regulated by Glc-6-P (activator) and Pi (inhibitor), which are antagonistic allosteric effectors (Doehlert and Huber, 1983a, b). The sensitivity of SPS to these allosteric effectors is altered by protein phosphorylation (discussed below). As a result, SPS activity is often measured under two conditions: (1) with rate-limiting substrate concentrations plus Pi ('selective' assay, V_{sel} activity), and (2) with saturating substrate concentrations in the absence of inhibitor ('nonselective' assay, V_{max} activity). Activation state of SPS is defined as V_{sel} activity expressed as a percentage of the V_{max} activity, and generally reflects the phosphorylation status of the enzyme. The activation state and enzymatic activity of SPS in spinach leaves are affected by phosphorylation of specific seryl residues in residues in response to light/dark conditions and osmotic stress (Huber *et al.*, 1992, 1996a, b; Huber and Huber, 1995). At the time of writing, three phosphorylation sites, which involve residues thought to be in loop regions of the protein (Salvucci *et al.*, 1995), have been suggested to be of regulatory significance (see Table I). Serine-158 (Ser158) is the residue thought to function in light/dark modulation of SPS activity (McMichael *et al.*, 1993), and this residue is conserved among species (Huber and Huber, 1996). One exception is the constitutive form of SPS recently cloned from sugar cane, which lacks a Ser residue at this position (Sugiharto *et al.*, 1997). Phosphorylation of Ser158 reduces the activation state of SPS by altering affinities for substrates and effectors (V_{sel} activity), without affecting maximum catalytic activity.

Site-directed mutagenesis of Ser158 of spinach leaf SPS has confirmed its role in the modulation of enzyme activity both *in vivo* and *in vitro* (Toroser *et al.*, 1999). The *S158A* mutant enzyme, expressed in transgenic tobacco plants, had a relatively high activation state in both light and dark conditions. In contrast, introduction of an acidic group at the 158 position in the double mutant, *S157F/S158E*, produced an enzyme that had a low activation state and was not activated by light. The results demonstrated the regulatory significance

TABLE I

Potential regulatory phosphorylation sites of spinach SPS

Residue	Sequence	Protein kinase	Effect	Physiological function
Ser158	RMRRISSVEMM	SnRK (PK_{III})	Inhibition	Light/dark modulation
Ser424	RMRRGVSCHGR	SnRK? (PK_{IV})	Activation	Osmotic stress activation
Ser229	LLTRQVSAPGV	CDPK? (PK_{I})	?	14-3-3 protein binding site

of Ser158 as the site responsible for dark inactivation of SPS *in vivo*, and suggest that the significance of phosphorylation is introduction of a negative charge at this position.

One of the major protein kinases that phosphorylates Ser158 in the native protein has been designated PK_{III} (protein kinase III) and tends to co-purify with SPS (McMichael *et al.*, 1995a). PK_{III} is strictly Ca^{2+} independent and has an apparent molecular mass of 150 kDa. The catalytic subunit is ~60 kDa (Douglas *et al.*, 1996) to 65 kDa (McMichael *et al.*, 1995a) and is recognized by antibodies against RKIN1, a member of the SnRK1 subfamily (Douglas *et al.*, 1996). The PK_{III} is equivalent to HRK-C, which has been demonstrated to be a member of the SNF1-related SnRK1 family of protein kinases (Sugden *et al.*, 1999). A characteristic property of the SnRK1 kinases is that they are often regulated by phosphorylation, and this has been demonstrated for PK_{III} (Douglas *et al.*, 1996; Sugden *et al.*, 1999). However, it is not known at present whether the phosphorylation state, and hence activity, of the SnRK1 kinases vary *in vivo*. If so, this could also contribute directly to the control of SPS activation state. The PK_{III} has been shown to target a Ser residue flanked by basic residues at P-3 and P-6, and with a hydrophobic group at P-5 (McMichael *et al.*, 1995; Toroser *et al.*, 1998). The sequence surrounding Ser158 of SPS matches this motif, and the positive recognition elements identified above are strictly conserved among species (Huber and Huber, 1996).

The second regulatory phosphorylation site to be identified on spinach SPS involves Ser424, which is phosphorylated when leaf tissue is subjected to osmotic stress (Toroser and Huber, 1997). This site is also highly conserved among species. Phosphorylation of Ser424 activates the enzyme, perhaps by attenuating the inhibitory effect of Ser158 phosphorylation. As a result, Suc synthesis can occur when it would otherwise be restricted (Huber *et al.*, 1999). In spinach SPS, phosphorylation of Ser424 is increased several-fold in response to osmotic stress, which correlated with increased activation state of the enzyme while maximum activity remained constant (Toroser and Huber, 1997). Control of the process apparently involves stress activation of PK_{IV} (via a presently unknown mechanism), which phosphorylates Ser424, and elevation of cytosolic [Ca^{2+}] that may accompany the environmental stress.

Recent results demonstrate that 14-3-3 proteins can associate with spinach leaf SPS (Toroser *et al.*, 1998; Moorhead *et al.*, 1999). Evidence for an association between the two proteins was obtained by co-immunoprecipitation and co-elution during gel filtration (Toroser *et al.*, 1998) and retention of SPS on a column containing immobilized 14-3-3 proteins (Moorhead *et al.*, 1999). Binding of 14-3-3s to SPS has been shown to require phosphorylation of SPS (Huber *et al.*, 1998; Moorhead *et al.*, 1999). The site of interaction was suggested to be Ser229, based on two lines of evidence. First, addition of a synthetic SPS-229 phosphopeptide was found to stimulate SPS activity, presumably by disrupting the SPS–14-3-3 complex. Second, the phosphorylated SPS-229 peptide was shown by surface plasmon resonance spectroscopy to bind a recombinant 14-3-3 protein (Toroser *et al.*, 1998). This finding was significant because the sequence surrounding Ser229 (RQVSAP) is a variant of the conserved motif RSXpSXP (where X is any amino acid and pS is phosphoserine) that serves as the binding site for 14-3-3s in many proteins (Muslin *et al.*, 1996). The protein kinase that seems to play a primary role in phosphorylation of Ser229 is the Ca^{2+}-dependent PK_I (Huber *et al.*, 1998), which has been shown to be a CDPK (Douglas *et al.*, 1998).

The effect of 14-3-3 binding to phospho-SPS has been suggested to either partially inhibit (Toroser *et al.*, 1998) or stimulate (Moorhead *et al.*, 1999) SPS activity. The basis for this discrepancy is not clear, but could be related to the source of the SPS (spinach versus cauliflower florets), the phosphorylation status of other regulatory sites, or specific assay conditions employed. However, it is clear that the two proteins can directly associate in a phosphorylation-dependent manner. Another important issue is the physiological significance of the SPS:14-3-3 association. It may function to regulate SPS activity directly, but if so, it is not yet clear under what conditions the association may be altered. Alternatively, the 14-3-3 protein may function as a scaffold protein (Jones *et al.*, 1995) to facilitate the interaction between SPS and another protein. One logical binding partner for SPS would be UDP-Glc pyrophosphorylase (UGPase), which forms the UDP-Glc substrate for SPS. Indeed, inspection of the UGPase primary structure indicates a putative 14-3-3 binding site (Toroser *et al.*, 1998) and raises the possibility of an SPS–UGPase complex that is bridged by a 14-3-3 protein. Recent results indicate that such a complex may exist (Toroser and Huber, manuscript in preparation), which suggests that UDP-Glc may be channeled between the two enzymes. However, further study will be required to establish the functional importance and physiological significance of SPS complexes.

B. HMG-CoA REDUCTASE

The enzyme HMG-CoA reductase converts HMG-CoA to mevalonic acid in an irreversible reaction utilizing two equivalents of NADPH. This reaction is the first committed step in isoprenoid biosynthesis, and in animals is known to be a regulatory point of cholesterol biosynthesis (Goldstein and Brown, 1990). In plants, the synthesis of total sterols appears to be limited by HMG-CoA reductase activity, but the accumulation of specific end-product sterols involves the activities of 'downstream' enzymes as well (Chappell *et al.*, 1995). Mammalian HMG-CoA reductases are inactivated by phosphorylation of a seryl residue close to their C-termini by the adenosine monophosphate (AMP)-activated protein kinase (Carling *et al.*, 1989). Plant HMG-CoA reductases contain a highly conserved C-terminal catalytic domain that can also be reversibly inactivated by phosphorylation of a single seryl residue – Ser577 of *Arabidopsis* isoform HMGR1 (Dale *et al.*, 1995). In both plants and animals, the single regulatory phosphorylation site is six residues C-terminal to a His residue that is essential for catalysis. Thus, phosphorylation is thought to inhibit enzyme activity by forming an electrostatic interaction between the negative charge of the phosphoserine and the imidazole ring of the protonated His residue, thereby preventing it from acting as a proton donor for catalysis (Omkumar and Rodwell, 1994). Early studies suggested that phosphorylation of the plant enzyme might be responsible for changes in activity in response to isoprenoid hormones (Russell *et al.*, 1985).

Studies with cauliflower florets identified a functional homolog in plants of the mammalian AMP-activated protein kinase that was involved in phosphorylation of HMG-CoA reductase (MacKintosh *et al.*, 1992). The plant kinase was shown to be part of a protein kinase cascade but, unlike the mammalian enzyme, was not regulated by AMP. Two distinct forms of the protein kinase were subsequently resolved by ion-exchange chromatography: a major form (HRK-A) and a minor form (HRK-B) (Ball *et al.*, 1994). Apparent molecular masses were 200 000 and 45 000, respectively, and HRK-A was reported to be Ca^{2+} independent. Subsequently, two distinct SNF1-related protein kinases (HRK-A and HRK-C) were identified in spinach leaf extracts and shown to phosphorylate NR, SPS and HMG-CoA reductase (Sugden *et al.*, 1999). All of the SNF1-related kinases could also phosphorylate the 'SAMS' synthetic peptide, which is based on the phosphorylation site sequence of rat acetyl-CoA carboxylase (Ball *et al.*, 1994), another substrate protein of the AMP-activated protein kinase. In subsequent studies with cauliflower florets, three peaks of protein kinase activity were resolved by fast protein liquid chromatography (FPLC)-Resource Q chromatography that could phosphorylate the SAMS peptide (Toroser and Huber, 1998). The peaks – listed in order of elution – were designated PK_I, PK_{IV} and PK_{III}, based on similarity to kinases resolved earlier from spinach leaves (Toroser and Huber, 1997). Interestingly, PK_{IV} was the predominant kinase to phosphorylate recombinant HMG-CoA reductase and it did so in a strictly Ca^{2+} dependent manner (Toroser and Huber, 1998). It is not known why the major HMG-CoA reductase kinase in cauliflower florets was Ca^{2+} independent in one study (Ball *et al.*, 1994) but strictly Ca^{2+} dependent in another (Toroser and Huber, 1998). Differences in plant material and protein extraction/purification conditions could be contributing factors.

In an attempt to develop a better synthetic peptide substrate to identify possible HMG-CoA reductase kinases, the SAMS* peptide was produced (Toroser and Huber, 1998), which is based on the phosphorylation site of plant HMG-CoA reductases. As shown in Table II, SAMS* was phosphorylated predominantly by PK_I and PK_{IV} and thus was a more specific substrate than the SAMS peptide. Interestingly, PK_{IV} was the most effective kinase in phosphorylating and inactivating the recombinant HMG-CoA reductase protein (see Table II). The PK_{IV} had a native molecular mass of ~150,000 – similar to SNF1-related kinases – but the family to which this Ca^{2+}-dependent enzyme belongs has not been determined.

TABLE II
Phosphorylation of synthetic peptides and recombinant HMG-CoA reductase by protein kinases resolved from cauliflower florets

Substrate	Peptide or protein kinase activity (cpm)		
	PK_I	PK_{IV}	PK_{III}
SAMS HMRSAM*S*GLHLVKRR	28×10^{-3}	16×10^{-3}	27×10^{-3}
SAMS* KSHMKYNR*S*TKDVK	11×10^{-3}	7.5×10^{-3}	1.7×10^{-3}
Recombinant HMG-CoA reductase	35	300	80
Ca^{2+} dependence of kinase	Partial	Complete	None

The HRK-A kinase was shown to be a member of the SNF1/AMP-activated protein kinase subfamily of kinases (Ball *et al.*, 1995), now known as SnRK1. Douglas *et al.* (1997) and Sugden *et al.* (1999) also identified several HMG-CoA reductase kinases in spinach leaf extracts. Two of them are distinct SNF1-related protein kinases: HRK-A and HRK-C [designated PK_{III} by Douglas *et al.* (1996)]. The kinases could phosphorylate NR, SPS and HMG-CoA reductase (Sugden *et al.*, 1999), thus establishing that SNF1-related kinases may function to regulate several major biosynthetic pathways.

C. NITRATE REDUCTASE

NADH:nitrate reductase (NR) catalyzes the reduction of nitrate to nitrite, which is the first step in the important process of inorganic N assimilation. This step is often considered to be limiting to plant growth and productivity. The steady-state level of NR protein is controlled at the level of gene expression (Vincentz *et al.*, 1993) and protein turnover (Somers *et al.*, 1983; Kaiser and Huber, 1997). The enzymatic activity of NR protein is controlled by covalent modification involving protein phosphorylation and binding of 14-3-3 proteins (Bachmann *et al.*, 1996a–c; Moorhead *et al.*, 1996; MacKintosh, 1998). Recent evidence suggests that phosphorylation and 14-3-3 protein binding not only reduces NR activity but may also increase susceptibility of the enzyme to proteolytic degradation (Kaiser and Huber, 1997). We will focus attention in this review on the control of NR activity by phosphorylation and 14-3-3 protein binding.

NR is known to contain a single major phosphorylation site that is located in hinge 1 (Douglas *et al.*, 1995; Bachmann *et al.*, 1996c; Su *et al.*, 1996), which links the cytochrome and molybdenum–molybdopterin (Mo–MPT) domains (Campbell, 1999) (see Fig. 1). Phosphorylation of Ser543 has no direct effect on NR activity, but completes the motif required for interaction with a 14-3-3 inhibitor protein (see Fig. 1). When 14-3-3s bind to phosphoSer543-NR, electron transfer between the cytochrome and molybdenum–molybdopterin

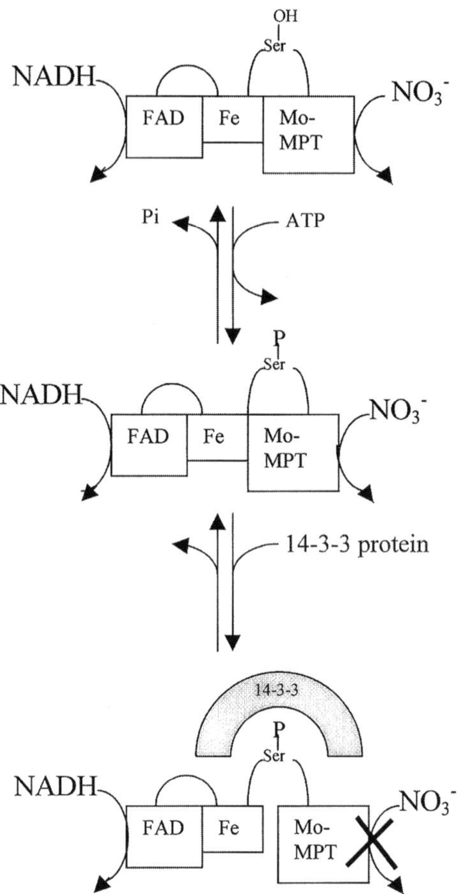

Fig. 1. Schematic representation of the two-stage regulation of NR by reversible protein phosphorylation and subsequent binding of a 14-3-3 inhibitor protein. The three domains of the NR polypeptide are shown as boxes that are connected by hinge regions. The regulatory phosphorylation site (Ser543) on hinge 1 is shown.

domains is blocked and nitrate reduction is inhibited (Huber *et al.*, 1992; Bachmann *et al.*, 1996c).

Two distinct protein kinases (PK_I and PK_{II}) can be resolved by ion-exchange chromatography from extracts of spinach leaves that can phosphorylate NR at Ser543 or synthetic peptides based on the phosphorylation site sequence (Bachmann *et al.*, 1995, 1996c; Douglas *et al.*, 1996). The protein kinases have apparent molecular masses of 45 kDa (PK_I) and 60 kDa (PK_{II}), and both are strictly Ca^{2+} dependent. However, they are immunochemically distinct – PK_{II} was identified as a CDPK whereas PK_I did not cross-react with a mixture of monoclonals raised against the catalytic domain of soybean CDPKα or polyclonals raised against the calmodulin-like domain of CDPKα (Bachmann *et al.*, 1996c). Subsequently, PK_I has been shown to contain a protein kinase with sequence similarity to *Arabidopsis* CDPK6 (Douglas *et al.*, 1998).

PK_I and PK_{II} have similar recognition motifs and require a Ser residue flanked by a basic residue at P − 3 with a hydrophobic residue at P − 5 (Bachmann *et al.*, 1996c). This recognition motif is a subset of that targeted by PK_{III} acting on SPS (see above). In other words, Ser158 of SPS should also be phosphorylated by PK_{II}, but the observation is that SPS is *not* a good substrate for PK_{II} (McMichael *et al.*, 1995a). Part of the answer to this question appears to rest with a 'negative recognition element' contained in the SPS sequence that inhibits phosphorylation by PK_{II}. As shown in Table III. PK_{II} activity with the NR7 synthetic peptide (based on NR-Ser543) is roughly four-fold higher than with the SP2 peptide (based on SPS-Ser158). At least part of the difference between the two peptides can be explained by the presence of an acidic residue at the P + 2 position in SP2. Substitution of an Ala for the Glu at this position (SP5 peptide) resulted in significantly greater kinase activity for PK_{II}. Similarly, substitution of an acidic residue for the neutral Pro at P + 2 in the NR peptide (NR19) reduced kinase activity and substitution with an Ala (NR11) increased activity (Table III). Thus, kinase specificity can be strongly influenced by negative elements, and this may explain why PK_{II} shows significant preference for NR over SPS (McMichael *et al.*, 1995a, b).

Synthetic phosphopeptides based on NR-Ser543 have been shown to bind to 14-3-3 proteins and to disrupt the phospho-NR–14-3-3 inactive complex by competing with native phospho-NR for the ligand binding site on the 14-3-3 protein (Bachmann *et al.*, 1996a, c; Moorhead *et al.*, 1996; Athwal *et al.*, 1998a, b). Because the 14-3-3s bind directly to the sequence containing the

TABLE III

Identification of an acidic residue at P + in synthetic peptides as a negative element that inhibits phosphorylation by spinach leaf PK_{II}. This may explain the preference of PK_{II} for NR as substrate over SPS (M. Bachmann, R. W. McMichael and S. C. Huber, unpublished results)

Peptide	Sequence	Peptide kinase activity (cpm × 10^{-3})
	−6 −5 −3 0 +4 +6	
SP2	K G R J R R I S <u>S</u> V E J J D K	12.2 (100)
SP5	——————————A——————	41.5 (340)
NR7	G P T L K R T A *S* T P F M N T T S K	38.0 (100)
NR19	——————————E——————	16.0 (42)
NR11	——————————A——————	51.2 (135)

phosphoserine residue, the ability of protein phosphatases to dephosphorylate the residue is reduced (Bachmann *et al.*, 1996a; Muslin *et al.*, 1996). Binding of 14-3-3 proteins to phospho-NR requires millimolar concentrations of a divalent cation such as Mg^{2+} (Kaiser and Brendle-Behnisch, 1991; Kaiser and Huber, 1994; Huber *et al.*, 1996a). Recent results suggest that divalent cations bind to the 14-3-3 proteins themselves and cause a conformational change that increases surface hydrophobicity and allows binding to target ligands, including synthetic phosphopeptides (Athwal *et al.*, 1998a, b). Binding in the absence of divalent cations can be induced by decreasing the pH from 7.5 to 6.5 (Athwal *et al.*, 1998), which explains the earlier observation that divalent cations were not required at pH 6.5 to maintain the inactive form of phospho-NR in crude leaf extracts (Huber *et al.*, 1995).

A metabolite that may control the binding of 14-3-3s to their target proteins is 5′-AMP (Athwal *et al.*, 1998b) (see Fig. 1). It will be important to identify other metabolites and factors that may regulate the binding of 14-3-3s and hence control accessibility of the phosphoserine residue to the requisite protein phosphatases.

D. SUCROSE SYNTHASE

SuSy is subject to post-translational modification by reversible protein phosphorylation and both SuSy isoforms in maize (SS1 and SS2) have been shown to be phosphorylated *in vivo* (Huber *et al.*, 1996b). Tryptic peptide mapping analysis suggested a single, similar phosphorylation site in both enzymes, and the site on maize SS2 protein was identified as Ser15 (Huber *et al.*, 1996b) and subsequently its homologue, Ser11, in soybean nodule-enhanced SuSy was also shown to be phosphorylated (Zhang and Chollet, 1997). The phosphorylation site is widely conserved in sequences deduced for SuSy complementary DNAs (cDNAs) cloned to date from mono- and dicotyledenous species. An endogenous SuSy kinase has been identified in extracts from elongating maize leaf tissue (Huber *et al.*, 1996b) and soybean nodules (Zhang *et al.*, 1999). Both kinases were strictly Ca^{2+} dependent and had estimated molecular masses of ~55 000 and 65 000, respectively. There is also a report that a PKC-like enzyme (phospholipid stimulated and Ca^{2+} dependent) can co-purify with and phosphorylate SuSy protein from maize seedlings (Lindblom *et al.*, 1997). Although the residue phosphorylated in that study still remains to be determined, the existence of a plant analog to PKC and its involvement in Suc metabolism is an exciting possibility.

The physiological significance of SuSy phosphorylation may be two-fold. First, phosphorylation of SuSy may stimulate cleavage activity. With extracts prepared from the elongation zone of maize leaves, pretreatment with ATP caused a slight (*ca* 25%) increase in SuSy cleavage activity with no change in synthetic activity (Huber *et al.*, 1996b; Winter *et al.*, 1997). Similarly, expression of mung bean SuSy in *Escherichia coli* yielded active recombinant enzyme (Nakai *et al.*, 1997, 1998) that had similar V_{max} and K_m values for UDP-Glc and Fru to the native enzyme. However, the K_m for Suc was 10-fold higher than native and was dramatically reduced by phosphorylation *in vitro* (Nakai *et al.*, 1997). Studies with site-directed mutants suggested that phosphorylation affects SuSy by introduction of a negative charge at Ser11, which specifically activates cleavage without affecting synthetic activity (Nakai *et al.*, 1998). It should also be noted that the sucrose-cleavage activity and kinetic properties of the recombinant soybean nodule SuSy were not altered by *in vitro* phosphorylation of Ser11, N-terminal truncation, or mutagenesis of the phosphorylated serine to an acidic group (S11D) or a neutral amino acid (S11A or S11C) (Zhang *et al.*, 1999). Thus, potential effects of phosphorylation on kinetic properties remain an important area for continued study.

The second effect of phosphorylation may be to influence the distribution of SuSy between the cytosol, plasma membrane and actin cytoskeleton (Winter *et al.*, 1997, 1998) (see Fig. 2). In particular, the membrane association of SuSy (Amor *et al.*, 1995) may be promoted by dephosphorylation of the enzyme (Winter *et al.*, 1997). It was shown that the membrane-associated form of SuSy was active *in vitro* and catalyzed the synthesis of cellulose and callose from

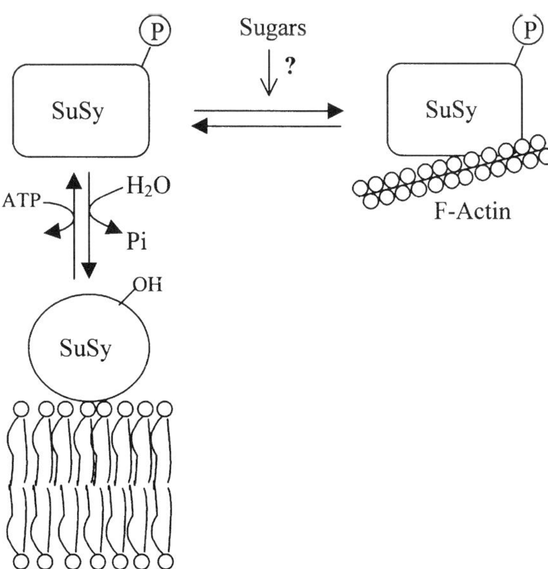

Fig. 2. Speculative schematic model of regulation of SuSy localization in a plant cell. The membrane association is postulated to be controlled, in part, by phosphorylation-dependent changes in surface hydrophobicity. The binding to F-actin is postulated to be under metabolic control, as binding is dependent upon Suc concentration.

$[^{14}C]$Suc in digitonin-permeabilized cotton fiber cells (Amor *et al.*, 1995). Thus, an important role for membrane-associated SuSy in channeling UDP-Glc into cell wall glucan synthesis was established. Both SuSy isozymes in maize SS1 and SS2 were found to be capable of associating with the plasma membrane (Carlson and Chourey, 1996). The relative amounts of SuSy association with the plasma membrane are variable and dynamic *in vivo*, due to developmental state (Amor *et al.*, 1995; Carlson and Chourey, 1996; Ruan *et al.*, 1997) and in response to environmental factors, such as gravity (Winter *et al.*, 1997). Although the mechanism of its membrane association is not well understood, phosphorylation-induced changes in surface hydrophobicity of SuSy could be at least part of the process (Winter *et al.*, 1997). It is unclear at present whether the membrane association of SuSy protein involves only an association of the protein with the lipid bilayer itself or also protein–protein interactions with one or more integral or peripheral membrane proteins. A specific interaction between SuSy and cellulose/callose synthase in the membrane is appearing to 'target' SuSy to specific sites but such interactions have not been shown to date.

The recent finding that the soluble form of SuSy protein is capable of binding to the actin cytoskeleton adds a third potential intracellular location for the enzyme (see Fig. 2). Evidence for an actin association is as follows. First, SuSy was found to co-precipitate with the detergent-insoluble cytoskeletal fraction when the actin cytoskeleton was stabilized *in situ* with phalloidin prior to extraction (Winter *et al.*, 1998). Second, several *SuSy* clones were identified as actin-interacting peptides in a yeast two-hybrid system screening a maize endosperm cDNA library (Carneiro and Larkins, unpublished results). Third, maize SuSy was shown to bind directly to F-actin *in vitro* (Winter *et al.*, 1998).

Several characteristics of SuSy binding to F-actin have been established. First, SuSy binding to actin was saturated at a monomer ratio of 1:5, respectively, suggesting that one SuSy tetramer can maximally bind to an actin filament consisting of 20 actin monomers. Second, binding of SuSy to F-actin was strictly dependent on Suc (H. Winter and S. C. Huber, unpublished data). Suc has been shown to induce a conformational change in phospho-SuSy, resulting in decreased surface hydrophobicity (Winter *et al.*, 1998), which may be necessary for actin binding. While the K_m(Suc) of purified maize SuSy in the cleavage direction was about 8 mM, half-maximal stimulation of actin binding was about an order of magnitude higher ($\sim$80 mM). This indicates that phospho-SuSy might have two different binding sites for sucrose: the catalytic site (high affinity) and a regulatory site (low affinity). We postulate that high concentrations of Suc bind to the regulatory site, which causes a conformational change exposing the actin-binding site. While the physiological significance of the interaction of SuSy and actin remains unknown, it may be that when Suc is abundant, SuSy is recruited to the actin cytoskeleton as part of a control mechanism to partition carbohydrate towards specific end products (e.g. starch biosynthesis) or to allow for accumulation of sucrose in the vacuole.

III. REGULATION OF PROTEIN KINASE ACTIVITY

As discussed above, SNF1-like protein kinases and the CDPKs undoubtedly play a major role in phosphorylating plant enzymes involved in primary C/N-metabolism. Factors that regulate kinase activity *in vitro* are of interest as they may help to identify mechanisms that control the phosphorylation status of enzymes *in vivo*.

A. REGULATION OF KINASE ACTIVITIES BY Ca^{2+} AND SECOND MESSENGERS

Ca^{2+} is recognized as an obligatory factor for many classes of plant protein kinases, in particular the CDPKs (Roberts and Harmon, 1992; Harmon *et al.*, 1997). NR can be phosphorylated *in vitro* by Ca^{2+}-dependent kinases (PK_I and PK_{II}) (Bachmann *et al.*, 1995, 1996c) and Ca^{2+}-independent kinases (HRK-C, which is equivalent to PK_{III}) (Douglas *et al.*, 1996, 1998; Sugden *et al.*, 1999). That Ca^{2+}-dependent kinases play a role in phosphorylation of NR was evidenced by the observation that NR could be activated in dark spinach leaves by depletion of cytosolic Ca^{2+} by a combination of ionophore plus extracellular chelator (J. L. Huber and S. C. Huber, unpublished results). Recent evidence suggests that cytosolic Ca^{2+} in leaf mesophyll cells may be reduced upon illumination as a result of energized uptake into chloroplasts (Plieth *et al.*, 1998). Consequently, part of the light signal that results in NR dephosphorylation/activation may be transduced to the cytosol via changes in $[Ca^{2+}]$. Light activation of other enzymes that do not involve Ca^{2+}-dependent kinases, e.g. SPS, may be mediated by other factors such as metabolites (see below).

The osmotic-stress activation of SPS that occurs in darkened leaves is thought to involve the Ca^{2+}-dependent phosphorylation of Ser424 catalyzed by PK_{IV} (Toroser and Huber, 1997). Increases in cytosolic $[Ca^{2+}]$ that often accompany stress (Price *et al.*, 1994; Sanders *et al.*, 1999) may thus be essential for this process to occur.

It is clear that not all Ca^{2+}-dependent phosphorylation events in leaves are light/dark modulated. For example, PK_I, a protein kinase that is thought to target SPS-Ser229, is dependent on micromolar concentrations of Ca^{2+} for activity (McMichael *et al.*, 1995a; Toroser and Huber, 1998). However, the SPS–14-3-3 interaction that is thought to require phospho-Ser229 apparently does not change appreciably during light–dark transitions (Toroser *et al.*, 1998). Thus, changes in cytosolic $[Ca^{2+}]$ may contribute to changes in enzyme phosphorylation state in some, but not all, cases even when Ca^{2+}-dependent kinases are thought to be involved.

In addition to Ca^{2+}, lipids and their derivatives are known to act as secondary messengers that can potentially control or provide specificity in protein kinase-mediated pathways. Many mammalian kinases bind phospholipids at specific sequences (Prescott, 1999) and it is likely that analogous mechanisms exist in plants. In fact, some CDPKs are stimulated by phospholipids (Karibe *et al.*, 1995). Preliminary data from the authors' laboratory suggest that certain intermediates in lipid metabolism may play a role in SNF1-related kinase regulation and thus potentially affect C/N metabolism. Specifically, low concentrations of lysophosphatidic acid (LPA; $\sim 50\,\mu M$) inhibited the phosphorylation of synthetic peptides by PK_{III}. Some specificity was observed because other lysophospholipids (e.g. lysophosphatidylcholine and lysophosphatidylserine) were not inhibitory. The ability of specific lipids to act at concentrations within physiological limits suggests that these results cannot be ascribed to a general lipophilic effect. We are presently characterizing the basis for the action of these bioactive compounds. Although preliminary, these studies raise the intriguing possibility that certain lipids and/or lysolipids may regulate the activity of certain CDPKs and SNF1-related kinases. The significance to lipid signalling, however, remains to be determined.

B. PHOSPHORYLATION CASCADES

Protein kinases that belong to the SNF1-like family are themselves regulated by site-specific phosphorylation of their catalytic subunit (Hawley *et al.*, 1996). In AMPK, in addition to the catalytic subunit, the regulatory β_1 subunit of the active heterotrimer is also phosphorylated (Mitchelhill *et al.*, 1997). In plants, recent evidence suggests that an SNF1-related kinase activity found in spinach (PK_{III}) can be inactivated via the action of mammalian PP2As or PP2Cs (Douglas *et al.*, 1996) and reactivated by mammalian kinase kinase (Sugden *et al.*, 1999). Several other lines of evidence indicate that protein kinases from spinach are themselves regulated at least in part by reversible protein phosphorylation. For example, recovery of kinase activities (in particular PK_I and PK_{III}) is enhanced when extractions and purifications are performed with a variety of phosphatase inhibitors (Douglas *et al.*, 1996, 1998). The authors have attempted to determine whether kinases can be dephosphorylated *in situ* using the phosphate sequestering agent d-mannose. Mannose treatment of spinach leaves has been shown to reduce dramatically phosphorylation of leaf proteins (Huber and Huber, 1992) and thus, dephosphorylation of protein kinases might be expected as well. Consistent with expectation, the authors have observed that recovery of activities associated with PK_I and PK_{III} was significantly reduced by mannose treatment (Toroser and Huber, unpublished results). However, there is as yet no evidence that the activities of these kinases change *in vivo* in response to an environmental stimulus or stress. For example, no change in extractable SP1-peptide kinase activity was observed when extracts were prepared from light versus dark leaves (McMichael *et al.*, 1995a). Nonetheless, reversible regulatory phosphorylation and the possibility of the widespread occurrence of phosphorylation cascades in plants are exciting possibilities.

C. METABOLITE REGULATION OF KINASE ACTIVITY

Metabolite regulation of kinase activity has been documented in some systems. For example, in yeast, Glc-6-P acts not only as a direct and potent activator of glycogen synthase but also as a specific inhibitor of glycogen synthase kinase (GSK-3) activity (Huang *et al.*, 1997). Another example is the mammalian SNF1 homolog, AMPK, which is directly regulated by 5′-AMP as the name implies, and also by phosphocreatine. AMPK phosphorylates and inactivates creatine kinase. Interestingly, inhibition of AMPK activity by phosphocreatine was antagonized by creatine, and was most pronounced at $pH < 7.5$ (Ponticos *et al.*, 1998). Because effects of creatine/creatine-P on AMPK activity were observed with synthetic peptides as substrates, these metabolite interactions were mediated by direct effects on the kinase (Ponticos *et al.*, 1998). This metabolic control of AMPK provides a potential mechanism for the co-ordinated regulation of energy metabolism in muscle tissue. Regulation by creatine-P and AMP provide a clear precedent for metabolite regulation of SNF1-related kinases. In plants, metabolites (e.g. Glc-6-P) have been shown to inhibit partially purified kinase activities from spinach that play a role in phosphorylation of SPS and NR.

1. *Phosphorylation of SPS on Ser158*

Glc-6-P inhibits the PK_{III}-catalyzed ATP-dependent inactivation of SPS *in vitro* (Weiner *et al.*, 1992). However, because Glc-6-P is an allosteric active form of SPS (Doehlert and Huber, 1983a, b; Huber and Huber, 1996), it is possible that the basis for Glc-6-P action is a change in the conformation of the substrate protein. However, several lines of evidence suggest that Glc-6-P may act to inhibit PK_{III} activity directly. First, Glc-6-P inhibited PK_{III}-catalyzed phosphorylation of a 26-kDa SPS fragment (Salvucci and Klein, 1993) that contains Ser158 (McMichael *et al.*, 1995b). Although the location of the allosteric site on SPS is not known, it is doubtful that the 26-kDa fragment would bind Glc-6-P and thus, an effect of Glc-6-P on the substrate polypeptide would seem unlikely. In addition, using a tryptophan fluorescence-based continuous kinase assay system, with a modified synthetic peptide ('SP46', RMKRKWS_{158}VEM), based on the SPS Ser158 sequence, the authors have obtained evidence that Glc-6-P inhibits PK_{III} activity. These results are consistent with those obtained with a standard [γ-^{32}P]ATP assay using numerous synthetic peptides based on the Ser158 sequence (Toroser, Plaut and Huber, unpublished results). The dual effect of Glc-6-P as an allosteric activator and also as an inhibitor of PK_{III} may function to amplify the effect of a change in cytosolic [Glc-6-P] that may occur in response to internal or external signals (see Fig. 3).

2. Phosphorylation of NR on Ser543

There is some evidence to suggest that P-esters may inhibit the phosphorylation and 14-3-3 protein-dependent inactivation of NR. Bachmann *et al.* (1996) found that NR inactivation *in vitro* was inhibited by dihydroxyacetone-P, Glc-6-P and Fru-1,6-P_2, and that the inhibition occurred at the phosphorylation step without affecting binding of the 14-3-3 inhibitor protein (Bachmann *et al.*, 1996). Specifically, the activities of PK_I and PK_{II} (Ca^{2+}-dependent kinases) were affected by P-esters *in vitro*. Evidence has also been obtained for a possible *in vivo* role of metabolites on NR activity modulation using a starchless mutant of *Nicotiana sylvestris*. The mutant tobacco plants hyperaccumulate P-esters during the second half of the photoperiod, and this was associated with a corresponding reduction in dark inactivation of NR *in vivo* (Bachmann *et al.*, 1996). Collectively, these results suggest that metabolic intermediates can inhibit the major protein kinases that act on NR and therefore could play a role in the post-translational regulation of NR activity *in vivo*.

3. Substrates Prevent the Phosphorylation of HR

HMG-CoA reductase activity is highly regulated by end-product inhibition (Bach *et al.*, 1990) and by down-modulation of catalysis by phosphorylation

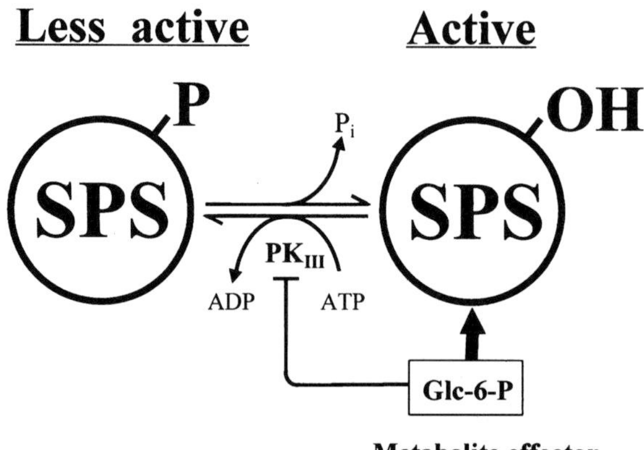

Fig. 3. Metabolite (Glc-6-P) control of SPS activity involves direct allosteric activation (solid arrow) of catalytic activity and inhibition of PK_{III}, which phosphorylates Ser158 and reduces activity. This dual action of Glc-6-P is postulated to amplify the metabolite signal.

(Hardie and MacKintosh, 1992; MacKintosh *et al.*, 1992; Dale *et al.*, 1995). Recent results suggest that substrate availability may influence the control of HMG-CoA reductase by phosphorylation. Toroser and Huber (1998) found that phosphorylation of recombinant *Nicotiana* HMG-CoA reductase by PK_{IV} was completely prevented in the presence of substrates NADPH plus HMG-CoA. The results demonstrated that metabolite supply can override signals that induce phosphorylation. In parallel experiments, assays were performed with PK_{IV} using the SAMS* synthetic peptide (KSHMKYNRSTKDVK), which is based on regulatory phosphorylation site of plant HMG-CoA reductase. There was no effect of substrates on peptide kinase activity (Toroser and Huber, 1998). Thus, substrates were affecting the phosphorylation of the regulatory Ser in the native protein molecule without affecting the requisite protein kinase per se. One possible explanation is that the phosphorylation motif targeted by PK_{IV} may involve residues involved in, or affected by, substrate binding or catalysis. In particular, the His residue located six residues N-terminal to the phosphorylated Ser (i.e. P-6 position) is known to be involved in catalysis and may also function as a recognition element for the kinase. It is reasonable to speculate that the His at P-6 may play a role in kinase targeting as other similar kinases, such as PK_{III} (a SNF1-like kinase), show a strong preference for peptide substrates with a basic residue at this position (McMichael *et al.*, 1995b; Toroser and Huber, 1998). It has been proposed (Omkumar and Rodwell, 1994) that the introduction of a negative charge at the regulatory phosphorylation site of HMG-CoA reductase impairs the function of the catalytic His residue at position P-6 and thus inhibits catalytic activity. Specifically, phosphorylation of the regulatory Ser is thought to result in an electrostatic interaction with the P-6 catalytic His residue, which blocks its ability to protonate the inhibitory $CoAS^-$ anion, thereby preventing its release from the active site and halting the overall reaction at the first reductive stage (Omkumar and Rodwell, 1994). Because substrates can attenuate phosphorylation of HMG-CoA reductase, it is clear that mevalonate production *in vivo* will not simply be determined by the activities of the protein kinase(s) that phosphorylate HMG-CoA reductase, but may be strongly impacted by substrate availability as well.

D. REGULATION OF KINASE ACTIVITIES BY PROTEIN FACTORS

Protein kinases are under intricate control involving changes in subcellular localization, phosphorylation events and protein–protein interactions. Many kinases interact with regulatory proteins that affect their activities. While CDPKs are thought to exist as monomers, it is likely that plant SNF1-like kinases are heterotrimeric proteins with the catalytic subunit complexed to one or more regulatory subunits (Dyck *et al.*, 1996; Gancedo, 1998), although this has yet to be conclusively demonstrated with plant SNF1-related kinases. In addition to homologs of these regulatory elements, other protein factors may be implicated in the regulation of the catalytic function of SNF1-like kinases, and some of these are described below.

1. 14-3-3 Proteins

14-3-3 proteins have been found to interact with at least one CDPK isoform in *Arabidopsis* (Camoni *et al.*, 1998; Moorhead *et al.*, 1999). Although a small stimulation of CDPK activity has been observed upon 14-3-3 binding to certain isoforms, it seems possible that an additional function of the interaction may be the bridging of the substrate protein and its kinase activity. In view of the functions in subcellular localization ascribed to 14-3-3s, a similar function may be ascribed to the interaction between 14-3-3s and certain CDPK isoforms.

2. Pleiotropic Regulatory Locus Protein 1 (PRL1)

SNF1-like kinases interact with a number of proteins, some of which may have inhibitory functions. The PRL1 protein is implicated in the transcriptional repression of glucose responsive genes in *Arabidopsis* (Németh *et al.*, 1998). Genetic and biochemical evidence (altered growth and carbohydrate partitioning) with the PRL1 mutation indicates that PRL1 probably represents a physiological inhibitor of plant SNF1-like protein kinases (Németh *et al.*, 1998; Bhalerao *et al.*, 1999). PRL1 has been found to interact with the *Arabidopsis* SNF1 homologs AKIN10 and AKIN11, and a comparison of the activation level of AKIN10 and AKIN11 immunocomplexes shows PRL1 to be a negative regulator of these *Arabidopsis* SNF1 homologs (Bhalerao *et al.*, 1999). Although most of the evidence for the regulatory interaction between PRL1 and SNF1 homologs centers around sugar-regulated gene expression in plants, analogous regulatory interactions between PRL1 and kinase activities may also have a major role in the regulation of enzyme activities involved in C/N metabolism.

3. Possible Interaction of Kinases with Calreticulin

Calreticulin (CRT) has long been considered to be restricted to the endoplasmic reticulum, but it is now thought that CRT can also localize to the nucleus and/or cytoplasm of some animal cells (Dedhar, 1994). CRT is known to interact with integrins and certain hormone receptors via the CRT-binding domain that is composed of the six amino acid motif KXFF[K,R]R (Dedhar, 1994). Interestingly, the putative CRT-binding motif is contained in several CDPK-related kinases (CRK) which, in contrast to the CDPKs, do not require Ca^{2+} for activity. Two CRKs from maize (ZmCRK1 and ZmCRK3) were recently cloned and the recombinant proteins characterized (Furomoto *et al.*, 1996). Both kinases contain the deduced sequence RRFFKR (residues 66–71 in ZmCRK1 and residues 90–95 in ZmCRK3). Similarly, an *Arabidopsis* CRK [European Molecular Biology Laboratory (EMBL) accession no.

O04290] contains the sequence KRFFKR (residues 87–92). All three are perfect matches for the CRT-binding motif; however, an interaction with CRT has not been demonstrated experimentally in any system and is only a possibility based on the occurrence of the binding motif. Other CRKs from carrot (accession no. P53681) and Virginia spiderwort (accession no. AAC24961) do not contain the binding motif, indicating that only some isoforms may be involved. In view of the possibility that Ca^{2+} may have a role in the modulation of numerous kinase activities that function to regulate kinases involved in C/N metabolism, putative CRT binding may be of interest in the study of the plant protein kinases.

REFERENCES

Amor, Y., Haigler, C. H., Johnson, S., Wainscott, M. and Delmer, D. P. (1995). A membrane-associated form of sucrose synthase and its potential role in synthesis of cellulose and callose in plants. *Proceedings of the National Academy of Sciences of the United States of America* **92**, 9353–9357.

Assmann, S. M. and Haubrick, L. L. (1996). Transport proteins of the plant plasma membrane. *Current Opinion in Cell Biology* **8**, 458–467.

Athwal, G. S., Huber, J. L. and Huber, S. C. (1998a). Biological significance of divalent cation binding to 14-3-3 proteins in relationship to nitrate reductase inactivation. *Plant Cell Physiology* **39**, 1065–1072.

Athwal, G. S., Huber, J. L. and Huber, S. C. (1998b). Phosphorylated nitrate reductase and 14-3-3 proteins. Site of interaction, effects of ions, and evidence for an AMP-binding site on 14-3-3 proteins. *Plant Physiology* **118**, 1041–1048.

Bach, T. J., Weber, T. and Motel, A. (1990). Some properties of enzymes involved in the biosynthesis and metabolism of 3-hydroxy-3-methyglutaryl-CoA in plants. *In* "Recent Advances in Phytochemistry" (G. H. N. Towers and H. A. Stafford, eds) pp. 1–82. Plenum Press, New York.

Bachmann, M., McMichael, R. W., Jr, Huber, J. L., Kaiser, W. M. and Huber, S. C. (1995). Partial purification and characterization of a calcium-dependent protein kinase and an inhibitor protein required for inactivation of spinach leaf nitrate reductase. *Plant Physiology* **108**, 1083–1091.

Bachmann, M., Huber, J. L., Athwal, G. S., Wu, K., Ferl, R. J. and Huber, S. C. (1996a). 14-3-3 proteins associate with the regulatory phosphorylation site of spinach leaf nitrate reductase in an isoform-specific manner and reduce dephosphorylation by endogenous protein phosphatases. *FEBS Letters* **398**, 26–30.

Bachmann, M., Huber, J. L., Liao, P.-C., Gage, D. A. and Huber, S. C. (1996b). The inhibitor protein of phosphorylated nitrate reductase from spinach (*Spinacia oleracea*) leaves is a 14-3-3 protein. *FEBS Letters* **387**, 127–131.

Bachmann, M., Shiraishi, N., Campbell, W. H., Yoo, B.-C., Harmon, A. C. and Huber, S. C. (1996c). Identification of Ser-543 as the major regulatory phosphorylation site in spinach leaf nitrate reductase. *Plant Cell* **8**, 505–517.

Ball, K. L., Dale, S., Weekes, J. and Hardie, D. G. (1994). Biochemical characterization of two forms of 3-hydroxy-3-methylglutaryl-CoA reductase kinase from cauliflower (*Brassica oleracea*). *European Journal of Biochemistry* **219**, 743–750.

Ball, K. L., Barker, J., Halford, N. G. and Hardie, D. G. (1995). Immunological evidence that HMG-CoA reductase kinase-A is the cauliflower homologue of the RKIN1 subfamily of plant protein kinases. *FEBS Letters* **377**, 189–192.

Bhalerao, R. P., Salchert, K., Bakó, L., Ökrez, L., Szabados, L., Muranaka, T., Machida, Y., Schell, J. and Koncz, C. (1999). Regulatory interaction of PRL1 WD protein with *Arabidopsis* SNF1-like protein kinases. *Proceedings of the National Academy of Sciences of the United States of America* **96**, 5322–5327.

Camoni, L., Harper, J. F. and Palmgren, M. G. (1998). 14-3-3 proteins activate a plant calcium-dependent protein kinase. *FEBS Letters* **430**, 381–384.

Campbell, W. H. (1999). Nitrate reductase structure, function and regulation: bridging the gap between biochemistry and physiology. *Annual Review of Plant Physiology and Plant Molecular Biology* **50**, 277–303.

Carling, D., Clarke, P. R., Zammit, V. A. and Hardie, D. G. (1989). Purification and characterization of the AMP-activated protein kinase. Copurification of acetyl-CoA carboxylase kinase and 3-hydroxy-3-methylglutaryl-CoA reductase kinase activities. *European Journal of Biochemistry* **186**, 129–136.

Carlson, S. J. and Chourey, P. S. (1996). Evidence for plasma membrane-associated forms of sucrose synthase in maize. *Molecular and General Genetics* **252**, 303–310.

Chappell, J., Wolf, F., Proulx, J., Cuellar, R. and Saunders, C. (1995). Is the reaction catalyzed by 3-hydroxy-3-methylglutaryl coenzyme A reductase a rate-limiting step for isoprenoid biosynthesis in plants? *Plant Physiology* **109**, 1337–1343.

Dale, S., Arró, M., Becerra, B., Morrice, N. G., Boronat, A., Hardie, D. G. and Ferrer, A. (1995). Bacterial expression of the catalytic domain of 3-hydroxy-3-methylglutaryl-CoA reductase (isoform HMGR1) from *Arabidopsis thaliana*, and its inactivation by phosphorylation at Ser577 by *Brassica oleracea* 3-hydroxyl-3-methylglutary-CoA reductase kinase. *European Journal of Biochemistry* **233**, 506–513.

Dedhar, S. (1994). Novel functions for calreticulin: interaction with integrins and modulation of gene expression? *Trends in Biochemical Sciences* **19**, 269–271.

Doehlert, D. C. and Huber, S. C. (1983a). Regulation of spinach leaf sucrose phosphate synthase by glucose-6-phosphate, inorganic phosphate and pH. *Plant Physiology* **73**, 989–994.

Doehlert, D. C. and Huber, S. C. (1983b). Spinach leaf sucrose phosphate synthase. Activation by glucose 6-phosphate and interaction with inorganic phosphate. *FEBS Letters* **153**, 293–297.

Douglas, P., Morrice, N. and MacKintosh, C. (1995). Identification of a regulatory phosphorylation site in the hinge 1 region of nitrate reductase from spinach (*Spinacea oleracea*) leaves. *FEBS Letters* **377**, 113–117.

Douglas, P., Pigaglio, E., Ferrer, A., Halford, N. G. and MacKintosh, C. (1996). Three spinach leaf nitrate reductase/3-hydroxy-3-methylglutaryl-CoA reductase kinases that are regulated by reversible phosphorylation and/or calcium ions. *Biochemical Journal* **325**, 101–109.

Douglas, P., Moorhead, G., Hong, Y., Morrice, N. and MacKintosh, C. (1998). Purification of a nitrate reductase kinase from *Spinacea oleracea* leaves, and its identification as a calmodulin-domain protein kinase. *Planta* **206**, 435–442.

Dyck, J. R. B., Gao, G., Widmer, J., Stapleton, D., Fernandez, C. S., Kemp, B. E. and Witters, L. A. (1996). Regulation of the 5′-AMP-activated protein kinase activity by the noncatalytic β and γ subunits. *Journal of Biological Chemistry* **271**, 17798–17803.

Furomoto, T., Ogawa, N., Hata, S. and Izui, K. (1996). Plant calcium-dependent protein kinase-related kinases (CRKs) do not require calcium for their activities. *FEBS Letters* **396**, 147–151.

Gancedo, J. M. (1998). Yeast carbon catabolite repression. *Microbiology and Molecular Biology Review* **62**, 334–361.

Geigenberger, P., Krause, K.-P., Hill, L. M., Reimholz, R., MacRae, E., Quick, P., Sonnewald, U. and Stitt, M. (1995). "The Regulation of Sucrose Synthesis in Leaves and Tubers of Potato Plants. Sucrose Metabolism, Biochemistry, Physiology and Molecular Biology". American Society of Plant Physiologists, Rockville, MD.

Goldstein, J. L. and Brown, M. S. (1990). Regulation of the mevalonate pathway. *Nature* **343**, 425–430.

Halford, N. G. and Hardie, D. G. (1998). SNF1-related protein kinases: global regulators of carbon metabolism in plants? *Plant Molecular Biology* **37**, 735–748.

Hardie, D. G. and MacKintosh, R. W. (1992). AMP-activated protein kinase – an archetypal protein kinase cascade? *BioEssays* **14**, 699–704.

Harmon, A. C., Lee, J.-Y. and Yoo, B.-C. (1997). CDPKs – a growing family of protein kinases involved in signal transduction. *In* "Protein Phosphorylation in Plants". (P. Gadal, M. Kreis and M. Dronet, eds) pp. 18–25. Université de Paris Sud Orsay, Paris.

Hawley, S. A., Davison, M., Woods, A., Davies, S. P., Beri, R. K., Carling, D. and Hardie, D. G. (1996). Characterization of the AMP-activated protein kinase kinase from rat liver and identification of threonine-172 as the major site at which it phosphorylates AMP-activated protein kinase. *Journal of Biological Chemistry* **271**, 27879–27887.

Huang, D., Wilson, W. A. and Roach, P. J. (1997). Glucose-6-P control of glycogen synthase phosphorylation in yeast. *Journal of Biological Chemistry* **272**, 22495–22501.

Huber, J. L., Huber, S. C., Campbell, W. H. and Redinbaugh, M. G. (1992). Reversible light/dark modulation of spinach leaf nitrate reductase activity involves protein phosphorylation. *Archives of Biochemistry and Biophysics* **296**, 58–65.

Huber, J. L. A. and Huber, S. C. (1992). Site-specific serine phosphorylation of spinach leaf sucrose-phosphate synthase. *Biochemical Journal* **283**, 877–882.

Huber, S. C. and Huber, J. L. (1995). Metabolic activators of spinach leaf nitrate reductase: effects on enzymatic activity and dephosphorylation by endogenous phosphatases. *Planta* **196**, 180–189.

Huber, S. C. and Huber, J. L. (1996). Role and regulation of sucrose-phosphate synthase in higher plants. *Annual Review of Plant Physiology and Plant Molecular Biology* **47**, 431–444.

Huber, S. C., McMichael, R. W., Jr, Huber, J. L., Bachmann, M., Yamamoto, Y. T. and Conkling, M. A. (1995). Light regulation of sucrose synthesis: role of protein phosphorylation and possible involvement of cytosolic [Ca^{2+}]. *In* "Carbon Partitioning and Source-Sink Interactions in Plants" (M. A. Madore and W. J. Lucas, eds) pp. 35–44. American Society of Plant Physiologists, Rockville, MD.

Huber, S. C., Bachmann, M. and Huber, J. L. (1996a). Post-translational regulation of nitrate reductase activity in higher plants: a role for Ca^{2+} and 14-3-3 proteins. *Trends in Plant Science* **1**, 432–438.

Huber, S. C., Huber, J. L., Liao, P.-C., Gage, D. A., McMichael, R. W., Chourey, P. S., Hannah, L. C. and Koch, K. (1996b). Phosphorylation of Ser-15 of maize leaf sucrose synthase: occurrence *in vivo* and possible regulatory significance. *Plant Physiology* **112**, 793–802.

Huber, S. C., Toroser, D., Winter, H., Athwal, G. S. and Huber, J. L. (1998). Regulation of plant metabolism by protein phosphorylation. Possible regulation of sucrose-phosphate synthase by 14-3-3 proteins. XIth International Photosynthesis Congress, Budapest, Hungary. Kluwer Academic, Dordrecht.

Huber, S. C., Kaiser, W. M., Toroser, D., Athwal, G. S., Winter, H. and Huber, J. L. (1999). Regulation of sucrose metabolism by protein phosphorylation. Stimulation of sucrose synthesis by osmotic stress and 5-aminoimidazole-4-carboxamide riboside. *In* "Carbohydrate Metabolism in Plants" (J. Bryant, M. Burrell and N. Kruger, eds) pp. 61–68. BIOS Scientific, Oxford.

Jones, D. H., Ley, S. and Aitken, A. (1995). Isoforms of 14-3-3 protein can form homo- and heterodimers *in vivo* and *in vitro*: implications for function as adapter proteins. *FEBS Letters* **368**, 55–58.

Kaiser, W. M. and Brendle-Behnisch, E. (1991). Rapid modulation of spinach leaf nitrate reductase activity by photosynthesis. I. Modulation *in vivo* by CO_2 availability. *Plant Physiology* **96**, 363–367.

Kaiser, W. M. and Huber, S. C. (1994). Post-translational regulation of nitrate reductase in higher plants. *Plant Physiology* **106**, 817–821.

Kaiser, W. M. and Huber, S. C. (1997). Correlation between apparent phosphorylation state of nitrate reductase (NR), NR hysteresis and degradation of NR protein. *Journal of Experimental Botany* **48**, 1367–1374.

Karibe, H., Komatsu, S. and Hirano, H. (1995). A calcium- and phospholipid-dependent protein kinase from rice (*Oryza sativa*) leaves. *Physiologia Plantarum* **95**, 127–133.

Li, J., Lee, Y.-R. and Assmann, S. M. (1998). Guard cells possess a calcium-dependent protein kinase that phosphorylates the KAT1 potassium channel. *Plant Physiology* **116**, 785–795.

Lindblom, S., Ek, P., Muszynska, G., Ek, B., Szczegielniak, J. and Engström, L. (1997). Phosphorylation of sucrose synthase from maize seedlings. *Acta Biochimica Polonica* **44**, 809–818.

MacKintosh, C. (1998). Regulation of cytosolic enzymes in primary metabolism by reversible protein phosphorylation. *Current Opinion in Plant Biology* **1**, 224–229.

MacKintosh, R. W., Davies, S. P., Clarke, P. R., Weekes, J., Gillespie, J. G., Gibb, B. J. and Hardie, D. G. (1992). Evidence for a protein kinase cascade in higher plants. 3-hydroxy-3-methylglutaryl-CoA reductase kinase. *European Journal of Biochemistry* **209**, 923–931.

McMichael, R. W., Jr, Klein, R. R., Salvucci, M. E. and Huber, S. C. (1993). Identification of the major regulatory phosphorylation site in sucrose-phosphate synthase. *Archives of Biochemistry and Biophysics* **307**, 248–252.

McMichael, R. W., Jr, Bachmann, M. and Huber, S. C. (1995a). Spinach leaf sucrose-phosphate synthase and nitrate reductase are phosphorylated/inactivated by multiple protein kinases *in vitro*. *Plant Physiology* **108**, 1077–1082.

McMichael, R. W., Jr, Kochansky, J., Klein, R. R. and Huber, S. C. (1995b). Characterization of the substrate specificity of sucrose-phosphate synthase protein kinase. *Archives of Biochemistry and Biophysics* **321**, 71–75.

Mitchelhill, K. I., Michell, B. J., House, C. M., Stapleton, D., Dyck, J., Gamble, J., Ullrich, C., Witters, L. A. and Kemp, B. E. (1997). Post-transcriptional modifications of the 5′-AMP-activated protein kinase β_1 subunit. *Journal of Biological Chemistry* **272**, 24475–24479.

Moorhead, G., Douglas, P., Morrice, N., Scarabel, M., Aitken, A. and MacKintosh, C. (1996). Phosphorylated nitrate reductase from spinach leaves is inhibited by 14-3-3 proteins and activated by fusicoccin. *Current Biology* **6**, 1104–1113.

Moorhead, G., Douglas, P., Cotelle, V. *et al.* (1999). Phosphorylation-dependent interactions between enzymes of plant metabolism and 14-3-3 proteins. *Plant Journal* **18**, 1–12.

Muslin, A. J., Tanner, J. W., Allen, P. M. and Shaw, A. S. (1996). Interaction of 14-3-3 with signaling proteins is mediated by the recognition of phosphoserine. *Cell* **84**, 889–897.

Nakai, T., Tonnouchi, N., Tsuchida, T., Mori, H., Sakai, F. and Hayashi, T. (1997). Synthesis of asymmetrically labeled sucrose by a recombinant sucrose synthase. *Bioscience Biotechnology and Biochemistry* **61**, 1955–1956.

Nakai, T., Konishi, T., Zhang, X. Q. *et al.* (1998). An increase in apparent affinity for sucrose of mung bean sucrose synthase is caused by *in vitro* phosphorylation or directed mutagenesis of Ser^{11}. *Plant Cell Physiology* **39**, 1337–1341.

Németh, K., Salchert, K., Putnoky, P. *et al.* (1998). Pleiotropic control of glucose and hormone responses by PRL1, a nuclear WD protein, in *Arabidopsis. Genes and Development* **12**, 3059–3073.

Omkumar, R. V. and Rodwell, V. W. (1994). Phosphorylation of Ser871 impairs the function of His865 of Syrian hamster 3-hydroxy-3-methylglutaryl-CoA reductase. *Journal of Biological Chemistry* **269**, 16862–16866.

Plieth, C., Sattelmacher, B. and Hansen, U.-P. (1998). Light-induced cytosolic calcium transients in green plant cells. I. Methodological aspects of chlorotetracycline usage in algae and higher-plant cells. *Planta* **207**, 42–51.

Ponticos, M., Lu, Q. L., Morgan, J. E., Hardie, D. G., Partridge, T. A. and Carling, D. (1998). Dual regulation of the AMP-activated protein kinase provides a novel mechanism for the control of creatine kinase in skeletal muscle. *EMBO Journal* **17**, 1688–1699.

Prescott, S. M. (1999). A thematic series on kinases and phosphatases that regulate lipid signalling. *Journal of Biological Chemistry* **274**, 8345.

Price, A. H., Taylor, A., Ripley, S. J., Griffiths, A., Trewavas, A. J. and Knight, M. R. (1994). Oxidative signals in tobacco increase cytosolic calcium. *Plant Cell* **6**, 1301–1310.

Roberts, D. M. and Harmon, A. C. (1992). Calcium-modulated proteins: targets of intracellular calcium signaling in higher plants. *Annual Review of Plant Physiology and Plant Molecular Biology* **43**, 375–414.

Roblin, G., Sakr, S., Bonmort, J. and Delrot, S. (1998). Regulation of a plant plasma membrane sucrose transporter by phosphorylation. *FEBS Letters* **424**, 165–168.

Ruan, Y.-L., Chourey, P. S., Delmer, D. P. and Perez-Grau, L. (1997). The differential expression of sucrose synthase in relation to diverse patterns of carbon partitioning in developing cotton seeds. *Plant Physiology* **115**, 325–385.

Russell, D. W., Knight, J. S. and Wilson, T. M. (1985). "Pea Seedling HMG-CoA Reductases: Regulation of Activity *In Vitro* by Phosphorylation and Ca^{2+}, and Post-translational Control *In Vivo* by Phytochrome and Isoprenoid Hormones". University of Missouri, Columbia.

Salvucci, M. E. and Klein, R. R. (1993). Identification of the uridine-binding domain of sucrose-phosphate synthase. Expression of a region of the protein that photoaffinity labels with 5-azidouridine diphosphate-glucose. *Plant Physiology* **102**, 529–536.

Salvucci, M. E., van de Loo, F. J. and Klein, R. R. (1995). The structure of sucrose-phosphate synthase. *In* "Sucrose Metabolism, Biochemistry, Physiology and Molecular Biology" (H. G. Pontis, G. L. Salerno and E. Echeverria, eds) pp. 1–10. American Society of Plant Physiologists, Rockville, MD.

Sanders, D., Brownlee, C. and Harper, J. F. (1999). Communicating with calcium. *Plant Cell* **11**, 691–706.

Schaller, G. E. and Sussman, M. R. (1988). Phosphorylation of the plasma-membrane H$^+$-ATPase of oat roots by a calcium-stimulated protein kinase. *Planta* **173**, 509–518.

Somers, D. A., Kuo, T.-M., Kleinhofs, A., Warner, R. L. and Oaks, A. (1983). Synthesis and degradation of barley nitrate reductase. *Plant Physiology* **72**, 949–952.

Stitt, M., Huber, S. C. and Kerr, P. S. (1987). Control of photosynthetic sucrose formation. *In* "The Biochemistry of Plants" (M. D. Hatch and N. K. Boardman, eds) pp. 327–409. Academic Press, New York.

Su, W., Huber, S. C. and Crawford, N. M. (1996). Identification *in vitro* of a post-translational regulatory site in the hinge 1 region of *Arabidopsis* nitrate reductase. *Plant Cell* **8**, 519–527.

Sugden, C., Donaghy, P. G., Halford, N. G. and Hardie, D. G. (1999). Two SNF1-related protein kinases from spinach leaf phosphorylate and inactivate 3-hydroxy-3-methylglutaryl-coenzyme A reductase, nitrate reductase, and sucrose phosphate synthase *in vitro*. *Plant Physiology* **120**, 257–274.

Sugiharto, B., Sakakibara, H., Sumadi and Sugiyama, T. (1997). Differential expression of two genes for sucrose-phosphate synthase in sugarcane: molecular cloning of the cDNAs and comparative analysis of gene expression. *Plant Cell Physiology* **38**, 961–965.

Toroser, D. and Huber, S. C. (1997). Protein phosphorylation as a mechanism for osmotic-stress activation of sucrose-phosphate synthase. *Plant Physiology* **114**, 947–955.

Toroser, D. and Huber, S. C. (1998). 3-Hydroxy-3-methylglutaryl-coenzyme A reductase kinase and sucrose-phosphate synthase kinase activities in cauliflower florets: Ca^{2+} dependence and substrate specificities. *Archives of Biochemistry and Biophysics* **355**, 291–300.

Toroser, D., Athwal, G. S. and Huber, S. C. (1998). Site-specific regulatory interaction between spinach leaf sucrose-phosphate synthase and 14-3-3 proteins. *FEBS Letters* **435**, 110–114.

Toroser, D., McMichael, R. W., Jr, Krause, K.-P., Kurreck, J., Sonnewald, U., Stitt, M. and Huber, S. C. (1999). Site-directed mutagenesis of serine 158 demonstrates its role in spinach leaf sucrose-phosphate synthase modulation. *Plant Journal* **17**, 407–413.

Vidal, J. and Chollet, R. (1997). Regulatory phosphorylation of C_4 PEP carboxylase. *Trends in Plant Science* **2**, 230–237.

Vincentz, M., Moureaux, T., Leydecker, M. T., Vaucheret, H. and Caboche, M. (1993). Regulation of nitrate and nitrite reductase expression in *Nicotiana plumbagnifolia* leaves by nitrogen and carbon metabolites. *Plant Journal* **3**, 315–324.

Weiner, H., McMichael, R. W., Jr and Huber, S. C. (1992). Identification of factors regulating the phosphorylation status of sucrose-phosphate synthase *in vivo*. *Plant Physiology* **99**, 1435–1442.

Winter, H., Huber, J. L. and Huber, S. C. (1997). Membrane association of sucrose synthase: changes during the graviresponse and possible control by protein phosphorylation. *FEBS Letters* **420**, 151–155.

Winter, H., Huber, J. L. and Huber, S. C. (1998). Identification of sucrose synthase as an actin binding protein. *FEBS Letters* **430**, 205–208.

Zhang, X.-Q. and Chollet, R. (1997). Seryl-phosphorylation of soybean nodule sucrose synthase (nodulin-100) by Ca^{2+}-dependent protein kinase. *FEBS Letters* **410**, 126–130.

Zhang, X.-Q., Lund, A. A., Sarath, G., Cerny, R. L., Roberts, D. M. and Chollet, R. (1999). Soybean nodule sucrose synthase (nodulin-100): analysis of its phosphorylation using recombinant and authentic nodule enzymes. *Archives of Biochemistry and Biophysics* **371**, 70–82.

Protein Phosphorylation and Ion Transport: A Case Study in Guard Cells

JIAXU LI and SARAH M. ASSMANN*

Department of Biology, 208 Mueller Laboratory, The Pennsylvania State University, University Park, PA 16802, USA

I. PROTEIN PHOSPHORYLATION

Protein phosphorylation is widely recognized as a common and important mechanism for regulation of cellular processes. Protein phosphorylation regulates proteins involved in a range of cellular processes, from membrane channels and pumps, through metabolic enzymes and cytoskeletal proteins, to transcription factors (Hardie, 1995). Even protein kinases and phosphatases themselves are subject to modulation by phosphorylation. The activities of protein kinases are highly regulated by activatory and inhibitory signals. Both extracellular ligands such as hormones and growth factors and key intracellular signaling molecules such as cyclic adenosine monophosphate (cAMP) or Ca^{2+}

*Author to whom correspondence should be addressed; e-mail: sma3@psu.edu

Advances in Botanical Research Vol. 32
incorporating Advances in Plant Pathology
ISBN 0-12-005932-0

can regulate the activities of protein kinases (Schenk and Snaar-Jagalska, 1999). In addition, for many protein kinases, autophosphorylation or phosphorylation by a separate kinase is required for activation (Hunter, 1995).

The co-ordination of protein kinases, which transfer the terminal phosphate group from adenosine triphosphate (ATP) to one or more amino acid residues within a protein, and protein phosphatases, which cleave the phosphate group(s) from the protein, endow protein phosphorylation with reversibility. The property of reversibility of protein phosphorylation provides an extraordinarily sensitive means by which the activity of a protein can be enhanced or diminished. Therefore, protein phosphorylation allows for switching of cellular processes from state to state in response to changing circumstances.

Protein kinases can be classified according to the amino acid residue(s) that they phosphorylate. For example, in eukaryotes, there are serine/threonine kinases, tyrosine kinases, and histidine kinases. However, this classification is not absolute, since protein kinases that can phosphorylate serine/threonine and tyrosine residues (dual-specificity protein kinases) have been reported (for review see Lindberg *et al.*, 1992). Similarly, protein phosphatases can be grouped into serine/threonine phosphatases, tyrosine phosphatases, and dual-specificity phosphatases (Hunter, 1995). Based on substrate specificity and sensitivity to endogenous inhibitors, serine/threonine phosphatases can be divided into type 1 protein phosphatases (PP1) and type 2 protein phosphatases (PP2). PP2s can be further divided into three subgroups, PP2A, PP2B, and PP2C, according to subunit structure, divalent cation requirement and regulatory properties (Cohen, 1989). For instance, the activities of PP2B and PP2C depend on the presence of Ca^{2+} and Mg^{2+}, respectively; PP1/PP2A can be selectively inhibited by okadaic acid, while PP2B can be specifically inhibited by cyclophilin–cyclosporin A and FKBP (FK506-binding protein)–FK506 complexes (see Table I for a summary of protein phosphatase inhibitors).

Serine/threonine kinases and tyrosine kinases contain a homologous region of about 250–300 amino acid residues which is responsible for the phosphotransferase reaction and is known as the catalytic domain (Hanks and Quinn, 1991). The catalytic domain can be further divided into 12 highly conserved subdomains which are invariant or nearly invariant among the members of certain subgroups of protein kinases (Hanks and Quinn, 1991). The conserved catalytic domains of protein kinases are flanked by highly variable regulatory domains, which contain signal-binding sites (e.g. for Ca^{2+}), regulatory subunit/protein-binding sites (e.g. for calmodulin), membrane-spanning segments (e.g. in receptor protein kinases), etc. These variable regulatory domains account for diverse regulation modes by different signals.

The specificity of protein kinases is essential for the accurate regulation of cellular processes by phosphorylation. The local amino acid sequence around the phosphorylation site may play an important role in recognition of the

TABLE I
Inhibitors of serine/threonine protein phosphatases

Inhibitor	Half-inhibitory concentration				Reference
	PP1	PP2A	PP2B	PP2C	
Okadaic acid	10–15 nM	~0.1 nM	~5 μM	NA[a]	Cohen (1989)
Calyculin A	2 nM	1 nM	NA	NA	Ishihara *et al.* (1989)
Cyclosporin A/ cyclophilin A	NA	NA	33 nM	NA	Liu *et al.* (1992)
FK506/FKBP12	NA	NA	32 nM	NA	Liu *et al.* (1992)
EDTA[b]	NA	NA	Chelate Ca^{2+}	Chelate Mg^{2+}	Hunter (1995)

[a]NA (not applicable) means that the inhibitor has no significant effect on the phosphatase.
[b]Ethylenediaminetetraacetic acid.

substrate by a protein kinase. For many protein kinases, specific consensus phosphorylation sites have been determined (Pearson and Kemp, 1991). Substrate specificity can also be achieved by intracellular translocation, targeting of kinases to a subcellular location through anchoring proteins, or association of kinases with their substrates via scaffold proteins (Inagaki *et al.*, 1994; Faux and Scott, 1996). The specificity, reversibility, and rapidity (minutes or even seconds) of protein phosphorylation provide an efficient means for regulation of complex and highly controlled and co-ordinated processes such as stomatal movement, the focus of this chapter.

II. INVOLVEMENT OF PROTEIN PHOSPHORYLATION IN STOMATAL APERTURE CHANGE

A. STOMATAL MOVEMENT

Stomata allow plants to take up CO_2 from the atmosphere for photosynthesis and to release water vapor to the air via transpiration. A pair of guard cells defines and regulates the aperture of each stomatal pore. Changes in the turgor of guard cells resulting from osmotically driven fluxes of water into or out of the cell cause guard cells to swell or shrink, which leads to opening or closure of stomata. Uptake of K^+ and Cl^- and production of organic solutes lower the water potential of guard cells, thereby stimulating osmotic uptake of water into guard cells, an increase in turgor and guard cell swelling. In contrast, release of K^+ and anions from guard cells and removal of organic osmotica lead to osmotic loss of water and guard cell shrinkage. Since some of the major ion transporters involved in stomatal opening are distinct from those functioning in stomatal closure, the two processes are not simply the reverse of each other (see Fig. 1 for a depiction of the major transporters involved in guard cell function).

During stomatal opening, the plasma membrane H^+-ATPase hyperpolarizes the plasma membrane (making the cytosol more negative relative to the outside of the cell) by pumping protons out of guard cells, thus generating the driving force for K^+ uptake. The uptake of K^+ into the guard cell cytosol is mediated by the plasma-membrane inward K^+ channels, which are activated by

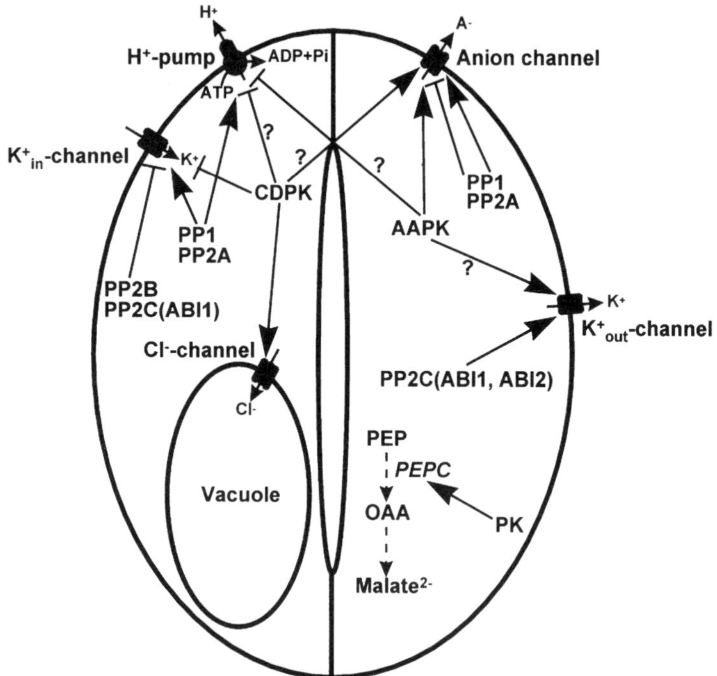

Fig. 1. Possible scheme for regulation of functionally important guard cell proteins by phosphorylation and dephosphorylation. Arrows with ion symbols indicate the direction of the movement of the ion through the channel or pump. Arrows followed by a protein kinase or phosphatase indicate an activation or enhancement by the protein kinase or phosphatase. Blocked arrows indicate an inactivation or inhibition by the protein kinase or phosphatase. The effects of PP1 and PP2A on the anion channel can be stimulatory or inhibitory, depending on the species (Grabov *et al.*,1997; Pei *et al.*, 1997). A, anions; AAPK, abscisic-acid-activated protein kinase from guard cells; ABI1 and ABI2, the products of the wild-type genes of the abscisic acid insensitive mutants *abi1* and *abi2*, respectively, whose deduced protein sequences are homologous to PP2C, CDPK, calcium-dependent protein kinase; OAA, oxaloacetate; PEP, phosphoenolpyruvate; PEPC, phosphoenolpyruvate carboxylase; PP1, PP2A, PP2B, and PP2C, type 1, 2A, 2B, and 2C serine/threonine protein phosphatase, respectively; PK, an unidentified protein kinase.

hyperpolarization. K^+ in the cytosol is further transported into the vacuole. Because of the unfavorable electrochemical gradient (the vacuole is positive relative to the cytosol), the uptake of K^+ into the vacuole is probably coupled to certain tonoplast energizing transporters, such as the vacuolar H^+-translocating pyrophosphatase (Darley *et al.*, 1998), which has been shown to facilitate K^+ accumulation in the vacuoles of *Beta vulgaris* storage tissue (Davies *et al.*, 1992). Proton extrusion by the plasma-membrane H^+-ATPase also energizes Cl^-/OH^- antiport or Cl^-/H^+ symport (Zeiger *et al.*, 1978). Cl^- entering the cytosol may be further taken up by the vacuole of the guard cell through a tonoplast Cl^-/malate^{2-} channel (Pei *et al.*, 1996). In addition to ion fluxes across the plasma membrane and the tonoplast, production of organic solutes such as malic acid also plays an important role in stomatal opening. The synthesis of malic acid from starch during stomatal opening provides both protons which replenish protons lost via proton extrusion by the plasma-membrane H^+-ATPase, and malate^{2-} which serves as a counter-ion for K^+ uptake, thus maintaining pH and charge balance in the guard cells. Sucrose is also an organic osmoticum for guard cell osmoregulation under some environmental conditions (for review see Talbott and Zeiger, 1998). Accumulation of sucrose in guard cells during stomatal opening may be due to import of sucrose from the apoplast, sucrose synthesis from the products of starch breakdown, and production of sucrose via the photosynthetic carbon reduction pathway (Assmann, 1993).

Removal of K^+, anions, and organic osmotica out of guard cells, which results in a decrease of turgor, is required for stomatal closing. Since most of the K^+ is stored in guard cell vacuoles, transport of K^+ from the vacuole to the cytosol must occur during stomatal closing. Tonoplast K^+-permeable channels such as the fast vacuolar (FV) channel, the slow vacuolar (SV) channel, and the vacuolar K^+-selective (VK) channel may mediate the efflux of K^+ from the guard cell vacuoles (Ward *et al.*, 1995; Allen *et al.*, 1998). K^+ efflux from the guard cell cytosol is mediated by plasma-membrane outward K^+ channels that are activated by membrane depolarization (making the cytosol less negative). Depolarization generates the driving force for K^+ efflux required for stomatal closure. The efflux of anions (e.g. Cl^- and malate^{2-}) from guard cells and inhibition of the H^+-ATPase both contribute to membrane depolarization. Efflux of anions from the guard cell cytosol is mediated by plasma membrane anion channels such as the S-type (slow) anion channel and R-type (rapid) anion channel (for review see Ward *et al.*, 1995). Removal of organic osmotica may be achieved by metabolism via the TCA cycle, conversion to starch (e.g. malate to starch), and export to the apoplast (Assmann, 1993).

B. PROTEIN PHOSPHORYLATION AND STOMATAL APERTURE CHANGES

Many environmental factors such as light, humidity, and CO_2 concentration, as well as endogenous signals such as hormones, regulate the apertures of stomata, and are believed to do so through modulation of ion transport activity and carbon metabolism (Assmann, 1993; Assmann and Shimazaki, 1999). Roles of G-proteins, Ca^{2+}, and pH in mediating regulation of stomata by environmental and hormonal signals have been well reviewed (Ward *et al.*, 1995; Assmann, 1996a, b; Thiel and Wolf, 1997; Grabov and Blatt, 1998; Assmann and Shimazaki, 1999). Protein phosphorylation is now also emerging as an important component of stomatal regulation.

Initial evidence for the involvement of protein phosphorylation in stomatal movements came from experiments showing that protein kinase and phosphatase inhibitors affected stimulus-induced stomatal responses in epidermal peels. In the leaf epidermis of *Commelina communis*, Lee and Assmann (1991) found that H-7, a very effective inhibitor of protein kinase C, inhibited light-induced stomatal opening. Furthermore, it was shown that synthetic diacylglycerols known to activate protein kinase C in animal systems stimulated stomatal opening, whereas a diacylglycerol that does not activate protein kinase C did not affect stomatal responses (Lee and Assmann, 1991). In another study with *Commelina benghalensis* epidermis, H-7 was reported to have no effect on light-induced stomatal opening, while ML-9 (an inhibitor of Ca^{2+}/calmodulin-dependent myosin light chain kinase) and W-7 (a calmodulin antagonist) inhibited light-induced stomatal opening (Shimazaki *et al.*, 1992). In contrast to the results of Shimazaki *et al.* (1992), Cousson *et al.* (1995) observed that ML-7 (an inhibitor of Ca^{2+}/calmodulin-dependent myosin light chain kinase) and W-7 induced an increase in stomatal aperture in *C. communis* under light. Although it is difficult to reconcile the results from the different groups, these studies do suggest that protein phosphorylation is involved in light-induced stomatal response. In addition, abscisic acid (ABA)-induced stomatal closure in *C. communis*, *Pisum sativum*, and *Vicia faba* could be blocked by K252a, a broad range inhibitor of serine/threonine protein kinases, and could be enhanced by the PP1/PP2A inhibitor okadaic acid (Schmidt *et al.*, 1995; Hey *et al.*, 1997), suggesting that phosphorylation may also play a role in stomatal response to ABA.

Because the effectiveness of pharmacological inhibitors in intact cells depends on many factors such as membrane permeability, residence time in the cell, and accessibility to targets, and these factors are difficult to bring under experimental control, similar experiments may give different and even controversial results, e.g. Shimazaki *et al.* (1992) versus Cousson *et al.* (1995). In addition, owing to the highly conserved nature of kinase active sites and lack of highly selective kinase inhibitors (Bishop *et al.*, 1998), studies that determine effects of inhibitors on a given stomatal response with intact cells and tissues cannot provide definitive information about the specific kinases involved and their targets.

III. GUARD CELL PROTEINS MODULATED BY PHOSPHORYLATION

A. INTRODUCTION

One approach to reveal the importance of protein phosphorylation in a response is to detect phosphoproteins whose phosphorylation status changes in response to the stimulus. Using this strategy, phosphorylation of a number of proteins from guard cells has been found to be stimulated by ABA, cAMP, or Ca^{2+} (Kinoshita and Shimazaki, 1995; Li and Assmann, 1996; Friedrich *et al.*, 1997; Li *et al.*, 1998). However, the identity of almost all of these guard cell phosphoproteins remains to be determined. Another approach to assess the role of protein phosphorylation in stomatal regulation is to examine the effects of inhibitors of protein kinases/phosphatases or purified protein kinases/phosphatases on the activity of a key guard cell protein whose activity can be assayed, e.g. an ion channel monitored by the patch clamp technique. As summarized in Fig. 1, much information regarding phosphorylation in stomatal regulation has been derived from studies using this approach.

B. TRANSPORT MOLECULES

1. H^+-ATPase

The plasma membrane H^+-ATPase in guard cells provides the driving force for K^+ and Cl^- uptake by pumping protons out of the cell. Many stomatal opening stimuli such as light, auxin, and fusicoccin activate the guard cell proton pump (Assmann and Shimazaki, 1999), while ABA and elevated cytosolic Ca^{2+}, which induce stomatal closure, inhibit this transporter (Kinoshita *et al.*, 1995; Goh *et al.*, 1996). Studies using activators or inhibitors of protein kinases or phosphatases suggest that phosphorylation and dephosphorylation may be involved in the regulation of the proton pump of guard cells. For example, in *V. faba*, synthetic diacylglycerols known to activate protein kinase C in other systems stimulated an ATP-dependent and voltage-independent current, indicating activation of an ATP-dependent pump such as the proton pump (Lee and Assmann, 1991). ML-9, an inhibitor of Ca^{2+}/calmodulin-dependent myosin light chain kinase, and the PP1/PP2A inhibitors calyculin A and okadaic acid inhibited blue light-dependent proton pumping in guard cells of *V. faba* (Shimazaki *et al.*, 1992; Kinoshita and Shimazaki, 1997). These studies fall short of determining that the guard cell plasma-membrane H^+-ATPase itself is a direct target of protein kinase(s) or phosphatase(s), but there is evidence from other plant systems that the plasma-membrane H^+-

ATPase can be phosphorylated by protein kinases and dephosphorylated by phosphatases (Schaller and Sussman, 1988; Xing *et al.*, 1996).

Dephosphorylation of the plasma-membrane H^+-ATPase from tomato cell cultures by an okadaic acid-sensitive protein phosphatase resulted in stimulation of its proton-pump activity (Vera-Estrella *et al.*, 1994). Strategies such as immunoprecipitation of the plasma-membrane proteins from [^{32}P]orthophosphate-incubated guard cell protoplasts with antibody to the plasma-membrane H^+-ATPase or *in vitro* phosphorylation assay by incubation of a purified guard cell kinase, e.g. CDPK or AAPK (see Section IV) with purified recombinant H^+-ATPase protein from a plasma-membrane H^+-ATPase cDNA of guard cells (Nakajima *et al.*, 1995; Henzen *et al.*, 1996), could be utilized to determine whether the guard cell plasma-membrane H^+-ATPase can be phosphorylated. Success in monitoring the activity of a purified H^+-ATPase incorporated into a planar bilayer system (Briskin *et al.*, 1995) also provides a system which conceivably could be used to determine which kinds of protein kinases and phosphatases can directly regulate activity of the guard cell H^+-ATPase. Such studies would be very useful to determine whether modulation of the guard cell proton pump by signals such as ABA and Ca^{2+} is mediated by direct phosphorylation of the H^+-ATPase.

2. K^+ *Channels*

It is well documented that elevated cytosolic Ca^{2+} inactivates the inward K^+ channels in the plasma membrane of guard cells (for reviews see Assmann, 1993; Maathuis *et al.*, 1997; Thiel and Wolf, 1997). Phosphorylation and dephosphorylation appear to be involved in Ca^{2+} inactivation of the inward K^+ channels. Inhibitors of the Ca^{2+}, calmodulin-dependent phosphatase calcineurin (PP2B) reversed Ca^{2+} inactivation of the inward K^+ channels in *V. faba* guard cells, while a constitutively active and Ca^{2+}-independent calcineurin fragment inhibited the activity of the inward K^+ channels in the absence of Ca^{2+} (Luan *et al.*, 1993), suggesting that a calcineurin-like phosphatase is involved in Ca^{2+} inactivation of the inward K^+ channels. In contrast, the cloned inward K^+ channel KAT1, whose gene is primarily expressed in guard cells (Nakamura *et al.*, 1995), could be phosphorylated *in vitro* by a guard cell calcium-dependent protein kinase (CDPK) from *V. faba* in a Ca^{2+}-dependent manner (Li *et al.*, 1998). Inward K^+ currents through KAT1 were greatly diminished by co-expression of KAT1 and CDPK in *Xenopus* oocytes (Kamasani *et al.*, 1997), indicating that phosphorylation of KAT1 by CDPK can inhibit inward K^+ currents through KAT1. Taken together, these data suggest that Ca^{2+}-induced inactivation of inward K^+ channels may be mediated by direct phosphorylation of the channels by CDPK and/or by dephosphorylation of either intermediary regulatory protein(s) or of the channels by calcineurin at amino acid residue(s) distinct from those CDPK targets. In addition, different types of protein phosphatases seem to exert different effects on the inward K^+ channels. For example, opposite to PP2B,

PP1 and PP2A appear to activate inward K^+ channels (Li *et al.*, 1994; Thiel and Blatt, 1994). Similar to Ca^{2+}, ABA inhibits inward K^+ channels in guard cells (for review see Grabov and Blatt, 1998). Studies on K^+ channels of tobacco transformed with the dominant mutant *abil–1* gene, which encodes a PP2C (Leung *et al.*, 1994; Meyer *et al.*, 1994), showed that ABA-induced inactivation of inward K^+ channels was suppressed, suggesting that phosphorylation and dephosphorylation play a role in the response of the inward K^+ channels to ABA (Armstrong *et al.*, 1995).

The outward K^+ channels of guard cells also seem to be modulated by phosphorylation and dephosphorylation. In *abil–1*-transformed tobacco plants, ABA was unable to activate the outward K^+ channels of guard cells, whereas ABA could activate the outward K^+ channels of guard cells in vector-transformed plants (Armstrong *et al.*, 1995), suggesting that a PP2C is involved in ABA activation of the outward K^+ channels. In addition, Thiel and Blatt (1994) found that the PP1/PP2A inhibitor, okadaic acid, blocked the activity of outward K^+ channels in *V. faba* guard cells. In contrast, Li *et al.* (1994) found that okadaic acid had no effect on the outward K^+ channel of *V. faba* guard cell. The different methods used in these two studies, for example, microelectrode-impalement of intact guard cells in one study and patch-clamping of guard cell protoplasts in the other study, might explain the different results.

From numerous studies, it is clear that protein phosphorylation and dephosphorylation is involved in modulation of K^+ channels in guard cells. However, in all cases except for the phosphorylation of KAT1 by CDPK (Li *et al.*, 1998), it is not clear whether the modulation of K^+ channels is due to direct phosphorylation and/or dephosphorylation of the channels themselves. Application of site-direct mutagenesis to cloned K^+ channel genes such as *KAT1* and *KST1* that are expressed in guard cells (for review see Maathuis *et al.*, 1997; Müller-Röber *et al.*, 1998) would allow the determination of not only whether the K^+ channel is a direct target for a given protein kinase or phosphatase but also which amino acid residue(s) is subject to the modification.

3. *Anion Channels*

Anion channels of guard cells are also subject to modulation by protein phosphorylation and dephosphorylation. For example, in *V. faba* guard cells, Schmidt *et al.* (1995) found that activity of plasma-membrane slow anion channels could be inactivated by protein kinase inhibitor K252a or by removal of cytosolic ATP, while the PP1/PP2A inhibitor, okadaic acid, prevented the inactivation of slow anion channel caused by depletion of ATP, suggesting that protein kinases are involved in channel activation. In *Nicotiana benthamiana*, recording of anion current in microelectrode-impaled intact guard cells showed that ABA activated slow anion channels (Grabov *et al.*, 1997). The ABA action on slow channels could be mimicked by calyculin A, an inhibitor of PP1/PP2A, whereas H-7 and staurosporine, both very effective inhibitors of protein kinase C, did not affect ABA activation of slow anion channels (Grabov *et al.*, 1997). One explanation of these data would be that activation of slow anion channels is downregulated by a calyculin A-sensitive protein phosphatase. Then activation of slow anion channels could be achieved by either inhibition of the phosphatase by calyculin A in the absence of ABA or by reversal of phosphatase action by a H-7/staurosporine-insensitive protein kinase activated by ABA. By contrast to *V. faba* and *N. benthamiana*, ABA activation of slow anion channels in *Arabidopsis* guard cells was inhibited by the PP1/PP2A inhibitor okadaic acid (Pei *et al.*, 1997). In addition, Pei *et al.* (1997) observed that ABA activation of slow anion channels was abolished in guard cells of the *Arabidopsis abi1–1* and *abi2–1* mutants [like *ABI1*, *ABI2* also encodes a PP2C homolog (see Leung *et al.*, 1997; Rodriguez *et al.*, 1998)], while Grabov *et al.* (1997) could not detect any difference in ABA activation of slow anion channels between wild-type and *abi1–1* transgenic *N. benthamiana*. Whether the different results are due to the presence of different regulatory components in different species, or different *abi1–1* background, i.e. *abi1–1 Arabidopsis* mutants versus *abi1–1*-transformed tobacco plants, or different methods, i.e. patch clamp of guard cell protoplasts versus microelectrode impalement of intact guard cell, remains to be investigated.

Like the plasma-membrane slow anion channels, guard cells' tonoplast anion channels may also be modulated by protein phosphorylation/dephosphorylation. For instance, Pei *et al.* (1996) observed that recombinant CDPK activated a Cl^-/malate^{2-} channel in the tonoplasts of *V. faba* guard cells in a Ca^{2+}-dependent manner while a constitutively active Ca^{2+}-independent mutant CDPK could activate the Cl^-/malate^{2-} channel in the absence of Ca^{2+}. Since the Cl^-/malate^{2-} channel could be a pathway for anion uptake into vacuoles of guard cells, these studies suggest that CDPK may play a role in anion uptake into guard cell vacuoles during stomatal opening (Pei *et al.*, 1996).

It is obvious from these studies that phosphorylation and dephosphorylation are involved in anion channel modulation at both the plasma membrane and tonoplast. However, it is at present unclear whether protein kinases and phosphatases involved in anion channel modulation act on the channels or on their regulatory proteins.

4. *Water Channels*

Besides ion channels, biological membranes also contain water channels that can facilitate and regulate water movement. Water channel proteins or 'aquaporins' have been found in a variety of organisms (Chrispeels and Agre, 1994). In plants, aquaporins are found in both the plasma membrane and the tonoplast (Maurel, 1997). The high water permeability of plant aquaporins has been demonstrated by swelling assays with *Xenopus* oocytes injected with aquaporin messenger RNA (mRNA) or with isolated protoplasts from transgenic plants expressing aquaporin genes (Kaldenhoff *et al.*, 1995; Maurel, 1997). Fundamentally, stomatal movement is achieved via transmembrane water flow between the apoplast, the cytosol, and the vacuole of guard cells. Therefore, it is not surprising that both plasma-membrane and tonoplast aquaporin genes are expressed in guard cells (Kaldenhoff *et al.*, 1995; Sarda *et al.*, 1997). The question is how the activity of aquaporins in guard cells is regulated, and protein phosphorylation appears to be a likely mechanism. For example, when the aquaporin α-TIP (tonoplast intrinsic protein) was expressed in *Xenopus* oocytes, a four- to eight-fold increase in osmotic water permeability was observed (oocyte swelling assay); mutation of the putative phosphorylation sites in α-TIP resulted in a decrease of α-TIP-mediated water transport in oocytes (Maurel *et al.*, 1995). In addition, α-TIP in isolated oocyte membranes could be phosphorylated by the catalytic subunit of bovine cAMP-dependent protein kinase and exposure of oocytes to agonists that stimulate the activity of cAMP-dependent protein kinase increased the water transport activity of a α-TIP (Maurel *et al.*, 1995). Similarly, plasma membrane aquaporins may also be regulated by phosphorylation. For instance, a 28-kDa plasma-membrane aquaporin from spinach leaves called PM28A could be phosphorylated *in vivo* at serine-274 (Johansson *et al.*, 1998). The water transport activity of PM28A expressed in oocytes (oocyte swelling assay) was inhibited by the protein kinase inhibitor K252a and enhanced by the PP1/PP2A inhibitor okadaic acid (Johansson *et al.*, 1998). Regulation of aquaporin activity by phosphorylation in other cell types suggests that guard cell aquaporins are probably also subject to modulation by phosphorylation.

C. NON-TRANSPORT MOLECULES

1. LHCPII

By incubation of guard cell protoplasts with [^{32}P]orthophosphate, Kinoshita *et al.* (1993) found that red light caused dephosphorylation of a 26-kDa protein. The red-light-induced dephosphorylation of the 26-kDa protein could be inhibited by 1 nM okadaic acid, suggesting that PP2A was involved. It was concluded that the 26-kDa protein is LHCPII (Kinoshita *et al.*, 1993) because:

1. The 26-kDa protein was localized in chloroplasts.
2. It has the same molecular mass as LHCPII (light-harvesting chlorophyll a/b protein complex of photosystem II).
3. DCMU (an inhibitor of electron transfer in photosystem II) stimulated dephosphorylation of the 26-kDa protein under light of 660 nm.
4. The strong reducing agent sodium hydrosulfite increased the phosphorylation of the 26-kDa protein under illumination.

However, the physiological significance of the red-light-induced dephosphorylation of this protein relative to stomatal function remains to be determined.

2. Metabolic Enzymes

In addition to membrane transport proteins, key enzymes involved in the production of malate^{2-} and sucrose also play important roles in stomatal movement. Phosphoenol-pyruvate carboxylase (PEPC), an enzyme involved in malate production during stomatal opening, has been partially purified from guard cells of *V. faba* (Denecke *et al.*, 1993). Like PEPC from other plant sources, the guard cell PEPC is subject to allosteric control by metabolites; for example, the activity of PEPC is inhibited by malate^{2-} and activated by glucose-6-phosphate (Denecke *et al.*, 1993). In addition to allosteric regulation, reversible phosphorylation is an important mechanism for controlling the activity of PEPC (Chollet *et al.*, 1996). Phosphorylation of PEPC by a Ca^{2+}-independent protein kinase (PEPC kinase) leads to an increase in the activity of leaf PEPC, while dephosphorylation of PEPC by a PP2A appears to decrease PEPC activity (Chollet *et al.*, 1996). PEPC from *V. faba* guard cells could be phosphorylated *in vitro* and *in vivo* (Schnabl *et al.*, 1992). However, unlike C_3-leaf PEPCs (Li *et al.*, 1996), phosphorylation of guard cell PEPC does not appear to be light regulated (Schnabl *et al.*, 1992). Fusicoccin, which induces stomatal opening, stimulated PEPC phosphorylation and malate accumulation in guard cells of *V. faba*, while ABA inhibited malate accumulation and suppressed PEPC phosphorylation (Du *et al.*, 1997). These data imply that signals regulating guard cell PEPC kinase may not be identical to those for other plant cell types. Intensive studies indicate that in C_4 and CAM plants phosphorylation of photosynthetic PEPC at a single serine residue near the N-terminus leads to an increase in its catalytic activity (Chollet *et al.*, 1996). In this regard, it is interesting to note that a PEPC cDNA clone isolated from potato guard cells has this highly conserved serine residue at the N-terminal region (Müller-Röber *et al.*, 1998). Molecular approaches such as site-directed mutagenesis could provide definite information on the regulatory sites of guard cell PEPC. Further identification of PEPC kinase from guard cells and elucidation of its mode of regulation are awaited.

Besides K^+ and its counter-ions Cl^- and malate^{2-}, sucrose is an important solute for guard cell osmoregulation (Talbott and Zeiger, 1998, and references therein). Sucrose synthesis is largely controlled by sucrose-phosphate synthase (SPS), whereas sucrose breakdown is mainly catalyzed by sucrose synthase (SuSy; for review see Huber and Huber, 1996). Plant SPS can be phosphorylated on multiple seryl residues *in vivo* (Huber and Huber, 1996). For example, in spinach leaves, phosphorylation of SPS at serine-158 inactivated SPS, while phosphorylation of the phosphoserine-158-SPS at serine-424 by a Ca^{2+}-dependent kinase led to activation of SPS (Toroser and Huber, 1997). Similar to SPS, SuSy is also subject to phosphorylation regulation. Phosphorylation of maize leaf SuSy at serine-15 by a Ca^{2+}-dependent kinase(s) resulted in activation of SuSy (Huber *et al.*, 1996). In *V. faba*, guard cells have very high levels of SPS and SuSy (Hite *et al.*, 1993). Determination of the modes of regulation of SPS and SuSy in guard cells would deepen our understanding concerning the role of sucrose in stomatal regulation.

IV. PROTEIN KINASES AND PHOSPHATASES IN GUARD CELLS

A. PROTEIN KINASES

Identification of protein kinases in guard cells is crucial to elucidation of the role of protein phosphorylation in stomatal movement. Data obtained from different approaches suggest that guard cells may have both protein kinases found in other cell types and guard cell-specific protein kinase(s). In addition to the pharmacological inhibitor studies discussed earlier, molecular studies on guard cells have also provided important clues about protein kinases in guard cells. By evaluation of 515 ESTs (expressed sequence tags) from a *Brassica campestris* guard cell cDNA library (Kwak *et al.*, 1997), ESTs were found that showed significant similarity to MAP kinase (mitogen-activated protein kinase) and receptor-like protein kinase. The MAP kinases have been found in all eukaryotes. In plants, MAP kinases have been suggested to be involved in several signal transduction pathways including those for drought stress (for review see Hirt, 1997). The receptor-like protein kinases (RLKs) are a group of serine/threonine protein kinases found in plants that are structurally similar to the animal receptor protein kinases of which most are tyrosine protein kinases (for review see Braun and Walker, 1996). Because RLKs have an extracellular domain, a transmembrane domain, and a cytoplasmic catalytic domain, it has been suggested that RLKs may function as receptors for transducing diverse extracellular signals into different cellular responses (Braun and Walker, 1996). Recently, it was found that ABA and dehydration could induce expression of a receptor-like protein kinase gene in *Arabidopsis* leaves within 1 h (Hong *et al.*, 1997). Therefore, cloning of MAP kinase and receptor-like protein kinase genes from guard cells and determination of the effects of ABA and drought stress on their kinase activity or gene expression may provide important clues about the roles of protein phosphorylation in stomatal responses to ABA and water stress.

Ca^{2+}-dependent protein kinases (CDPKs) comprise a major group of protein kinases found in plants and are believed to be one of the most important mediators for the second messenger Ca^{2+} (Roberts and Harmon, 1992). Recently, a CDPK has been identified in guard cells of *V. faba* using biochemical approaches (Li *et al.*, 1998). Both autophosphorylation and catalytic activity of this kinase were Ca^{2+} dependent; the Ca^{2+}-dependent catalytic activity of this kinase could be inhibited by the calmodulin antagonists TFP and W-7. Furthermore, this kinase could be recognized by an affinity-purified soybean CDPK antibody (Li *et al.*, 1998). Guard cell CDPK may participate in the regulation of the plasma membrane inward K^+

channel, the tonoplast Cl^-/malate^{2-} channel or other Ca^{2+}-regulated activities of guard cells (Pei *et al.*, 1996; Li *et al.*, 1998).

Besides several protein kinases which are well known to exist in other sources, guard cells may have unique protein kinases. For instance, an ABA-activated and Ca^{2+}-independent protein kinase known as AAPK (ABA-activated protein kinase) has been found in guard cells of *V. faba* (Li and Assmann, 1996; Mori and Muto, 1997). In-gel activity assays of kinases from guard cells, mesophyll cells, epidermal cells, and root tissues of *V. faba* demonstrated that the ABA-activated and Ca^{2+}-independent autophosphorylation or kinase activity was only present in guard cells, suggesting that AAPK may be guard cell specific (Li and Assmann, 1996; Mori and Muto, 1997). The activation of AAPK by ABA occurs within minutes and at physiologically relevant ABA concentrations. Two-dimensional phosphoamino acid analysis demonstrated that AAPK is a serine/threonine protein kinase (Li and Assmann, 1996). Therefore, AAPK may represent a unique serine/threonine protein kinase found in guard cells. Determination of AAPK function in ABA signaling in guard cells would provide a better understanding of the role of protein phosphorylation in ABA regulation of stomatal aperture.

B. PROTEIN PHOSPHATASES

A number of pharmacological inhibitor studies suggest that guard cells have protein phosphatases akin to PP1 and PP2A (Li *et al.*, 1994; Thiel and Blatt, 1994; Armstrong *et al.*, 1995; Schmidt *et al.*, 1995; Hey *et al.*, 1997; Kinoshita and Shimazaki, 1997; Pei *et al.*, 1997; Grabov and Blatt, 1998). Protein phosphatase activity assays using mammalian phosphoprotein substrates further support this notion. In protein extract from blended *V. faba* leaf epidermal peels (which contain guard cells as the only intact cell type), about 50% of total protein phosphatase activity (assays in the presence of Ca^{2+} and Mg^{2+}) was sensitive to okadaic acid (see Table I for a summary of protein phosphatase inhibitors), indicating that PP1 and PP2A accounted for about 50% of total protein phosphatase activity (Luan *et al.*, 1993). Interestingly, Luan *et al.* (1993) also observed a Ca^{2+}-dependent protein phosphatase activity in epidermal peel protein extract that could be inhibited by cyclophilin–cyclosporin A and FKBP–FK506 complexes, suggesting that epidermal peels (most likely guard cells) may have a PP2B-like phosphatase. Recently, a phosphatase activity stimulated by 1 μM Ca^{2+} and inhibited by cyclosporin A has been detected in guard cells of *V. faba* (Kinoshita and Shimazaki, 1999a), confirming that guard cells do have a PP2B-like activity.

Consistent with this conclusion, in guard cells of *V. faba*, PP2B inhibitors blocked Ca^{2+}-induced inactivation of inward K^+ channels, reversed the inhibitory effect of PP2B on a Ca^{2+}-permeable slow vacuolar channel, and enhanced phosphorylation of several proteins in a Ca^{2+}-dependent manner

(Luan *et al.*, 1993; Allen and Sanders, 1995; Li *et al.*, 1998). The Ca^{2+}-dependent PP2B-like phosphatase and a Ca^{2+}-independent and okadaic-acid-insensitive phosphatase activity accounted for the other 50% of the total protein phosphatase activity in blended *V. faba* epidermal peels (Luan *et al.*, 1993). Since the protein phosphatase activity assays were performed in the presence of Mg^{2+} (Luan *et al.*, 1993), the Ca^{2+}-independent and okadaic-acid-insensitive phosphatase activity probably represents activity of PP2C in the epidermal peels of *V. faba*.

Studies with the *Arabidopsis* ABA-insensitive mutant *abi1*, whose deduced protein sequence is homologous to PP2C (Leung *et al.*, 1994; Meyer *et al.*, 1994), indicate that guard cells have a PP2C which is required for stomatal response to ABA. For example, the *abi1–1* mutation eliminated the abilities of inward K^+ channels, outward K^+ channels, and anion channels of guard cells to respond to ABA (Armstrong *et al.*, 1995; Pei *et al.*, 1997). Biochemical assay of the recombinant *ABI1* protein demonstrated that ABI1 does have phosphatase activity characteristic of a PP2C, while the phosphatase activity of abi1 was reduced (Bertauche *et al.*, 1996). Furthermore, Kwak *et al.* (1997) found an EST from guard cells of *B. campestris* showing 93% identity to *ABI1* at the peptide level, supporting that PP2C exists in guard cells. Therefore, it appears that all the major classes of protein serine/threonine phosphatases (PP1, PP2A, PP2B and PP2C) are present in guard cells.

V. CONCLUSIONS AND PERSPECTIVES

Since the first report on this topic (Lee and Assmann, 1991), the study of protein phosphorylation in stomatal regulation has expanded. Phosphorylation/dephosphorylation modulation of ion channels has been revealed and some guard cell protein kinases and phosphatases have been identified. The functional effects of phosphorylation on ion channels appear to be complex (see Fig. 1). A given channel may be regulated by several different protein kinases and phosphatases, each modulating channel activity in a distinct way. However, the same protein kinase or phosphatase may exert opposite effects on stomatal movement by modulating different effectors. For instance, CDPK probably exerts an inhibitory effect on the inward plasma membrane K^+ channel of guard cells (retarding stomatal opening), while CDPK may activate a tonoplast Cl^-/malate^{2-} channel of guard cells (contributing to stomatal opening). How regulation is achieved such that these antagonistic effects do not occur simultaneously is not known.

Although ion channels in guard cells have been shown to be modified by phosphorylation and dephosphorylation, in almost all cases it is still unclear whether the modulation is due to direct phosphorylation of the channel itself or to phosphorylation of an upstream component (regulatory proteins or enzymes) in a cascade, which, in turn, leads ultimately to modulation of channel activity. Moreover, in many cases, the identity of the kinases and phosphatases responsible is not known. Like ion channels, the guard cell H^+-ATPase and metabolic enzymes involved in synthesis of malate^{2-} or sucrose are also very likely to be subject to modulation by protein phosphorylation, but studies on these aspects are just beginning. Protein phosphorylation may also be involved in regulation of gene expression in guard cells, as implied by the observations that ABA-induced dehydrin gene expression in guard cells of *V. faba* (Shen *et al.*, 1995) and ABA-induced dehydrin gene expression in the leaf epidermis of *Pisum sativum* are affected by inhibitors of protein kinases and phosphatases (Hey *et al.*, 1997). Finally, it must be kept in mind that while protein phosphorylation as a mechanism for stomatal regulation is highlighted in this chapter, other covalent protein modifications, e.g. farnesylation (Pei *et al.*, 1998), are also important in the control of stomatal action.

In conclusion, despite progress in the studies of protein phosphorylation in stomatal regulation, the field is still at an early stage. The merging of electrophysiological, biochemical, cell biological, and molecular biological approaches will be required to generate a synthetic picture of the role of protein phosphorylation in stomatal regulation.

NOTE ADDED IN PROOF

Several advances have recently been made in characterizing the role of protein phosphorylation in guard cell signalling cascades. Kinoshita and Shimazaki (1999b) demonstrated that blue light activation of the guard cell plasma membrane H^+-ATPase is mediated by phosphorylation of the C-terminus of the H^+-ATPase protein. Li *et al.* (2000) used peptide sequence information derived by mass spectrometry to clone the cDNA encoding AAPK, and then demonstrated that this kinase mediates ABA activation of anion channels during stomatal closure.

ACKNOWLEDGEMENTS

Research in the authors' laboratory on protein phosphorylation and stomatal action is supported by an NSF grant (MCB-9874438) to S. M. Assmann.

REFERENCES

Allen G. J. and Sanders, D. (1995). Calcineurin, a type 2B protein phosphatase, modulates the Ca^{2+}-permeable slow vacuolar ion channel of stomatal guard cells. *Plant Cell* **7**, 1473–1483.

Allen, G. J., Amtmann, A. and Sanders, D. (1998). Calcium-dependent and calcium-independent K^+ mobilization channels in *Vicia faba* guard cell vacuoles. *Journal of Experimental Botany* **49**, 305–318.

Armstrong, F., Leung, J., Grabov, A., Brearley, J., Giraudat, J. and Blatt, M. R. (1995). Sensitivity to abscisic acid of guard cell K^+ channels is suppressed by *abi-1*, a mutant *Arabidopsis* gene encoding a putative protein phosphatase. *Proceedings of the National Academy of Sciences of the United States of America* **92**, 9520–9524.

Assmann, S. M. (1993). Signal transduction in guard cells. *Annual Review of Cell Biology* **9**, 345–375.

Assmann, S. M. (1996a). Guard cell G proteins. *Trends in Plant Science* **1**, 73–74.

Assmann, S. M. (1996b). G-protein regulation of plant K^+ channels. *In* "Signal Transduction in Plant Growth and Development" (D. P. S. Verma, ed.) pp. 39–61. Springer, New York.

Assmann, S. M. and Shimazaki, K. (1999). The multisensory guard cell. Stomatal responses to blue light and abscisic acid. *Plant Physiology* **119**, 809–815.

Bertauche, N., Leung, J. and Giraudat, J. (1996). Protein phosphatase-activity of abscisic-acid insensitive-1 (ABI1) protein from *Arabidopsis thaliana*. *European Journal of Biochemistry* **241**, 193–200.

Biermann, B., Johnson, E. M. and Feldman, L. J. (1990). Characterization and distribution of a maize cDNA encoding a peptide similar to the catalytic region of second messenger dependent protein kinases. *Plant Physiology* **94**, 1609–1615.

Bishop, A. C., Shah, K., Liu, Y., Witucki, L., Kung, C. and Shokat, K. M. (1998). Design of allele-specific inhibitors to protein kinase signaling. *Current Biology* **8**, 257–266.

Braun, D. M. and Walker, J. C. (1996). Plant transmembrane receptors: new pieces in the signaling puzzle. *Trends in Biochemical Sciences* **21**, 70–73.

Briskin, D. P., Basu, S. and Assmann, S. M. (1995). Characterization of the red beet plasma membrane H^+-ATPase reconstituted in a planar bilayer system. *Plant Physiology* **108**, 393–398.

Chollet, R., Vidal, J. and O'Leary, M. H. (1996). Phosphoenolpyruvate carboxylase: a ubiquitous, highly regulated enzyme in plants. *Annual Review of Plant Physiology and Plant Molecular Biology* **47**, 273–298.

Chrispeels, M. J. and Agre, P. (1994). Aquaporins: water channel proteins of plant and animal cells. *Trends in Biochemical Sciences* **19**, 421–425.

Cohen, P. (1989). Structure and regulation of protein phosphatases. *Annual Review of Biochemistry* **58**, 453–508.

Cousson, A., Cotelle, V. and Vavasseur, A. (1995). Induction of stomatal closure by vanadate or a light/dark transition involves Ca^{2+}–calmodulin-dependent protein phosphorylations. *Plant Physiology* **109**, 491–497.

Darley, C. P., Skiera, L. A., Northrop, F. D., Sanders, D. and Davies, J. M. (1998). Tonoplast inorganic pyrophosphatase in *Vicia faba* guard cells. *Planta* **206**, 272–277.

Davies, J. M., Poole, R. J., Rea, P. A. and Sanders, D. (1992). Potassium transport into vacuoles energized directly by a proton-pumping inorganic pyrophosphatase. *Proceedings of the National Academy of Sciences of the United States of America* **89**, 11701–11705.

Denecke, M., Schulz, M., Fischer, C. and Schnable, H. (1993). Partial purification and characterization of stomatal phosphoenolpyruvate carboxylase from *Vicia faba*. *Physiologia Plantarum* **87**, 96–102.

Du, Z., Aghoram, K. and Outlaw, W. H., Jr (1997). *In vitro* phosphorylation of phosphoenolpyruvate carboxylase in guard cells of *Vicia faba* L. is enhanced by fusicoccin and suppressed by abscisic acid. *Archives of Biochemistry and Biophysics* **337**, 345–350.

Faux, M. C. and Scott, J. D. (1996). Molecular glue: kinase anchoring and scaffold proteins. *Cell* **85**, 9–12.

Friedrich, P. G., Darjania, L., Curvetto, N. R. and Giusto, N. M. (1997). Cyclic AMP, its agonist Sp-cAMPS and forskolin affect the protein phosphorylation pattern in guard cell protoplasts of *Vicia faba* L. *Plant Physiology* **114**, 270.

Goh, C-H., Kinoshita, T., Oku, T. and Shimazaki, K. (1996). Inhibition of blue-light-dependent H^+ pumping by abscisic acid in *Vicia* guard-cell protoplasts. *Plant Physiology* **111**, 433–440.

Grabov, A. and Blatt, M. R. (1998). Co-ordination of signalling elements in guard cell ion channel control. *Journal of Experimental Botany* **49**, 351–360.

Grabov, A., Leung, J., Giraudat, J. and Blatt, M. R. (1997). Alteration of anion channel kinetics in wild-type and abi1-1 transgenic *Nicotiana benthamiana* guard cells by abscisic acid. *Plant Journal* **12**, 203–213.

Hanks, S. K. and Quinn, A. M. (1991). Protein kinase catalytic domain sequence database: identification of conserved features of primary structure and classification of family members. *Methods in Enzymology* **200**, 38–61.

Hardie, D. G. (1995). Cellular functions of protein kinases. *In* "The Protein Kinase FactsBook" (G. Hardie and S. Hanks, eds) pp. 48–56. Academic Press, London.

Hayashida, N., Mizoguchi, T. and Shinozaki, K. (1993). Cloning and characterization of a plant gene encoding a protein kinase. *Gene* **124**, 251–255.

Henzen, A. E., Smart, L. B., Wimmers, L. E., Fang, H. H., Schroeder, J. I. and Bennett, A. B. (1996). Two plasma membrane H^+-ATPase genes expressed in guard cells of *Vicia faba* are also expressed thought the plant. *Plant and Cell Physiology* **37**, 650–659.

Hey, S. J., Bacon, A., Burnett, E. and Neill, S. J. (1997) Abscisic acid signal transduction in epidermal cells of *Pisum sativum* L. *Argenteum*: both dehydrin mRNA accumulation and stomatal responses require protein phosphorylation and dephosphorylation. *Planta* **202**, 85–92.

Hirt, H. (1997). Multiple roles of MAP kinases in plant signal transduction. *Trends in Biochemical Sciences* **2**, 11–15.

Hite, D. R. C., Outlaw, W. H., Jr and Tarczynski, M. C. (1993). Elevated levels of both sucrose-phosphate synthase and sucrose synthase in *Vicia* guard cells indicate cell-specific carbohydrate interconversions. *Plant Physiology* **101**, 1217–1221.

Hong, S. W., Jon, J. H., Kwak, J. M. and Nam, H. G. (1997). Identification of a receptor-like protein kinase gene rapidly induced by abscisic acid, dehydration, high salt, and cold treatments in *Arabidopsis thaliana*. *Plant Physiology* **113**, 1203–1212.

Huber, S. C. and Huber, J. L. (1996). Role and regulation of sucrose-phosphate synthase in higher plants. *Annual Review of Plant Physiology and Plant Molecular Biology* **47**, 431–444.

Huber, S. C., Huber, J. L., Liao, P. C., Gage, D. A., McMichael, R. W. Jr, Chourey, P. S., Hannah, L. C. and Koch, K. (1996). Phosphorylation of serine-15 of maize leaf sucrose synthase. Occurrence *in vivo* and possible regulatory significance. *Plant Physiology* **112**, 793–802.

Hunter, T. (1995). Protein kinases and phosphatases: the yin and yang of protein phosphorylation and signaling. *Cell* **80**, 225–236.

Inagaki, N., Ito, M., Nakano, T. and Inagaki, M. (1994). Spatiotemporal distribution of protein kinase and phosphatase activities. *Trends in Biochemical Sciences* **19**, 448–452.

Ishihara, H., Martin, B. L., Brautigan, D. L., Karaki, H., Ozaki, H., Kato, Y., Fusetani, N., Watabe, S., Hashimoto, K., Uemura, D. and Hartshorne, D. J. (1989). Calyculin and okadaic acid: inhibitors of protein phosphatase activity. *Biochemical and Biophysical Research Communications* **159**, 871–877.

Johansson, I., Karlsson, M., Shukla, V. K., Chrispeels, M. J., Larsson, C. and Kjellbom, P. (1998). Water transport activity of the plasma membrane aquaporin PM28A is regulated by phosphorylation. *Plant Cell* **10**, 451–459.

Kaldenhoff, R., Kolling, A., Meyers, J., Karmann, U., Ruppel, G. and Richter, G. (1995). The blue-light-responsive *AthH2* gene of *Arabidopsis thaliana* is primarily expressed in expanding as well as in differentiating cells and encodes a putative channel protein of the plasmalemma. *Plant Journal* **7**, 87–95.

Kamasani, U., Zhang, X., Lawton, M. and Berkowitz, G. A. (1997). Ca^{2+}-dependent protein kinase modulates activity of the K^+ channel KAT1. *Plant Physiology* **114**, 197.

Kinoshita, T. and Shimazaki, K. (1995). Evidence for Ca^{2+}-dependent protein phosphorylation *in vitro* in guard cells from *Vicia faba* L. *Plant Science* **110**, 173–180.

Kinoshita, T. and Shimazaki, K. (1997). Involvement of calyculin A- and okadaic acid-sensitive protein phosphatase in the blue light response of stomatal guard cells. *Plant and Cell Physiology* **38**, 1281–1285.

Kinoshita, T. and Shimazaki, K. (1999a). Characterization of cytosolic cyclophilin from guard cells of *Vicia faba* L. *Plant and Cell Physiology* **40**, 53–59.

Kinoshita, T. and Shimazaki, K. (1999b). Blue light activates the plasma membrane H^+-ATPase by phosphorylation of the C-terminus in stomatal guard cells. *EMBO Journal* **18**, 5548–5558.

Kinoshita, T., Shimazaki, K. and Nishimura, M. (1993). Phosphorylation and dephosphorylation of guard cell proteins from *Vicia faba* L. in response to light and dark. *Plant Physiology* **102**, 917–923.

Kinoshita, T., Nishimura, M. and Shimazaki, K. (1995). Cytosolic concentration of Ca^{2+} regulates the plasma membrane H^+-ATPase in guard cells of fava bean. *Plant Cell* **7**, 1333–1342.

Kwak, J. M., Kim, S. A., Hong, S. W. and Nam, H. G. (1997). Evaluation of 515 expressed sequence tags obtained from guard cells of *Brassica campestris*. *Planta* **202**, 9–17.

Lawton, M. A., Yamamoto, R. T., Hanks, S. K. and Lamb, C. J. (1989). Molecular cloning of plant transcripts encoding protein kinase homologs. *Proceedings of the National Academy of Sciences of the United States of America* **86**, 3140–3144.

Lee, Y. and Assmann, S. M. (1991). Diacylglycerols induce both ion pumping in patch-clamped guard-cell protoplasts and opening of intact stomata. *Proceedings of the National Academy of Sciences of the United States of America* **88**, 2127–2131.

Leung, J., Bouvier-Durand, M., Morris, P.-C., Guerrier, D., Chefdor, F. and Giraudat, J. (1994). *Arabidopsis* ABA-response gene *ABI1*: features of a calcium-modulated protein phosphatase. *Science* **264**, 1448–1452.

Leung, J., Merlot, S. and Giraudat, J. (1997). The *Arabidopsis ABSCISIC ACID-INSENSITIVE2 (ABI2)* and *ABI1* genes encode homologous protein phosphatases 2C involved in abscisic acid and signal transduction. *Plant Cell* **9**, 759–771.

Li, B., Zhang, X.-Q. and Chollet, R. (1996). Phospho*enol*pyruvate carboxylase kinase in tobacco leaves is activated by light in a similar but not identical way as in maize. *Plant Physiology* **111**, 497–505.

Li, J. and Assmann, S. M. (1996). An abscisic acid-activated and calcium-independent protein kinase from guard cells of fava bean. *Plant Cell* **8** 2359–2368.

Li, J., Lee, Y.-R. and Assmann, S. M. (1998). Guard cells possess a calcium-dependent protein kinase that phosphorylates the KAT1 K^+ channel. *Plant Physiology* **116**, 785–795.

Li, J., Wang, X.-Q., Watson, M. B. and Assmann, S. M. (2000). Regulation of abscisic acid-induced stomatal closure and anion channels by guard cell AAPK kinase. *Science* **287**, 300-303.

Li, W., Luan, S., Schreiber, S. L. and Assmann, S. M. (1994). Evidence for protein phosphatase 1 and 2A regulation of K^+ channels in two types of leaf cells. *Plant Physiology* **106**, 963–970.

Lindberg, R. A., Quinn, A. M. and Hunter, T. (1992). Dual-specificity protein kinases: will any hydroxyl do? *Trends in Biochemical Sciences* **17**, 114–119.

Liu, J., Albers, M. W., Wandless, T. J., Luan, S., Alberg, D. G., Belshaw, P. J., Cohen, P., MacKintosh, C., Klee, C. B. and Schreiber, S. L. (1992). Inhibition of T cell signaling by immunophilin-ligand complexes correlates with loss of calcineurin phosphatase activity. *Biochemistry* **31**, 3896–3901.

Luan, S., Li, W., Rusnak, F., Assmann, S. M. and Schreiber, S. L. (1993). Immunosuppressants implicate protein phosphatase regulation of K^+ channels in guard cells. *Proceedings of the National Academy of Sciences of the United States of America* **90**, 2202–2206.

Maathuis, F. J. M., Ichida, A. M., Sanders, D. and Schroeder, J. I. (1997). Roles of higher plant K^+ channels. *Plant Physiology* **114**, 1141–1149.

Maurel, C. (1997). Aquaporins and water permeability of plant membranes. *Annual Review of Plant Physiology and Plant Molecular Biology* **48**, 399–429.

Maurel, C., Kado, R. T., Guern, J. and Chrispeels, M. J. (1995). Phosphorylation regulates the water channel activity of the seed-specific aquaporin α-TIP. *EMBO Journal* **14**, 3028–3035.

Meyer, K., Leube, M. P. and Grill, E. (1994). A protein phosphatase 2C involved in ABA signal transduction in *Arabidopsis thaliana. Science* **264**, 1452–1455.

Mori, I. C. and Muto, S. (1997). Abscisic acid activates a 48-kilodalton protein kinase in guard cell protoplasts. *Plant Physiology* **113**, 833–839.

Müller-Röber, B., Ehrhardt, T. and Plesch, G. (1998). Molecular features of stomatal guard cells. *Journal of Experimental Botany* **49**, 293–304.

Nakajima, N., Saji, H., Aono, M. and Kondo, N. (1995). Isolation of cDNA for a plasma membrane H^+-ATPase from guard cells of *Vicia faba* L. *Plant and Cell Physiology* **36**, 919–924.

Nakamura, R. L., McKendree, W. L., Jr, Hirsch, R. E., Sedbrook, J. C., Gaber, R. F. and Sussman, M. R. (1995). Expression of an *Arabidopsis* potassium channel gene in guard cells. *Plant Physiology* **109**, 371–374.

Pearson, R. B. and Kemp, B. E. (1991). Protein kinase phosphorylation site sequences and consensus specificity motifs: tabulations. *Methods in Enzymology* **200**, 62–81.

Pei, Z. M., Ward, J. M., Harper, J. F. and Schroeder, J. I. (1996). A novel chloride channel in *Vicia faba* guard cell vacuoles activated by the serine/threonine kinase, CDPK. *EMBO Journal* **15**, 6564–6574.

Pei, Z. M., Kuchitsu, K., Ward, J. M., Schwarz, M. and Schroeder, J. I. (1997). Differential abscisic acid regulation of guard cell slow anion channels in *Arabidopsis* wild-type and *abi1* and *abi2* mutants. *Plant Cell* **9**, 409–423.

Pei, Z. M., Ghassemian, M., Kwak, C. M., McCourt, P. and Schroeder, J. I. (1998). Role of farnesyltransferase in ABA regulation of guard cell anion channels and plant water loss. *Science* **282**, 287–290.

Roberts, D. M. and Harmon, A. C. (1992). Calcium-modulated proteins: targets of intracellular calcium signals in higher plants. *Annual Review of Plant Physiology and Plant Molecular Biology* **43**, 375–414.

Rodriguez, P. L., Benning, G. and Grill, E. (1998). AB12, a second protein phosphatase 2C involved in abscisic acid signal transduction in *Arabidopsis. FEBS Letters* **421**, 185–190.

Sarda, X., Tousch, D., Ferrare, K., Legrand, E., Dupuis, J. M., Casse-Delbart, F. and Lamaze, T. (1997). Two TIP-like genes encoding aquaporins are expressed in sunflower guard cells. *Plant Journal* **12**, 1103–1111.

Schaller, G. E. and Sussman, M. R. (1988). Phosphorylation of the plasma-membrane H^+-ATPase of oat roots by a calcium-stimulated protein kinase. *Planta* **173**, 509–518.

Schenk, P. W. and Snaar-Jagalska, B. E. (1999). Signal perception and transduction: the role of protein kinases. *Biochimica et Biophysica Acta* **1449**, 1–24.

Schmidt, C., Schelle, I., Liao, Y. and Schroeder, J. I. (1995). Strong regulation of slow anion channels and abscisic acid signaling in guard cells by phosphorylation and dephosphorylation events. *Proceedings of the National Academy of Sciences of the United States of America* **92**, 9535–9539.

Schnabl, H., Denecke, M. and Schulz, M. (1992). *In vitro* and *in vivo* phosphorylation of stomatal phosphoenolpyruvate carboxylase from *Vicia faba* L. *Botanica Acta* **105**, 367–369.

Shen, L., Outlaw, W. H., Jr and Epstein, L. M. (1995). Expression of an mRNA with sequence similarity to pea dehydrin (Psdhn 1) in guard cells of *Vicia faba* in response to exogenous abscisic acid. *Physiologia Plantarum* **95**, 99–105.

Shimazaki, K., Kinoshita, T. and Nishimura, M. (1992). Involvement of Ca^{2+}/calmodulin-dependent myosin light chain kinase in blue-light-dependent H^+ pumping of guard cell protoplasts from *Vicia faba* L. *Plant Physiology* **99**, 1416–1421.

Talbott, L. D. and Zeiger, E. (1998). The role of sucrose in guard cell osmoregulation. *Journal of Experimental Botany* **49**, 329–337.

Thiel, G. and Blatt, M. R. (1994). Phosphatase antagonist okadaic acid inhibits steady-state K^+ currents in guard cells of *Vicia faba*. *Plant Journal* **5**, 727–733.

Thiel, G. and Wolf, A. H. (1997). Operation of K^+-channels in stomatal movement. *Trends in Plant Science* **2**, 339–345.

Toroser, D. and Huber, S. C. (1997). Protein phosphorylation as a mechanism for osmotic stress activation of sucrose-phosphate synthase in spinach leaves. *Plant Physiology* **114**, 947–955.

Vera-Estrella, R., Barkla, B. J., Higgins, V. J. and Blumwald, E. (1994). Plant defense response to fungal pathogens: activation of host plasma membrane H^+-ATPase by elicitor-induced enzyme dephosphorylation. *Plant Physiology* **104**, 209–215.

Ward, J. M., Pei, Z.-M. and Schroeder, J. I. (1995). Roles of ion channels in initiation of signal transduction in higher plants. *Plant Cell* **7**, 833–844.

Xing, T., Higgins, V. J. and Blumwald, E. (1996). Regulation of plant defense response to fungal pathogens: two types of protein kinases in the reversible phosphorylation of the host plasma membrane H^+-ATPase. *Plant Cell* **8**, 555–564.

Zeiger, E., Bloom, A. J. and Hepler, P. K. (1978). Ion transport in stomatal guard cells: a chemiosmotic hypothesis. *What's New in Plant Physiology* **9**, 29–31.

AUTHOR INDEX

Numbers in *italic* refer to pages on which full references are listed

H

L

T

SUBJECT INDEX

P

bchaudhuri@dmu.ac.uk.